Combustion Chemistry

Combustion Chemistry

Edited by
W. C. Gardiner, Jr.

With Contributions by

A. Burcat G. Dixon-Lewis M. Frenklach
W. C. Gardiner, Jr. R. K. Hanson S. Salimian
J. Troe J. Warnatz R. Zellner

With 164 Figures

Springer-Verlag
New York Berlin Heidelberg Tokyo

William C. Gardiner, Jr.
Department of Chemistry
University of Texas at Austin
Austin, Texas 78712
U.S.A.

Library of Congress Cataloging in Publication Data
Gardiner, William C. (William Cecil), 1933–
Combustion chemistry.
Bibliography: p.
1. Combustion. I. Title.
QD516.G24 1984 541.3′61 84-5355

Typeset by Polyglot Pte Ltd., Singapore.
Printed and bound by R. R. Donnelley & Sons, Harrisonburg, Virginia.
Printed in the United States of America.

9 8 7 6 5 4 3 2 1

ISBN 0-387-90963-X Springer-Verlag New York Berlin Heidelberg Tokyo
ISBN 3-540-90963-X Springer-Verlag Berlin Heidelberg New York Tokyo

Preface

Detailed study of the rates and mechanisms of combustion reactions has not been in the mainstream of combustion research until the recent recognition that further progress in optimizing burner performance and reducing pollutant emission can only be done with fundamental understanding of combustion chemistry. This has become apparent at a time when our understanding of the chemistry, at least of small-molecule combustion, and our ability to model combustion processes on large computers have developed to the point that real confidence can be placed in the results. This book is an introduction for outsiders or beginners as well as a reference work for people already active in the field. Because the spectrum of combustion scientists ranges from chemists with little computing experience to engineers who have had only one college chemistry course, everything needed to bring all kinds of beginners up to the level of current practice in detailed combustion modeling is included.

It was a temptation to include critical discussions of modeling results and computer programs that would enable outsiders to start quickly into problem solving. We elected not to do either, because we feel that the former are better put into the primary research literature and that people who are going to do combustion modeling should either write their own programs or collaborate with experts. The only exception to this is in the thermochemical area, where programs have been included to do routine fitting operations. For reference purposes there are tables of thermochemical, transport-property, and rate-coefficient data.

The material is organized as follows. In Chapters 1 and 2 the basic chemistry and physics of reactive flow are described, first simply and then in the sophisticated form required for dealing with steady flames. The theory of elementary reaction rate constants is presented in Chapters 3 and 4, subdivided into bimolecular and unimolecular types and with particular emphasis on the temperature dependence of rate coefficients and procedures for estimating rate coefficients when little or no experimental information is available. Surveys of elementary reaction rate coefficients are given in Chapters 5 and 6 for reactions of molecules containing the elements carbon, hydrogen and oxygen and nitrogen, hydrogen and oxygen respectively. In Chapter 7 the mathematical theory of modeling is reviewed, with the particular purpose of showing how the predictions and parameters of combustion models can be subjected to critical statistical and sensitivity analysis. The concluding Chapter 8 gives methods of thermochemical property estimation and polynomial fitting for combustion-related purposes.

Our current knowledge of combustion chemistry can be rather well characterized by stating what cannot yet be presented in a book like this. First, the fuel molecules whose combustion can be described with confidence have at most only two carbon atoms; propane and butane combustion models are topics of exploratory research and the mechanism of octane combustion can only be surmised. Second, when hydrocarbon fuels burn with too little oxygen for complete combustion, solid carbon is formed by mechanisms that are still not understood in enough detail to permit models to be constructed. Similarly, the interaction of physical and chemical processes at surfaces of burning solids such as coal is known only in outline. In the area of pollutant-formation mechanisms, while we can explain why oxides of nitrogen form when combustion temperatures get very high, we cannot explain how they are formed from nitrogen-containing fuel molecules or from attack of carbon-containing radicals on molecular nitrogen, nor how nitrogen oxides produced early in flames get reduced back to molecular nitrogen within the main reaction zone. We know little about the chemistry of sulfur in flames. On the modeling side, it is still usually necessary to oversimplify either the fluid dynamics or the chemistry: For chemically detailed combustion modeling, one must treat the fluid dynamics with restrictive approximations that usually amount to an assumption of steady one-dimensional adiabatic flow, which can only be achieved in practice using special laboratory burners; when unsteady or two- or three-dimensional flow is to be modeled, the demands upon computer time are already so great that chemical details must be suppressed, the flow model must be simplified, or computational accuracy must be sacrificed. This list of shortcomings in our understanding of and ability to apply combustion chemistry is not a declaration that real-world combustion is too complex to understand at the level of elementary reactions and realistic fluid dynamics. On the contrary, it is a program for combustion research that we hope readers of this book will help us to carry out.

Austin, Texas William C. Gardiner, Jr.

Contents

Contributors

Alexander Burcat, Department of Aeronautical Engineering, Technion—Israel Institute of Technology, Haifa, Israel.

Graham Dixon-Lewis, Department of Fuel and Energy, The University, Leeds LS2 9JT, England.

Michael Frenklach, Department of Chemical Engineering, Louisiana State University, Baton Rouge, Louisiana 70803, U.S.A.

William C. Gardiner, Jr., Department of Chemistry, University of Texas, Austin, Texas 78712, U.S.A.

Ronald K. Hanson, Department of Mechanical Engineering, Stanford University, Stanford, California 94305, U.S.A.

Siamak Salimian, Department of Mechanical Engineering, Stanford University, Stanford, California 94305, U.S.A.

Jürgen Troe, Institute for Physical Chemistry, University of Göttingen, D-3400 Göttingen, Federal Republic of Germany.

Jürgen Warnatz, Institute for Physical Chemistry, University of Heidelberg, D-6900 Heidelberg, Federal Republic of Germany.

Reinhard Zellner, Institute for Physical Chemistry, University of Göttingen, D-3400 Göttingen, Federal Republic of Germany.

Chapter 1

Introduction to Combustion Modeling

William C. Gardiner, Jr.

1. Terminology of reaction kinetics

It was known from the very beginning of combustion science that the chemical reactions converting fuel to combustion products are complex, because the directly observed phenomena of flame propagation, explosions, slow ignition, critical fuel/oxidizer ratios, and so forth depend in subtle ways on the conditions under which combustion takes place. For most of the history of the subject this complexity could be described only in a patchwork fashion, much like describing the character of a vast country by showing a handful of snapshots from different parts of it. It was always appreciated that an understanding of combustion chemistry could only be attained by identifying the individual molecular events responsible for it and knowing how rapidly these events take place.

Until recently, two hindrances to achieving this could not be overcome. The molecular events and their speeds could not be determined, and there was no practical way to combine known or estimated molecular-level knowledge into descriptions of the complete combustion process. Continuous development and application of sophisticated research techniques gradually increased the amount of molecular-level information available, and development of large digital computers and programs to use them for studying combustion reactions gradually made it possible to deal with this information in a comprehensive manner. It is now possible, for simple fuel molecules at least, to give computer-generated descriptions of combustion reactions that we think are accurate reflections of what happens at the molecular level and are in agreement with manifold experiments that have accumulated over the years.

The languages in which all of this is discussed are the languages of chemical kinetics and computer modeling, as these two subjects pertain to combustion.

In this chapter we introduce these languages in order to provide the background needed to appreciate the technical presentations of later chapters. First some basic terms: a *chemical reaction* is the conversion of one kind of matter into another chemically different form; a *combustion reaction* is a chemical reaction in which a *fuel* combines with an *oxidizer* (usually oxygen from air) to form *combustion products.* Combustion reactions are described by writing chemical symbols for the starting materials—the *reactants*—and the final *products* connected by an arrow that means "react with one another to form." Thus, for the combustion of propane in air the combustion reaction is

$$C_3H_8 + 5O_2 \rightarrow 3CO_2 + 4H_2O$$

where the *stoichiometric coefficients* 5, 3, and 4 denote the numbers of moles of oxygen, carbon dioxide, and water, that participate in the combustion of one mole of propane. The terms chemical reaction and chemical equation are both used to designate shorthand descriptions like this.

Chemical equations only name the observed reactants and final products and their relative molar amounts. At the molecular level a similar form of expression is used to describe molecular events that are responsible for the observed changes. An example is the attack of hydrogen atoms on oxygen molecules to form hydroxyl radicals and oxygen atoms:

$$H + O_2 \rightarrow OH + O$$

Chemical transformations that are also real molecular events are called *elementary reactions.* (Alternate terms are *elementary steps* or simply *steps.*)

It is important to distinguish between these two uses of identical-appearing chemical shorthand expressions: Combustion reactions occur because large numbers of different elementary reactions combine to produce the transformation of fuel and oxidizer to combustion products described in the chemical equation. The whole set of elementary reactions is called the *reaction mechanism.* In summary, combustion reactions can be described at the molecular level by giving the elementary reactions that comprise the combustion mechanism of the fuel of interest. In Section 8 of Chapter 5 can be found a selection of elementary reactions that one might assemble into a mechanism for the combustion of propane, for example.

2. Rate laws and reaction mechanisms

Given that combustion reactions occur as a consequence of a number of elementary reactions taking place in the burning gas, how does one identify these and find out how each of them contributes to the progress of reaction? These questions belong to the area of chemistry known as *reaction kinetics*, because they involve knowing how fast each elementary reaction proceeds. In chemical kinetics one defines the *rate* of a chemical reaction as the derivative of the concentration of a product species with respect to time due to the reaction involved; thus for the elementary reaction $H + O_2 \rightarrow OH + O$ the increases in the concentrations of OH and O per unit time due to this reaction would be measures of its rate. Since the concentrations of H and O_2 are decreased as a result of this reaction, their concentration changes attributable to this reaction would be multiplied by -1 to get the rate.

Elementary reactions may proceed in either direction; thus an encounter between OH and O may lead to formation of H and O_2 in what is called a *reverse reaction*. The *net rate r* of an elementary reaction is then the difference between the forward and reverse rates:

$$r = \text{forward rate} - \text{reverse rate} \tag{2.1}$$

At any time, the net rate must be either positive or negative. As the composition of a reacting system changes in time, it is possible that the direction of net reaction (that is, the sign of r) changes. For purposes of modeling combustion reactions, the two opposing directions are considered separately; if one direction is negligible under all conditions of interest, then it may be omitted from consideration.

The higher the concentrations of the reacting molecules, the faster the reaction will proceed. Since the rate in either direction is proportional to the concentrations of each of the reactant molecules, a *mass action rate law* connects the rate of reaction to the reactant concentrations. Thus for the $H + O_2 \rightarrow OH + O$ elementary reaction we have

$$\text{forward rate} = k_f[H][O_2]$$

$$\text{reverse rate} = k_r[OH][O]$$

$$r = k_f[H][O_2] - k_r[OH][O]$$

where the proportionality constants k_f and k_r are called the *rate coefficients* or *rate constants* in the forward and reverse directions and the brackets are used to denote concentrations. For gas-phase reactions, the latter may be expressed in mol/cm^3 or molecule/cm^3 units, with the rate coefficients then having cm^3/mol s or cm^3/s units. Elementary reactions like this in which there are two reactant molecules (where atoms and radicals also count as molecules) are called *bimolecular* elementary reactions. In combustion reactions there are also *unimolecular* elementary reactions such as

$$C_3H_8 \rightarrow CH_3 + C_2H_5$$

for which the forward rate is given by

$$\text{forward rate} = k_f[C_3H_8]$$

and *termolecular* elementary reactions such as

$$H + CO + H_2O \rightarrow HCO + H_2O$$

for which the forward rate is given by

$$\text{forward rate} = k_f[\text{H}][\text{CO}][\text{H}_2\text{O}]$$

In Chapters 3 and 4, physical theories underlying the values which are measured for the rate coefficients of unimolecular, bimolecular, and termolecular rate coefficients are described.

Rate coefficients of bimolecular and termolecular gas reactions prove to depend on one variable only, namely temperature. Unimolecular reaction rate coefficients also depend on the total molar concentration $[M]$ (Chapter 4). One could be explicit and write $k(T)$ or $k(T, [M])$ rather than k alone. Indeed, in combustion processes rate coefficients do change by orders of magnitude from the cool to the hot parts of a flame. It is not customary to write these dependences explicitly, however, and so one must just keep in mind that rate coefficients are not constants even though they often bear that name.

In combustion processes, the concentration of a given *species* (atom, molecule, or radical) may be affected by a large number of elementary reactions. The rate of change of each species concentration must therefore be found by adding up the effects of all elementary reactions producing the species and subtracting the effects of all elementary reactions that consume it. Since one does not know before doing the calculation whether the net rate of an elementary reaction is positive or negative, the most straightforward way to compute the net effects of all elementary reactions on the concentration of a given species is simply to add up the forward and reverse rates of all reactions that involve the species, prefixing each contribution by a positive or negative sign depending on whether the species is produced or consumed in that direction of the elementary reaction. (In a few cases, such as for H in $CH_2 + O_2 \rightarrow CO_2 + 2H$, an additional factor of 2 is required to account for double production or consumption.) In subsequent sections we will express this summation mathematically to provide the right-hand sides of equations for the time derivatives of species concentrations.

For one special case, isothermal reaction at constant density, the set of differential equations comprising the time derivatives of all species concentrations is sufficient to determine the evolution of a system described by a given reaction mechanism for any assumed starting concentrations. While this special case does apply for some experiments of interest in combustion research, it does not pertain to the conditions under which most combustion processes occur. Usually we must expand our set of differential equations so as to describe the effects of chemical reaction on the physical conditions and the effects of changes in the physical conditions on the chemistry. In either case, the evolution of the system is found by numerical integration of the appropriate set of ordinary differential equations with a computer. This procedure is known in the language of numerical analysis as the solution of an *initial value problem.*

3. Physical constraints on gas-phase combustion reactions

We noted in the preceding section that concentrations and temperature are the variables needed to describe the rates of chemical reactions. Temperature and concentration changes in gases interact with changes of other variables; among them, pressure and density—which may mean either the *molar density* (total moles/cm^3) or the *mass density* (g/cm^3)—are coupled to temperature through the ideal gas law $PV = nRT$ rearranged to $P = (n/V)RT$, where R is the universal gas constant 82.057 atm cm^3 mol^{-1} K^{-1} or 8.314 in SI units J mol^{-1} K^{-1} or Pa m^3 mol^{-1} K^{-1} and n is the total number of moles of gas; thus (n/V) is the molar density. If the mass density ρ is to appear in the equations, then since $n = \text{mass}/\overline{MW}$, where $\overline{MW}$ is the average molecular weight, $P = (\text{mass}/\overline{MW})RT/V = (\text{mass}/V)RT/\overline{MW} = \rho(R/\overline{MW})T = \rho R'T$. Here the *specific gas constant*, $R' = R/\overline{MW}$, is a function of the composition of the gas and may change appreciably in a combustion process as the average number of atoms per molecule is changed by the combustion reactions.

Gas-phase reactions may take place in closed vessels, which leads to the physical constraint that the density of the gas is not changed as a result of chemical reactions. The nature of the vessel wall then determines how chemical reactions affect the pressure and temperature. If the walls conduct heat well (are *diabatic*) and the vessel is in a thermostat, then the temperature remains constant and the pressure in the vessel is affected only by changes in the total number of molecules present; that is, the pressure is inversely proportional to the average molecular weight through $P = \rho(R/\overline{MW})T = \text{constant}/\overline{MW}$. If the walls do not conduct heat at all (are *adiabatic*), then heat released by chemical reactions results in a temperature increase. The pressure, temperature, and number of molecules are all free to change, but coupled to one another and to the extent of reaction by the ideal gas law.

Another common constraint is for reaction taking place at constant pressure, the volume being free to change as a result of reaction. In this case, work of expansion moderates the temperature and pressure increases that would result from the same reaction taking place in a closed vessel. The volume, temperature, and number of molecules change with extent of reaction in a manner determined by the ideal gas law and the first law of thermodynamics.

So far we have ignored the spatial coordinates of combustion processes, assuming that the composition and physical properties of the reacting gas are the same everywhere. Such *homogeneous* systems can be realized in some laboratory situations, but for many combustion experiments and virtually all practical situations the gas composition and physical properties change with position, and the gas is usually in motion. This implies that *transport properties*, including heat conduction and diffusion, contribute to the evolution of the system, and that forces generated by chemical reaction will cause the *fluid dynamics* of the reacting system to become important.

In the following sections we develop the equations needed to describe the evolution of homogeneous static systems, under several sets of constraints, and of one-dimensional flowing systems for the case that diffusion and heat transport can be neglected. The more complicated equations describing processes where these transport processes are important are developed in Chapter 2.

4. Differential equations of homogeneous reaction without transport

Our first objective is to provide a compact notation for the rates of chemical reaction in a many-species, many-reaction situation. Let the number of species be $\mathscr{S}$ and the number of elementary reactions be $\mathscr{R}$; let the concentration of species i be c_i and the net rate of elementary reaction j (as defined above) be r_j. Define β_{ij} to be the *net* stoichiometric coefficient of species i in elementary reaction j, i.e., the stoichiometric coefficient of i on the product side minus its stoichiometric coefficient on the reactant side. (For most elementary reactions one of these is zero.) Thus, for the $CH_2 + O_2 \rightarrow CO_2 + 2H$ example the four β_{ij} values are -1, -1, 1, and 2 for the species CH_2, O_2, CO_2, and H, respectively. In problems where $\mathscr{R}$ and $\mathscr{S}$ are large, most of the β_{ij} are zero, giving a severe computational problem that we discuss in detail in Section 5.

We introduce at this point a simple example mechanism that will be used in Sections 4 and 6 to illustrate important features of combustion mechanisms. Suppose that nitrous oxide N_2O is heated rapidly to 1500 K at atmospheric pressure, whereupon it decomposes in an exothermic reaction. The species that one might assume to be present as the decomposition proceeds are to include N_2O itself, N_2, O_2, NO, NO_2, N_2O_4, and O atoms, a total of $\mathscr{S} = 7$ species. Suppose that the forward and reverse directions of the following eight reactions are to be tested as a possible reaction mechanism:

$$N_2O + M \rightleftharpoons N_2 + O + M \qquad 2 \times 10^6 \qquad (1)$$

$$O_2 + M \rightleftharpoons O + O + M \qquad 8 \times 10^{-3} \qquad (2)$$

$$N_2O + O \rightleftharpoons NO + NO \qquad 9 \times 10^9 \qquad (3)$$

$$N_2O + O \rightleftharpoons N_2 + O_2 \qquad 8 \times 10^{12} \qquad (4)$$

$$NO_2 + M \rightleftharpoons NO + O + M \qquad 3 \times 10^6 \qquad (5)$$

$$NO_2 + O \rightleftharpoons NO + O_2 \qquad 8 \times 10^{12} \qquad (6)$$

$$NO_2 + NO_2 \rightleftharpoons NO + NO + O_2 \qquad 3 \times 10^8 \qquad (7)$$

$$NO_2 + NO_2 \rightleftharpoons N_2O_4 \qquad 1 \times 10^{13} \qquad (8)$$

In this mechanism the species "M" (standing for "molecule") is included as a representation of all of the species that may participate in reactions (1), (2), and (5) as "collision partners." (See Chapter 4.) The rate coefficients listed are for concentration units of mol/cm^3 for the forward directions of reaction at 1500 K. (Expressions giving these rate coefficients as a function of temperature may be found in Chapter 6.) The β matrix for this trial reaction mechanism is

$j =$	1	2	3	4	5	6	7	8
N_2O	−1	0	−1	−1	0	0	0	0
N_2	+1	0	0	+1	0	0	0	0
O_2	0	−1	0	+1	0	+1	+1	0
NO	0	0	+2	0	+1	+1	+2	0
NO_2	0	0	0	0	−1	−1	−2	−2
N_2O_4	0	0	0	0	0	0	0	+1
O	+1	+2	−1	−1	+1	−1	0	0

Each of the net rates r_j is given by a mass-action expression

$$r_j = k_f \prod c_i - k_r \prod c_i \tag{4.1}$$

where the first product is taken over all i belonging to the reactant species of reaction j and the second product is taken over all i belonging to product species. (If a species appears more than once in the reaction, it also appears more than once in these products.) Introduce the operator notation $\Delta = d/dt$ for time derivative. Then the total rate of change of concentration c_i due to all elementary reactions is given by a sum over all reactions j affecting c_i

$$\Delta c_i = \sum \beta_{ij} r_j \tag{4.2}$$

The set of $\mathscr{S}$ differential equations consisting of $\mathscr{S}$ equations like Eq. (4.2) defines the chemical part of kinetic modeling. The functions on the right-hand side of Eqs. (4.2) contain only the concentrations c_i as variables, with the rate coefficients k_{fj} and k_{rj} as parameters depending on the temperature T.

We may note at this point that k_f and k_r are not independent of one another, but are instead constrained to obey the principle of *detailed balance*

$$k_f / k_r = K_{\text{eq}} \tag{4.3}$$

where K_{eq} is the equilibrium constant in concentration units, a function of temperature only that can be computed from the thermochemical properties of the species participating in the elementary reaction (Chapter 8).

4.1. Constant-density isothermal reaction

The simplest form that combustion modeling can take is that for which the set of $\mathscr{S}$ differential equations (4.2) completely defines the evolution of the system. Because rate coefficients depend upon temperature, and concentrations are affected by changes of volume even if the numbers of molecules of each species remain constant, this simplicity can only be attained if the constraint of reaction at constant temperature and volume is assumed. Since we are not now concerned with the size of the system, what we really assume is constancy of the volume per unit mass (the inverse of the density) or the *specific volume* of the reacting system.

The solution of the initial value problem defined by Eqs. (4.2) is straightforward. For any composition, the right-hand side of Eqs. (4.2) can be computed for the forward and reverse rate coefficients appropriate for the specified temperature. Increments and decrements to the starting concentrations are computed repeatedly by the integration algorithm to generate the solution to the initial value problem (Section 5).

For fixed temperature and density the only state variable free to change is the pressure. There are two ways to keep track of the pressure during the course of reaction. The first is to use the ideal gas law in the form

$$P = (n/V)RT = RT \sum c_i \tag{4.4}$$

whenever the pressure is to be computed. The second way will introduce the procedures used later for other constraints. The total derivative of pressure ΔP with respect to all of the concentration changes is the derivative of Eq. (4.4)

$$\Delta P = RT \sum \Delta c_i \tag{4.5}$$

Changes in the pressure can be followed in differential form by adding the pressure derivative Eq. (4.5) to the set (4.2) to give $\mathscr{S} + 1$ differential equations to be solved by the numerical integration algorithm.

4.2. Constant-density adiabatic reaction

If the heat liberated in the combustion reaction is not lost from the system, then the temperature will rise as reaction proceeds. To solve for the evolution in this case we need in addition to the ideal gas law an appropriate form of the law of conservation of energy.

The total derivative of the ideal gas law divided by the ideal gas law itself is

$$(1/P)\,\Delta P - (1/\textstyle\sum c_i)\sum \Delta c_i - (1/T)\,\Delta T = 0 \tag{4.6}$$

At constant density, the volume is constant, no work can be done on the system, and since in an adiabatic reaction no heat flows into or out of the system, the thermodynamic energy equation *per unit mass* is

$$\Delta U = q + w = 0 \tag{4.7}$$

For thermochemistry one usually keeps track of enthalpy rather than internal energy, in which case Eq. (4.7) is

$$\Delta U = \Delta(h - Pv) = 0 \tag{4.8}$$

where h is the enthalpy of the system per unit mass (*specific enthalpy*) and $v = 1/\rho$ is the specific volume. The specific enthalpy can be calculated from the molar enthalpies H_i of each species by

$$h = (1/\rho)\sum c_i H_i \tag{4.9}$$

Substituting Eq. (4.9) for h and the inverse density for v in Eq. (4.8) gives

$$(1/\rho)\,\Delta\sum c_i H_i - (1/\rho)\,\Delta P = 0 \tag{4.10}$$

or

$$\Delta\sum c_i H_i - \Delta P = 0 \tag{4.11}$$

The products $c_i H_i$ in Eq. (4.11) can be differentiated as

$$\Delta(c_i H_i) = c_i \Delta H_i + H_i \Delta c_i \tag{4.12}$$

Since the molar enthalpy H_i is a function of the temperature only

$$\Delta H_i = C_{Pi}\,\Delta T \tag{4.13}$$

Combining Eqs. (4.11), (4.12), and (4.13) casts the energy conservation law into

$$\sum H_i\,\Delta c_i + \sum c_i C_{Pi}\,\Delta T - \Delta P = 0 \tag{4.14}$$

Eliminating P between Eq. (4.6) and Eq. (4.14) and factoring gives

$$\Delta T = \sum (H_i - P/\sum c_i)\,\Delta c_i/(P/T - \sum c_i C_{Pi}) \tag{4.15}$$

We see that once the concentration changes Δc_i have been calculated from Eq. (4.2), the temperature change can be found from Eq. (4.15). Then the pressure change can be found using the rearranged form of Eq. (4.6)

$$\Delta P = P(\Delta T/T + \sum \Delta c_i/\sum c_i) \tag{4.16}$$

Equations (4.2), (4.15), and (4.16) then provide $\mathscr{S} + 2$ equations for the $\mathscr{S} + 2$ derivatives to be found. The evolution of the system is computed in principle just as for the constant-temperature case; in practice, however, we do have the additional complication that the rate coefficients change as the temperature changes, and so subsequent evaluations of Eq. (4.2) have to be done with k_f and k_r values appropriate to the current temperature.

As an example of the results that one may obtain in modeling an adiabatic constant-density reaction, the evolution of the N_2O thermal decomposition system is shown in Fig. 1 in semilog form and in Fig. 2 in linear form for a time interval of 3 ms. Several features of the reaction are immediately evident: the strong increases in pressure and temperature due to reaction, and the fact that so little N_2O_4 is produced that even including it in the list of species can be viewed as a mistake for these conditions. Other characteristics of the reactions emerge on further study, some using alternate presentations of the results of a single numerical integration like this and others using repeated integrations for different assumptions about parameters such as rate coefficients, thermochemical properties, or initial conditions.

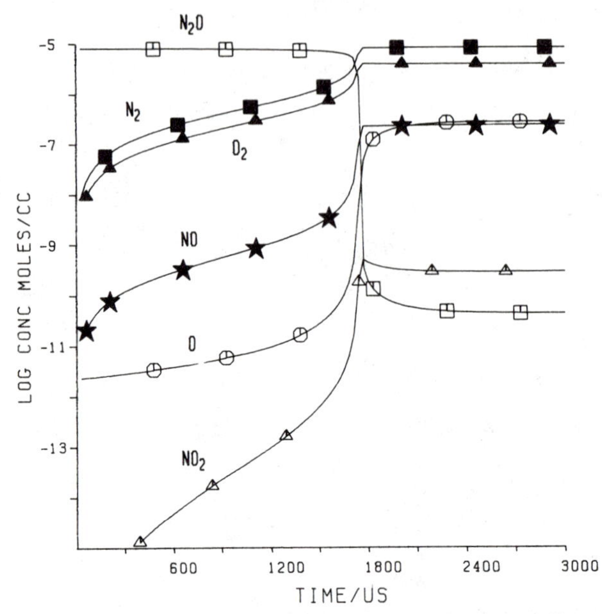

Figure 1. Semilogarithmic presentation of species profiles for thermal decomposition of N_2O starting at $P = 1$ atm and $T = 1500$ K under constant-density, adiabatic constraints. The concentration of N_2O_4 is off scale. (See Fig. 2.)

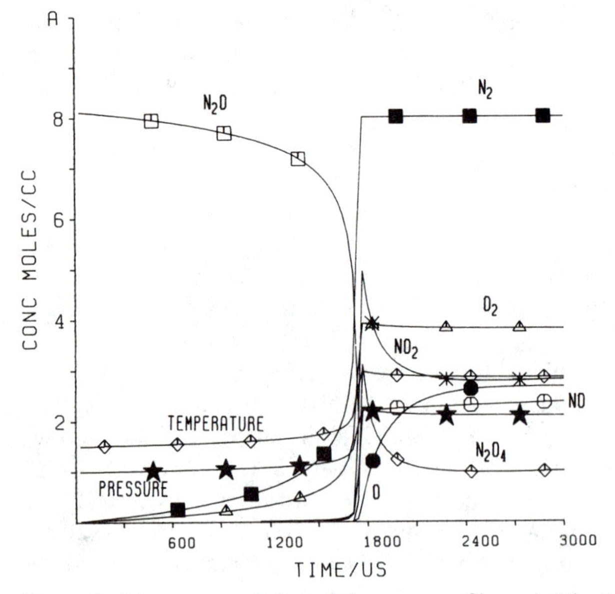

Figure 2. Linear presentation of the same profiles as in Fig. 1. The temperature profile (in kK) and pressure profile (in atm) are also shown. The A values for the species concentrations in mol/cm^3 units are N_2O 10^{-5}, NO 10^{-6}, NO_2 10^{-9}, N_2O_4 10^{-20}, N_2 10^{-5}, O 10^{-6}, O_2 10^{-5}. The A values for T and P are 10 kK and 10 atm.

4.3. Constant-pressure adiabatic reaction

If a burning gas mixture can expand as it gets hot, the constant-density constraint considered in the previous two sections is no longer appropriate, for concentrations then also change because of the expansion. For situations like this it may be a suitable approximation to account for the effect of the expansion while still neglecting the motion of the gas and transport properties. An example would be the ignition process of a fuel/oxidizer mixture heated to ignition temperature by compression. In this section we derive the differential equations needed to describe homogeneous adiabatic reaction at constant pressure.

The energy conservation equation for adiabatic ($dq = 0$) reaction of unit mass of gas is

$$\Delta U = \Delta(h - Pv) = dw = -P\,dv \tag{4.17}$$

Expanding the derivative of Pv gives

$$\Delta h = v\,dP = 0 \tag{4.18}$$

at constant pressure. Taking the derivative of $h = (1/\rho)\sum c_i H_i$ and multiplying by ρ converts Eq. (4.18) into

$$\sum c_i \Delta H_i + \sum H_i \Delta c_i - (1/\rho)\sum H_i c_i \Delta\rho = 0 \tag{4.19}$$

The molar enthalpy depends on temperature only, i.e., $\Delta H_i = C_{Pi}\Delta T$, so that the basic energy conservation equation for constant-pressure adiabatic reaction is

$$\sum c_i C_{Pi}\Delta T + \sum H_i \Delta c_i - \sum H_i c_i \Delta\rho/\rho = 0 \tag{4.20}$$

The derivative of the ideal gas law [Eq. (4.6)] for $\Delta P = 0$ gives the second governing equation

$$(1/\sum c_i)\sum \Delta c_i + (1/T)\Delta T = 0 \tag{4.21}$$

The concentration changes must now be found as a total derivative

$$\Delta c_i = \frac{\partial c_i}{\partial t} + \frac{\partial c_i}{\partial \rho}\Delta\rho \tag{4.22}$$

The partial derivative with respect to density can be found by noting that the sum of all concentrations gives the density divided by the average molecular weight

$$\sum c_i = \frac{\rho}{\overline{MW}} \tag{4.23}$$

The partial derivative of Eq. (4.23) is

$$\frac{\partial \sum c_i}{\partial \rho} = \frac{1}{\overline{MW}} = \frac{\sum c_i}{\rho} \tag{4.24}$$

Since the c_i are independent variables their partial derivatives must each be given by

$$\frac{\partial c_i}{\partial \rho} = \frac{c_i}{\rho} \tag{4.25}$$

The total derivatives of the concentrations are thus

$$\Delta c_i = \sum \beta_{ij} r_j + (c_i/\rho)\,\Delta\rho \tag{4.26}$$

where substitution of Eq. (4.2) gives the reactive contribution of the first term and the second term (which is negative for an expanding heated gas) is the contribution caused by density change.

Eliminating ΔT between Eqs. (4.20) and (4.21) gives

$$\sum H_i\,\Delta c_i - \left(T \sum \Delta c_i / \sum c_i\right) \sum c_i C_{Pi} = \sum c_i H_i\,\Delta\rho/\rho \tag{4.27}$$

which contains only $\Delta\rho$ and the Δc_i. Substituting Eq. (4.26) into Eq. (4.27) gives, after some algebra,

$$\Delta\rho = \left(\rho/T \sum c_i C_{Pi}\right) \sum_i \sum_j H_i \beta_{ij} r_j - \left(\rho/\sum c_i\right) \sum_i \sum_j \beta_{ij} r_j \tag{4.28}$$

which contains only known quantities on the right-hand side. The last equation needed is obtained by rearranging Eq. (4.21) to

$$\Delta T = -T \sum \Delta c_i / \sum c_i \tag{4.29}$$

The procedure needed to evaluate the $\mathcal{S} + 2$ derivatives of ρ, T, and the $\mathcal{S}$ species concentrations is as follows: (1) evaluate the sums on the right-hand side of Eq. (4.28) to find $\Delta\rho$; (2) evaluate the right-hand sides of the S equations (4.26) for the Δc_i's; evaluate the right-hand side of Eq. (4.29) to find ΔT. These equations generate the evolution of the constant-pressure adiabatic reacting system.

4.4. Reactive steady flow

Many combustion-related experiments as well as practical devices involve reaction in flowing gas streams. When reacting gas is moving, the kinetic energy of motion must be considered as part of the total energy that is conserved when chemical reaction releases heat energy. In essence, the flow velocity changes when reaction occurs. The differential equations that we use to describe the process must take this into account.

As in the static reaction cases treated in the preceding sections, the equations have to be derived for an idealized physical model. To begin with, we assume that the flow is one dimensional and that the gas moves without encountering frictional forces. Transport of matter and energy through diffusion and heat transport are again assumed to be negligible. These idealizations do indeed restrict the accuracy with which computed reaction profiles represent the real world, but on the other hand they permit us to see the essential physics and chemistry of reaction in flowing gases both in the governing equations and in the computed profiles. The walls constraining the flow to be one dimensional could be diabatic or adiabatic; we consider here only the latter case. In the language of fluid mechanics, our model is that of *inviscid* (no friction), *isentropic* (no heat loss) constant area duct flow with chemical reaction.

There are four physical constraints: the ideal gas law, conservation of matter expressed as a *continuity equation*, conservation of momentum expressed in one form of *Euler's equation*, and conservation of energy expressed as the sum of internal energy and kinetic energy. These take the following mathematical forms.

The ideal gas law requires

$$P - \sum c_i RT = 0 \tag{4.30}$$

Conservation of matter requires that the mass of matter flowing into any segment of the duct in a given time be equal to the mass flowing out of it. Designating the flow velocity as u we have

$$\rho u = m \tag{4.31}$$

where the constant m is known as the *mass flow*. Momentum conservation is expressed in differential form (see any textbook of fluid dynamics, e.g., Liepmann and Roshko, 1957) as

$$\frac{dP}{dx} + \rho u \frac{du}{dx} = 0 \tag{4.32}$$

Conservation of energy per unit mass requires

$$h + u^2/2 = h^* \tag{4.33}$$

where the constant h^* is called the *stagnation enthalpy*.

Equations (4.30) through (4.33) express the physical constraints under which reaction takes place, each of them being expressions that must remain constant along the flow profile. In order to get a set of ordinary differential equations, total derivatives of Eqs. (4.30), (4.31), and (4.33) are formed and then manipulated by a tedious algebraic procedure to generate $S + 4$ equations for the derivatives of P, T, ρ, u, and the $\mathscr{S}$ c_i. Matrix manipulations can be avoided by first finding the derivative of one of the concentrations, then the remaining concentrations and the other derivatives. The derivations can be found elsewhere in the same notation as used here. (Gardiner *et al.*, 1981) The final equations are simplified by introducing the definitions

$$\Sigma c = \sum c_i \tag{4.34}$$

$$\Sigma C_P = \sum c_i C_{Pi} \tag{4.35}$$

$$D_i = \frac{1}{u} \sum \beta_{ij} r_j \tag{4.36}$$

$$\Sigma D = \sum D_i \tag{4.37}$$

$$\Sigma HD = \sum H_i D_i \tag{4.38}$$

We let the operator symbol Δ now stand for the derivative with respect to distance along the flow direction rather than the time derivative as before. The derivative with respect to x for c_1 is

$$\Delta c_1 = \frac{\rho u^2(T\Sigma C_P/P - 1)D_1 - c_1\Sigma HD + T\Sigma C_P(c_1\Sigma D/\Sigma c - D_1)}{\rho u^2(T\Sigma C_P/P - 1) - T\Sigma C_P} \tag{4.39}$$

The other concentration derivatives are

$$\Delta c_i = D_i + \frac{c_i(\Delta c_1 - D_1)}{c_1} \tag{4.40}$$

and the four state variable derivatives are

$$\Delta \rho = \frac{\rho(\Delta c_1 - D_1)}{c_1} \tag{4.41}$$

$$\Delta u = \frac{u(D_1 - \Delta c_1)}{c_1} \tag{4.42}$$

$$\Delta P = -\rho u \Delta u \tag{4.43}$$

$$\Delta T = T\left(\frac{\Delta P}{P} - \frac{\Sigma \Delta c_i}{\Sigma c}\right) \tag{4.44}$$

One must choose species number 1 to be one whose concentration (especially at the start of the integration) is never zero, because c_1 appears in the denominators of Eqs. (4.40) through (4.42). If one is interested in the elapsed reaction time rather than the distance along the flow direction, then since $dt/dx = 1/u$, the set of differential equations can be expanded to $\mathscr{S} + 5$ and the reaction time computed as one of the dependent variables.

Equations (4.39) through (4.44) are the $\mathscr{S} + 4$ equations needed to obtain the solution of the initial value problem for adiabatic chemical reaction in constant-area duct flow. One sees that while there is more computing to do, it is really more complex than the static cases only in that there is one more variable, the flow velocity, to keep track of.

This concludes our presentation of introductory sets of differential equations describing chemical reaction in gases. We now consider how best to go about the numerical integrations.

5. Methods of numerical integration

The simplest way to integrate a differential equation of the form $dx/dt = f(x)$ from $t = 0$ to $t = t_1$ is to choose a small time increment h, called the *step size*, and repeatedly add the product $hf(x)$ to the current value of x until one has repeated the process t_1/h times. The graph of $x(t)$ so generated is the *profile* of x over the time interval 0 to t_1. If we have not one but N differential equations, such that each dependent variable x_i may be a function of up to $N + 1$ variables $x_1, x_2, \ldots, x_N, t$, then the very same procedure can be used to generate all N profiles. This rudimentary method, called Euler integration, has indeed been used to solve many initial value problems in chemical kinetics.

Long before electronic computers were invented, it was realized that mathematical sophistication could be introduced into numerical integration in order to save computational effort and improve accuracy. Textbooks of numerical analysis are full of ways to do this. The most popular of them, the Runge-Kutta and predictor-corrector algorithms, once were standard methods for numerical solution of the initial value problems of chemical kinetics. They have been replaced, however, by more suitable methods invented for the specific purpose of dealing with chemical kinetics problems.

The more suitable integration methods address problems stemming from two characteristics of the differential equations of chemical kinetics, such as those developed in Section 4, that eventually lead to severe difficulties in solving them by conventional integration methods. The first is that in virtually all chemical reactions some of the species are held at nearly stationary concentrations by very rapid elementary reactions while other concentrations change slowly by elementary reactions that determine the overall net rate of reaction. This is illustrated in Fig. 3 for the N_2O example. Sets of differential equations describing situations like this are called "stiff" because of their analogy to the mechanical equation of motion of a stiff rod. The existence of a wide range of rates means that for ordinary numerical integration methods the length of the integration is determined by the slowest rates while the step size is determined by the fastest ones. This conflict leads indeed not only to impossibly long computing times but can, for some integration algorithms, even guarantee that the method will not work at all. In Fig. 3 one can see that the rate of Reaction 8 oscillates in a peculiar and unphysical manner. This was deliberately caused to appear in this case (by adjusting the tolerance demanded of the integration subroutine) by way of illustrating (a) that the instability is invisible in the profiles themselves (Figs. 1 and 2) and (b) that the integration program actually used was able to continue the integration and produce meaningful profiles despite the instability.

There is an extensive literature on stiff differential equations (see, for example, Shampine and Gear, 1979; Gear, 1969), and general-purpose subroutines for solving them can be found in the libraries of most computation centers (e.g., the DVOGER subroutine of the IMSL library) or obtained in transportable FORTRAN form from software specialists (Hindmarsh, 1972; Hindmarsh and Byrne, 1977). For combustion modeling, use of standard subroutines for solving stiff sets of differential equations is common practice.

The second problem has already been noted in Section 4 in connection with the beta matrix: most species concentrations are affected by only a small fraction of the elementary reactions assumed to take place, which leads to a characteristic called "sparseness" in two important matrices, the beta matrix itself and a Jacobian matrix—the matrix of partial derivatives of the species

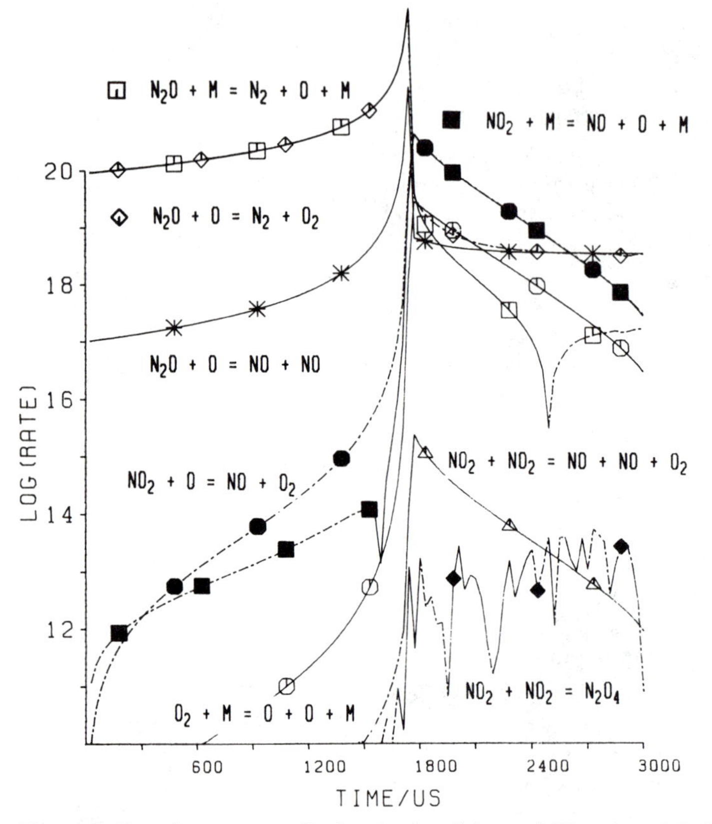

Figure 3. Reaction rate profile for the conditions of Figs. 1 and 2. Reactions proceeding in the reverse direction from that given in the text are shown as dashed lines. The oscillations in the rate of $NO_2 + NO_2 \rightarrow N_2O_4$ are not real, but result from relaxing the accuracy demand placed on the integration subroutine.

concentration derivatives with respect to species concentrations—required in the numerical integration procedures for stiff differential equations. A sparse matrix is one in which there are many more zero elements than nonzero elements. For small mechanisms such as the N_2O thermal decomposition example introduced above, this abundance of zeros is not such an aggravating problem that one needs to face it at all. Indeed, the solution shown in Figs. 1 and 2 was obtained in less than a second of computer time. The size of the beta matrix, however, grows as the product $\mathscr{R}\mathscr{S}$ and the size of the Jacobian matrix grows as $\mathscr{S}^2$ while the number of nonzero elements increases only linearly. For larger reaction mechanisms, therefore, one must confront the problem of the abundance of zero matrix elements that contribute to the computational burden but not to the solution of the equations. This is accomplished with the aid of "sparse matrix" subroutines that process matrices—in particular the Jacobian matrix—in such a manner that a large fraction of the zeros disappear. (Hindmarsh, 1979; Sherman and Hindmarsh, 1980) For typical chemical reactions, the break-even point where ignoring the zeros is no longer reasonable is at about 30 species, corresponding to combustion of C_3 or larger hydrocarbons.

No matter what integration subroutine is used, there are several important ways that a programmer can save on computer cost in modeling combustion reactions:

(1) The accuracy of the computed results usually does not need to be better than say 1 part in 10^3 or even less. Making no more than the necessary demand for integration accuracy permits the integration algorithm to use larger—sometimes much larger—step sizes and thus complete the integration with fewer operations. In modern subroutines it is even possible to assign different accuracy requirements to each dependent variable, so that only the important ones need to be computed accurately.
(2) While most of the mathematical operations of integration procedures are multiplications, additions, and subtractions, evaluating rate coefficients and thermochemical properties entails computing exponential and logarithm functions. Since these are costly of computer resources, one should minimize recomputation of rate coefficients and thermochemical properties by doing so only when the temperature has increased or decreased by an amount sufficient to make computationally significant changes in them.
(3) Output operations are significant cost items in many computer installations. It may bring a substantial saving to use output buffers, either within the program or in computer center local programs, rather than printing or graphing profiles as they are computed.
(4) The equations for the derivatives contain numerous summations over the $\mathscr{S}$ species and $\mathscr{R}$ reactions. Since the derivatives have to be computed over and over again during the integration, it is important to suppress multiplications by 1 and additions of 0 in these summations. In particular, the sum over j of $\beta_{ij}r_j$ really only needs to be done for the reactions which have nonzero matrix elements. One way to accomplish this (there are others, including some general-purpose methods) is to compute the Δc_i functions in a special-purpose subroutine specific to the particular mechanism under study. For the N_2O thermal decomposition example one could write derivative evaluations like

```
C     FIND DERIVATIVE OF N2 CONCENTRATION
      DELC(2) =
     .KF(1) * C(1) * C(8)
     .-KR(1) * C(2) * C(7) * C(8)
     .+KF(4) * C(1) * C(7)
     .-KR(4) * C(2) * C(3)
```

where the coding can be understood by comparing it with the mechanism listed at the start of Section 4 and noting that the extra concentration c_8 has been used for the "M" concentration $\sum c_i$.

(5) Frequent detailed inspection of computed results may show that useless reactions and/or species can be eliminated from a trial mechanism.

One purpose of this section has been to suggest that the bridge from the differential equation sets developed in Section 4 to implementation in a computer program should involve use of a standard library subroutine for solving stiff sets of differential equations. We conclude by noting that there also exist a number of prepackaged computer programs ready for use in combustion modeling (under the same constraints considered in this chapter) by users who prefer trusting other people's programs to writing their own. Without making a recommendation one way or the other on this point we simply note that such packages are described by Turner, Emanuel, and Wilkins (1970), Bittker and Scullin (1972), and Kee, Miller, and Jefferson (1980).

6. Interpretation of combustion modeling profiles

The expositions of basic terminology and computation methods given so far are concerned with explaining how to compute concentration and physical conditions profiles for combustion reactions taking place under simple physical constraints. Now we consider what can be done with such profiles.

Obviously, the ideal thing to do with computed species profiles would be to compare them with experimental ones and to figure out—by studying how the profiles and the comparisons are affected by varying the mechanism and the assumed rate-coefficient expressions—just how combustion reactions occur. This is indeed only an ideal, however, because it is rarely possible to measure enough profiles and even more rarely possible to understand the effects of variation in rate-coefficient expression without substantial help from the computer itself. Somehow it is required to make the computation meet the experiment in a meaningful way. The mathematical aspects of this are discussed in Chapter 7. Here we briefly mention three chemical aspects that users of detailed combustion modeling programs should keep in mind.

First, it should be recognized that there is a hierarchy of confidence that can be placed in rate-coefficient expressions. Some are accurate and some are only more or less educated guesses. In variation of rate-coefficient parameters for sensitivity analysis, a user must bear in mind that the ranges of uncertainty studied must go from effectively zero to the theoretical maximum values (Chapters 3 and 4) for some reactions. The elementary combustion reactions that presently stand high in this hierarchy are discussed in Chapters 5 and 6.

Second, it should be kept in mind that comparisons between theory and experiment can be no more accurate than the degree to which the physical constraints adopted for the modeling equations are faithful representations of the constraints that actually pertain in the laboratory. Some experiments lend themselves, by choice of reaction conditions, better to modeling than others.

Finally, it is unlikely that species profiles themselves will be effective objects for contemplation while undertaking a modeling study. Instead, one can come more rapidly to an understanding of the model predictions by *parameterizing* the profiles in ways that facilitate comparisons with experiment. (Gardiner, 1977, 1979) An important direction of thought at the outset of a modeling investigation is into selection of the most suitable parameters for characterizing the computed and experimental results, and the main expression of the final conclusions from the modeling study should be in terms of how well these parameters agree with one another. Agreement—or lack thereof—between rate-coefficient parameters used in the study and elsewhere likewise demands the most careful consideration before, during, and after a modeling study of combustion chemistry.

7. References

Bittker, D. A. & Scullin, V. J. (1972). *General chemical kinetics computer program for static or flow reactors, with application to combustion and shock-tube kinetics*, NASA Technical Note D-6586.

Gardiner, W. C., Jr. (1977). *J. Phys. Chem.* **81,** 2367.

Gardiner, W. C., Jr. (1979). *J. Phys. Chem.* **83,** 37.

Gardiner, W. C., Jr., Walker, B. F., & Wakefield, C. B. (1981). In *Shock Waves in Chemistry*, A. Lifshitz, Ed., Dekker, New York, p. 319.

Gear, C. W., 1969. *Numerical Initial Value Problems in Ordinary Differential Equations*, Prentice-Hall, Englewood Cliffs, N.J.

Hindmarsh, A. C. (1972). *GEAR: Ordinary differential equation system solver*, Technical Report No. UCID-30001, Rev. 2, Lawrence Livermore Laboratory, Livermore, California.

Hindmarsh, A. C. (1979). *A collection of software for ordinary differential equations*, Lawrence Livermore Laboratory preprint UCRL-82091.

Hindmarsh, A. C. & Byrne, G. D. (1977). *EPISODE*, Report No. UCID-30112, Rev. 1, Lawrence Livermore Laboratory, Livermore, California.

Sherman, A. H. & Hindmarsh, A. C. (1980). *GEARS: A package for the solution of sparse, stiff ordinary differential equations*, Lawrence Livermore Laboratory preprint UCRL-84012.

Kee, R. J., Miller, J. A., & Jefferson, T. H. (1980). *CHEMKIN: A general-purpose, problem-independent, transportable, fortran chemical kinetics code package*, Sandia National Laboratory Report SANBO-8003.

Liepmann, H. W. & Roshko, A. (1957). *Elements of Gasdynamics*, Van Nostrand, New York, Chapter 2.

Shampine, L. F. & Gear, C. W. (1979). *SIAM Review* **21,** 1.

Turner, E. B., Emanuel, G., & Wilkins, R. L. (1970). *The NEST chemistry computer program*, Vol. 1, Aerospace Report No. TR-0059(6420-20)-1, Aerospace Corporation, El Segundo, California.

Chapter 2

Computer Modeling of Combustion Reactions in Flowing Systems with Transport

Graham Dixon-Lewis

1. Introduction

In combustion systems, the strongly exothermic processes of fuel oxidation may give rise to localized reaction zones which propagate themselves into the unreacted material near them. There are two distinct mechanisms of propagation, deflagration and detonation. Deflagrations, travelling through the unburnt material at subsonic velocities, depend for their propagation on activation of adjacent material to a reactive condition by diffusive transport processes. Detonations, on the other hand, propagate at supersonic velocities by virtue of gasdynamic (shock) compression and heating of adjacent material, the shock wave itself being sustained by the energy release from the combustion process. In both cases the reaction zone propagates as a consequence of strong coupling between the combustion chemistry and the appropriate fluid mechanical process.

Despite its title, and although it contains discussion of relevant numerical techniques, this article is not a comprehensive survey of the numerical methods currently employed in detailed combustion modeling. For that, the reader is referred to the reviews by McDonald (1979) and Oran and Boris (1981). Rather, the aim here is to provide an introduction that will stimulate interest and guide the enthusiastic and persistent amateur. The discussion will center mainly about low-velocity, laminar, premixed flames, which form a substantial group of reactive flow systems with transport. Present computational capabilities virtually dictate that such systems be studied as quasi-one-dimensional flows. We also consider two-dimensional boundary layer flows, in which the variation of properties in the direction of flow is small compared with the variation in the cross-stream direction. The extension of the numerical methods to multidimensional flows is straightforward in principle, but implementation at acceptable cost is much more difficult.

2. Conservation or continuity equations, and other useful relations

In order to analyze flames and other reacting systems in terms of basic flow principles and chemical kinetics, it is necessary to consider the conditions in a differential control volume within the reacting fluid. To this control volume we must apply the four fundamental conservation principles which are the basis of all physical and chemical problems: conservation of mass, momentum, energy, and atoms. For each of these, the fundamental continuity equation for a volume element states that the rate of accumulation of the quantity within the element is equal to the rate of gain due to flow plus the rate of gain due to reaction, i.e., for a quantity denoted by the subscript k, and referring to unit volume

$$\partial \rho_k/\partial t = -\nabla \cdot \mathbf{F}_k + q_k \tag{2.1}$$

where ρ_k is the density of the quantity within the element, $\mathbf{F}_k$ is its flux (a vector quantity), and q_k is its rate of formation per unit volume. All three are functions both of position and time. The term $\nabla \cdot \mathbf{F}_k$, the divergence of the flux, represents the rate of loss of the quantity per unit volume due to the flow. The expression to be used for the divergence depends on the coordinate system employed, which in turn is chosen so as best to fit the flow problem under consideration. The coordinate systems commonly used in flame work are the Cartesian, cylindrical polar (for flows with symmetry about an axis), and spherical polar systems. The properties of these systems are as follows, where vector notation has been discarded in relation to the components and appropriate symmetry has been assumed

(1) *Cartesian coordinates* x, y, z
Volume element: parallelepiped with sides $\delta x, \delta y, \delta z$
Flux components along axes: F_x, F_y, F_z

$$\nabla \cdot \mathbf{F} = \partial F_x/\partial x + \partial F_y/\partial y + \partial F_z/\partial z \tag{2.2a}$$

(2) *Cylindrical polar coordinates* Radial coordinate r
Axial coordinate z
Volume element: annulus with radial thickness δr and axial height δz
Flux components: F_r, F_z

$$\nabla \cdot \mathbf{F} = \partial F_z/\partial z + (1/r)\partial/\partial r(rF_r) \tag{2.2b}$$

(3) *Spherical polar coordinates* Radial coordinate r
Volume element: spherical shell with thickness δr.

$$\nabla \cdot \mathbf{F} = (1/r^2)\,\partial/\partial r\,(r^2 F_r) \tag{2.2c}$$

The divergences are obtained from analysis of the flow into and out of the volume element at each face. The spherical polar system is naturally one dimensional. The other systems become effectively so if the distance derivatives of an appropriate number of the flux components become zero. Thus, a flat flame lying in the x-z plane of Cartesian coordinates and fed by a nondivergent gas flow will have both $\partial F_x/\partial x$ and $\partial F_z/\partial z = 0$ [Eq. (2.2a)]. Similarly, a flame front developing from a semi-infinite line source of ignition will have $\partial F_z/\partial z = 0$ [Eq. (2.2b)].

Computer modeling with detailed reaction mechanisms has so far been confined essentially to such one-dimensional flow systems, and particularly to studies of flat, laminar, premixed flames for which profiles of temperature and composition as a function of distance y through the flame front may be measured. Figure 1 illustrates the main aerodynamic features of such a flame. In addition to the convective or advective fluxes through the flame there are transport fluxes along the directions of the temperature and composition gradients, and it is the coupling of these with the combustion chemistry which causes the flame to propagate. If the flame is in the x-z plane of Cartesian coordinates, so that all derivatives in this plane are zero, the generalized continuity equation for the one-dimensional system becomes

$$\partial\rho_k/\partial t + \partial F_{ky}/\partial y = q_k \tag{2.3}$$

We shall further assume for convenience that both F_{kx} and F_{kz} are zero, in which case $\mathbf{F}_k = F_{ky}$.

In reality, a true one-dimensional flame with constant stream tube area is not stable to flow perturbations. For stability of a stationary flame on a burner the flow must be somewhat divergent towards the hot side. In such a situation we may still treat the system as pseudo-one-dimensional by including an allowance for the change in the stream tube area A normal to the flow. The continuity equation then becomes

$$\partial\rho_k/\partial t + (1/A)\,\partial/\partial y\,(AF_{ky}) = q_k \tag{2.4}$$

Equation (2.4) will be used in the following sections as the basic continuity equation, with the understanding that in a true one-dimensional Cartesian system $A = 1.0$ for all y.

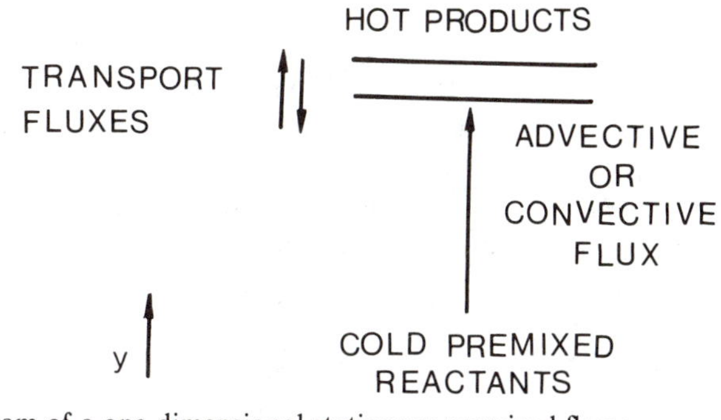

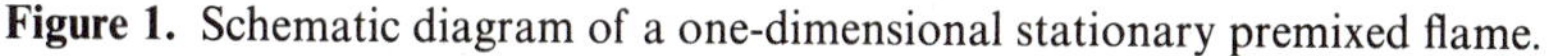
Figure 1. Schematic diagram of a one-dimensional stationary premixed flame.

It is necessary next to develop the individual conservation equations. We continue to use the one-dimensional premixed flame as the practical system exhibiting the basic features to be modeled. The principles of extension to two or three dimensions are not difficult. For a comprehensive review the reader is referred to the standard texts in fluid dynamics, e.g., Landau and Lifshitz (1959) or Williams (1965).

2.1. Conservation of total mass

Here ρ_k and F_{ky} are the density ρ (g cm^{-3}) and the overall mass flux M_y (g cm^{-2} s^{-1}), respectively, while q_k is the rate of formation of mass, i.e., zero. Conservation of mass thus leads to

$$\partial\rho/\partial t + (1/A)\,\partial/\partial y\,(AM_y) = 0 \tag{2.5}$$

When the rate of accumulation $\partial\rho_k/\partial t$ of all quantities in the system is zero, the condition is known as the steady state. All other systems are time dependent. Stationary flames supported on burners are steady-state phenomena, and so for a quasi-one-dimensional stationary flame $d(AM_y)/dy = 0$ and AM_y = constant. In the hypothetical case of a true one-dimensional adiabatic flame the constant M_y is the adiabatic mass burning velocity. It is an eigenvalue solution of the physical problem, equal to the product of the density and linear velocity of the gas at any position in the flame. Thus

$$M_y = \rho_u S_u = \rho_b S_b = \text{constant} \tag{2.6}$$

where the subscripts u and b refer to unburnt and burnt gases, respectively, and S is the linear burning velocity.

2.2. Conservation of *y*-direction momentum

With the introduction of $\rho_k \equiv \rho v_y$ and $F \equiv M_y v_y$, Eq. (2.4) becomes

$$\partial/\partial t\,(\rho v_y) + (1/A)\,\partial/\partial y\,(AM_y v_y) = -\partial/\partial y\,(P + Q) \tag{2.7}$$

Here v_y is the scalar value of the y-direction velocity, P is the pressure, and Q is a viscous pressure, which may be real or artificial. In the context of the equation $(P + Q)$ is the component $(P + Q)_{yy}$ of the pressure tensor $\mathscr{P}$ defined by Eq. (3.28). This component represents the stress normal to the x-z plane. The right-hand side of Eq. (2.7) represents the body force acting on unit volume of the fluid, and is equal by Newton's second law to a rate of formation of momentum. In the absence of the viscosity term, Eq. (2.7) is a version of Euler's equation for the motion of an ideal (nonviscous) fluid. By differentiating out the products in Eq. (2.7) and then substituting from Eq. (2.5) the momentum equation may be reduced to

$$\partial v_y/\partial t + v_y \partial v_y/\partial y = -(1/\rho)\,\partial/\partial y\,(P + Q) \tag{2.7a}$$

The viscous pressure Q is given in laminar flow by $Q = -\{(4/3)\eta + \zeta\}\,\partial v_y/\partial y$, where η and ζ are the coefficients of shear and volume viscosity, respectively. Volume viscosity effects are negligible in combustion systems.

An *artificial* viscosity is introduced into the equations for systems where strong shocks are present (Richtmyer and Morton, 1967) in order to give stability to computation. It does not specifically concern us here. Further, it is only when treating flows approaching sonic velocities that it is necessary to consider the momentum equations at all. In slower-burning laminar flames, viscous effects and the change of velocity of the gases as they pass through the flame front are *not* large enough to require a significant pressure drop. Open flames, that is, flames not confined in containers, can therefore be regarded as *constant-pressure* systems, and we do not need to consider the force-momentum balance at all.

Specifically, the condition for neglect of the momentum equation can be seen by integrating the steady-state version of Eq. (2.7a) with (ρv_y) constant and viscous pressure neglected. The integration leads to

$$v_y^2 + P/\rho = \text{constant} \tag{2.7b}$$

Hence, if $v_y^2 \ll P/\rho$, meaning that the ordered kinetic energy of the flow is much less than the random thermal energy, the fractional pressure changes that can occur remain negligibly small. The condition $v_y^2 \ll P/\rho$ simply implies low Mach number flow.

2.3. Species equations. Conservation of atoms

The species continuity equations ensure conservation of atoms by balancing the chemical equations governing the reaction system. The concentration of an individual species i will be expressed either as a product of total molar density and mole fraction X_i, or as a product of density and mass fraction w_i. In an N-component system mass fraction and mole fraction are related by

$$w_i = \frac{m_i X_i}{\sum_{j=1}^{N} (m_j X_j)} \tag{2.8}$$

$$X_i = \frac{w_i/m_i}{\sum_{j=1}^{N} (w_j/m_j)} \tag{2.8a}$$

where m denotes molecular weight. Further

$$\rho = \rho_{\text{molar}} \sum_j (m_j X_j) = \frac{\rho_{\text{molar}}}{\sum_j (w_j/m_j)} \tag{2.9}$$

The species continuity equation then becomes

$$\partial/\partial t\,(\rho w_i) + (1/A)\,\partial/\partial y\,(AF_{iy}) = q_i \tag{2.10}$$

where F_{iy} is the (y-direction) flux of the species i and q_i is its mass chemical rate of formation. There is one equation [Eq. (2.10)] for each species present.

Because of the species concentration gradients in flames the fluxes F_{iy} consist of two parts: a convective flux ($M_y w_i$) due to the overall mass flow and a mass diffusive flux j_i. The expressions for the diffusive fluxes will be discussed in Section 3. In the one-dimensional case we may write

$$F_{iy} = M_y w_i + j_i \tag{2.11}$$

$$= M_y G_i \tag{2.11a}$$

where G_i is the mass flux fraction, introduced into the treatment of one-dimensional flames by Hirschfelder, Curtiss, and Bird (1954), and used below. For a stationary, truly one-dimensional flame with M_y constant

$$q_i = M_y(dG_i/dy) \tag{2.12}$$

The situation with variable stream tube area (and hence M_y varying with y in the steady state) is slightly more complex. For an appropriately curved reaction zone in cylindrical or spherical coordinates, Eq. (2.11) may again be substituted directly into Eq. (2.10) to give

$$\partial/\partial t\,(\rho w_i) + (1/A)\,\partial/\partial y\,\{A(M_y w_i + j_i)\} = q_i \tag{2.10a}$$

If, however, a planar reaction zone is fed by a divergent flow, Eq. (2.10) should be replaced by

$$\partial/\partial t\,(\rho w_i) + (1/A)\,\partial/\partial y\,(AM_y w_i) + \partial j_i/\partial y = q_i \tag{2.10b}$$

For simplicity, the further development will treat only the first of these alternatives.

The chemistry is represented in Eqs. (2.10) by q_i, which may be expressed as

$$q_i = m_i R_i \tag{2.13}$$

where R_i is the net molar rate of formation of the species. For the elementary reaction

$$A + B \underset{k_{-1}}{\overset{k_1}{\rightleftharpoons}} C + D$$

the net molar forward rate is given by

$$\begin{aligned} R &= k_1 \rho_{\text{molar}}^2 X_A X_B - k_{-1} \rho_{\text{molar}}^2 X_C X_D \\ &= k_1 \rho_{\text{molar}}^2 (X_A X_B - X_C X_D/K_1) \\ &= k_1 \rho^2 (\sigma_A \sigma_B - \sigma_C \sigma_D/K_1) \end{aligned} \tag{2.14}$$

where k_1 and k_{-1} are rate coefficients, $K_1 = k_1/k_{-1}$ is the equilibrium constant of the reaction, $\sigma_i \equiv w_i/m_i$, and the last line of the equation follows from Eqs. (2.8) and (2.9). Both the rate coefficients and the equilibrium constant are functions of temperature, and sometimes also of pressure. The temperature dependence is usually expressed by

$$k = AT^B \exp(-C/T) \tag{2.15}$$

or

$$K = DT^E \exp(-F/T) \tag{2.16}$$

The foundations of these expressions are discussed in Chapters 3, 4 and 8 along with the various types of elementary reactions which may be encountered. In the case of a chain reaction, the overall rate q_i for each species is obtained by summation of the rates of its formation in each elementary reaction in which it occurs.

The last line of Eq. (2.14) has introduced the variables $\sigma_i \equiv w_i/m_i$. Being simply related to both the mass fraction and the mole fraction, these variables are useful in the discussion of reactive flows. In terms of σ_i, Eq. (2.10a) becomes, after dividing throughout by m_i,

$$\partial/\partial t\,(\rho\sigma_i) + (1/A)\,\partial/\partial y\,\{A(M_y\sigma_i + j_i/m_i)\} = R_i \tag{2.17}$$

Differentiating out the products and substituting from Eq. (2.5) then leads to

$$\rho(\partial\sigma_i/\partial t) + M_y(\partial\sigma_i/\partial y) = -(1/A)\,\partial/\partial y\,(Aj_i/m_i) + R_i \qquad (i = 1, 2, \ldots, N) \tag{2.18}$$

2.4. Conservation of energy

Let $\hat{U}_i$ be the specific internal energy (internal energy per unit mass) of species i, and Q_T and Q_D be the energy fluxes in the y-direction due to thermal conduction and diffusion, respectively. If account is taken also of the change in ordered kinetic energy consequent upon the acceleration of the gases as they flow through the quasi-one-dimensional system, then the conservation of energy may be expressed as

$$\partial/\partial t\left\{\sum_i \rho w_i(\hat{U}_i + \tfrac{1}{2}v_y^2)\right\} + (1/A)\,\partial/\partial y\left\{A\left[\sum_i M_y w_i(\hat{U}_i + \tfrac{1}{2}v_y^2 + (P + Q)/\rho) + Q_D + Q_T\right]\right\} = q_E \tag{2.19}$$

where q_E is the rate of receipt of energy per unit volume by heat transfer from the surroundings and the term $(P + Q)/\rho$ arises from the work done against pressure and viscous forces by the fluid within the control volume. Since $(\hat{U}_i + P/\rho)$ is equal to the specific enthalpy h_i, and since $\sum_i w_i\hat{U}_i = \hat{U}$ and

$\sum_i w_i h_i = h$, where $\hat{U}$ and h refer to the overall mixture, then Eq. (2.19) may be rewritten as

$$\partial/\partial t\{\rho(\hat{U} + \tfrac{1}{2}v_y^2)\} + (1/A)\partial/\partial y\{A[M_y(h + \tfrac{1}{2}v_y^2) + v_y Q + Q_D + Q_T]\} = q_E \tag{2.19a}$$

or, by differentiation of the products and substitution from Eq. (2.5)

$$\rho\{\partial/\partial t\,(h + \tfrac{1}{2}v_y^2)\} + M_y\{\partial/\partial y(h + \tfrac{1}{2}v_y^2)\}$$
$$= q_E - (1/A)\,\partial/\partial y\,\{A(v_y Q + Q_D + Q_T)\} + \partial P/\partial t \tag{2.20}$$

The kinetic energy terms in this equation may be modified by rewriting Eq. (2.7a) as

$$\rho\{\partial/\partial t\,(\tfrac{1}{2}v_y^2)\} + M_y\{\partial/\partial y\,(\tfrac{1}{2}v_y^2)\} = \rho v_y\{(\partial v_y/\partial t) + v_y(\partial v_y/\partial y)\}$$
$$= -v_y\{\partial/\partial y\,(P + Q)\} \tag{2.7c}$$

With this substitution Eq. (2.20) becomes

$$\rho(\partial h/\partial t) + M_y(\partial h/\partial y) = q_E - (1/A)\,\partial/\partial y\,\{A(Q_D + Q_T)\} + DP/Dt$$
$$- (Q/A)\,\partial/\partial y\,(v_y A) \tag{2.20a}$$

where $DP/Dt \equiv \partial P/\partial t + v_y(\partial P/\partial y)$ in the one-dimensional system.

At constant pressure the final term on the right-hand side of Eq. (2.20) is zero. If at the same time, as is usually possible, we can neglect the contributions of the kinetic energy terms and the viscous term, Eq. (2.20) becomes, for constant pressure,

$$\rho(\partial h/\partial t) + M_y(\partial h/\partial y) = q_E - (1/A)\,\partial/\partial y\,\{A(Q_D + Q_T)\} \tag{2.20b}$$

The constant-pressure subgroup of systems is frequently encountered because of its relevance to laminar flames. Letting H_i be the molar enthalpy of the species i, then since $\sum_i j_i = 0$, we have $Q_D = \sum_i j_i h_i = \sum_i j_i H_i/m_i$. For the same conditions that apply to Eq. (2.20b) the total energy flux in the y-direction of the one-dimensional flame corresponding with Eq. (2.12) is given by $F_E = M_y \sum_i(G_i h_i) + Q_T$, and for the stationary flame

$$d/dy\left\{M_y \sum_i (G_i h_i) + Q_T\right\} = q_E \tag{2.21}$$

For an adiabatic flame (i.e., a flame not losing energy by radiation or to walls) $q_E = 0$, and since $Q_T = 0$ also in the unburnt gas of the free flame, where all the gradients are zero, Eq. (2.21) becomes

$$M_y \sum_i (G_i h_i - G_{iu} h_{iu}) = -Q_T \tag{2.22}$$

where the additional subscript u refers to the unburnt gas.

Finally, it should be noted that Eqs. (2.19) to (2.22) refer to the conservation of *total* (i.e., chemical *plus* thermal) energy. To illustrate the parallel between the release of sensible heat and the formation of other combustion products, we note first that for the conditions of Eq. (2.20b) the energy equation for the adiabatic flame ($q_E = 0$) may be written as

$$\partial/\partial t\left(\sum_i \rho\sigma_i H_i\right) + (1/A)\,\partial/\partial y\left\{A\left(\sum_i M_y\sigma_i H_i + Q_D + Q_T\right)\right\} = 0 \quad (2.23)$$

Again noting that $Q_D = \sum_i j_i H_i/m_i$, differentiating out the products and substitution from Eq. (2.17) leads to

$$\sum_i \{\rho\sigma_i(\partial H_i/\partial t)\} + \sum_i \{(F_{iy}/m_i)\,\partial H_i/\partial y\} = -(1/A)\,\partial/\partial y(AQ_T) - \sum_i (R_i H_i) \quad (2.24)$$

Here $-\sum_i(R_i H_i)$ is the rate of heat release by the chemical reaction.

2.5. Auxiliary equations

In addition to the continuity equations and the expressions for rate coefficients and equilibrium constants, several further relationships are required in order to close the system of equations. Apart from the expressions for the transport fluxes, these are

(1) The ideal gas equation of state

$$P/\rho = \left(\sum_i \sigma_i\right) RT \quad (2.25)$$

where R is the molar gas constant and $(\sum_i \sigma_i)^{-1}$ is the average molecular weight.

(2) A relation between enthalpy and temperature for each species present. These relations may be provided either by linear interpolation in stored tables of thermal data, or by polynomial fits of such data. (Chapter 8) We then have

$$h = \sum_i (\sigma_i H_i) \quad (2.26)$$

(3) Rather than set up a species continuity equation for all N components of the reacting system, the physical nature of the problem is often preserved against numerical rounding errors by replacing one of the equations by

$$\sum_i w_i = 1 \quad (2.27)$$

2.6. Eulerian and Lagrangian coordinate reference frames

The continuity equations of Sections 2.1 to 2.4 have been derived for a reference frame which is stationary in space. This is known as an Eulerian frame of reference, and in it at the steady state the time derivatives vanish, as in

Eqs. (2.6), (2.12), and (2.21). This reference frame is particularly useful for examination of the properties of steady-state reaction systems such as flames on burners, which are themselves stationary in space. The convection terms remain in the equations.

On the other hand, the development of a flame from a localized ignition source is more conveniently studied as a time-dependent phenomenon that eventually reaches a steady propagation rate in space. Such a phenomenon is more conveniently studied in a coordinate system in which the observer moves with the fluid and the properties at a point change as the reaction zone passes by. Such a coordinate system is known as a Lagrangian reference frame, and the time derivatives in it become the substantial derivatives, denoted by D/Dt, where $D/Dt \equiv \partial/\partial t + v_y \partial/\partial y$ in the one-dimensional system. (In general $D/Dt \equiv \partial/\partial t + \mathbf{v} \cdot \text{grad}$). Remembering that $M_y = \rho v_y$, Eqs. (2.5), (2.7a), (2.18), and (2.20b) transform in the Lagrangian reference frame to

$$D\rho/Dt = -(\rho/A)\,\partial/\partial y\,(v_y A) \tag{2.28}$$

$$Dv_y/Dt = -(1/\rho)\,\partial/\partial y\,(P + Q) \tag{2.29}$$

$$D\sigma_i/Dt = -(1/\rho)\{(1/A)\,\partial/\partial y\,(Aj_i/m_i) - R_i\} \tag{2.30}$$

$$Dh/Dt = -(1/\rho)\{(1/A)\,\partial/\partial y\,[A(Q_D + Q_T)] - q_E\} \tag{2.31}$$

The convection terms have now disappeared from the equations.

2.7. Space integral rate

When the unsteady-state equations are being integrated forward in time either towards the Eulerian steady state or towards the Lagrangian steady propagation rate, a criterion is required for deciding when the steady situation is achieved. In both cases this is provided by the space integral reaction rates $\int_{-\infty}^{+\infty} q_i A\, dy$, which for sufficiently small finite difference intervals may be replaced by the summations $\sum_{n=-\infty}^{+\infty} (q_i A'\,\delta y)_n$ or an equivalent quantity, where $(A'\,\delta y)_n$ is the volume of the nth element and A' a weighted mean area within the element. For the quasi-one-dimensional steady-state flame we have from Eq. (2.12)

$$\int_{y_1}^{y_2} q_i A\, dy = (M_y A)\{(G_i)_2 - (G_i)_1\} \tag{2.32}$$

and in particular

$$\int_{-\infty}^{+\infty} q_i A\, dy = (M_y A)(G_{ib} - G_{iu}) \tag{2.32a}$$

Since both G_{iu} and G_{ib} are known from the boundary conditions defining the specific flame problem (Section 4), the space integral rate provides a means of extracting the current value of the product $(M_y A)$ from an unsteady-state calculation and of investigating its variation with time. Provided that inaccuracy is not introduced by near equality of G_{iu} and G_{ib}, the computed steady-state value of the product $(M_y A)$ should also be independent of the species selected for its computation.

3. Formulation of transport fluxes

The next problem is the formulation of the transport fluxes. The precision with which it must be done for multicomponent mixtures is a matter on which there is diversity of opinion. This arises because the more precise formulations rapidly become more expensive in computer time but have uncertain returns in terms of accuracy because of shortcomings in the primary data. The author is of the opinion that a detailed transport flux model is a necessary reference representation on which approximations may be based so as to give the most satisfactory compromise for a specific problem. Fortunately, the various degrees of approximation result in formally similar continuity equations which can be treated by the same numerical techniques. Readers interested in the general aspects of modeling reactive flows are advised to omit this section on first reading and to accept the results given at the commencement of Section 4.

For gases having no internal degrees of freedom, the classical Chapman-Enskog procedure provides the most satisfactory means of obtaining the fundamental transport relationships. The procedure has been described by Chapman and Cowling (1939, 1970) and Hirschfelder *et al.* (1954), and is further discussed for systems containing more than two components by Hellund (1940), Waldmann (1947, 1958), Curtiss and Hirschfelder (1949), Hirschfelder *et al.* (1954), and others. For the contribution of ordinary diffusion to the diffusive flux, the expressions derived from this theory may be used without change to derive first approximations for systems of polyatomic gases. However, for thermal conduction and thermal diffusion, application to polyatomic gases requires modification of the original theory to include effects of (a) the diffusive flux of internal energy and (b) inelastic collisions. A formal theory incorporating the necessary modifications for polyatomic gases was proposed by Wang Chang and Uhlenbeck (1951), by de Boer (unpublished), and by Taxman (1958), and is discussed by Wang Chang, Uhlenbeck, and de Boer (1964). The procedure follows the Chapman-Enskog method, but instead of considering only the translational velocity distribution function for each species in the gas, it regards each quantum state of each species as a separate entity, and thus considers internal energy states also. The present section begins by outlining this modification of the Chapman-Enskog procedure and the subsequent development by Mason, Monchick and coworkers (1961 to 1966), and then goes on to discuss its implementation. The notation used is that of Mason, Monchick *et al.*, and Hirschfelder *et al.* (1954). The initial account follows closely that by Dixon-Lewis (1968).

3.1. Transport processes in mixtures of nonpolar gases

(a) Extended Chapman-Enskog procedure for polyatomic gas mixtures

In molecular terms, the local values of the diffusion velocity $\bar{\mathbf{V}}_i(\mathbf{r}, t)$ of the ith component and the local value of the heat flux vector $\mathbf{q}'(\mathbf{r}, t)$ in a volume element $d\mathbf{r}$ in a nonuniform N-component mixture of polyatomic gases are given by

$$\bar{\mathbf{V}}_i = (1/n_i)\sum_I \int V_i f_{iI}\, d\mathbf{v}_i \tag{3.1}$$

$$\begin{aligned}\mathbf{q}' &= \mathbf{q}'_{\text{tr}} + \mathbf{q}'_{\text{int}} \\ &= \sum_{i=1}^{N}\sum_I \int \{(1/2)m'_i V_i^2 + E_{iI}\}\mathbf{V}_i f_{iI}\, d\mathbf{v}_i\end{aligned} \tag{3.2}$$

Here $n_i(\mathbf{r}, t)$ is the local number density of species i, the index I denotes the Ith quantum state, m'_i is the molecular mass, and $f_{iI}(\mathbf{v}_i, E_{iI}, \mathbf{r}, t)$ is the local value of the distribution function for the molecular velocity $\mathbf{v}_i(\mathbf{r}, t)$ and internal energy E_{iI}, normalized so that

$$n_i = \sum_I \int f_{iI}\, d\mathbf{v}_i \tag{3.3}$$

and $\mathbf{V}_i(\mathbf{r}, t) = \mathbf{v}_i(\mathbf{r}, t) - \mathbf{v}_0(\mathbf{r}, t)$, where $\mathbf{v}_0(\mathbf{r}, t)$ is the local mass average velocity given by

$$\mathbf{v}_0 = \left(1\Big/\sum_i m'_i n_i\right)\sum_i \left\{m'_i \sum_I \int f_{iI}\mathbf{v}_i\, d\mathbf{v}_i\right\} \tag{3.4}$$

In order to integrate Eqs. (3.1) and (3.2), it is necessary to know the distribution function f_{iI}. In the absence of external forces, the rate of decay of a nonequilibrium velocity distribution to an equilibrium one is given by the generalized Boltzmann equation

$$\begin{aligned}Df_{iI}/Dt = \sum_I \sum_{J,K,L} \int \cdots \int \{&f'_{iK} f'_{1jL} g' I_{KL}^{IJ}(\mathbf{g}' \to \mathbf{g}, \psi, \phi) \\ &- f_{iI} f_{1jJ} g I_{IJ}^{KL}(\mathbf{g} \to \mathbf{g}', \psi, \phi)\} \sin(\psi)\, d\psi\, d\phi\, d\mathbf{v}_{1j}\end{aligned} \tag{3.5}$$

which states that the velocity distribution in a fluid element changes as a result of collisions. There is one equation (3.5) for each quantum state of each component in the mixture. The indices I and J denote the Ith and Jth quantum states, respectively, of components i and j before collision, and K and L the corresponding states after collision. The subscript 1 is an extra index used to distinguish one of the collision partners from the other in case i and j are the same. The primes indicate values after collision. $I_{IJ}^{KL}(\mathbf{g} \to \mathbf{g}', \psi, \phi)$ is the differential scattering cross section for the process ($\mathbf{g} \equiv \mathbf{v} - \mathbf{v}_1 \rightarrow \mathbf{v}' - \mathbf{v}_1' \equiv \mathbf{g}'$, $E_{iI} \rightarrow E_{iK}$, $E_{jJ} \rightarrow E_{jL}$), where ψ and ϕ are the polar and azimuthal angles, respectively, describing the orientation of $\mathbf{g}'$ relative to $\mathbf{g}$.

The *equilibrium* Boltzmann distribution function $f_{iI}^{(0)}$ is given by

$$f_{iI}^{(0)} = (n_i/Q_i)(m_i'/2\pi k_B T)^{3/2} \exp(-W_i^2 - \varepsilon_{iI}) \tag{3.6}$$

where k_B is the Boltzmann constant and

$$\mathbf{W}_i = (m_i'/2k_B T)^{1/2}\mathbf{V}_i$$

$$\varepsilon_{iI} = E_{iI}/k_B T$$

$$Q_i = \sum_I \exp(-\varepsilon_{iI})$$

and now

$$n_i = \sum_I \int f_{iI}^{(0)}\, d\mathbf{v}_i \tag{3.7a}$$

$$\mathbf{v}_0 = \left(1\Big/\sum_i m_i' n_i\right)\sum_i \left\{m_i' \sum_I \int f_{iI}^{(0)}\mathbf{v}_i\, d\mathbf{v}_i\right\} \tag{3.7b}$$

$$\hat{U}^{(0)}(\mathbf{r}, t) = \hat{U}_{\mathrm{tr}}^{(0)}(\mathbf{r}, t) + \hat{U}_{\mathrm{int}}^{(0)}(\mathbf{r}, t)$$

$$= \left(1\Big/\sum_i m_i' n_i\right)\sum_i \sum_I \int f_{iI}^{(0)}\{(1/2)m_i' V_i^2 + E_{iI}\}\, d\mathbf{v}_i \tag{3.7c}$$

The temperature T, defined by Eqs. (3.6) and (3.7c), implies that a single temperature is sufficient to describe the distribution of energy among the translational *and* internal degrees of freedom. While this is true in the equilibrium case, it is only strictly true in nonequilibrium cases (with which we are concerned) when the exchange between internal and translational energy is rapid.

If the *nonequilibrium* distribution function f_{iI} is expressed as a perturbation expansion

$$f_{iI} \approx f_{iI}^{(0)}(1 + \phi_{iI}) \tag{3.8}$$

it can be shown that ϕ_{iI} is of the form

$$\phi_{iI} = -\mathbf{A}_{iI} \cdot \nabla_r \ln(T) - \mathscr{B}_{iI}{:}\nabla_r \mathbf{v}_0 + n \sum_{h=1}^{N} \mathbf{C}_{iI}^h \cdot \mathbf{d}_h - D_{iI}\nabla_r \cdot \mathbf{v}_0 \tag{3.9}$$

where

$$\mathbf{d}_h = \nabla_r X_h + (X_h - w_h)\nabla_r \ln(P) \tag{3.10}$$

and the vector functions $\mathbf{A}_{iI}$ and $\mathbf{C}_{iI}^h$, the tensor function $\mathscr{B}_{iI}$, and the scalar function D_{iI} are to be determined. Substitution of Eqs. (3.8) and (3.9) into the Boltzmann equation (3.5) and evaluation of the result then separates the single integral equation into separate equations for $\mathbf{A}_{iI}$, $\mathscr{B}_{iI}$, $\mathbf{C}_{iI}^h$, and D_{iI}. The equation in D_{iI} gives rise to the volume viscosity, and since such effects are negligible in the context of reactive flows, this equation will not be developed further. The equations for $\mathbf{A}_{iI}$, $\mathscr{B}_{iI}$, and $\mathbf{C}_{iI}^h$ become, after some manipulation,

$$f_{iI}^{(0)}\{(W_i^2 - 5/2) + (\varepsilon_{iI} - \bar{\varepsilon}_i)\}\mathbf{V}_i = \sum_j \sum_{J,K,L} \int \cdots \int \{(\mathbf{A}_{iI} + \mathbf{A}_{1jJ})gI_{IJ}^{KL} - (\mathbf{A}'_{iK} + \mathbf{A}'_{1jL})g'I_{KL}^{IJ}\} f_{iI}^{(0)} f_{1jJ}^{(0)} \sin(\psi) d\psi\, d\phi\, d\mathbf{v}_{1j} \tag{3.11}$$

$$\bar{\varepsilon}_i = (1/Q_i) \sum_I \varepsilon_{iI} \exp(-\varepsilon_{iI}) \tag{3.12}$$

$$f_{iI}^{(0)}\{2(\mathbf{W}_i\mathbf{W}_i - (1/3)W_i^2\mathscr{U})\} = \sum_j \sum_{J,K,L} \int \cdots \int \{(\mathscr{B}_{iI} + \mathscr{B}_{1jJ})gI_{IJ}^{KL} - (\mathscr{B}'_{iK} + \mathscr{B}'_{1jJ})g'I_{KL}^{IJ}\} \times f_{iI}^{(0)} f_{1jJ}^{(0)} \sin(\psi) d\psi\, d\phi\, d\mathbf{v}_{1j} \tag{3.13}$$

$$(1/n_{iI}) f_{iI}^{(0)} (\delta_{ih} - \delta_{ik})\mathbf{V}_i = -\sum_j \sum_{J,K,L} \int \cdots \int \{(\mathbf{C}_{iI}^h + \mathbf{C}_{1jJ}^h - \mathbf{C}_{iI}^k - \mathbf{C}_{1jJ}^k)gI_{IJ}^{KL} - (\mathbf{C}'^h_{ik} + \mathbf{C}'^h_{1jL} - \mathbf{C}'^k_{iK} - \mathbf{C}'^k_{1jL})g'I_{KL}^{IJ}\} f_{iI}^{(0)} f_{1jJ}^{(0)} \sin(\psi)\, d\psi\, d\phi\, d\mathbf{v}_{1j} \tag{3.14}$$

together with the auxiliary relations

$$\sum_i \sqrt{(m_i')} \sum_I \int (\mathbf{A}_{iI} \cdot \mathbf{W}_i) f_{iI}^{(0)}\, d\mathbf{v}_i = 0 \tag{3.15}$$

$$\sum_i \sqrt{(m_i')} \sum_I \int \{(\mathbf{C}_{iI}^h - \mathbf{C}_{iI}^k) \cdot \mathbf{W}_i\} f_{iI}^{(0)}\, d\mathbf{v}_i = 0 \tag{3.16}$$

which arise from the normalization conditions (3.7). $\mathscr{U}$ in Eq. (3.13) is the unit tensor. Equations (3.11) and (3.14) may also be expressed in terms of the reduced velocity $\mathbf{W}_i$ instead of $\mathbf{V}_i$. The form of Eqs. (3.11) to (3.14) then demands that $\mathbf{A}_{iI}$, $\mathscr{B}_{iI}$, and $\mathbf{C}_{iI}^h$ are of the form

$$\mathbf{A}_{iI} = A_{iI}(W_i)\mathbf{W}_i \tag{3.17}$$

$$\mathscr{B}_{iI} = B_{iI}(W_i)\{\mathbf{W}_i\mathbf{W}_i - (1/3)W_i^2\mathscr{U}\} \tag{3.18}$$

$$\mathbf{C}_{iI}^h = C_{iI}^h(W_i)\mathbf{W}_i \tag{3.19}$$

where A_{iI}, B_{iI}, and C_{iI}^h are scalar functions of the absolute value of W_i (or V_i) which depend parametrically on temperature, pressure, and composition. In order to solve for the vector functions, the latter are expanded into a double series of orthogonal polynomials

$$\mathbf{A}_{iI} = \mathbf{W}_i \sum_{m,n}^{M} \{a_{imn} S_{3/2}^{(m)}(W_i^2) P^{(n)}(\varepsilon_{iI})\} \tag{3.20a}$$

$$\mathcal{B}_{iI} = \{\mathbf{W}_i\mathbf{W}_i - (1/3)W_i^2\mathcal{U}\} \sum_{m,n}^{M} \{b_{imn}S_{5/2}^{(m)}(W_i^2)P^{(n)}(\varepsilon_{iI})\} \quad (3.20b)$$

$$\mathbf{C}_{iI}^h - \mathbf{C}_{iI}^k = \mathbf{W}_i \sum_{m,n}^{M} \{c_{imn}^{hk}S_{3/2}^{(m)}(W_i^2)P^{(n)}(\varepsilon_{iI})\} \quad (3.20c)$$

where the $S_{3/2}^{(m)}(W_i^2)$ and $S_{5/2}^{(m)}(W_i^2)$ are the Sonine polynomials (Chapman and Cowling, 1970; Hirschfelder *et al.*, 1954), and the $P^{(n)}(\varepsilon_{iI})$ are the polynomials used by Wang Chang and Uhlenbeck (1951). The first two of these are

$$\begin{aligned} P^{(0)} &= 1 \\ P^{(1)} &= \varepsilon_{iI} - \bar{\varepsilon}_i \end{aligned} \quad (3.21)$$

Application of the variational procedure for mixtures (Enskog, 1917; Hellund and Uehling, 1939; Hirschfelder *et al.*, 1954) and subsequent use of the auxiliary relations (3.15) and (3.16) in the case of $\mathbf{A}_{iI}$ and $\mathbf{C}_{iI}$ then leads to three sets of $M \times N$ linear algebraic equations which determine the various expansion coefficients uniquely

$$\sum_{j=1}^{N} \sum_{m,n}^{M} \tilde{Q}_{ij}^{rs,mn} a_{jmn} = -n_i(2k_BT/m_i')^{1/2}\{(15/4)\,\delta_{r1}\,\delta_{s0} - (3/2)(c_{i,\text{int}}/k_B)\,\delta_{r0}\,\delta_{s1}\} \quad (3.22a)$$

$$\sum_{j=1}^{N} \sum_{m,n}^{M} \tilde{H}_{ij}^{rs,mn} b_{jmn} = X_i \quad (3.22b)$$

$$\sum_{j=1}^{N} \sum_{m,n}^{M} \tilde{Q}_{ij}^{rs,mn} c_{jmn}^{hk} = -(3/2)(2k_BT/m_i')^{1/2}\,\delta_{r0}\,\delta_{s0}\,(\delta_{ih} - \delta_{ik}) \quad (3.22c)$$

In Eq. (3.22a), $c_{i,\text{int}}$ is the internal heat capacity per molecule of the ith component. M is the total number of terms in the expansions (3.20). In the limit as M approaches infinity, the expansions are presumed to become exact, and the coefficients a_{jmn}, b_{jmn}, and c_{jmn}^{hk} are obtained exactly as solutions of an infinite set of linear simultaneous equations. However, the only coefficients actually needed for the calculation of the transport properties are a_{j00}, a_{j01}, b_{j00}, and c_{j00}. To get numerical values of these to a first approximation, the set of equations is truncated by taking only three terms ($mn = 00$, 10, 01) in Eq. (3.20a), which controls thermal conduction and thermal diffusion, and only one term ($mn = 00$) in each of Eqs. (3.20b) and (3.20c).

The terms $\tilde{Q}_{ij}^{rs,mn}$ and $\tilde{H}_{ij}^{rs,mn}$ in Eqs. (3.22) are linear combinations of a number of collision integrals which arise from the right-hand sides of Eqs. (3.11), (3.13), and (3.14). These integrals will be discussed below. For numerical calculation relating to Eqs. (3.22a) and (3.22c), some simplification is obtained if the $\tilde{Q}_{ij}^{rs,mn}$ are replaced by the coefficients $L_{ij}^{rs,mn}$, where for rs and mn equal to 00, 10 and 01,

$$\tilde{Q}_{ij}^{rs,mn} = -(1/12)k_B(g_{irs}g_{jmn}/X_iX_j)L_{ij}^{rs,mn} \quad (3.23)$$

$$g_{irs} = n_i(2k_BT/m_i')^{1/2}\{(15/4)\,\delta_{s0} - (3/2)(c_{i,\text{int}}/k_B)\,\delta_{r0}\,\delta_{s1}\} \quad (3.24)$$

This leads after some cancellation to

$$\sum_{j=1}^{N} \sum_{m,n}^{M} L_{ij}^{rs,mn} a_{jmn}^{1} = (\delta_{r1} + \delta_{s1}) X_i$$

$$(rs = 00, 10, 01; i = 1, 2, \ldots, N) \quad (3.25a)$$

$$\sum_{j=1}^{N} \sum_{m,n}^{M} L_{ij}^{rs,mn} c_{jmn}^{1hk} = (\delta_{ih} - \delta_{ik}) \delta_{r0} \delta_{s0}$$

$$(rs = 00, 10, 01; i, h, k = 1, 2, \ldots, N) \quad (3.25c)$$

where

$$a_{jmn}^{1} = (k_B g_{imn}/12X_j) a_{jmn} \quad (3.26a)$$

$$c_{jmn}^{1hk} = (5P/24T)(g_{imn}/X_j) c_{jmn}^{hk} \quad (3.26c)$$

Formal expressions for the $L_{ij}^{rs,mn}$ and $\tilde{H}_{ij}^{rs,mn}$ are given by Monchick, Yun, and Mason (1963). The working definition and evaluation of the coefficients are discussed in Sections 3.1(c) and 3.2.

(b) Formulation of transport fluxes and coefficients

The diffusion velocities $\mathbf{V}_i$, the pressure tensor $\mathscr{P}$, and the heat flux vector $\mathbf{q}'$ in a mixture of polyatomic gases may now be expressed in terms of the ϕ_{iI}, and hence in terms of the functions $\mathbf{A}_{iI}$, $\mathscr{B}_{iI}$, and $\mathbf{C}_{iI}$.

The pressure tensor becomes

$$\begin{aligned}
\mathscr{P} &= \sum_{i=1}^{N} m_i' \sum_{I} \int \mathbf{V}_i \mathbf{V}_i f_{iI} \, d\mathbf{v}_i \\
&= \sum_{i=1}^{N} m_i' \sum_{I} \left\{ \int \mathbf{V}_i \mathbf{V}_i f_{iI}^{(0)} \, d\mathbf{v}_i + \int \mathbf{V}_i \mathbf{V}_i f_{iI}^{(0)} \phi_{iI} \, d\mathbf{v}_i \right\} \\
&= P\mathscr{U} + \sum_{i=1}^{N} m_i' \sum_{I} \int \mathbf{V}_i \mathbf{V}_i f_{iI}^{(0)} \phi_{iI} \, d\mathbf{v}_i
\end{aligned} \quad (3.27)$$

where P is the scalar pressure and $\mathscr{U}$ is the unit tensor. It can be shown by symmetry arguments that in this case the tensor function $\mathscr{B}_{iI}$ is the only nonzero contributor to ϕ_{iI} in the integral of Eq. (3.27). Then, making use of the expression for $\mathscr{B}_{iI}$ from Eq. (3.18) and neglecting volume viscosity effects, we find (Hirschfelder *et al.*, 1954)

$$\mathscr{P} = P\mathscr{U} - 2\eta\mathscr{S} \quad (3.28)$$

where $\mathscr{S}$ is the rate of shear tensor defined by

$$\mathscr{S}_{\alpha\beta} = (1/2)\{(\partial v_{0\beta}/\partial x_\alpha) + (\partial v_{0\alpha}/\partial x_\beta)\} - (1/3)(\nabla \cdot \mathbf{v}_0)\, \delta_{\alpha\beta} \quad (3.29)$$

and Eq. (3.28) also defines the coefficient of shear viscosity [cf. Eqs. (2.7) and

(2.7a)]. Then, using the Sonine polynomial expansion (3.20b) together with the appropriate orthogonality relation, we obtain the first approximation $[\eta]_1$ as

$$[\eta]_1 = \sum_{j=1}^{N} X_j b_{j00} \tag{3.30}$$

The diffusion velocity $\bar{\mathbf{V}}_i$ and the heat flux vector $\mathbf{q}'$ may be treated by similar methods. In dealing with the concentration gradients, it should be remembered that since $\sum_{h=1}^{N} \mathbf{d}_h = 0$ there are only $(N-1)$ independent vectors $\mathbf{d}_h$. Hence, one of the $\sum_I \mathbf{C}_{iI}^h$ in each of the Eqs. (3.19) may be set equal to zero. The term involving $\mathscr{B}_{iI}$ vanishes in the integrations for both the diffusion velocity and the heat flux vector, so that taking $\sum_I \mathbf{C}_{iI}^i = 0$ we obtain, for example, for $\bar{\mathbf{V}}_i$,

$$\begin{aligned}
\bar{\mathbf{V}}_i &= (1/n_i) \sum_I \int \mathbf{V}_i f_{iI} \, d\mathbf{v}_i \\
&= (1/n_i) \sum_I \int \mathbf{V}_i f_{iI}^{(0)} \phi_{iI} \, d\mathbf{v}_i \\
&= (1/n_i) \sum_I \left\{ n \sum_{j=1}^{N} (\mathbf{C}_{iI}^j - \mathbf{C}_{iI}^i) \cdot \mathbf{d}_j - \mathbf{A}_{iI} \cdot \nabla_r \ln(T) \right\} \mathbf{V}_i f_{iI}^{(0)} \, d\mathbf{v}_i
\end{aligned} \tag{3.31}$$

Using the polynomial expansions (3.20a) and (3.20c) with appropriate orthogonality relations we then obtain (Hirschfelder *et al.*, 1954)

$$\bar{\mathbf{V}}_i = (n^2/n_i\rho) \sum_{j=1}^{N} m'_j D_{ij} \mathbf{d}_j - (1/n_i m'_i) D_i^T \nabla_r \ln(T) \tag{3.32}$$

where D_{ij} and D_i^T are the multicomponent diffusion coefficients and multicomponent thermal diffusion coefficients, respectively, defined by

$$D_{ij} = X_i(16T/25P) \left(\sum_k m_k X_k / m_j \right) c_{i00}^{1\,ji} \tag{3.33}$$

$$D_i^T = (8/5)(m_i X_i/R) a_{i00}^1 \tag{3.34}$$

The diffusive mass flux $\mathbf{j}_i$ is given in terms of molar quantities by

$$\begin{aligned}
\mathbf{j}_i &= n_i m'_i \bar{\mathbf{V}}_i \\
&= \left(\rho_{\text{molar}} / \sum_k m_k X_k \right) \sum_{j=1}^{N} m_i m_j D_{ij} \mathbf{d}_j - D_i^T \nabla_r \ln(T)
\end{aligned} \tag{3.35}$$

By commencing at Eq. (3.2), it can be shown similarly (Monchick, Yun, and Mason, 1963) that the heat flux vector is given by

$$\mathbf{q}' = k_B T \sum_i (5/2 + \bar{\varepsilon}_i) n_i \bar{\mathbf{V}}_i - \lambda_0 \nabla_r T - n k_B T \sum_i \{(1/n_i m'_i) D_i^T \mathbf{d}_i\} \tag{3.36}$$

where λ_0 is a sort of thermal conductivity, but not the usual one. λ_0 is given in terms of the coefficients a_{i10}^1 and a_{i01}^1 by

$$\lambda_0 = \lambda_{0,\mathrm{tr}} + \lambda_{0,\mathrm{int}}$$

$$\lambda_{0,\mathrm{tr}} = -4 \sum_i X_i a_{i10}^1 \tag{3.37}$$

$$\lambda_{0,\mathrm{int}} = -4 \sum_i X_i a_{i01}^1$$

In terms of the mass flux vector and molar quantities, the heat flux vector is given by

$$\mathbf{q}' = \sum_i \mathbf{j}_i h_i - \lambda_0 \nabla_r T - \sum_i (RT/m_i X_i) D_i^T \,\mathbf{d}_i \tag{3.38}$$

and for the one-dimensional system discussed in Section 2 we have

$$j_i = \left(\rho_{\mathrm{molar}} \Big/ \sum_k m_k X_k \right) \sum_{j=1}^{N} m_i m_j D_{ij} \partial X_j / \partial y - (D_i^T/T)\, \partial T/\partial y \tag{3.35a}$$

$$Q_T = -\lambda_0\, \partial T/\partial y - \sum_i (RT/m_i X_i) D_i^T \,\partial X_i/\partial y \tag{3.38a}$$

where pressure-gradient diffusion has been neglected. The final terms in Eqs. (3.35) and (3.38) represent the contributions of thermal diffusion and its reciprocal effect (Dufour effect) on the respective fluxes.

(c) Evaluation of the collision integrals, the L_{ij}, and the H_{ij}

For the practical implementation of Eqs. (3.22b) and (3.25), it is necessary to evaluate the collision integrals in the formal expressions given by Monchick *et al.* (1963) for the L_{ij}. These integrals arise from the right-hand sides of Eqs. (3.11), (3.13), and (3.14). In the notation of Monchick *et al.* (1963, 1965), a pair of pointed brackets around the operand denotes integration

$$\begin{aligned} \langle F \rangle_{ij} &\equiv (k_B T/2\pi\mu_{ij})^{1/2} \sum \int d\Omega_{ij}\, F \\ &\equiv (k_B T/2\pi\mu_{ij})^{1/2} (Q_i Q_j)^{-1} \sum_{I,J,K,L} \int_0^\infty d\gamma \int_0^{2\pi} d\phi \int_0^\pi d\psi \\ &\quad \times \{F\gamma^3 \exp(-\gamma^2 - \varepsilon_{iI} - \varepsilon_{iJ}) I_{IJ}^{KL} \sin(\psi)\} \end{aligned} \tag{3.39}$$

where

$$\mu_{ij} = m_i' m_j' / (m_i' + m_j') \tag{3.40}$$

$$\gamma = (\mu_{ij}/2k_B T)^{1/2} g_{ij} \tag{3.41}$$

and F may be any function of the internal energy changes $\Delta\varepsilon_i$, $\Delta\varepsilon_j$, or $\Delta\varepsilon_{ij}$, given by

$$\Delta\varepsilon_i = \varepsilon_{iK} - \varepsilon_{iI}$$

$$\Delta\varepsilon_j = \varepsilon_{jL} - \varepsilon_{jJ} \tag{3.42}$$

$$\Delta\varepsilon_{ij} = \Delta\varepsilon_i + \Delta\varepsilon_j$$

For *elastic collisions* ($\Delta\varepsilon_i = \Delta\varepsilon_j = 0$) the only collision integrals required for the complete set of first approximations are the integrals $\Omega_{ij}^{(1,1)}(T)$, $\Omega_{ij}^{(1,2)}(T)$, $\Omega_{ij}^{(1,3)}(T)$, and $\Omega_{ij}^{(2,2)}(T)$, given by

$$\Omega_{ij}^{(l,s)} \equiv \langle \gamma^{2s}\{1 - \cos^l(\psi)\}\rangle_{ij} \tag{3.43}$$

These integrals depend on the forces of interaction between the colliding molecules and can be calculated for the elastic-collision case. Tables of the integrals are available for the most widely used intermolecular potential functions. For nonpolar molecules the quantities $\Omega^{(l,s)*}$, equal to the collision integrals for the Lennard-Jones (12 : 6) potential divided by the corresponding rigid sphere values, are quoted by Hirschfelder *et al.* (1954) for a wide range of reduced temperatures $T^* = k_B T/(eps)_{ij}$. The Lennard-Jones parameters $(eps)_{ij}/k_B$ and $(si)_{ij}$ (written in this way instead of the more usual ε/k_B and σ so as to distinguish from other uses of ε and σ here) are characteristic of the colliding species, and thus in the elastic-collision case all the necessary integrals may be found at any temperature if these parameters are known. The parameters are tabulated by Svehla (1962) for a variety of pure substances and for some mixtures. For gases or gas pairs whose transport properties have been measured experimentally, they are obtained by fitting the temperature dependence of the observed data. For other pairs of nonpolar molecules the combining rules (3.44) are usually employed in calculation

$$\begin{aligned}(eps)_{ij} &= \{(eps)_i \times (eps)_j\}^{1/2} \\ (si)_{ij} &= (1/2)\{(si)_i + (si)_j\}\end{aligned} \tag{3.44}$$

A list of values of $(eps)_i$ and $(si)_i$ for species of interest in combustion is given in Table 1.

The expressions used in practice to calculate the L_{ij} are not normally written directly in terms of the Ω integrals either in the elastic or the inelastic case. In terms of these integrals it can be shown that in the first approximation (only the term in Eq. (3.20c) having $mn = 00$ considered) the binary diffusion coefficient $[\mathscr{D}_{ij}]_1$ is given by

$$\begin{aligned}[\mathscr{D}_{ij}]_1 &= \frac{3}{16n\mu_{ij}} \frac{k_B T}{\Omega_{ij}^{(1,1)*}(T^*)} \\ &= 0.002628 \frac{\sqrt{(T^3/2\mu_{ij})}}{P(si)_{ij}^2 \Omega_{ij}^{(1,1)*}(T^*)}\end{aligned} \tag{3.45}$$

where in the second equation the units are: $\mathscr{D}$, cm^2 s^{-1}; P, atm; T, K; and $(si)_{ij}$, Å(= 10^{-8} cm). Thus, $[\mathscr{D}_{ij}]_1$ has usually been used in place of $\Omega^{(1,1)}$. For the same reason three further quantities are commonly used

$$\begin{aligned}A_{ij}^* &= \Omega_{ij}^{(2,2)*}/\Omega_{ij}^{(1,1)*} \\ B_{ij}^* &= \{5\Omega_{ij}^{(1,2)*} - 4\Omega_{ij}^{(1,3)*}\}/\Omega_{ij}^{(1,1)*} \\ C_{ij}^* &= \Omega_{ij}^{(1,2)*}/\Omega_{ij}^{(1,1)*}\end{aligned} \tag{3.46}$$

Table 1. Lennard-Jones (12:6) potential parameters (mostly from Svehla, 1962), rotational collision numbers ζ_{ij}, and polarizabilities α_n of "nonpolar" species of interest in combustion systems, together with Stockmayer parameters of H_2O (Mason and Monchick 1962).

Species	$(eps)/k_B(K)$	(si)(nm)	ζ_{ij} [a]	$10^{24}\alpha_n$ (cm^3)
H[b]	37.0	0.2070	—	0.55
O	106.7	0.3050	—	0.6
OH	79.8	0.3147	4.5	0.6
N_2	71.4	0.3798	4.5	1.76
O_2	106.7	0.3467	4.5	1.60
H_2	59.7	0.2827	200.0	0.79
CO	91.7	0.3690	4.6	1.95
CO_2	195.2	0.3941	2.5	2.65
H_2O^*	260.0	0.2800	4.0	—
H_2O_2	289.3	0.4196	—	1.50
CH_4	148.6	0.3758	2.5	2.60
C_2H_6	215.7	0.4443	2.5	4.47
C_2H_4	224.7	0.4163	2.5	4.26
C_2H_2	231.8	0.4033	2.5	3.33
CH_3	312.0	0.3644	—	2.0
C_2H_5	220.2	0.4303	—	4.36
CH_2O	312.0	0.3758	2.5	3.0
CHO	187.0	0.3465	—	2.2
HO_2	168.0	0.3068	—	1.5

[a] The assumption is made that the ζ_{ij} depend only on i. Many of these values are estimates only, e.g., for hydrocarbons.

[b] This combination of parameters for atomic hydrogen uses $(eps)_H/k_B$ from Svehla (1962). $(si)_H$ has been chosen so as to fit $\mathscr{D}_{H,N_2} = 1.35 \pm 0.3$ cm^2 s^{-1} at 294 K and 1 atm, measured by Clifford *et al.* (1982).

For the elastic-collision case the quantities in Eqs. (3.46) are again available in tables as a function of the reduced temperature T^*. From the point of view of computation, it is convenient also to have the collision integrals and their ratios expressed as polynomials in T^*. The appropriate polynomial coefficients for a fourth-order fit of $\Omega^{(1,1)*}$ and second-order fits of A^*, B^*, and C^* over limited ranges of T^* are given in Table 2.

For *inelastic collisions* the Ω integrals cannot yet be calculated from first principles. The procedure adopted with polyatomic gases is therefore to treat them as if the collisions were elastic, that is, to use tables of *elastic*-collision integrals and to derive effective values of $(eps)_{ij}/k_B$ and $(si)_{ij}$ from the temperature variation of the transport properties. The first approximations to the binary diffusion coefficients $[\mathscr{D}_{ij}]_1$, as well as A^*, B^*, and $(1.2C^* - 1)$, may then be calculated as for elastic collisions.

In the inelastic case, the new quantities relating to the diffusion of internal energy are the internal specific heats per molecule, $c_{i,\mathrm{int}}$, and the diffusion

Table 2. Coefficients of fourth-order polynomial fits of reduced collision integrals $\Omega^{(1,1)*}$ and $A^*_{H_2O}$ [a] (see Eq. (3.43) and Section 3.2) and second-order fits of A^*, B^*, and C^* [see Eq. (3.46)]. Integrals are expressed as $I(T^*) = a_0 + a_1 T^* + a_2 T^{*2} + a_3 T^{*3} + a_4 T^{*4}$.

I	T^*	a_0	a_1	a_2	a_3	a_4
$\Omega^{(1,1)*}$	<5.0	$0.23527333E+01$	$-0.13589968E+01$	$0.52202460E+00$	$-0.94262883E-01$	$0.64354629E-02$
	5.0–10.0	$0.12660308E+01$	$-0.16441443E+00$	$0.22945928E-01$	$-0.16324168E-02$	$0.45833672E-04$
	>10.0	$0.85263337E+00$	$-0.13552911E-01$	$0.26162080E-03$	$-0.24647654E-05$	$0.86538568E-08$
A^*	<5.0	$0.11077725E+01$	$-0.94802344E-02$	$0.16918277E-02$		
	5.0–10.0	$0.10871429E+01$	$0.31964282E-02$	$-0.89285689E-04$		
	>10.0	$0.11059000E+01$	$0.65136364E-03$	$-0.34090910E-05$		
B^*	<5.0	$0.12432868E+01$	$-0.78288929E-01$	$0.99572721E-02$		
	5.0–10.0	$0.11091429E+01$	$-0.53107128E-02$	$0.37499990E-03$		
	>10.0	$0.10938667E+01$	$0.43636359E-04$	$-0.30303027E-06$		
C^*	<5.0	$7.83721415E-01$	$6.24667525E-02$	$-6.72439798E-03$		
	5.0–10.0	$8.95485714E-01$	$9.78107143E-03$	$-4.87500000E-04$		
	>10.0	$9.44930000E-01$	$1.10560600E-04$	$--8.10606061E-07$		
[a] $A^*_{H_2O}$	<10.0	$0.10764205E+01$	$0.46037515E-01$	$-0.13506975E-01$	$0.15404522E-02$	$-0.60887567E-04$
	>10.0	$0.11141689E+01$	$0.48711959E-03$	$-0.44570091E-05$	$0.99643413E-08$	$0.68639118E-10$

[a] Table of coefficients of $A^*_{H_2O}$ is for $\bar{\mu}^*_{H_2O} = \sqrt{5}$, $(eps)_{H_2O} = 260.0$ K, and $(si)_{H_2O} = 0.2800$ nm.

coefficients $\mathscr{D}_{i\,\mathrm{int},j}$ and $\mathscr{D}_{i\,\mathrm{int},i}$ for internal energy. Monchick, Pereira, and Mason (1965) approximate the $\mathscr{D}_{i\,\mathrm{int},j}$ by the ordinary binary diffusion coefficients $[\mathscr{D}_{ij}]_1$, and the $\mathscr{D}_{i\,\mathrm{int},i}$ for nonpolar gases by the self-diffusion coefficients $\mathscr{D}_{ii}$ obtained from experimental viscosity data by means of

$$\begin{aligned}\mathscr{D}_{ii} &= 6A_{ii}^{*}(T^{*})\eta_i/5nm_i' \\ &= 98.4708\ TA_{ii}^{*}(T^{*})\eta_i/Pm_i' \end{aligned}\tag{3.47}$$

where the units are: $\mathscr{D}$, cm^2 s^{-1}; η, g cm^{-1} s^{-1}; and P, atm.

The remaining experimental quantities relevant to the inelastic case are the rotational specific heats per molecule, $c_{i,\mathrm{rot}}$, and the relaxation times τ_{ij} for transfer of internal energy of the ith species into translational energy on colliding with species j. Only rotational energy transfer need be considered, since transfer of vibrational energy is much slower. These relaxation times are expressed as collision numbers ζ_{ij} defined by

$$\begin{aligned}\zeta_{ii} &= (4/\pi)(P\tau_{ii}/\eta_i) \\ \zeta_{ij} &= (12k_B T/5\pi)(\tau_{ij}A_{ij}^{*}/\mu_{ij}\mathscr{D}_{ij}) \end{aligned}\tag{3.48}$$

It is assumed that complex collisions involving more than a single quantum jump can be neglected, that is, integrals such as $\langle \Delta\varepsilon_i\, \Delta\varepsilon_j \rangle_{ij}$ can be neglected.

Using the quantities now defined, and setting certain other collision integrals which should be very small equal to zero, Monchick *et al.* (1965) were able to express the $L_{ij}^{rs,mn}$ of Eqs. (3.25a) and (3.25c) entirely in terms of experimental or calculable quantities as follows[1]:

$$L_{ij}^{00,00} = \frac{16T}{25P}\sum_k \frac{X_k}{m_i[\mathscr{D}_{ik}]_1}\{m_j X_j(1-\delta_{ik}) - m_i X_i(\delta_{ij}-\delta_{jk})\}$$

$$L_{ij}^{00,10} = \frac{8T}{5P}\sum_k \left\{X_j X_k(\delta_{ij}-\delta_{ik})\frac{m_k(1.2C_{jk}^{*}-1)}{(m_j+m_k)[\mathscr{D}_{jk}]_1}\right\}$$

$$L_{ij}^{10,00} = L_{ji}^{00,10}$$

$$L_{ij}^{01,00} = L_{ji}^{00,01} = 0$$

$$\begin{aligned}L_{ij}^{10,10} = {}& \frac{16T}{25P}\sum_k \frac{m_i}{m_j}\frac{X_i X_k}{(m_i+m_k)^2[\mathscr{D}_{ik}]_1} \\ & \times \{(\delta_{jk}-\delta_{ij})[(15/2)m_j^2 + (25/4)m_k^2 - 3m_k^2 B_{ik}^{*}] - 4m_j m_k A_{ik}^{*}(\delta_{jk}+\delta_{ij}) \\ & \times [1 + (5/3\pi)\{(c_{i,\mathrm{rot}}/k_B\zeta_{ik}) + (c_{k,\mathrm{rot}}/k_B\zeta_{ki})\}]\}\end{aligned}$$

$$L_{ij}^{10,01} = \frac{32T}{5\pi P c_{j,\mathrm{int}}}\sum_k \left\{\frac{m_j A_{jk}^{*}}{(m_j+m_k)[\mathscr{D}_{jk}]_1}(\delta_{ik}+\delta_{ij})X_j X_k(c_{j,\mathrm{rot}}/\zeta_{jk})\right\}$$

---

[1] Units: Units of $L_{ij}^{rs,mn}$ are those of $(T/P\mathscr{D})$. If T is in K, P in atm, and $\mathscr{D}$ in cm^2/s, then the units of a_{i00}^1, a_{i10}^1, a_{i01}^1, and c_{jmn}^{1hk} in Eqs. (3.25), (3.33), (3.34), and (3.37) are (cm^3 atm) cm^{-1} K^{-1} s^{-1}, where 1 cm^3 atm = 1.01325 × 10^3 J. If P is in N/m^2 and $\mathscr{D}$ in m^2/s, then $a_{i00}^1 \cdots$ are in W m^{-1} K^{-1}.

$$L_{ij}^{01,10} = L_{ji}^{10,01}$$

$$L_{ii}^{01,01} = -\frac{4k_B T}{c_{i,\text{int}} P} \sum_k \left\{ \frac{X_i X_k}{\mathscr{D}_{i\,\text{int},k}} + \frac{12 X_i X_k}{5\pi c_{i,\text{int}}} \frac{m_i A_{ik}^* c_{i,\text{rot}}}{m_k [\mathscr{D}_{ik}]_1 \zeta_{ik}} \right\}$$

$$L_{ij}^{01,01} = 0 (j \neq i) \tag{3.49}$$

For mixtures which have both polyatomic and monatomic components, the rows which refer to the monatomic components in the submatrices of **L** having $rs = 01$ and the corresponding columns in the submatrices having $mn = 01$ are omitted.

If Eq. (3.47) is used to substitute for the self-diffusion coefficients in the expressions for $L_{ii}^{10,10}$, $L_{ii}^{10,01}$, $L_{ii}^{01,10}$, and $L_{ii}^{01,01}$, these expressions become, in terms of η_i,

$$L_{ii}^{10,10} = -\frac{16}{15} \frac{m_i X_i^2}{R\eta_i} \left\{ 1 + \frac{10}{3\pi} \frac{c_{i,\text{rot}}}{k_B \zeta_{ii}} \right\} - \frac{16T}{25P} \sum_{k \neq i} \left\{ \frac{X_i X_k}{(m_i + m_k)^2 [\mathscr{D}_{ik}]_1} \right.$$

$$\times \left[(15/2) m_i^2 + (25/4) m_k^2 - 3 m_k^2 B_{ik}^* \right.$$

$$\left. \left. + 4 m_i m_k A_{ik}^* \left(1 + \frac{5}{3\pi} \left\{ \frac{c_{i,\text{rot}}}{k_B \zeta_{ik}} + \frac{c_{k,\text{rot}}}{k_B \zeta_{ki}} \right\} \right) \right] \right\}$$

$$L_{ii}^{10,01} = L_{ii}^{01,10}$$

$$= \frac{16}{3\pi} \frac{m_i X_i^2 k_B}{R \eta_i c_{i,\text{int}}} \frac{c_{i,\text{rot}}}{k_B \zeta_{ii}} + \frac{32 T k_B}{5\pi P c_{i,\text{int}}} \sum_{k \neq i} \left\{ \frac{m_i A_{ik}^* X_i X_k}{(m_i + m_k)[\mathscr{D}_{ik}]_1} \frac{c_{i,\text{rot}}}{k_B \zeta_{ik}} \right\}$$

$$L_{ii}^{01,01} = -\frac{8k_B^2}{\pi (c_{i,\text{int}})^2} \frac{m_i X_i^2}{R\eta_i} \frac{c_{i,\text{rot}}}{k_B \zeta_{ii}} - \frac{4k_B T}{c_{i,\text{int}} P}$$

$$\times \left\{ \sum_k \frac{X_i X_k}{\mathscr{D}_{i\,\text{int},k}} + \sum_{k \neq i} \frac{12 X_i X_k}{5\pi c_{i,\text{int}}} \frac{m_i}{m_k} \frac{A_{ik}^*}{[\mathscr{D}_{ik}]_1} \frac{c_{i,\text{rot}}}{k_B \zeta_{ii}} \right\} \tag{3.49a}$$

Experimental values of the viscosities η_i of the pure components may then be used in Eqs. (3.49a) instead of calculated self-diffusion coefficients in the corresponding Eqs. (3.49). The approach by Eqs. (3.49a) is preferred since the combining rules (3.44) do not always allow the best fit of both viscosity and diffusion data when the same force constants are used for the pure gases. With the approach via the viscosities, uncertainties only enter through A^*, giving a weak dependence on the force constants.

For practical use Eqs. (3.49) and (3.49a) take the place of the formal expressions for the L_{ij} given by Monchick, Yun, and Mason (1963). If the inelastic-collision effects can be neglected, the expressions for $L_{ij}^{10,10}$, $L_{ij}^{10,01}$, and $L_{ij}^{01,01}$ further simplify to

$$L_{ij}^{10,10} = \frac{16T}{25P} \sum_k \frac{m_i}{m_j} \frac{X_i X_k}{(m_i + m_k)^2 [\mathscr{D}_{ik}]_1}$$

$$\times \{ (\delta_{jk} - \delta_{ij}) [(15/2) m_j^2 + (25/4) m_k^2 - 3 m_k^2 B_{ik}^*] - 4 m_j m_k A_{ik}^* (\delta_{jk} + \delta_{ij}) \}$$

$$L_{ii}^{10,10} = -\frac{16}{15}\frac{m_i X_i^2}{R\eta_i} - \frac{16T}{25P}\sum_{k\neq i}\frac{X_i X_k}{(m_i+m_k)^2[\mathscr{D}_{ik}]_1}$$
$$\times \{(15/2)m_i^2 + (25/4)m_k^2 - 3m_k^2 B_{ik}^* + 4m_i m_k A_{ik}^*\}$$
$$L_{ij}^{10,01} = L_{ij}^{01,10} = 0$$
$$L_{ij}^{01,01} = 0\ (j \neq i)$$
$$L_{ii}^{01,01} = -\frac{4k_B T}{c_{i,\mathrm{int}}P}\sum_k \frac{X_i X_k}{\mathscr{D}_{i\,\mathrm{int},k}} \tag{3.49b}$$

The only nonzero terms remaining in all of the submatrices having either rs or $mn = 01$ are the leading diagonal terms in $\mathbf{L}^{01,01}$. The equations with both rs and $mn = 01$ then separate out from the set (3.25a) to give

$$\left(-\frac{4k_B T}{c_{i,\mathrm{int}}P}\sum_k \frac{X_i X_k}{\mathscr{D}_{i\,\mathrm{int},k}}\right) a_{i01}^1 = X_i \tag{3.50}$$

and hence, from Eqs. (3.37)

$$\lambda_{0,\mathrm{int}} = \frac{P}{T}\sum_{i=1}^{N}\frac{c_{i,\mathrm{int}}X_i}{k_B\left(\sum_k X_k/\mathscr{D}_{i\,\mathrm{int},k}\right)} \tag{3.51}$$

For a pure gas Eq. (3.51) reduces, as it should, to the form of the Eucken internal energy correction term.

The situation regarding viscosity of gas mixtures is somewhat simpler. Here the coefficients $\tilde{H}_{ij}^{00,00}$ of Eq. (3.22b) may be expressed similarly to the L_{ij} as

$$H_{ij}^{00,00} = -\frac{2X_i X_j}{(m_i+m_j)}\frac{RT}{P[\mathscr{D}_{ij}]_1}\{1 - (3/5)A_{ij}^*\}$$
$$H_{ii}^{00,00} = \frac{X_i^2}{\eta_i} + \sum_{k\neq i}\frac{2X_i X_k}{(m_i+m_k)}\frac{RT}{P[\mathscr{D}_{ik}]_1}\left\{1 + \frac{3m_k A_{ik}^*}{5m_i}\right\} \tag{3.52}$$

In terms of the collision integrals, the viscosity η_i of a pure component is given to first approximation by

$$[\eta_i]_1 = 2.6693 \times 10^{-5}\frac{\sqrt{(m_i T)}}{(si)_i^2\Omega_{ii}^{(2,2)*}(T^*)} \tag{3.53}$$

where the units are η, g cm^{-1} s^{-1} and T, K.

3.2. Mixtures containing one polar component

Combustion products usually contain water vapor, and it is necessary to consider the effect of the polar nature of the water molecule on transport properties. For mixtures containing one polar component, the same general methods are used as for mixtures of nonpolar molecules, except that the collision integrals involving the polar molecule need modification, and that allowance must be made for the fact that the dipole field induces relatively easy exchange of rotational energy in resonant collisions between polar molecules.

Two types of interaction require consideration in the collision integrals—one between polar and nonpolar molecules and the other between pairs of polar molecules. For pairs of polar molecules the interaction potential ϕ is usually represented by the Stockmayer potential

$$\begin{aligned}\phi &= 4(eps)_0\{[(si)_0/r]^{12} - [(si)_0/r]^6\} \\ &\quad - (\bar{\mu}_1\bar{\mu}_2/r^3)\{2\cos(\theta_1)\cos(\theta_2) - \sin(\theta_1)\sin(\theta_2)\cos(\phi)\} \qquad (3.54)\end{aligned}$$

where $\bar{\mu}_1$ and $\bar{\mu}_2$ are the dipole moments of the two molecules, r is the distance between centers, $(eps)_0$ and $(si)_0$ are the Stockmayer potential parameters, θ_1 and θ_2 are the angles of inclination of the two dipoles to the line joining the centers of the molecules, and ϕ is the azimuthal angle between them.

In the limit as $\bar{\mu}_1$ or $\bar{\mu}_2 \to 0$, Eq. (3.54) takes the same form as that for the Lennard-Jones (12 : 6) potential for pairs of nonpolar molecules. Hence, for the interaction between a polar and a nonpolar molecule the Lennard-Jones potential may be used. However, in order to allow for small induced dipole effects, the combining laws (3.44) are replaced (Hirschfelder *et al.*, 1954) by

$$\begin{aligned}(eps)_{np} &= \{(eps)_n \times (eps)_p\}^{1/2}\xi^2 \\ (si)_{np} &= (1/2)\{(si)_n + (si)_p\}\xi^{(-1/6)} \qquad (3.55)\end{aligned}$$

where

$$\xi = 1 + \alpha_n^* t_p^* \{(eps)_p/2(eps)_n\}^{1/2} \qquad (3.56)$$

the subscripts n and p refer to nonpolar and polar molecules, respectively, α_n^* is the reduced polarizability of the nonpolar molecule $(= \alpha_n/(si)_n^3)$, $t_p^* = \bar{\mu}_p^{*2}/\sqrt{8}$, and $\bar{\mu}_p^*$ is the reduced dipole moment of the polar molecule $(= \bar{\mu}_p/\sqrt{\{(eps)_p(si)_p^3\}})$.

For pairs of polar molecules the situation is made much more difficult by the angle-dependent nature of the Stockmayer potential, since orientation effects must then be considered. The problem has been discussed by Monchick and Mason (1961), who calculated special sets of orientation-averaged collision integrals to be used for collisions between pairs of polar molecules. The integrals are functions both of the reduced temperature T^* and the reduced dipole moment $\bar{\mu}_p^*$ of the polar molecule. Polynomial expressions for the case of $H_2O - H_2O$ collisions are given in Table 2 on the assumption that $\bar{\mu}_{H_2O}^* = \sqrt{5}$ (Monchick and Mason, 1961). This value of $\bar{\mu}_{H_2O}^*$ is much higher than would be expected from the actual dipole moment of the water molecule. However, in combination with the force constants $(eps)_{H_2O}$ and $(si)_{H_2O}$ in Table 1, it gives the best fit for the viscosity of water vapor as measured by Kestin and Wang (1960), Shifrin (1959), and Bonilla, Wang, and Weiner (1956). The same value of $\bar{\mu}_{H_2O}^*$ is also used in Eq. (3.54) and the modified combining laws (3.55). Reduced polarizabilities of a number of nonpolar gases of interest in combustion systems are given in Table 1. It should be noted that since the theory can only deal with one polar component, all species other than H_2O are regarded as nonpolar.

It remains to consider the relatively easy rotational energy transfer in resonant collisions between polar molecules. These are collisions in which energy transfer takes place in a single mode between molecules of the same species. They are complex collisions in which $\Delta\epsilon_{pp} = 0$, and they do not contribute to the inelastic-collision integrals that determine relaxation times. Resonant collisions must, however, be taken into account when assessing the diffusion coefficients for internal energy. For nonpolar gases, where it is assumed that relatively close collisions are needed to cause transfer of internal energy, the binary diffusion coefficients for internal energy have been approximated by the ordinary binary diffusion coefficients. For collisions between polar molecules, where the exchange is energetically resonant, a large correction is necessary, of the form

$$\mathscr{D}_{p\,\mathrm{int},\,p} = \mathscr{D}_{pp}/(1 + \delta'_{pp}) \tag{3.57}$$

where $\mathscr{D}_{pp}$ is the ordinary self-diffusion coefficient for the polar species and δ'_{pp} may be evaluated using the method given by Mason and Monchick (1962). For water vapor, $\delta'_{H_2O} = 2958/T^{(3/2)}$.

3.3 Application of the extended Chapman-Enskog procedure to reactive flow systems

(a) General outline

The practical implementation of the detailed transport property theory just outlined is formidable and very expensive in computational effort if performed frequently as part of an iterative process. It is therefore natural to search for means of simplifying this part of the computation. It is necessary to examine each problem closely in relation (i) to the background of the theory employed and (ii) to the requirements of the results. Initial considerations must include the assumptions inherent in the Chapman-Enskog kinetic theory, which to some extent limit its applicability. Three of these assumptions are the consideration only of binary collisions, the treatment of translational energy by classical mechanics, and the condition that the gradients of the physical quantities are small. These assumptions, which are common to both the elastic- and inelastic-collision theories, are discussed by Hirschfelder *et al.* (1954). Reactive flow systems at all but very high pressures easily satisfy the required conditions.

Also involved in the case of inelastic collisions is the question of interconversion of translational and internal energy. In order that the distribution of energy among the translational and internal degrees of freedom may be described by a single temperature, as has been assumed (see Section 3.1), it is necessary for there to be easy exchange of translational and internal energy.

The vibrational relaxation is the more critical, since the appropriate collision numbers are high, often ranging from 10^3 to 10^7 (Herzfeld and Litovitz, 1959) as compared with only a few collisions for the attainment of translation-rotation equilibrium. The mean free path in a gas at a pressure of 1 atm is about 10^{-5} cm, and for this mean free path the number of collisions undergone by a molecule during a diffusive displacement of 10^{-2} cm is about 10^6. Hence, if appreciable changes in the physical quantities in flames at atmospheric pressure are confined to distances greater than 10^{-2} cm, the condition of easy interchange among all degrees of freedom is generally satisfied. Faster steady-state flames may not fall into this category, however, and neither will the boundaries of freshly ignited combustion kernels (around a localized ignition source) where the momentary gradients are extremely steep. Because of relaxation problems in the reaction zones of such systems it may not be possible to use a single temperature to define both the translation and internal energy distributions within a control volume. No firm theory yet exists for this situation, and so an approximate approach to the transport processes must be accepted.

In our experience, on the final approach to the steady state in an Eulerian type of time-dependent reactive flow calculation (see Section 4), when the properties at a point in the flow field vary slowly with time and fresh calculations of the transport coefficients are only required at infrequent intervals, use of the detailed transport model presents little problem. A sound basis then exists for studies of the effect of variations in basic transport parameters on the steady-state flow. Unsteady flows, on the other hand, are most economically treated by less demanding, though more approximate, methods. Some approaches along these lines are discussed below and in Section 3.4.

Perhaps the most time-consuming part of the detailed transport property computation is the solution of Eqs. (3.25a), which involves the inversion of approximately a $3N \times 3N$ matrix. The solution of this set of equations is necessary for the evaluation of the thermal conductivity and the thermal diffusion coefficients in the mixture. However, in hydrocarbon combustion systems N may be 20 or more, and if all these components are included the cost of the calculation becomes prohibitive. One approach to overcoming the problem is to exclude from this part of the calculation minor reaction intermediates which are unlikely to contribute appreciably either to the thermal conductivity or to thermal diffusion. This reduces the mixture to seven or eight effective components which are normally H, N_2 or Ar, O_2, H_2, CO, CO_2, H_2O, and the initial hydrocarbon.

In considering the terms involving internal energy in Eqs. (3.49), we note that these terms are of two types. First, there are terms involving $\mathcal{D}_{i\,\mathrm{int},k}$, occurring in $L_{ii}^{01,01}$ only, which are due to the diffusional flux of internal energy; and, second, there are the terms containing $c_{\mathrm{rot}}/\zeta_{ik}$ which arise from distortion of the elastic-collision trajectories when $\Delta\varepsilon \neq 0$. Although in the language of the discussion above there must be "easy" interchange of translational and internal energy, so that the diffusional terms are always important, it is only when the number of inelastic collisions becomes an appreciable fraction of the total number (ζ_{ik} less than about five) that the inelastic terms begin to exert an effect on the thermal conductivity. If the thermal conductivity of a pure gas is written as

$$\lambda = (\eta/m)(f_{\text{tr}}C_{v,\text{tr}} + f_{\text{int}}C_{v,\text{int}}) \tag{3.58}$$

where $C_{\text{v,tr}} = 1.5R$ and $C_{\text{v,int}} = C_{\text{v}} - C_{\text{v,tr}}$, it can be shown that in the absence of effects of inelastic collisions on the trajectories $f_{\text{tr}} = 2.5$ and $f_{\text{int}} = \rho\mathscr{D}_{ii}/\eta$. The effect of inelastic collisions is to decrease f_{tr} and simultaneously to increase f_{int}. The two effects oppose each other, and except at temperatures below about 300–400 K (where $\zeta_{ik,\text{rot}}$ for most simple molecules becomes quite small), the overall effect of the inelastic-collision terms is small (Mason and Monchick, 1962). For flame work, this prediction was confirmed by numerical experimentation (Dixon-Lewis, unpublished) on three hydrogen–air flames containing 20, 41 and 70% hydrogen in the unburnt gas, each with an initial temperature of 298 K. Results of computations in which ζ_{ik} took the values given in Table 1 were compared with results when ζ_{ik} for all the components took the value 200.0. The maximum change in burning velocity (in the 20% hydrogen–air flame) was 1.5%. On the basis of this result it is clearly permissible to use Eqs. (3.50) and (3.51) to give $\lambda_{0,\text{int}}$, then to solve the truncated set (3.25a) in which $rs = 00{,}10$ only, in order to obtain the a_{j10}^{1} and a_{j00}^{1}, and finally to obtain $\lambda_{0,\text{tr}}$ and the D_i^T by Eqs. (3.34) and (3.37).

Thermal diffusion effects in hydrogen–oxygen–nitrogen flames have been investigated by Dixon-Lewis (1968, 1979a,b) and by Warnatz (1978b). The inclusion of thermal diffusion among the transport fluxes makes a considerable difference to the molecular hydrogen profiles (particularly in hydrogen-rich flames), but not to the profiles of the other species. As a somewhat extreme example, Fig. 2 shows the magnitudes of the molecular

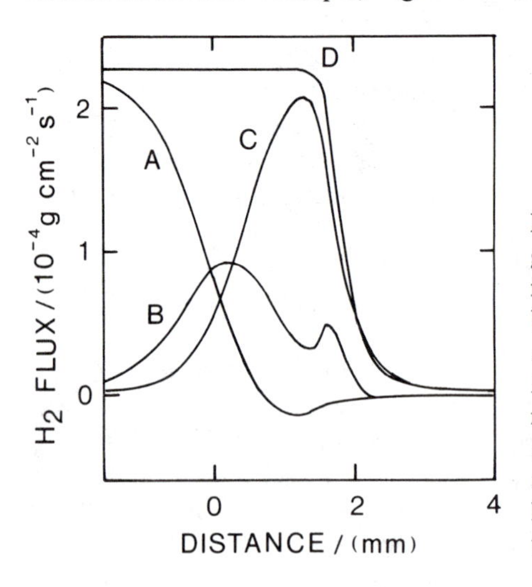

Figure 2. Fluxes of molecular hydrogen in a 60% hydrogen–air flame inhibited by addition of 4% hydrogen bromide (from Dixon-Lewis, 1979b). *A*, convective flux, $M_y(w_{\text{H}_2} - w_{\text{H}_2,b})$; *B*, ordinary diffusive flux; *C*, thermal diffusive flux; *D*, overall flux, $M_y(G_{\text{H}_2} - G_{\text{H}_2,b})$. Positive values denote fluxes from left to right (cold to hot in flame) and vice versa. (By courtesy of The Combustion Institute.)

hydrogen fluxes in a 60% hydrogen–air flame containing 4% hydrogen bromide. However, because the major effects on the molecular hydrogen profiles are in hydrogen-rich flames, these effects are unlikely to be kinetically very significant. The more important, though less directly observable, effect of thermal diffusion is associated with hydrogen atoms. Dixon-Lewis (1979a) found that the inclusion specifically of hydrogen-atom thermal diffusion effects in premixed hydrogen–air flames (other thermal diffusion contributions being present the whole time) reduced the computed burning velocity by 5–6%. Warnatz (1978b) made a similar observation regarding thermal diffusion in a hydrogen-rich flame. It seems therefore that thermal diffusion effects should be included in more precise modeling studies.

In the case of the hydrogen–air flames containing 20, 30, and 41% hydrogen, Dixon-Lewis (1979a) found including the hydrogen-atom thermal diffusion effect to be approximately equivalent to increasing $(si)_H$ by 15%.

(b) Organization of computer programs

The expressions required for pressure tensor, heat flux, and diffusion calculations following the extended Chapman-Enskog procedure without further approximation are given by Eqs. (3.28) to (3.30) and (3.33) to (3.37) inclusive. The evaluation of these expressions involves in turn the determination of the b_{j00}, a^1_{jmn}, and c^1_{j00} by means of Eqs. (3.22b), (3.25a), and (3.25c). The first step of each flux calculation thus involves setting up the matrix **H** or **L** of the coefficients $H^{00,00}_{ij}$ or $L^{rs,mn}_{ij}$ respectively. The steps involved for the $L^{rs,mn}_{ij}$ may be set out as follows. The procedure for the $H^{00,00}_{ij}$ is similar,

(i) Input of tables or expressions which give $\Omega^{(1,1)*}$, A^*, B^*, and C^* as a function of reduced temperature for nonpolar interactions, and an appropriate table or expression giving A^* for any polar–polar interaction. The table or expression used in the last case depends on the reduced dipole moment of the polar molecule (Section 3.2). Table 2 gives a set of polynomial coefficients satisfying these requirements.

(ii) Input of the molecular weight, reduced dipole moment or reduced polarizability, force constants $(si)_i$ and $(eps)_i/k_B$, and mole fraction of each component in the mixture. Table 1 gives this information for a number of species. A polynomial expression is also required for the molar specific heat $C_{p,i}$ of each component as a function of temperature (Chapter 8). The total internal specific heat $C_{i,\text{int}}$ may then be found at any temperature, since $C_{i,\text{int}} = C_{p,i} - (5/2)R$. The specific heats may also be expressed nondimensionally as $C_{i,\text{int}}/R$ $(= c_{i,\text{int}}/k_B)$. If inelastic-collision terms are considered, the rotational specific heat of each component is also required, again expressed as $C_{i,\text{rot}}/R$. The rotational specific heat remains constant in most reactive flow problems, since it is already fully developed at the lowest temperatures likely to be of interest.

(iii) The off-diagonal elements $\mathscr{D}_{ij}$ of the symmetric Nth order matrix $[\mathscr{D}_{ij}]$ of the binary diffusion coefficients are derived by using Eq. (3.45). The collision integrals are found after using the combination rules (3.44) or (3.55) to obtain $(si)_{ij}$ and the reduced temperature $T_{ij}^* = k_B T/(eps)_{ij}$. Values of T_{ij}^* are stored for later use when finding A_{ij}^*, B_{ij}^*, and $(1.2C_{ij}^* - 1)$ for the formation of the elements of **L**.

(iv) Provision is made for the program to have access, either from tables or from equations, to the viscosities η_i of each of the pure components of the thermal conductivity matrix [see Section 3.3(a)]. For N_2, Ar, O_2, H_2, CO, and CO_2, values used by the author are taken from Hilsenrath *et al.* (1960), and for water the formula (3.59) due to Shifrin (1959) is used

$$\eta_{H_2O} = 4.07 \times 10^{-7} T - 3.07 \times 10^{-5} \tag{3.59}$$

For other species, Lennard-Jones (12 : 6) potential parameters are taken from Table 1. The resulting viscosities are used in Eqs. (3.49a) and (3.49b).

(v) The remaining required data are the collision numbers ζ_{ij} [if inelastic collisions are included—but see Section 3.3(a)] and the quantity δ'_{pp} of Eq. (3.57) if a polar species is present. For water vapor, evaluation using the method given by Mason and Monchick (1962) leads to $\delta'_{H_2O} = 2958/T^{(3/2)}$. By means of Eqs. (3.49) and (3.57) the matrix **L** of coefficients in Eq. (3.25) may then be set up without difficulty.

(vi) Solution of Eqs. (3.25a) for the coefficients a_{jmn}^1 is now straightforward. In the case of diffusion, however, the necessary c_{i00}^{1ji} for the first approximations to the ordinary multicomponent diffusion coefficients are derived from a truncated set of Eqs. (3.25c) obtained by taking only $rs, mn = 00, 00$. Initially, this gives N^2 sets (corresponding with the possible combinations hk) of N equations each. A full solution of each set is not required. It can be shown that if the inverse matrix **P** is defined such that $\mathbf{P}\mathbf{L}^{00,00} = \mathbf{I}$, where **I** is the unit matrix, then the D_{ij} are given to first approximation by Eq. (3.60), where the P_{ij} are the elements of **P**

$$[D_{ij}]_1 = X_i \frac{16T}{25P} \frac{\sum_k m_k X_k}{m_j} (P_{ij} - P_{ii}) \tag{3.60}$$

Inversion of $\mathbf{L}^{00,00}$ thus provides the means of evaluation of all the ordinary multicomponent diffusion coefficients.

3.4. Approximate equations for transport fluxes in multicomponent mixtures

Despite the complicated nature of the formulations of the transport fluxes discussed above, the results of computations which use these equations do not differ much (e.g., less than 5–10% in the steady-state burning velocity of hydrogen–oxygen-nitrogen premixed flames) from the results of models which use more approximate and much simpler formulations. Since there is also some uncertainty about the combining rules and the extrapolation of the "primary" expressions such as Eqs. (3.45) to temperatures over 1000 K, then clearly for unsteady-state systems at least, and for the initial stages of approach to a steady state in an Eulerian calculation, a more approximate formulation of the transport fluxes will suffice. A mumber of approximate equations relating to transport coefficients in multicomponent systems are given by Chapman and Cowling (1970), Hirschfelder *et al.* (1954), and others; and several such formulations have recently been investigated in relation to reactive flow problems by Picone and Oran (1980), Oran and Boris (1981), and Coffee and Heimerl (1981). The most satisfactory compromise between complexity and accuracy is probably given by the following formulas and approaches.

(a) Viscosity

The equation (3.28) for the pressure tensor remains unchanged. The coefficient of viscosity is given by the following equations:

(i) For a pure gas

$$[\eta_i]_1 = 266.93 \times 10^{-7} \frac{(m_i T)^{1/2}}{(si)_i^2 \Omega_{ii}^{(2,2)*}(T^*)} \tag{3.53}$$

(ii) For multicomponent mixtures Eq. (3.61) due to Buddenberg and Wilke (1949) provides a good approximation

$$[\eta_{\text{mix}}]_1 = \sum_{i=1}^{N} \frac{[\eta_i]_1}{1 + 1.385 \dfrac{[\eta_i]_1 RT}{X_i P m_i} \displaystyle\sum_{\substack{k=1 \\ k \neq i}}^{N} \frac{X_k}{[\mathscr{D}_{ik}]_1}} \tag{3.61}$$

Alternatively, if we define ϕ_{ik} as

$$\phi_{ik} = \frac{1}{2\sqrt{2}} \{1 + (m_i/m_k)\}^{(-1/2)} \{1 + (\eta_i/\eta_k)^{(1/2)} (m_k/m_i)^{(1/4)}\}^2 \tag{3.62}$$

then we may also write (Wilke, 1950)

$$[\eta_{\text{mix}}]_1 = \sum_{i=1}^{N} \frac{[\eta_i]_1}{1 + \displaystyle\sum_{\substack{k=1 \\ k \neq i}}^{N} (X_k/X_i)\phi_{ik}} \tag{3.63}$$

This ϕ_{ik} is an approximation to the ratio $(\mathscr{D}_{ii}/\mathscr{D}_{ik})$ for mixtures of nonpolar gases (Mason and Saxena, 1958; Cheung, Bromley, and Wilke, 1960). Equations (3.62) and (3.63) may be used in conjunction with either experimental or calculated values of the viscosities of the pure gases.

(b) Diffusion

The approximations to the diffusive fluxes involve a slightly different formulation from Eq. (3.32) for the diffusion velocities. Equation (3.32) itself, together with the solutions of Eqs. (3.25c), can be rearranged to give the Stefan-Maxwell diffusion equations (Muckenfuss and Curtiss, 1958; Monchick, Munn, and Mason, 1966; Dixon-Lewis, 1968). If the thermal diffusion terms are neglected these equations become, in the first approximation,

$$\mathbf{d}_i = \sum_{j=1}^{N} \frac{X_i X_j}{[\mathscr{D}_{ij}]_1} (\bar{\mathbf{V}}_j - \bar{\mathbf{V}}_i) \qquad (i = 1, 2, \ldots, N) \tag{3.64}$$

where the vector $\mathbf{d}_i$ is defined in Eq. (3.10). If it is now assumed that all the $\bar{\mathbf{V}}_j$ take a common value $\bar{\mathbf{V}}$, Eq. (3.64) becomes

$$\mathbf{d}_i = \sum_{j=1}^{N} \frac{X_i X_j}{[\mathscr{D}_{ij}]_1} (\bar{\mathbf{V}} - \bar{\mathbf{V}}_i) \tag{3.65}$$

with the additional constraint (by definition) that

$$w_i \bar{\mathbf{V}}_i + (1 - w_i)\bar{\mathbf{V}} = 0 \tag{3.66}$$

Substituting back for $\bar{\mathbf{V}}$ in Eq. (3.65) then leads to the result used by Hirschfelder and Curtiss (1949)

$$\bar{\mathbf{V}}_i = -\frac{(1 - w_i)}{X_i \sum_{j \neq i} (X_j/[\mathscr{D}_{ij}]_1)} \mathbf{d}_i \tag{3.67}$$

When using Eq. (3.67) it must be remembered that there are only $N - 1$ independent diffusive fluxes, since $\sum_i (w_i \bar{\mathbf{V}}_i) = 0$. It is therefore necessary to calculate the Nth flux by means of this last relation. This may result in considerable error. Strictly speaking, Eq. (3.67) only applies for a trace component in a mixture of gases.

Oran and Boris (1981) have recently used a form very close to Eq. (3.67)

$$\bar{\mathbf{V}}_i = -\frac{(1 - w_i)D_{im}}{(1 - X_i)X_i} \mathbf{d}_i \tag{3.68}$$

where

$$\frac{1 - X_i}{D_{im}} = \sum_{j \neq i} \frac{X_j}{[\mathscr{D}_{ij}]_1} \tag{3.69}$$

as the first term in a more precise series solution for the $\bar{\mathbf{V}}_i$. The remaining terms, obtained by assuming all the velocities to be independent but subject to the constraint $\sum_i (w_i \bar{\mathbf{V}}_i) = 0$, lead to the series

$$\bar{\mathbf{V}}_i = -\frac{(1 - w_i)D_{im}}{(1 - X_i)X_i} \sum_{j=1}^{N} \left\{ \delta_{ij} + A_{ij} + \sum_{k=1}^{N} A_{ik} A_{kj} + \cdots \right\} \mathbf{d}_j \tag{3.70}$$

where

$$A_{ij} = w_i \delta_{ij} + \frac{X_i}{[\mathscr{D}_{ij}]_1} \frac{(1 - w_j) D_{jm}}{(1 - X_j)} (1 - \delta_{ij}) \tag{3.71}$$

Evaluation using all three terms given in the series gives results accurate to a few percent, while scaling approximately as N^2 in computer operations instead of N^3. The method is computationally efficient.

(c) Thermal conductivity

Equations (3.38) for the heat flux vector remain unchanged. However, in the approximate formulation it becomes necessary to calculate the thermal conductivities of the separate components of a multicomponent mixture, and then to combine these in an appropriate manner. The equations are:

(i) For a pure monatomic gas

$$[\lambda_{i,\mathrm{tr}}]_1 = 8.322 \times 10^{-4} \frac{\sqrt{(T/m_i)}}{(si)_i^2 \Omega_{ii}^{(2,2)*}(T^*)} \ \mathrm{W\,cm^{-1}\,K^{-1}} \tag{3.72}$$

$$= \frac{15}{4} \frac{R}{m_i} [\eta_i]_1 \tag{3.72a}$$

(ii) For a polyatomic gas, a modified Eucken correction is used

$$\begin{aligned} \lambda_i &= \lambda_{i,\mathrm{tr}} + \lambda_{i,\mathrm{int}} \\ &= (\eta/m_i)\{(15/4)R + f_{\mathrm{int}} C_{i,\mathrm{int}}\} \\ &= \lambda_{i,\mathrm{tr}}(0.115 + 0.354 C_{p,i}/R) \qquad (\text{for } f_{\mathrm{int}} = 1.3275) \end{aligned} \tag{3.72b}$$

(iii) For multicomponent mixtures the equation (3.73) derived by Mason and Saxena (1958) is probably most satisfactory

$$[\lambda_{\mathrm{mix}}]_1 \simeq \sum_{i=1}^{N} \frac{\lambda_i}{1 + \dfrac{1}{X_i} \sum\limits_{\substack{k=1 \\ k \neq i}}^{N} (X_k G_{ik})} \tag{3.73}$$

where $G_{ik} = 1.065 \phi_{ik}$ and ϕ_{ik} is given by Eq. (3.62). However, Coffee and Heimerl (1981) claim that the much simpler formula (3.74) gives almost as accurate results

$$\lambda_{\mathrm{mix}} = 0.5 \left\{ \sum_{i=1}^{N} X_i \lambda_i + \left(\sum_{i=1}^{N} X_i/\lambda_i \right)^{-1} \right\} \tag{3.74}$$

Thermal diffusion is not normally considered in approximate treatments of reactive flows unless there are very large differences between the molecular weights of the participating species. However, thermal diffusion of hydrogen atoms can be very important in flames [see Section 3.3(a)].

4. One-dimensional premixed laminar flame properties by solution of the time-dependent equations

4.1. Preliminary transformations

The forward integration of the appropriate time-dependent conservation equations by finite difference methods not only provides a means of following nonstationary combustion phenomena, but also by more prolonged integration offers a method of entry into the steady-state flame properties. While nonstationary processes such as ignitions or flames traveling in initially quiescent gas mixtures must be studied in a Lagrangian reference frame, either a Lagrangian or an Eulerian reference frame may be used for entry into the steady state. Still further computations may employ a mixed Lagrange-Euler approach in certain circumstances. Since the practical implementation of both schemes requires a number of similar operations, they will be discussed together at first. Simplifications which are possible with particular formulations will be discussed at a later stage.

We consider first systems in which the pressure is effectively constant, the simplest case of reactive flow with transport. Armed with the expressions for the transport fluxes, we return to the continuity equations for the one-dimensional flow and consider an adiabatic system, neglecting the contributions of both the viscous and the kinetic energy terms in the energy conservation equation. In Eulerian coordinates, the species and energy continuity equations become

$$\rho(\partial\sigma_i/\partial t) + M_y(\partial\sigma_i/\partial y) = -(1/A)\,\partial/\partial y\,(Aj_i/m_i) + R_i \qquad (i = 1, 2, \ldots, N) \tag{2.18}$$

$$\rho(\partial h/\partial t) + M_y(\partial h/\partial y) = -(1/A)\,\partial/\partial y\,\{A(Q_D + Q_T)\} \tag{2.20b}$$

where, in terms of σ_i and h, the transport fluxes defined by Eqs. (3.35a) and (3.38) become

$$j_i = -\left\{\beta_i^h(\partial h/\partial y) + \sum_{j=1}^{N} \beta_{ij} m_j(\partial\sigma_j/\partial y)\right\} \tag{4.1}$$

$$\begin{aligned} q' &= Q_D + Q_T \\ &= -\left\{\gamma^h(\partial h/\partial y) + \sum_{j=1}^{N} \gamma_j m_j(\partial\sigma_j/\partial y)\right\} \end{aligned} \tag{4.2}$$

with

$$\beta_{ij} = \sum_{k=1}^{N} \{(1 - \delta_{ik})\rho\sigma_k D_{ik}(m_i m_k/m_j)\} - (1 - \delta_{ij})\rho D_{ij}(m_i/m_G) - \beta_i^h h_j \qquad (j = 1, 2, \ldots, N) \tag{4.3}$$

$$\beta_i^h = D_i^T/c_p T \tag{4.4}$$

$$\gamma_j = \frac{RT}{m_j}\left\{\frac{D_j^T}{m_j\sigma_j} - \sum_{k=1}^{N} \frac{D_k^T m_G}{m_k}\right\} - \frac{\lambda_o h_j}{c_p} + \sum_{k=1}^{N} h_k \beta_{ki}$$

$$(j = 1, 2, \ldots, N) \tag{4.5}$$

$$\gamma^h = \frac{\lambda_o}{c_p} + \sum_{k=1}^{N} \beta_k^h h_k \tag{4.6}$$

where m_G denotes average molecular weight $(\sum_j \sigma_j)^{-1}$.

The detailed transport flux formulation has been used here. If approximations like those outlined in Section 3.4 are introduced, the modifications are confined to the expressions for the coefficients β_{ij}, β_i^h, γ_j, and γ^h. There is no change in the form of the flux terms. The complete set of Eqs. (2.18) and (2.20b) must be solved subject to boundary conditions which define the specific problem. For a "free flame" these can be written

$$\begin{aligned} y_b &= -\infty, \quad \partial\sigma_i/\partial y = 0, \quad \partial h/\partial y = 0 \\ y_u &= +\infty, \quad \sigma_i = \sigma_{iu}, \quad h = h_u(T_u, \sigma_{iu}), \\ &\qquad\quad \partial\sigma_i/\partial y = 0, \quad \partial h/\partial y = 0 \end{aligned} \tag{4.7}$$

where the subscripts u and b refer to unburnt and burnt gas, respectively. The object of the calculation is either to observe the evolution of the flame profiles with time or to determine the steady-state value of $(M_y A)$ and the corresponding profiles of the dependent variables.

The computational procedure outlined below follows in principle the approach initially described by Patankar and Spalding (1970) for handling boundary layer flows and later applied to the solution of stationary flame problems by Spalding and Stephenson (1971), Spalding, Stephenson, and Taylor (1971), and Stephenson and Taylor (1973). Other procedures, for example that of Lund (1978), follow the same general outline but differ in detail.

As a first step Eqs. (2.18) and (2.20b) are transformed into the von Mises system of coordinates. The transformation is defined for the present system by

$$\begin{aligned} d\psi/dy &= \rho A \\ d\psi/dt &= -M_y A \end{aligned} \tag{4.8}$$

It leads to the use of mass (a conserved quantity) rather than distance as the independent variable. A nondimensional stream function ω is next introduced, defined by

$$\omega = \frac{\psi - \psi_b}{\psi_u - \psi_b} \qquad (0 \leqslant \omega \leqslant 1) \tag{4.9}$$

The difference $(\psi_u - \psi_b)$ defines a finite total mass of fluid in the computational grid, and in practice this must be estimated at the start of an integration so as to cover economically the region of major reactive flow interest within the theoretically infinite "free flame" boundaries. According to our definition we have

$$\begin{aligned} d\psi_u/dt &= -(M_yA)_u \\ d\psi_b/dt &= -(M_yA)_b. \end{aligned} \tag{4.10}$$

In the steady state

$$(M_yA)_u = (M_yA)_b \tag{4.11}$$

However, in the unsteady state, when the flame is expanding or contracting towards its steady-state thickness and shape, $(M_yA)_u \neq (M_yA)_b$, the value of $(\psi_u - \psi_b)$ varies somewhat, and (M_yA) is a function of ω. If it is assumed that this function is linear, Eqs. (2.18) and (2.20b) become

$$\partial\sigma_i/\partial t + (a + b\omega)\,\partial\sigma_i/\partial\omega = -\{1/(\psi_u - \psi_b)\}\,\partial/\partial\omega\,(Aj_i/m_i) + R_i/\rho \qquad (i = 1, 2, \ldots, N) \tag{4.12}$$

$$\partial h/\partial t + (a + b\omega)\,\partial h/\partial\omega = -\{1/(\psi_u - \psi_b)\}\,\partial/\partial\omega\,\{A(Q_D + Q_T)\} \tag{4.13}$$

where

$$a = (M_yA)_b/(\psi_u - \psi_b) \tag{4.14}$$

$$b = \{(M_yA)_u - (M_yA)_b\}/(\psi_u - \psi_b) \tag{4.15}$$

and $(M_yA)_u$ and $(M_yA)_b$ are the convection terms or "entrainment rates" at the cold and hot ends of the grid, respectively.

The boundary conditions for the free flame then become

$$\begin{aligned} &\omega = 0, 1 \quad \partial\sigma_i/\partial\omega = 0, \quad \partial h/\partial\omega = 0 \\ &\omega = 1, \quad\;\; \sigma_i = \sigma_{iu}, \quad\;\; h = h_u(T_u, \sigma_{iu}) \end{aligned} \tag{4.16}$$

Other types of boundary conditions will be discussed later. The condition that the spatial gradients of all dependent variables are zero at the hot boundary ensures no diffusive loss of energy or material through this boundary. The boundary condition demands judicious choice of the quantity $(\psi_u - \psi_b)$ so that the approximations of zero gradient, together with zero gradient at the cold boundary, are not unreasonable.

4.2. Finite-difference formulation

(a) Setting up the computational grid

Equations (4.12) and (4.13) now have to be integrated forward in time subject to the appropriate boundary conditions such as Eq. (4.16). Finite-difference methods are used. At $t = 0$, sigmoid-shaped starting profiles of species concentrations and enthalpy must be set up to represent the transfer from cold to hot boundary conditions. These are input as vectors $\mathbf{\Phi}$ of values of each dependent variable at discrete values of ω. In order to concentrate the grid spacing at the position where most reaction occurs, the grid points are distributed nonuniformly over the total interval $0 \leqslant \omega \leqslant 1$. In computations carried out by the author, typically 33 grid points (but sometimes more) are used, with a uniform value of $\delta\omega$ over the central 14 intervals, outside of which each value of $\delta\omega$ increases by a constant factor (usually 1.4 or higher) relative to its nearest inside neighbor. The ω distribution usually stays constant throughout the integration. The dependent variables which are input are the molar composition, pressure, and temperature of the initial mixture, together with rough estimates of the hot boundary values. The mole fractions are converted to values of the working composition variables σ_i, and the specific enthalpy of the unburnt mixture is calculated. The final specific enthalpy takes the same value. A few grid points near each boundary also are given the boundary values of the dependent variables, and a suitable curve-generating algorithm is used to initialize values in the central region of the grid.

At the start of the integration, the specific enthalpy is assumed constant over the whole flame. For interconversion between enthalpy and temperature, the program must be provided with polynomial coefficients which allow the molar enthalpies of the pure components to be expressed as functions of temperature (Chapter 8). If the molar enthalpy of component i is expressed as

$$H_i = a_{i,0} + a_{i,1}T + a_{i,2}T^2 + \cdots \tag{4.17}$$

then in terms of the temperature the specific enthalpy of the mixture is

$$h = \sum_{i=1}^{N} \sigma_i(a_{i,0} + a_{i,1}T + a_{i,2}T^2 + \cdots) \tag{4.18}$$

The temperature at each grid point may be calculated by solution of the polynomial expression.[1] The mixture density may then be found from the equation of state (2.25).

(b) The finite-difference equations in the central region of the grid

The numerical integrations are performed by means of finite-difference techniques using a rectangular grid in the (ω, t) plane having (nonuniform) intervals $\delta\omega$ and δt. We denote by $\mathbf{\Phi}_{n',j'}$ the vector of dependent variables at grid point $n'(\delta t), j'(\delta\omega)$. The equations to be integrated [(4.12) and (4.13)] may then be written as the matrix equation

$$\partial\mathbf{\Phi}/\partial t + (a + b\omega)\partial\mathbf{\Phi}/\partial\omega = -\{1/(\psi_u - \psi_b)\}\, \partial/\partial\omega\,(A\mathbf{Q}) + \mathbf{d} \tag{4.19}$$

---

[1] If Eq. (4.17) is of order greater than three, then solving Eq. (4.18) for the temperature has to be done numerically. As an alternative, Eq. (4.17)—equivalent to the equations given in Table 2 of Chapter 8—can be replaced by a quadratic form valid for a small range of T near that estimated for the grid point.

where a and b are defined by Eqs. (4.14) and (4.15), respectively, $(A\mathbf{Q})$ represents the transport flux terms in Eqs. (4.12) and (4.13), and $\mathbf{d}$ represents the terms R_i/ρ, which do not contain ω. For economy of notation the terms of $\mathbf{Q}$ are represented as

$$Q_i = -\sum_{k=1}^{N+1} \beta'_{ik}(\partial\phi_k/\partial y) \qquad (i = 1, 2, \ldots, N+1) \tag{4.20}$$

where the first N integers i and k refer to the chemical components and the $(N + 1)$th integer refers to the specific enthalpy. With reference to Eqs. (4.1) to (4.6) and (4.19)

$$\begin{aligned} \phi_k &\equiv \sigma_k && (k \le N) \\ \phi_{N+1} &\equiv h && \\ \beta'_{ik} &\equiv \beta_{ik}m_k/m_i && (i, k \le N) \\ \beta'_{i,N+1} &\equiv \beta_i^h/m_i && (i \le N) \\ \beta'_{N+1,k} &\equiv \gamma_k m_k && (k \le N) \\ \beta'_{N+1,N+1} &\equiv \gamma^h && \end{aligned} \tag{4.21}$$

The finite-difference integration at time $n'(\delta t)$ involves computing the quantities $\partial\mathbf{\Phi}/\partial t \equiv (\mathbf{\Phi}_{n'+1,j'} - \mathbf{\Phi}_{n',j'})/\delta t$ by evaluating the remaining terms in Eqs. (4.19). There are two extreme possibilities for this. One is to compute these terms completely at time $n'(\delta t)$. This is the *explicit* method, or method of forward differences. It is easy to use, since the values of all dependent variables at the start of the time step are known. Unfortunately, the explicit approach by itself is unsuitable for dealing with coupled equations for rate processes with very different time constants. The chain reactions of combustion chemistry have just this "stiff" property: rapid reactions reach equilibrium long before the system as a whole does. Stability of computation demands a small time step δt appropriate to the rates of the faster reactions, with a corresponding prohibitive increase in the cost of the complete calculation.

The other extreme is to evaluate the remaining terms at time $(n' + 1)(\delta t)$, the *fully implicit* or backward differencing approach. It leads to a set of algebraic equations from which the dependent variables at time $(n' + 1)(\delta t)$ can be calculated. This approach is unconditionally stable (Richtmyer and Morton, 1967), and is the approach used here. We may of course also use other schemes in which intermediate weights are given to the forward and backward differences. These partially implicit schemes lead to improved accuracy. However, if attempts are made to use them on systems of stiff equations, the latter must be treated by asymptotic techniques. In chemical situations such techniques are equivalent to the use of the chemical quasi-steady-state or partial equilibrium assumptions at long times. They will be considered again in Section 9.

Figure 3 shows a control volume (shaded) in the center of the computational grid. The technique used by Patankar and Spalding (1970), and

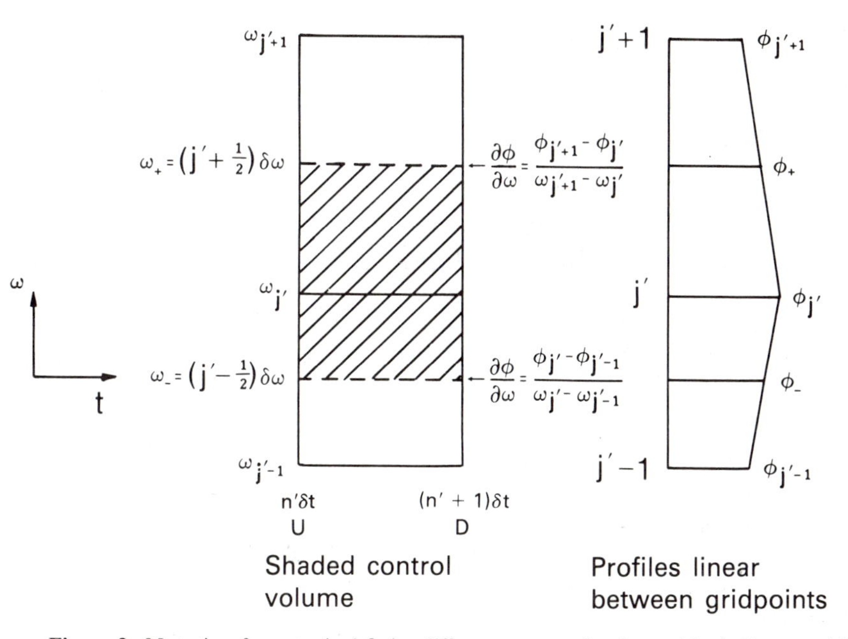

Figure 3. Notation for a typical finite difference control volume (shaded) over which the microintegration is carried out.

followed here, is to perform a microintegration over the control volume on the assumption of linear profiles of the dependent variables between each pair of grid points. The control volume boundaries of cell j', at $(j' + \frac{1}{2})(\delta\omega)$ and $(j' - \frac{1}{2})(\delta\omega)$, are denoted by "+" and "−," respectively. Also, for the purpose of integration the interval $n' \to n' + 1$ will be denoted by $U \to D$ (upstream → downstream), and the suffix n' will be discarded. Since the method is implicit, terms other than $\partial\mathbf{\Phi}/\partial t$ in Eqs. (4.19) are all evaluated at time D. The subscript D can therefore be omitted, and only the subscript U will be included. It is understood that the remaining terms take their downstream (D) values.

On integration over the control volume and multiplication throughout by $(\psi_u - \psi_b)$, Eq. (4.19) gives

$$\left[\int_-^+ \mathbf{\Phi}\, d\omega - \int_-^+ \mathbf{\Phi}_U\, d\omega\right](\psi_u - \psi_b)/\delta t$$

$$+ (\psi_u - \psi_b)[\{(a + b\omega)\mathbf{\Phi}\}_+ - \{(a + b\omega)\mathbf{\Phi}\}_-] - (\psi_u - \psi_b)b\int_-^+ \mathbf{\Phi}\, d\omega$$

$$= -\{(A\mathbf{Q})_+ - (A\mathbf{Q})_-\} + \int_-^+ (\psi_u - \psi_b)\,\mathbf{d}\cdot d\omega \tag{4.22}$$

where the integration limits are signified by the defined positions in the grid.

Referring to Eqs. (4.14) and (4.15) we introduce

$$P \equiv (\psi_u - \psi_b)/\delta t \tag{4.23}$$

$$\begin{aligned} G &\equiv (M_y A)_b - (M_y A)_u \\ &= -(\psi_u - \psi_b)b \end{aligned} \tag{4.24}$$

$$\begin{aligned} L_+ &= (\psi_u - \psi_b)(a + b\omega_+) \\ &= (M_y A)_b + \omega_+\{(M_y A)_u - (M_y A)_b\} \\ &= (M_y A)_b - G\omega_+ \end{aligned} \tag{4.25}$$

$$L_- \equiv (M_y A)_b - G\omega_- \tag{4.26}$$

Equation (4.22) then becomes

$$(P + G)\int_-^+ \mathbf{\Phi}\, d\omega - P\int_-^+ \mathbf{\Phi}_U\, d\omega + L_+\mathbf{\Phi}_+ - L_-\Phi_- = -\{(A\mathbf{Q})_+ - (A\mathbf{Q})_-\} + \int_-^+ (\psi_u - \psi_b)\mathbf{d}\cdot d\omega \tag{4.27}$$

Further introducing

$$\Omega = \omega_{j'+1} - \omega_{j'-1} \tag{4.28}$$

$$\Omega_+ = \omega_{j'+1} - \omega_{j'} \tag{4.29}$$

$$\Omega_- = \omega_{j'} - \omega_{j'-1} \tag{4.30}$$

then on the assumption that the $\mathbf{\Phi}$ profiles are linear in ω between the grid points, twice the left-hand side of Eq. (4.27) can be written

$$\begin{aligned} 2 \times \text{L.H.S.} &= (1/4)(P + G)(\Omega_+\mathbf{\Phi}_{j'+1} + 3\Omega\mathbf{\Phi}_{j'} + \Omega_-\mathbf{\Phi}_{j'-1}) \\ &\quad - (1/4)P(\Omega_+\mathbf{\Phi}_{U,j'+1} + 3\Omega\mathbf{\Phi}_{U,j'} + \Omega_-\mathbf{\Phi}_{U,j'-1}) \\ &\quad + L_+(\mathbf{\Phi}_{j'} + \mathbf{\Phi}_{j'+1}) - L_-(\mathbf{\Phi}_{j'} + \mathbf{\Phi}_{j'-1}) \\ &= \{(1/4)(P + G)\Omega_- - L_-\}\mathbf{\Phi}_{j'-1} \\ &\quad + \{(3/4)(P + G)\Omega + L_+ - L_-\}\mathbf{\Phi}_{j'} \\ &\quad + \{(1/4)(P + G)\Omega_+ + L_+\}\mathbf{\Phi}_{j'+1} \\ &\quad - (1/4)P(\Omega_+\mathbf{\Phi}_{U,j'+1} + 3\Omega\mathbf{\Phi}_{U,j'} + \Omega_-\mathbf{\Phi}_{U,j'-1}) \end{aligned} \tag{4.31}$$

We now turn to the right-hand side of Eq. (4.27).

(i) *The transport flux terms* $\mathbf{Q}$ are given in differential form by Eqs. (4.20) and (4.21). In finite-difference form they become to first order

$$Q_{i+} = -\sum_{k=1}^{N+1} T_{ik+}\{(\phi_k)_{j'+1} - (\phi_k)_{j'}\} \tag{4.32}$$

$$Q_{i-} = -\sum_{k=1}^{N+1} T_{ik-}\{(\phi_k)_{j'} - (\phi_k)_{j'-1}\} \tag{4.33}$$

where

$$T_{ik+} = (1/2)\{(\beta'_{ik})_{j'+1} + (\beta'_{ik})_{j'}\}/\delta y_{j'} \quad (4.34)$$

$$T_{ik-} = (1/2)\{(\beta'_{ik})_{j'} + (\beta'_{ik})_{j'-1}\}/\delta y_{j'-1} \quad (4.35)$$

$$\delta y_{j'} = y_{j'+1} - y_{j'} \quad (4.36)$$

$$\delta y_{j'-1} = y_{j'} - y_{j'-1} \quad (4.37)$$

(ii) *Source terms.* On back-substituting R_i/ρ for d_i [Eqs. (4.19)], the final terms on the right-hand sides of Eqs. (4.27) become the space integral rates $\int_-^+ R_i A dy$. On the assumption that the product $(R_i A)$ is linear in y between the grid points, these space integral rates can be expressed in finite-difference form as

$$S_i = \int_-^+ R_i A dy$$
$$= (1/8)[\{(R_i A)_{j'+1} + 3(R_i A)_{j'}\}\ \delta y_{j'} + \{(R_i A)_{j'-1} + 3(R_i A)_{j'}\}\ \delta y_{j'-1}] \quad (4.38)$$

Expressions for the net molar rates of formation R_i are given as Eqs. (2.14) to (2.16). They are nonlinear in the dependent variables. In order to linearize the equations, $R_i(\sigma_1, \sigma_2, \ldots, \sigma_N, \rho, T)$ is expanded to first order around an approximate solution $R_{i,B}$ as

$$R_i = R_{i,B} + \sum_{k=1}^{N} (\partial R_i/\partial \sigma_k)_B \delta\sigma_k + (\partial R_i/\partial \rho)_B\ \delta\rho + (\partial R_i/\partial T)_B\ \delta T \quad (4.39)$$

$$= R_{i,B} + \sum_{k=1}^{N} (\partial R_i/\partial \sigma_k)_B(\sigma_k - \sigma_{k,B}) + (\partial R_i/\partial \rho)_B(\rho - \rho_B) + (\partial R_i/\partial T)_B(T - T_B) \quad (4.39a)$$

The subscript B denotes values of the variables *before updating*, and at the end of a time step they may be close approximations to a true solution. As an initial approximation to a solution, however, they may be much further away, and could, for example, be identical to the values at the beginning of the time step.

Generalizing further along the lines of Eq. (4.20), we introduce three further dependent variables, making $(N + 4)$ in all for the N-component system. The new variables are ρ, T and δy, and the whole set of variables in order becomes $\sigma_1, \sigma_2, \ldots, \sigma_N, h, \rho, T$, and δy. With this larger set we may write

$$R_i A = R_{i,B} A + \sum_{\substack{k=1 \\ k \neq N+1}}^{N+3} A S_{D,ik}\ \delta\phi_k \quad (4.40)$$

where $\mathbf{S}_D$ represents the Jacobian matrix whose elements are $(\partial R_i/\partial \phi_k)_B$. The "chaperon" factors in third-order reactions are considered to remain constant during a time interval and do not therefore contribute to these derivatives.

The source term or space integral rate S_i may now be expressed sufficiently accurately as

$$(S_i)_j = (S_{i,B})_j + (1/2) \sum_{\substack{k=1 \\ k \neq N+1}}^{N+3} (AS_{D,ik})_{j'} (\delta\phi_k)_{j'} (\delta y_{j'} + \delta y_{j'-1}) \tag{4.41}$$

$$= (S_{i,B})_j + (1/2) \sum_{\substack{k=1 \\ k \neq N+1}}^{N+3} (AS_{D,ik})_{j'} (\phi_k - \phi_{k,B})_{j'} \Delta y_{j'} \tag{4.41a}$$

where

$$\begin{aligned}\Delta y_{j'} &= \delta y_{j'} + \delta y_{j'-1} \\ &= y_{j'+1} - y_{j'-1}\end{aligned} \tag{4.42}$$

Then, gathering together the results expressed in Eqs. (4.32), (4.33), and (4.41a), multiplying by two, and equating with expression (4.31), Eq. (4.27) becomes

$$\begin{aligned}
&\sum_{k=1}^{N+1} [\{(1/4)(P+G)\Omega_- - L_-\}\delta_{ik} - 2T_{ik-}A_-](\phi_k)_{j'-1} \\
&\qquad + \sum_{k=1}^{N+3} [\{(3/4)(P+G)\Omega + L_+ - L_-\}\delta_{ik} \\
&\qquad\qquad + 2(T_{ik+}A_+ + T_{ik-}A_-)(1-\delta_{k,N+2})(1-\delta_{k,N+3}) \\
&\qquad\qquad - (AS_{D,ik}\Delta y)_{j'}(1-\delta_{k,N+1})](\phi_k)_{j'} \\
&\qquad + \sum_{k=1}^{N+1} [\{(1/4)(P+G)\Omega_+ + L_+\}\delta_{ik} - 2T_{ik+}A_+](\phi_k)_{j'+1} \\
&= (1/4)P\{\Omega_+(\phi_{i,U})_{j'+1} + 3\Omega(\phi_{i,U})_{j'} + \Omega_-(\phi_{i,U})_{j'-1}\} \\
&\quad + 2(S_{i,B})_{j'} - \sum_{\substack{k=1 \\ k \neq N+1}}^{N+3} (AS_{D,ik}\phi_{k,B}\Delta y)_{j'}
\end{aligned}$$

$$(i = 1, 2, \ldots, N+1) \tag{4.43}$$

Equation (4.43) arises as a result of expressing the conservation conditions in finite-difference form, there being one equation for each species i and one for the enthalpy. However, the closure of the set of equations at grid point j' requires three further relations. Two of these are provided by the auxiliary equations (2.25) and (2.26). In certain computations directed towards an Eulerian steady state these relations may be used as they stand, as will be indicated in Section 4.4(b). Otherwise, both equations are expanded to first order in the dependent variables to

$$\delta\rho = -(\rho/T)_B\,\delta T - \sum_{i=1}^{N} (\rho/\textstyle\sum_i \sigma_i)_B\,\delta\sigma_i \tag{4.44}$$

$$\begin{aligned}\delta h &= \sum_{i=1}^{N} (H_i)_B\,\delta\sigma_i + \sum_{i=1}^{N} (\sigma_i)_B\,\delta H_i \\ &= \sum_{i=1}^{N} (H_i)_B\,\delta\sigma_i + c_{p,B}\,\delta T\end{aligned} \tag{4.45}$$

where c_p is the specific heat of the mixture. Additionally, one of the species conservation equations is normally discarded in favor of Eq. (2.27). Similarly expanded, Eq. (2.27) becomes

$$\sum_{i=1}^{N} m_i \,\delta\sigma_i = 1 - \left(\sum_{i=1}^{N} m_i\sigma_i\right)_B \tag{4.46}$$

The right-hand side of Eq. (4.46) is expressed in this form in order to maintain relation (2.27) strictly despite the possible effects of rounding errors on the σ_i in the extended calculation. We note in Eqs. (4.44) to (4.46) that $\delta\mathbf{\Phi} \equiv \mathbf{\Phi} - \mathbf{\Phi}_B$. The discarded species conservation equation should belong to a component which is either present in excess or inert, preferably both.

The third extra relationship is required for the calculation of the physical distances δy.

(i) For *true one-dimensional flames* ($A = 1.0$ everywhere) the distances δy are calculated from the finite-difference analog of Eq. (4.8)

$$\begin{aligned}\delta y_{j'} &= y_{j'+1} - y_{j'} \\ &= \frac{2(\omega_{j'+1} - \omega_{j'})(\psi_u - \psi_b)}{(\rho_{j'+1} + \rho_{j'})}\end{aligned} \tag{4.47}$$

Here $(y_{j'+1} - y_{j'})$ is the distance between the centers of control volumes j' and $j' + 1$ and $(1/2)(\rho_{j'} + \rho_{j'+1})$ represents the average density between the two grid points. A more satisfactory physical expression for the average density would be $2\{(1/\rho)_{j'+1} + (1/\rho)_{j'}\}^{-1}$, but this formulation can have disadvantages since an improperly low value of ρ at a grid point then assumes undue importance (Patankar and Spalding, 1970). It is only when the two densities are very nearly equal that either formula is really satisfactory, and this is perhaps one of the weaker aspects of the calculation. However, results obtained from computations which use the approximation (4.47) are found to be in satisfactory agreement with results obtained by a completely different approach to the calculation of the properties of certain hydrogen–oxygen–nitrogen flames (Dixon-Lewis, 1979). The approximation is therefore adequate. The alternative calculation will be briefly discussed in Section 9.

Expanding $\delta y_{j'}$ to first order in the dependent variables gives

$$\delta(\delta y_{j'}) = -(\delta y_{j'})_B \frac{\delta\rho_{j'+1} + \delta\rho_{j'}}{\rho_{j'+1} + \rho_{j'}} \tag{4.48}$$

(ii) In *divergent flow* situations an expression relating stream tube area A with normal distance y must be provided. In such cases the cell centers j' and $(j' + 1)$ are symmetrically placed about the boundary $(j' + \frac{1}{2})$ in terms of ω, but not in terms of y. It then becomes necessary to calculate the physical thickness of each half-cell separately, so as to evaluate the areas of the cell boundaries. The boundary surface areas $A_{j'+1/2}$, are of direct relevance to the calculation of the transport ($A\mathbf{Q}$) terms in Eq. (4.27). In the case of divergent flow it is necessary first to evaluate

$$I_{1,j'} = \int_{j'}^{j'+1/2} A\,dy = \frac{2(\omega_{j'+1} - \omega_{j'})(\psi_u - \psi_b)}{(3\rho_{j'} + \rho_{j'+1})} \tag{4.49}$$

and

$$I_{2,j'} = \int_{j'+1/2}^{j'+1} A\,dy = \frac{2(\omega_{j'+1} - \omega_{j'})(\psi_u - \psi_b)}{(\rho_{j'} + 3\rho_{j'+1})} \tag{4.49a}$$

and then, given the relation $A = f(y)$ and the value of y at one end of the interval, to compute the appropriate thickness $\delta y_{1,j'}$ or $\delta y_{2,j'}$. The stream tube area at each half-interval is then readily calculated from the given relation.

We may wish to look at changes in δy_1 and δy_2 directly during an iteration. This is probably best done by combining the two operations just discussed in an approximate manner to

$$\delta y_{1,j'} = \frac{1}{(A_{j'+1/2})_B} \frac{2(\omega_{j'+1} - \omega_{j'})(\psi_u - \psi_b)}{(3\rho_{j'} + \rho_{j'+1})} \tag{4.50}$$

$$\delta y_{2,j'} = \frac{1}{(A_{j'+1})_B} \frac{2(\omega_{j'+1} - \omega_{j'})(\psi_u - \psi_b)}{(\rho_{j'} + 3\rho_{j'+1})} \tag{4.50a}$$

and hence

$$\delta(\delta y_{1,j'}) = -(\delta y_{1,j'})_B \frac{3\delta\rho_{j'} + \delta\rho_{j'+1}}{3\rho_{j'} + \rho_{j'+1}} \tag{4.51}$$

$$\delta(\delta y_{2,j'}) = -(\delta y_{2,j'})_B \frac{\delta\rho_{j'} + 3\delta\rho_{j'+1}}{\rho_{j'} + 3\rho_{j'+1}} \tag{4.51a}$$

The areas A are then assumed constant during each iteration, but must be updated between iterations.

The set of equations for the $N + 4$ dependent variables ($N + 5$ in divergent flow) is now complete with the exception that both the transport terms and the source terms in Eq. (4.43) are functions of δy. Referring to Eqs. (4.32) to (4.37), the normal practice in the evaluation of the transport terms is to treat the β'_{ik} as constant during an iteration, so that for variable δy Eqs. (4.34) and (4.35) take the additional modification

$$T_{ik+} = (T_{ik+})_B \left\{1 - \frac{\delta(\delta y_{j'})}{\delta y_{j'}}\right\} \tag{4.52}$$

$$T_{ik-} = (T_{ik-})_B \left\{1 - \frac{\delta(\delta y_{j'-1})}{\delta y_{j'-1}}\right\} \tag{4.53}$$

In a corresponding manner, $(S_i)_{j'}$ in Eq. (4.41) should more properly be written, to first order, as

$$(S_i)_{j'} = (S_{i,B})_{j'} + (1/8)\{(R_i A)_{j'+1} + 3(R_i A)_{j'}\}\,\delta(\delta y_{j'}) + (1/8)\{(R_i A)_{j'-1} + 3(R_i A)_{j'}\}\,\delta(\delta y_{j'-1}) + (1/2)\sum_{\substack{k=1\\k\neq N+1}}^{N+3} (AS_{D,ik}\,\delta\phi_k)_{j'}\,(\delta y_{j'} + \delta y_{j'-1}) \tag{4.54}$$

As a preliminary to a final rearrangement, we next assume that the $N + 4$ element line vector $\{\mathbf{\Phi}_B\}_{j'}$ represents an approximate solution of the complete set of equations for the cell around j', so that $\mathbf{\Phi}_{j'} = \mathbf{\Phi}_{B,j'} + \delta\mathbf{\Phi}_{j'}$. Then, from Eqs. (4.43) to (4.48) the relations which allow us to calculate $\delta\mathbf{\Phi}_{j'}$ in the true one-dimensional case (with $A = 1.0$) become

$$\mathbf{B}_{j'}\,\delta\mathbf{\Phi}_{j'-1} + \mathbf{C}_{j'}\,\delta\mathbf{\Phi}_{j'} + \mathbf{D}_{j'}\,\delta\mathbf{\Phi}_{j'+1} = \mathbf{W}_{j'} \tag{4.55}$$

where, omitting the subscript j' on the elements of $\mathbf{B}$, $\mathbf{C}$, $\mathbf{D}$, and $\mathbf{W}$, and on quantities other than $\mathbf{\Phi}$, $\mathbf{R}_i$ and δy,

$$B_{ik} = \{(1/4)(P + G)\Omega_- - L_-\}\delta_{ik} - 2T_{ik-}$$
$$(i = 1, 2, \ldots, N-1, N+1;\ k = 1, 2, \ldots, N+1)$$

$$B_{i,N+4} = -2\sum_{k=1}^{N+1} T_{ik-}\{(\phi_{k,B})_{j'} - (\phi_{k,B})_{j'-1}\}/\delta y_{j'-1} - (1/4)\{(R_i)_{j'-1} + 3(R_i)_{j'}\}$$
$$(i = 1, 2, \ldots, N-1, N+1)$$

$$C_{ik} = \{(3/4)(P + G)\Omega + L_+ - L_-\}\delta_{ik} + 2(T_{ik+} + T_{ik-}) - S_{D,ik}\,\Delta y(1 - \delta_{k,N+1})$$
$$= -(B_{ik} + D_{ik}) - S_{D,ik}\,\Delta y(1 - \delta_{k,N+1}) + P\Omega\delta_{ik}$$
$$(i = 1, 2, \ldots, N-1, N+1;\ k = 1, 2, \ldots, N+1)$$

$$C_{ik} = -S_{D,ik}\,\Delta y$$
$$(i = 1, 2, \ldots, N-1, N+1;\ k = N+2, N+3)$$

$$C_{i,N+4} = 2\sum_{k=1}^{N+1} T_{ik+}\{(\phi_{k,B})_{j'+1} - (\phi_{k,B})_{j'}\}/\delta y_{j'} - (1/4)\{(R_i)_{j'+1} + 3(R_i)_{j'}\}$$
$$(i = 1, 2, \ldots, N-1, N+1)$$

$$C_{N,k} = m_k \qquad (k = 1, 2, \ldots, N)$$

$$C_{N+2,k} = \rho \Big/ \sum_{j=1}^{N} \sigma_j \qquad (k = 1, 2, \ldots, N)$$

$$C_{N+2,N+2} = 1.0$$

$$
\begin{aligned}
C_{N+2,N+3} &= \rho/T \\
C_{N+3,k} &= H_k \qquad (k = 1, 2, \ldots, N) \qquad\qquad (4.56) \\
C_{N+3,N+1} &= -1.0 \\
C_{N+3,N+3} &= c_p \\
C_{N+4,N+2} &= (\delta y_{j'})/(\rho_{j'} + \rho_{j'+1}) \\
C_{N+4,N+4} &= 1.0 \\
D_{ik} &= \{(1/4)(P+G)\Omega_+ + L_+\}\delta_{ik} - 2T_{ik+} \\
&\qquad (i = 1, 2, \ldots, N-1, N+1; k = 1, 2, \ldots, N+1) \\
D_{N+4,N+2} &= (\delta y_{j'})/(\rho_{j'} + \rho_{j'+1}) \\
W_i &= (1/4)P\{\Omega_+(\phi_{i,U})_{j'+1} + 3\Omega(\phi_{i,U})_{j'} + \Omega_-(\phi_{i,U})_{j'-1}\} \\
&\quad + 2(S_{i,B})_{j'} - \sum_{k=1}^{N+1} B_{ik}(\phi_{k,B})_{j'-1} \\
&\quad - \sum_{k=1}^{N+1} [\{(3/4)(P+G)\Omega + L_+ - L_-\}\delta_{ik} \\
&\quad + 2(T_{ik+} + T_{ik-})](\phi_{k,B})_{j'} - \sum_{k=1}^{N+1} D_{ik}(\phi_{k,B})_{j'+1} \\
&\qquad (i = 1, 2, \ldots, N-1, N+1) \\
W_N &= 1 - \sum_{k=1}^{N} m_k\sigma_k \\
W_k &= 0 \qquad (k = N+2, N+3, N+4)
\end{aligned}
$$

When calculating these elements, the subscript B is to be understood except when the subscript U is explicitly stated. The remaining elements in the coefficient matrices **B**, **C**, and **D** are everywhere zero. In these equations, the subscript order is $\sigma_1, \sigma_2, \ldots, \sigma_N, h, \rho, T, \delta y$. Equation (4.46) replaces the normal conservation equation for the species N. As already indicated, the discarded conservation condition should in practice belong to a component whose concentration does not critically influence the reaction velocities.

The extension of the above manipulation to the equations for divergent flow is straightforward, with one further equation being required to allow for treatment of the divided $\delta y_{j'}$ terms.

(c) Representation of boundary conditions

In the von Mises system of coordinates the cell boundaries of the finite-difference grid are located half-way between the grid points, and the intervals at each end of the grid are similarly divided. We now consider the pair of end half-intervals. Instead of only one set of dependent variables being defined at each of the extreme points $\omega = 0$ and $\omega = 1$, we use two sets, so that a grid with M finite-difference intervals has in all $M + 3$ defined sets of variables, with positions as shown in Fig. 4. It is then assumed that the gradients of the dependent variables are linear in y over the end pair of half-intervals, taking the values $2\{(1/2)(\phi_1 + \phi_2) - \phi_0\}/\delta y_1$ between $j' = 1$ and 1.5, $(\phi_2 - \phi_1)/\delta y_1$ between $j' = 1.5$ and 2, $(\phi_{M+1} - \phi_M)/\delta y_M$ between $j' = M$ and $(M + 0.5)$, and $2\{\phi_{M+2} - (1/2)(\phi_M + \phi_{M+1})\}/\delta y_M$ between $j' = M + 0.5$ and $(M + 1)$. The values of the dependent variables at $j' = 0$ and $j' = M + 2$ are normally determined by the boundary conditions, and the extra grid points at $j' = 1$ and $j' = M + 1$ provide buffers between the boundaries and the central grid where the $\phi - \omega$ profiles are linear over each whole interval. By admitting changes in the gradients at $j' = 1.5$ and $j' = M + 0.5$, the buffer effect is able to allow, if necessary, for significant curvatures in the profiles of the dependent variables near the boundaries.

In the flame system which has been described, Eq. (4.9) defines two imaginary surfaces, $b(\omega = 0)$ and $u(\omega = 1)$, which bound the flow region of interest. In the particular context of laminar premixed flames, three types of boundary are of practical importance:

(i) The "free" boundary, where the grid material merges into the surrounding uniform medium (burnt or unburnt gas) and the gradients of the profiles of the dependent variables become effectively zero.

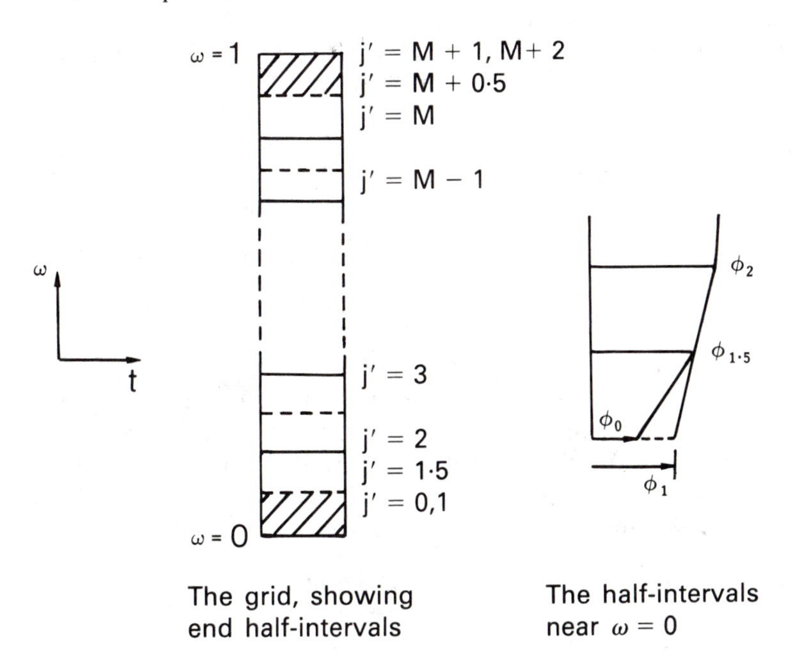

Figure 4. Arrangements at the ends of the finite-difference grid.

(ii) A boundary at a solid surface, where the gradients of the profiles of the dependent variables may be high. Examples are a flame approaching a cold, nonporous wall (end-on quenching) or approaching a cooled porous plate through which fresh gas is supplied (cooled porous plate burner).

(iii) The third situation occurs when the boundary coincides with a line or point of symmetry in the flow, such as the center line of a jet in boundary layer flow or the center of a spherically expanding flame. In such cases the flow across the boundary and the gradients normal to the symmetry position are all zero.

To solve the differential equations, we need information about the u and b boundaries so as to define the specific problem. This information may be given in terms of (i) values of the dependent variables along the boundary (for example, the unburnt gas composition); (ii) the gradients of the dependent variables at the boundary; or (iii) mass transfer rates of unburnt or burnt components through a porous wall.

Our next task is thus to set up the microintegral equations corresponding with expression (4.55) for the u and b boundary half-intervals. We commence at Eq. (4.22) and proceed as before. In the flame context, it is assumed that no chemical reaction takes place in the volume of either of the end half-intervals, that is, that the space integral rates there are zero. This is reasonable, since there should be no major reaction near free boundaries, while walls are presumed to be perfect sinks for free radicals.

In the general case, the microintegral equation at the b boundary becomes

$$\begin{aligned}(1/4)(P+G)\Omega_{+}\{3(\phi_i)_1+(\phi_i)_2\} &- (1/4)P\Omega_{+}\{3(\phi_{i,U})_1+(\phi_{1,U})_2\}\\ &+ L_{+}\{(\phi_i)_1+(\phi_i)_2\} - 2L_0(\phi_i)_0 - 2\sum_{k=1}^{N+1} T_{ik+}\{(\phi_k)_2-(\phi_k)_1\}\\ &+ 2\sum_{k=1}^{N+1} T_{ik}^{0}[(1/2)\{(\phi_k)_1+(\phi_k)_2\}-(\phi_k)_0]=0\end{aligned}$$

$$(i = 1, 2, \ldots, N-1, N+1) \qquad (4.57)$$

where L_0 refers to the advection term $(M_yA)_b$ at the hot boundary and $\Omega_+ = \omega_2 = 2\omega_{1.5}$, in accordance with the definition (4.29). The terms T_{ik}^0 are calculated in the normal manner for conditions at grid point 1, except that it must be remembered that the distance over which the gradient is estimated is now $(1/2)\delta y_1$ and not δy_1 as in Eqs. (4.34) and (4.35).

We distinguish three separate situations:

(i) *When the dependent variables at the boundary are prescribed*, the corrections $\delta\mathbf{\Phi}_0$ in the appropriate equation (4.55) all become zero, and the equation itself becomes

$$\mathbf{C}_1\,\delta\mathbf{\Phi}_1 + \mathbf{D}_1\,\delta\mathbf{\Phi}_2 = \mathbf{W}_1 \qquad (4.58)$$

where

$$(C_{ik})_1 = \{(3/4)(P+G)\Omega_+ + L_+\}\delta_{ik} + 2T_{ik_+} + T^0_{ik}$$
$$= -\{(B_{ik})_1 + (D_{ik})_1\} + P\Omega_+\delta_{ik}$$
$$(i = 1, 2, \ldots, N-1, N+1; k = 1, 2, \ldots, N+1)$$

$$(C_{i,N+4})_1 = 2\sum_{k=1}^{N+1} \{T_{ik_+}[(\phi_k)_2 - (\phi_k)_1]$$
$$- T^0_{ik}[(1/2)\{(\phi_k)_1 + (\phi_k)_2\} - (\phi_k)_0]\}/\delta y_1$$
$$(i = 1, 2, \ldots, N-1, N+1)$$

$$(D_{ik})_1 = \{(1/4)(P+G)\Omega_+ + L_+\}\delta_{ik} - 2T_{ik_+} + T^0_{ik} \tag{4.59}$$
$$(i = 1, 2, \ldots, N-1, N+1; k = 1, 2, \ldots, N+1)$$

$$(B_{ik})_1 = -2(T^0_{ik} + L_0\delta_{ik})$$
$$(i = 1, 2, \ldots, N-1, N+1; k = 1, 2, \ldots, N+1)$$

$$(W_i)_1 = (1/4)P\Omega_+\{3(\phi_{i,U})_1 + (\phi_{i,U})_2\}$$
$$- \sum_{k=1}^{N+1} \{(B_{ik})_1(\phi_{k,B})_0 + (C_{ik})_1(\phi_{k,B})_1 + (D_{ik})_1(\phi_{k,B})_2\}$$
$$(i = 1, 2, \ldots, N-1, N+1)$$

and the remaining nonzero elements of **C** and **D** take values as in Eq. (4.56).

We may also now express the effective diffusive fluxes of material at the boundary as

$$Q_{i,b} = -\sum_{k=1}^{N+1} T^0_{ik}[(1/2)\{(\phi_k)_1 + (\phi_k)_2\} - (\phi_k)_0] \tag{4.60}$$

These fluxes become of particular importance when, for example, radical species are removed at a wall boundary.

(ii) *When the gradients of the dependent variables are known*, these may be equated with the quantities $\{(1/2)(\mathbf{\Phi}_1 + \mathbf{\Phi}_2) - \mathbf{\Phi}_0\}/\delta y_1$. In particular, for a free boundary where all the gradients are zero, as at the b boundary of a flame [cf. Eq. (4.7)]

$$\mathbf{\Phi}_0 = (1/2)(\mathbf{\Phi}_1 + \mathbf{\Phi}_2) \tag{4.61}$$

(iii) *When the total flux of a species is prescribed at the boundary*, we formulate

$$J_{\text{tot}\,i,b} = L_0(\phi_i)_0 - \sum_{k=1}^{N+1} T^0_{ik}[(1/2)\{(\phi_k)_1 + (\phi_k)_2\} - (\phi_k)_0] + \delta J_{i,b} \tag{4.62}$$

Substitution into Eq. (4.57) then leads to

$$(1/4)(P+G)\Omega_+\{3(\phi_i)_1 + (\phi_i)_2\} - (1/4)P\Omega_+\{3(\phi_{i,U})_1 + (\phi_{i,U})_2\}$$
$$+ L_+\{(\phi_i)_1 + (\phi_i)_2\} - 2(J_{\text{tot}\,i,b} - \delta J_{i,b}) - 2\sum_{k=1}^{N+1} T_{ik_+}\{(\phi_k)_2 - (\phi_k)_1\} = 0$$
$$(i = 1, 2, \ldots, N-1, N+1)$$

The terms δJ_i are here added to the right-hand sides of Eqs. (4.62) in order to allow for possible catalytic formation of the species at a wall boundary. In flame systems, a free radical arriving at a solid surface is assumed to be immediately destroyed there; that is, we assume that all radical concentrations are zero at such surfaces. Computationally, radicals which diffuse to the surface are assumed to react at "infinite" rate with one of the major components of the mixture in such a way as to return a stable species to the gas phase. The terms δJ are necessary to maintain conservation at the boundary while allowing for this. The diffusive fluxes of the radicals are evaluated from Eq. (4.60), and the effect on other fluxes of their conversion to stable species at the boundary is calculated to give the δJ. The assumption of zero radical concentrations at the boundary (or at least zero penetration into the porous wall) is essential if reaction is to be confined to the gas phase.

The rearrangement of Eq. (4.63) leads to the following coefficients in Eq. (4.58) for the case where the total wall fluxes are prescribed:

$$\begin{aligned}
(C_{ik})_1 &= \{(3/4)(P + G)\Omega_+ + L_+\}\delta_{ik} + 2T_{ik+} \\
&= -(D_{ik})_1 + P\Omega_+\delta_{ik} + 2L_0\delta_{ik} \\
&\qquad (i = 1, 2, \ldots, N-1, N+1;\ k = 1, 2, \ldots, N+1) \\
(C_{i,N+4})_1 &= 2\sum_{k=1}^{N+1} T_{ik+}\{(\phi_k)_2 - (\phi_k)_1\}/\delta y_1 \\
&\qquad (i = 1, 2, \ldots, N-1, N+1) \qquad (4.64) \\
(D_{ik})_1 &= \{(1/4)(P + G)\Omega_+ + L_+\}\delta_{ik} - 2T_{ik+} \\
&\qquad (i = 1, 2, \ldots, N-1, N+1;\ k = 1, 2, \ldots, N+1) \\
(W_i)_1 &= (1/4)P\Omega_+\{3(\phi_{i,U})_1 + (\phi_{i,U})_2\} + 2(J_{\text{tot}\,i,b} - \delta J_{i,b}) \\
&\quad - \sum_{k=1}^{N+1}\{(C_{ik})_1(\phi_{k,B})_1 + (D_{ik})_1(\phi_{k,B})_2\} \\
&\qquad (i = 1, 2, \ldots, N-1, N+1)
\end{aligned}$$

Having obtained values of $\delta\mathbf{\Phi}_1$ and $\delta\mathbf{\Phi}_2$ (and hence the new $\mathbf{\Phi}_1$ and $\mathbf{\Phi}_2$) by solution of these equations, back-substitution into Eq. (4.61) or (4.62) gives the new boundary values $\mathbf{\Phi}_0$. These new values becomes the B values for the next iteration.

It is of course possible to have a situation where some of the variables k have prescribed values ϕ_k at the boundary, while others have prescribed total fluxes. The treatment of the free hot boundary of an adiabatic flame provides an interesting and important example of this type. The condition of zero gradient of each of the concentration variables is first enforced at such a boundary [cf. Eqs. (4.7) and (4.61)]. In these circumstances the specific enthalpy of the combustion products must be equal to that of the gases entering the adiabatic flame, so that h_0 is automatically defined. Equations (4.62) become

$$J_{\text{tot}\,i,b} = L_0(\phi_i)_0 - T^0_{ih}\{(1/2)(h_1 + h_2) - h_0\} \tag{4.62a}$$

and by keeping h_0 constant (or $\delta h_0 = 0$), it is then possible to update the $(\phi_k)_0$ associated with the material species iteratively as above. At convergence of the solution the enthalpy gradient at the boundary also becomes zero.

Equations (4.57) to (4.64) define the incorporation of "b" or hot boundary conditions. At the cold or "u" boundary it is necessary to specify both the initial composition and the nature of the boundary. Thus, for example, in the "free flame" situation it is necessary to specify values of $\mathbf{\Phi}_{M+2}$, and then to ensure zero gradients of the dependent variables by equating $\mathbf{\Phi}_{M+1}$ with $\mathbf{\Phi}_{M+2}$. It is further necessary at the end of the computation to check that one or two sets of dependent variables belonging to grid points near, but not at, the boundary do not differ appreciably from $\mathbf{\Phi}_{M+1}$ and $\mathbf{\Phi}_{M+2}$, and to adjust the grid accordingly if this is not the case.

For reference, the cold or "u" boundary equations corresponding with Eqs. (4.57) to (4.64) will be given without further comment. The matrix equation corresponding with Eq. (4.58) becomes

$$\mathbf{B}_{M+1}\,\delta\mathbf{\Phi}_M + \mathbf{C}_{M+1}\,\delta\mathbf{\Phi}_{M+1} = \mathbf{W}_{M+1} \tag{4.58a}$$

The expressions for the coefficients become, when $\mathbf{\Phi}_{M+2}$ is prescribed,

$$(B_{ik})_{M+1} = \{(1/4)(P+G)\Omega_- - L_-\}\delta_{ik} + T^{M+2}_{ik} - 2T_{ik-}$$
$$(i = 1, 2, \ldots, N-1, N+1;\ k = 1, 2, \ldots, N+1)$$

$$\begin{aligned}(C_{ik})_{M+1} &= \{(3/4)(P+G)\Omega_- - L_-\}\delta_{ik} + T^{M+2}_{ik} + 2T_{ik-}\\ &= -\{(B_{ik})_{M+1} + (D_{ik})_{M+1}\} + P\Omega_-\delta_{ik}\end{aligned}$$
$$(i = 1, 2, \ldots, N-1, N+1;\ k = 1, 2, \ldots, N+1)$$

$$\begin{aligned}(C_{i,N+4})_{M+1} = 2\sum_{k=1}^{N+1} \{&T^{M+2}_{ik}[(\phi_k)_{M+2} - (1/2)\{(\phi_k)_M + (\phi_k)_{M+1}\}]\\ &- T_{ik-}[(\phi_k)_{M+1} - (\phi_k)_M]\}/\delta y_M\end{aligned}$$
$$(i = 1, 2, \ldots, N-1, N+1) \tag{4.59a}$$

$$(D_{ik})_{M+1} = 2(L_{M+2}\delta_{ik} - T^{M+2}_{ik})$$
$$(i = 1, 2, \ldots, N-1, N+1;\ k = 1, 2, \ldots, N+1)$$

$$\begin{aligned}(W_i)_{M+1} = (1/4)P\Omega_-\{(\phi_{i,U})_M + 3(\phi_{i,U})_{M+1}\} - \sum_{k=1}^{N+1}\{&(B_{ik})_{M+1}(\phi_{k,B})_M\\ &+ (C_{ik})_{M+1}(\phi_{k,B})_{M+1} + (D_{ik})_{M+1}(\phi_{k,B})_{M+2}\}\end{aligned}$$
$$(i = 1, 2, \ldots, N-1, N+1)$$

(Note that the T^{M+2}_{ik} are evaluated on the basis of a gradient over the distance $(1/2)\delta y_M$.)

When the total flux at the "u" boundary is prescribed, the flux and coefficients become

$$J_{\text{tot}\,i,u} = L_{M+2}(\phi_i)_{M+2} - \sum_{k=1}^{N+1} T_{ik}^{M+2}[(\phi_k)_{M+2} - (1/2)\{(\phi_k)_M + (\phi_k)_{M+1}\}] + \delta J_{i,u} \quad (4.62a)$$

and

$$(B_{ik})_{M+1} = \{(1/4)(P+G)\Omega_- - L_-\}\delta_{ik} - 2T_{ik_-}$$
$$(i = 1, 2, \ldots, N-1, N+1;\ k = 1, 2, \ldots, N+1)$$

$$(C_{ik})_{M+1} = \{(3/4)(P+G)\Omega_- - L_-\}\delta_{ik} + 2T_{ik_-}$$
$$= -(B_{ik})_{M+1} + P\Omega_-\delta_{ik} - 2L_{M+2}\delta_{ik}$$
$$(i = 1, 2, \ldots, N-1, N+1;\ k = 1, 2, \ldots, N+1)$$

$$(C_{i,N+4})_{M+1} = -2\sum_{k=1}^{N+1} T_{ik_-}\{(\phi_k)_{M+1} - (\phi_k)_M\}/\delta y_M$$
$$(i = 1, 2, \ldots, N-1, N+1) \quad (4.64a)$$

$$(W_i)_{M+1} = (1/4)P\Omega_-\{(\phi_{i,U})_M + 3(\phi_{i,U})_{M+1}\} + 2(J_{\text{tot}\,i,u} - \delta J_{i,u}) - \sum_{k=1}^{N+1}\{(B_{ik})_{M+1}(\phi_{k,B})_M + (C_{ik})_{M+1}(\phi_{k,B})_{M+1}\}$$
$$(i = 1, 2, \ldots, N-1, N+1)$$

4.3. Solution of equations

The procedures outlined in Section 4.2 lead to a set of simultaneous linear equations, the solution of which gives the corrections $\delta\mathbf{\Phi}$ to be applied to the starting (or B) approximations for the dependent variables throughout the flame. When represented by the matrices of Eqs. (4.55), (4.58), and (4.58a), the coefficients corresponding with successive grid points through the flame form the tridiagonal matrix which is characteristic of one-dimensional transport problems. The finite-difference equations through the whole flame may be represented as

$$\begin{vmatrix} \mathbf{C}_1 & \mathbf{D}_1 & & & & \\ \mathbf{B}_2 & \mathbf{C}_2 & \mathbf{D}_2 & & & \\ & \mathbf{B}_3 & \mathbf{C}_3 & \mathbf{D}_3 & & \\ & & \cdot & \cdot & \cdot & \\ & & & \cdot & \cdot & \mathbf{D}_M \\ & & & & \mathbf{B}_{M+1} & \mathbf{C}_{M+1} \end{vmatrix} \begin{vmatrix} \delta\mathbf{\Phi}_1 \\ \delta\mathbf{\Phi}_2 \\ \delta\mathbf{\Phi}_3 \\ \cdot \\ \delta\mathbf{\Phi}_M \\ \delta\mathbf{\Phi}_{M+1} \end{vmatrix} = \begin{vmatrix} \mathbf{W}_1 \\ \mathbf{W}_2 \\ \mathbf{W}_3 \\ \cdot \\ \mathbf{W}_M \\ \mathbf{W}_{M+1} \end{vmatrix} \quad (4.65)$$

The equations are solved by Gaussian elimination as described by Richtmyer and Morton (1967) and by Roache (1972). We perform a standard LU decomposition of the coefficient matrix $\mathbf{V}$ on the left-hand side

$$\mathbf{V} = \mathbf{LU} \tag{4.66}$$

where

$$\mathbf{L} = \begin{vmatrix} \mathbf{E}_1 & & & & \\ \mathbf{B}_2 & \mathbf{E}_2 & & & \\ & \mathbf{B}_3 & \mathbf{E}_3 & & \\ & & \cdot & \cdot & \\ & & & \cdot & \cdot \end{vmatrix}$$

and

$$\mathbf{U} = \begin{vmatrix} 1 & \mathbf{F}_1 & & & \\ & 1 & \mathbf{F}_2 & & \\ & & 1 & \mathbf{F}_3 & \\ & & & \cdot & \cdot \\ & & & & \cdot \end{vmatrix}$$

Putting $\mathbf{U}\,\delta\mathbf{\Phi} = \mathbf{Y}$, we then solve successively, by back-substitution, the equations

$$\mathbf{LY} = \mathbf{W} \tag{4.67}$$

$$\mathbf{U}\,\delta\mathbf{\Phi} = \mathbf{Y} \tag{4.68}$$

to give the sought-after corrections. An appropriate algorithm for performing the LU decomposition and solving Eq. (4.67) is

$$\begin{aligned} \mathbf{E}_1 &= \mathbf{C}_1 \\ \mathbf{X}_1 &= \mathbf{W}_1 \\ \mathbf{F}_k &= \mathbf{E}_k^{-1}\mathbf{D}_k \\ \mathbf{Y}_k &= \mathbf{E}_k^{-1}\mathbf{X}_k \\ \mathbf{E}_k &= \mathbf{C}_k - \mathbf{B}_k\mathbf{F}_{k-1} \quad \text{for} \quad k > 1 \\ \mathbf{X}_k &= \mathbf{W}_k - \mathbf{B}_k\mathbf{Y}_{k-1} \end{aligned} \tag{4.69}$$

This solution allows us to proceed from the "b" to the "u" boundary of the flame, after which a similar procedure is carried out in reverse from "u" to "b" to give the $\delta\mathbf{\Phi}$ and so complete the cycle of computation for the updating of the values of the dependent variables. The algorithm is

$$\begin{aligned} \delta\mathbf{\Phi}_{M+1} &= \mathbf{Y}_{M+1} \\ \delta\mathbf{\Phi}_k &= \mathbf{Y}_k - \mathbf{F}_k\,\delta\mathbf{\Phi}_{k+1} \quad \text{for} \quad k < M + 1 \end{aligned} \tag{4.70}$$

4.4. The convection term. Lagrangian and Eulerian calculations

Apart from including them in the equations, nothing has been said yet about the convective flux terms L. It is to these that attention is now directed. Including the modeling of flames stabilized on water-cooled burners, three distinct cases arise:

(a) Pure Lagrangian calculations

Pure Lagrangian computations use Eqs. (2.28) to (2.31) as the governing equations. The convective transfer terms are zero everywhere, that is, G, L_+, and L_- are zero in the finite-difference equations

To perform this type of calculation it is necessary to determine accurately the final state after each interval δt. This is achieved by using Eqs. (4.65) to update initial estimates of $\mathbf{\Phi}$ iteratively by a Newton-Raphson process until a prescribed precision is reached, for example, $\delta\phi/\phi = 10^{-6}$ for all variables. Because of the nonlinear nature of the governing differential equations, the iteration will not converge if too large a time step is attempted, and provision must be made for, say, halving the latter if convergence is not achieved after a prescribed number of Newton-Raphson iterations. On the other hand, computational economy demands maximization of the time step within these constraints. It is therefore convenient to commence with a small value of δt (say 10^{-9} s) and to allow this to increase gradually, by perhaps 10% per time step, until a good compromise is achieved.

Convergence to a steady rate of propagation may be monitored by the approach of the space integral rate over the whole flame to a constant value (cf. Section 2.6). The steady-state burning velocity M_y [or the product $(M_y A)$] may be obtained directly from Eq. (4.71)

$$(M_y A)_{m,i} = \frac{\int_0^1 (R_i/\rho)(\psi_u - \psi_b)\, d\omega}{(1/m_i)(G_{i,b} - G_{i,u})} \qquad (G_{i,b} \neq G_{i,u}) \tag{4.71}$$

$$= \frac{\int_0^1 (R_i/\rho)(\psi_u - \psi_b)\, d\omega}{\sigma_{i,b} - \sigma_{i,u}} \qquad (\sigma_{i,b} \neq \sigma_{i,u}) \tag{4.71a}$$

where Eq. (4.71a) applies specifically to the "free" flame.

In order to maintain the high concentration of grid points in the region of primary reactive flow interest, it is necessary with this approach to advance the grid with the flame in a separate regridding stage if maximum computational efficiency is sought. By this means the technique can be made to yield a two-stage approach to an Eulerian calculation.

The pure Lagrangian approach without regridding has the theoretical advantage (over the Eulerian approaches to be described below) that it is demonstrably free from the "numerical diffusion" effects (Boris and Book, 1973) which are associated with the convection terms of the latter. However, despite the fact that it is not necessary to reevaluate the coefficient matrix in Eq. (4.65) at *every* iteration, the nonconstancy of properties at a grid point even when the propagation rate is steady makes the Lagrangian method considerably more computationally demanding for steady-state problems than the alternative Eulerian approach. Further, the regridding process may itself introduce numerical diffusion effects.

(b) Eulerian calculations

In the Eulerian steady-state, stationary flame the time derivatives in Eqs. (4.12) and (4.13) become zero, and the computational problem becomes one of determining convective fluxes and flame profiles that will allow this to occur everywhere.

In general, the convective fluxes at different positions in the flame are represented in Eqs. (4.12) and (4.13) by the terms $(a + b\omega)$, with the proviso that in the steady state $b = 0$. In the approach to the steady state both a and b may vary with time, and our objective is to relate these parameters to the reaction rates in the flame. In the case of the "free" flame, we wish at the same time to locate the flame in the grid so that the greatest concentration of grid points is at the position of maximum change. These two objectives are achieved by means of "entrainment formulas," which fix the rates of entrainment across the cold and hot boundaries of the grid. For a free flame, satisfactory entrainment formulas are

$$(M_y A)_u = (M_y A)_{m,i} \left[\left| \frac{\sigma_{i,uu} - \sigma_{i,u}}{\sigma_{i,b} - \sigma_{i,u}} \right| F_u \right]^{0.1} \tag{4.72}$$

$$(M_y A)_b = (M_y A)_{m,i} \left\{ 2.0 - \left[\left| \frac{\sigma_{i,b} - \sigma_{i,bb}}{\sigma_{i,b} - \sigma_{i,u}} \right| F_b \right]^{0.1} \right\} \tag{4.73}$$

where $(M_y A)_{m,i}$ is given by Eq. (4.71a) and uu and bb refer to fixed points near, but not at, the cold and hot boundaries, respectively. The specific component i used in the entrainment formulas should normally be a rate-controlling reactant. New values of $(M_y A)_u$ and $(M_y A)_b$ are evaluated at the commencement of each time step, and at the end of each step the value of $(\psi_u - \psi_b)$ is updated by means of

$$\psi_u - \psi_b = (\psi_u - \psi_b)_U - \{(M_y A)_u - (M_y A)_b\}\, \delta t \tag{4.74}$$

Equation (4.11) is satisfied and the steady state is reached when the products in Eqs. (4.72) and (4.73), which are raised to the power 0.1, are each equal to unity. The steady-state condition, towards which the calculation is directed, also entails that the multipliers F_u and F_b form a pair so that the changes in concentration between u and uu, and between b and bb, are equal to fractions $(1/F_u)$ and $(1/F_b)$, respectively, of the total concentration change in the flame. Arbitrary values of F_u and F_b may not be consistent with the selected distance scale for the computation, that is, with the selected $(\psi_u - \psi_b)$ and distribution of mesh points. Thus, only one of F_u and F_b can be chosen arbitrarily. This will fix the final position of the flame reaction zone within the grid. After an initial settling period of say 30% of the total number of cycles expected for convergence, the second parameter must then be adjusted from time to time during the calculation (say every further 10%) so that $(M_y A)_u = (M_y A)_b$ at that time. Only when the steady state has been actually reached will this equality continue to hold during subsequent time steps.

Normally, for stable integration the grid points *uu* and *bb* should each be about eight points inside their respective boundaries, and they may conveniently immediately surround the central region of high concentration of grid points. The indices 0.1 in Eqs. (4.72) and (4.73) are also present for stability reasons (indices other than 0.1 may of course be used if preferred for specific problems). It should be added that even with these safeguards minor catastrophes are not impossible; and in order to counteract possible violent changes in $(M_y A)_u$ and $(M_y A)_b$ early in a computation, an override condition is usefully imposed by insisting that the contents of the parentheses in Eqs. (4.72) and (4.73) remain within the range $0.1 \leq [\quad] \leq 10$, for example. The moduli of the quotients within these parentheses are also employed for obvious reasons, in case the change from σ_u to σ_b is not monotonic at some stage. If a nonmonotonic situation does arise to the extent that the quotient becomes negative, then F_u (or F_b as the case may be) should be replaced by its reciprocal in order to maintain the "spirit" of the formulas.

The modeling of flames stabilized on water-cooled porous plate burners follows a similar pattern, with the exception that the convective flux at the cold end of the flame is already fixed by the experiment. For water cooling, a plate temperature of between 300 and 370 K at the cold boundary may be assumed, the precise temperature not being critical. The grid spacing is best arranged so that the intervals $\delta\omega$ are equal and give good coverage from the burner into the flame for some 20 to 25 points out of 30 (cf. Section 4.2(a) for the "free" flame). The parameter F_b must then be adjusted at intervals as above so that the equality $(M_y A)_u = (M_y A)_b$ continues to hold during subsequent time steps. Because of diffusion effects at the "u" boundaries of these flames, the $\sigma_{i,u}$ at the porous plate surface are not the same as in the initial feedstock. Equation (4.71) must therefore be used for the evaluation of $(M_y A)_{m,i}$ in connection with the entrainment formula (4.73), as well as in the evaluation (for verification purposes) of the nonadiabatic burning velocities via the space integral rates of all the reactants and products at output stages of the computation.

To set against the drawback of possible uncertainty about effects of the numerical diffusion associated with the convective terms, the Eulerian approach has very considerable operational advantages if the computational objective is solely the determination of steady-state properties. Because the properties at each grid point are time-invariant in the steady state, Eulerian time steps are in themselves successive approximations to the end result. They do not therefore individually require Newton-Raphson iteration, nor do they require consideration of variations of density, distance interval, or temperature actually during the time step. Instead, and along with the convection terms, these variables may be reevaluated explicitly between time steps by direct application of the auxiliary equations (2.25), (2.26), and (4.48) or (4.50). This reduces by three the number of variables to be determined implicitly. Equation (2.27) may also be used explicitly between time steps. Conservation during a time step may then be checked if desired by solving the continuity equations for all N species and the enthalpy, followed by checking via Eq. (2.27) that errors are not excessive.

We now aim during a time step at a single solution $\mathbf{\Phi}$ for Eq. (4.43), slightly modified to

$$\begin{aligned}
&\sum_{k=1}^{N+1} [\{(1/4)(P+G)\Omega_- - L_-\}\delta_{ik} - 2T_{ik-}A_-](\phi_k)_{j'-1} \\
&\quad + \sum_{k=1}^{N+1} [\{(3/4)(P+G)\Omega + L_+ - L_-\}\delta_{ik} + 2(T_{ik+}A_+ + T_{ik-}A_-) \\
&\qquad - (AS_{D,ik}\Delta y)_{j'}(1-\delta_{k,N+1})](\phi_k)_{j'} \\
&\quad + \sum_{k=1}^{N+1} [\{(1/4)(P+G)\Omega_+ + L_+\}\delta_{ik} - 2T_{ik+}A_+](\phi_k)_{j'+1} \\
&= (1/4)P\{\Omega_+(\phi_{i,U})_{j'+1} + 3\Omega(\phi_{i,U})_{j'} + \Omega_-(\phi_{i,U})_{j'-1}\} \\
&\quad + 2(S_{i,U})_{j'} - \sum_{k=1}^{N+1} (AS_{D,ik}\phi_{k,U}\,\Delta y)_{j'}
\end{aligned}$$

$$(i = 1, 2, \ldots, N+1) \quad (4.75)$$

Compared with Eq. (4.43), the subscripts B and U have now become identical, and all summations are from 1 to $(N + 1)$ only. The coefficients are also all evaluated explicitly at the start of the time step. Compared with Eqs. (4.56), (4.59), and (4.64), the coefficients and other terms having suffices greater than $N + 1$ now become redundant, as also do the expressions for $C_{N,k}$ and W_N. Otherwise, only the right-hand sides of the equations are modified, leading in the general case $(W_i)_{j'}$, which includes reaction terms, to

$$\begin{aligned}
(W_i)_{j'} &= (1/4)P\{\Omega_+(\phi_{i,U})_{j'+1} + 3\Omega(\phi_{i,U})_{j'} + \Omega_-(\phi_{i,U})_{j'-1}\} \\
&\quad + 2(S_{i,U})_{j'} - \sum_{k=1}^{N} (AS_{D,ik}\phi_{k,U}\,\Delta y)_{j'} \qquad (4.76)
\end{aligned}$$

The modifications to $(W_i)_1$ and $(W_i)_{M+1}$ merely involve discarding the terms in $(\phi_{k,B})_1$ and $(\phi_{k,B})_2$, or $(\phi_{k,B})_M$ and $(\phi_{k,B})_{M+1}$, from the right-hand side of the appropriate equation (4.59) or (4.64).

The solutions discussed so far, both Eulerian and Lagrangian, are known as *block implicit* solutions. Eulerian integrations into the steady state may, however, be further simplified by *uncoupling* the separate equations, so that each dependent variable is considered separately during each time step, under the assumption that the remaining variables stay constant. The assumption is correct in the steady state itself. There are then $N + 1$ solutions of successive single conservation equations per time interval, but the submatrices **B**, **C**, **D**, and **W** of Eq. (4.65) become single terms in each solution. (The equation is of course modified to deal with a vector $\mathbf{\Phi}$ of unknowns instead of $\delta\mathbf{\Phi}$.) The computational efficiency of the block implicit vis-à-vis the uncoupled implicit approach will be discussed below. The uncoupled modification returns essentially to the scheme of Spalding *et al.* (1971, 1973), who used a cruder representation of the transport processes.

Specifically, the changes introduced in this further modification, which was first used with detailed transport properties by Tsatsaronis (1978), are that the reaction rate coefficients and the $k \neq i$ components of the transport fluxes for each species are all evaluated explicitly at the beginning of the time interval. The remaining terms of Eq. (4.75) are calculated implicitly, leading for each species to

$$\begin{aligned}
&\{(1/4)(P + G)\Omega_- - L_- - 2T_{ii-}A_-\}(\phi_i)_{j'-1} + \{(3/4)(P + G)\Omega + L_+ - L_- \\
&\qquad + 2(T_{ii+}A_+ + T_{ii-}A_-) - (AS_{D,ii}\,\Delta y)_{j'}(1 - \delta_{i,N+1})\}(\phi_i)_{j'} \\
&\qquad + \{(1/4)(P + G)\Omega_+ + L_+ - 2T_{ii+}A_+\}(\phi_i)_{j'+1} \\
&= (1/4)P\{\Omega_+(\phi_{i,U})_{j'+1} + 3\Omega(\phi_{i,U})_{j'} + \Omega_-(\phi_{i,U})_{j'-1}\} \\
&\quad + 2(S_{i,U})_{j'} - (AS_{D,ii}\phi_{i,U}\,\Delta y)_{j'}(1 - \delta_{i,N+1}) \\
&\quad + 2\sum_{\substack{k=1\\k\neq i}}^{N+1}[T_{ik+}\{(\phi_{k,U})_{j'+1} - (\phi_{k,U})_{j'}\} - T_{ik-}\{(\phi_{k,U})_{j'} - (\phi_{k,U})_{j'-1}\}] \quad (4.77)
\end{aligned}$$

There are two possible methods of approach. Either the space integral rates $S_{i,U}$ and the associated derivative terms $S_{D,ii}$ are evaluated at the beginning of the time interval, and the dependent variables are updated together at the end of the interval; or the appropriate $S_{i,U}$ and $S_{D,ii}$ are evaluated immediately prior to the solution of each equation and each dependent variable is updated immediately after the solution of its equation. When using the latter method, the order of solution of the equations becomes important. The most satisfactory order treats the most reactive free radicals first, then less reactive intermediates, initial reactants, reaction products and inert species in that order, and finally the enthalpy. The latter method is potentially the more economical computationally, but it forfeits precise chemical conservation during single time steps on the approach to the steady state. This has been found to cause instability in certain diffusion flame problems to be outlined in Section 7.2(b). The former method does not suffer from this disadvantage.

In the interest of computational economy, experience with the application of the method has shown that it is only necessary when using the Eulerian technique to recompute the transport coefficients at infrequent intervals (say every 20 cycles at most), and then only if, for example, the temperature at a grid point has changed by more than some specified amount since the previous reevaluation.

It remains to consider once more the questions of the distance intervals within the grid, the time intervals to be chosen for the integration, and the overall computational efficiency.

(i) *Distance intervals.* In their treatment of boundary layer flows, Patankar and Spalding (1970) drew attention to the likely occurrence of implausible effects, for example, saw-toothed irregularities in the profiles of the dependent variables, in flow situations where the transport terms have declined sufficiently in importance relative to the convection terms. Specifically in relation to the present calculation, they apply a correction to the transport flux at a cell boundary if $|L| > 2T$. The need for such a correction should not be allowed to arise in realistic representations of those reactive flows that depend on the transport processes for their very occurrence. The appropriate remedy is to increase the coefficients T of Eqs. (4.34) and (4.35) by decreasing the distance intervals δy in the important regions of the flow.

Increased resolution in regions of steep gradients also reduces numerical diffusion effects. We shall return to this aspect below.

(ii) *Time intervals.* It was remarked in Section 4.2(b) that the implicit approach described here is unconditionally stable. Nevertheless, because of the nonlinear nature of the governing differential equations, there is a practical upper limit to the integration time interval which may be employed, and with too large time intervals the method may fail to converge. The closer the system is to convergence at a given instant, the larger the time interval which may be employed. The most satisfactory practical procedure is to commence integration with a low value of δt (say 10^{-6} s in flame systems) and to allow this to increase by 10% per time step until some prescribed maximum value is reached. A convenient formula for defining the maximum is given by

$$\delta t_{\max} = \frac{\xi_t(\psi_u - \psi_b)}{|(M_y A)_u| + |(M_y A)_b|} \tag{4.78}$$

where the factor ξ_t in the numerator, a matter for programming decision, is discussed below.

(iii) *Overall computational efficiency.* Within a given block or uncoupled implicit program structure, optimization of computational efficiency demands maximization of ξ_t subject to the constraint that satisfactory convergence is achieved. In comparing block with uncoupled implicit approaches, it is presumed that the block structure normally gives convergence with fewer iterations at the expense of increased computational effort per iteration. The balance between these two conflicting factors can only be found by experience. For a 70% hydrogen-air flame, a system with eight species (H, O, OH, N_2, O_2, H_2, H_2O, and HO_2), the processing time required for convergence in the uncoupled case is compared with corresponding times for several block implicit solutions in Table 3. The completely uncoupled approach is clearly the most efficient

Table 3. Operational features of computation of 70% hydrogen–air flame, for grid with $M = 30$ distance intervals.

Type of implicit approach	Unknowns in block matrix	Size of matrices to be inverted [algorithm (4.69)]	Solutions per time step	ξ_t	Time steps for convergence	Processor time (Amdahl seconds)	S_u (cm/s)
1. Uncoupled	All treated separately	1×1	12	2×10^{-3}	1200	184	80.0
				2×10^{-2}	400	56	79.9
2. Block	All species, enthalpy	9×9	1 (plus three solutions of auxiliary eq. for T, ρ, δy)	2×10^{-2}	300	100	80.0
3. Block	All species, enthalpy, temperature	10×10	1 (plus two solutions of auxiliary eq. for ρ, δy)	2×10^{-2}	300	117	80.2
4. Block	All species, ethalpy, temperature, density	11×11	1 (plus one solution of auxiliary eq. for δy)	2×10^{-2}	320	140	79.2
				5×10^{-2}	200	92	78.5
				1×10^{-1}	200	89	78.2
				2×10^{-1}	computation broke down		
5. Block	All species, enthalpy, T, ρ, δy	12×12	1	2×10^{-2}	600	300 (convergence not good)	79.3
				1×10^{-1}	400	200	78.4

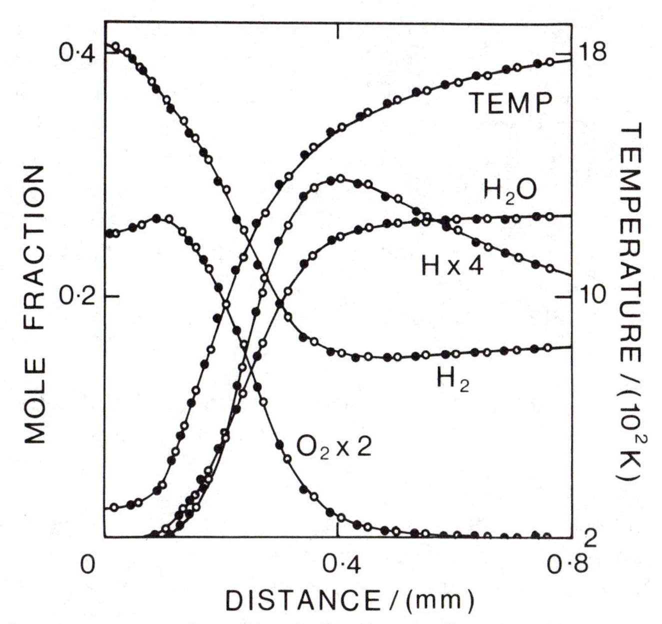

Figure 5. Computed temperature profile and mole fraction profiles of stable species and hydrogen atoms in 41% hydrogen-59% air flame at atmospheric pressure and $T_u = 298$ K. ○ values computed after Eulerian calculation alone; ● after 500 subsequent Lagrangian cycles of 10^{-7} s each, after which the distance scale was moved uniformly by 0.784 mm to bring profiles back to coincidence.

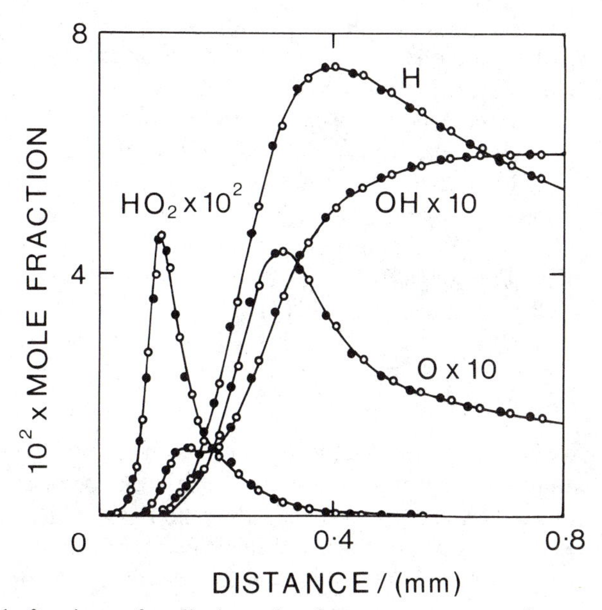

Figure 6. Computed mole fractions of radical species, from same computation as Fig. 5.

in this case, because of its much smaller computational demands per time step. The computed burning velocities resulting from the block implicit approaches 4 and 5 decrease somewhat as ξ_t is increased. Similar overall results were obtained in the case of the 41% hydrogen-air flame, but with the additional feature that it became necessary to use a smaller ξ_t ($= 2 \times 10^{-3}$), with a correspondingly increased number of cycles, for convergence. The value $\xi_t = 0.002$ appears to give satisfactory convergence for the whole range of hydrogen-carbon monoxide-methane-air flames, though it may be possible by trial and error to establish satisfactory larger values of ξ_t in certain composition ranges.

(c) Comparison of flame properties computed by Eulerian and Lagrangian approaches

The computational advantages of Eulerian methods must be weighed against the possibility of appreciable errors caused by the "numerical" diffusion which accompanies inclusion of advection terms. We carried out computations on two true one-dimensional flames to compare results from Eulerian and Lagrangian approaches. The two flames were hydrogen-air flames containing 41 and 70% hydrogen respectively. The Eulerian and Lagrangian codes performed the operations described above, including a detailed transport property formulation. The solution was first made to converge to the steady state by use of the uncoupled Eulerian approach. This steady-state solution was then continued in a Lagrangian manner for 500 cycles of 10^{-7} s each for the 41% hydrogen-air flame and 600 cycles of 4×10^{-7} s each for the 70% hydrogen–air flame. The computed burning velocity of the 41% hydrogen–air flame decreased from 264.3 to 263.2 cm s^{-1}, while that of the 70% hydrogen flame remained at 80.0 cm s^{-1} to the third significant digit. Figures 5 and 6 show the composition and temperature profiles for the 41% hydrogen–air flame before and after the Lagrangian part of the calculation, with the end profiles moved backwards by 0.784 mm. Alternate points only are shown to the left of the distance of 0.15 mm in Fig. 5. Clearly there is no appreciable effect of numerical diffusion errors. It is nevertheless necessary to be eternally vigilant with respect to this problem.

4.5. Gasdynamic effects

We have so far treated the quasi-one-dimensional flame as a constant pressure system with an interaction solely between chemical events and diffusion of matter and energy. Provided that the flame Mach number is low, this is a perfectly legitimate means of investigating the interaction per se, and it leads to a solution which includes the steady state, one-dimensional, laminar burning velocity. As an eigenvalue of the steady-state differential equations, the laminar burning velocity is a fundamental property which depends on the composition, temperature, and pressure of the initial mixture feeding the flame, and it provides the means of representing the progress of the flame in a larger flow field. For higher-velocity flames, however, and for flames freshly ignited by powerful localized energy sources, gasdynamic effects associated with the ignition or propagation of the reaction zone also become important; indeed, even for low-velocity flames the interaction of the flame with an external flow field may necessitate consideration of quite small pressure gradient effects. In these situations, pressure must be introduced into the equations as an additional dependent variable, accompanied by the momentum equation (2.7a) as the additional required relation. However, the momentum source term includes the pressure gradient across the computational cell, and, since the pressure nodes are at the centers of normal cells, it is most convenient to stagger the velocity nodes with respect to the others, so that the momentum cell centers are at the normal cell boundaries and their boundaries at the positions of the normal cell nodes. On the assumption of linear variation of the velocity with ω between velocity nodes, a microintegration across a momentum cell between the boundaries at $\omega_{j'}$ and $\omega_{j'+1}$ leads, with notation similar to that of Eqs. (4.23) to (4.26), to

$$\begin{aligned}
&\frac{\Omega_{j'}}{\Omega_{j'-1}+\Omega_{j'}}\{(1/2)(P+G)\Omega_{j'}-L_{j'-1}\}(v_y)_{j'-1/2} \\
&\quad+\left\{(1/2)(P+G)\Omega_{j'}\left(2+\frac{\Omega_{j'-1}}{\Omega_{j'-1}+\Omega_{j'}}+\frac{\Omega_{j'+1}}{\Omega_{j'}+\Omega_{j'+1}}\right)\right. \\
&\quad\left.+L_{j'+1}\left(\frac{\Omega_{j'+1}}{\Omega_{j'}+\Omega_{j'+1}}\right)-L_{j'-1}\left(\frac{\Omega_{j'-1}}{\Omega_{j-1}+\Omega_{j'}}\right)\right\}(v_y)_{j'+1/2} \\
&\quad+\frac{\Omega_{j'}}{\Omega_{j'}+\Omega_{j'+1}}\{(1/2)(P+G)\Omega_{j'}+L_{j'+1}\}(v_y)_{j'+3/2} \\
&\quad-(1/2)P\Omega_{j'}\left\{\left(\frac{\Omega_{j'}}{\Omega_{j'-1}+\Omega_{j'}}\right)(v_y)_{U,j'-1/2}\right. \\
&\quad+\left(2+\frac{\Omega_{j'-1}}{\Omega_{j'-1}+\Omega_{j'}}+\frac{\Omega_{j'+1}}{\Omega_{j'}+\Omega_{j'+1}}\right)(v_y)_{U,j'+1/2} \\
&\quad\left.+\left(\frac{\Omega_{j'}}{\Omega_{j'}+\Omega_{j'+1}}\right)(v_y)_{U,j'+3/2}\right\} \\
&=\{(P'+Q')_{j'+1}-(P'+Q')_{j'}\}(I_{1,j'}+I_{2,j'})/(\delta y_{1,j'}+\delta y_{2,j'})
\end{aligned}\tag{4.79}$$

where

$$\begin{aligned}
\Omega_{j'+1}&=(1/2)(\omega_{j'}-\omega_{j'-1}) \\
\Omega_{j'}&=(1/2)(\omega_{j'+1}-\omega_{j'}) \\
\Omega_{j'+1}&=(1/2)(\omega_{j'+2}-\omega_{j'+1})
\end{aligned}$$

and $I_{1,j'}$, $I_{2,j'}$, $\delta y_{1,j'}$, and $\delta y_{2,j'}$ are defined by Eqs. (4.49) and (4.50). For Eq. (4.79) only, P' and Q' denote the pressure and viscous pressure respectively, of Eq. (2.7a). The multiplying factor $(I_{1,j'} + I_{2,j'})/(\delta y_{1,j'} + \delta y_{2,j'})$ on the right-hand side of Eq. (4.79) is the mean area of the computational cell.

For slow flames, Eq. (4.79) may be uncoupled from the remainder of the calculation (as has been done so far), and Eq. (2.7b) may be used to determine the steady-state pressure profile at the end of the integration. For much faster flames where there are appreciable gasdynamic effects and associated density changes, the momentum equation must be coupled directly into the system, and the energy equations (2.19), (2.20) or (2.20a) must be used in place of Eq. (2.20b). In the finite-difference formulation discussed in Section 4.2, it then also becomes necessary to modify Eq. (4.44) to include the effect of variable pressure on the density and to introduce the condition

$$\delta(\delta y_{j'}) = \delta t\{(v_y)_{j'+1} - (v_y)_{j'}\} \tag{4.80}$$

in order to close the system in combination with Eq. (4.48) or (4.51). Laminar premixed flames do not fall into this "fast" category.

We have been discussing the pressure gradients that are sustained by the reaction zone of the flame, which can be done retaining a time scale characteristic of the flame reaction. If it is necessary to follow pressure dissipation ahead of a freely propagating flame, or in the early stages of a powerful ignition, then the time scale must be compatible with sonic velocities and the Courant stability criterion must be obeyed: the (Lagrangian) model must not be advanced in a time step by an interval greater than the sound travel time between successive grid points.

5. Premixed laminar flames and kinetic studies

The implicit numerical solution of the time-dependent conservation equations provides the most powerful general method of solving premixed laminar flame problems in systems of (in principle) arbitrary chemical complexity. Indeed, with the simultaneous development of improved diagnostic techniques for the measurement of flame profiles, the possibility of obtaining such solutions has opened the way to realistic studies of reaction mechanisms even in hydrocarbon flames. The choice of solution method and transport flux formulation involves compromise between precision and cost, which becomes a matter of considerable import when modeling hydrocarbon oxidation in flames, which may involve some 25 chemical species and 80 or so elementary reactions.

The implicit solution approach has been used successfully for one-dimensional premixed flame modeling of flames in hydrogen–bromine (Spalding and Stephenson, 1971), hydrogen–air (Stephenson and Taylor,

1973; Warnatz, 1978b,c; Dixon-Lewis, 1979, 1983), hydrogen–carbon monoxide–air (Warnatz, 1979; Cherian *et al.*, 1981a,b), methane–air (Smoot, Hecker, and Williams, 1976; Tsatsaronis, 1978; Warnatz, 1981; Dixon-Lewis, 1981; Dixon-Lewis and Islam, 1983), methanol–air (Westbrook and Dryer, 1980), and other lower hydrocarbon–air (Warnatz, 1981) systems, as well as for studies of the hydrogen–fluorine flame (Warnatz, 1978c), the hydrazine (Spalding, Stephenson, an Taylor, 1971) and ozone (Warnatz, 1978a) decomposition flames, and the inhibitory effect of hydrogen bromide on methane–air flames (Westbrook, 1980). Several further papers appear in Combustion Science and Technology, **34** (1983), which is a special issue on "Modeling of Laminar Flame Propagation in Premixed Gases." An eventual objective of such modeling is validation of reaction mechanisms and rate parameters, which may ultimately be of service for the solution of more practical flame problems.

Following a series of flame computations over an extended range of initial compositions and conditions, the validity of an assumed kinetic mechanism and set of rate parameters may be tested by comparison of computed properties with experiment. In the case of the flame profiles there must be comparison not only for major reactants and products, but also for reactive intermediates. It is necessary to be circumspect as far as minor intermediates are concerned (for example, with mole fractions less than about 10^{-4}), since there may be large uncertainties in the measurements themselves. When assessing rate parameters it is also necessary to carry out sensitivity analyses (Chapter 7) to determine the relative importance of each elementary reaction and intermediate in the mechanism. The final mechanism and rate parameters must, further, be consistent with results from studies of other combustion phenomena, such as data obtained from shock tube, explosion limit, or fast flow studies.

For flame modeling it is necessary to define reaction-rate coefficients over extended ranges of temperature. Rate coefficients are rarely known to better than $\pm 25\%$ over extended temperature ranges (Chapters 5 and 6). Thus, when groups of well-characterized and not-so-well-characterized reactions are combined into a hydrocarbon–air flame mechanism, there is much room for error. For this reason it is wise to consider a hydrocarbon–air flame mechanism as three or four groups of reactions which can be arranged hierarchically from hydrogen–air through hydrogen–carbon monoxide–air and formaldehyde–air flames to the hydrocarbon system itself. From studies of the intermediate systems, a self-consistent set of rate parameters can be built up for the whole ensemble.

To illustrate the degree of agreement with experiment which is achieved by current methane–air kinetic models, Fig. 7 compares predicted and measured burning velocities of methane–air flames at atmospheric pressure, and Figs. 8 and 9 compare predicted species profiles for a stoichiometric methane–air flame at atmospheric pressure with measurements by Bechtel *et al.* (1981).

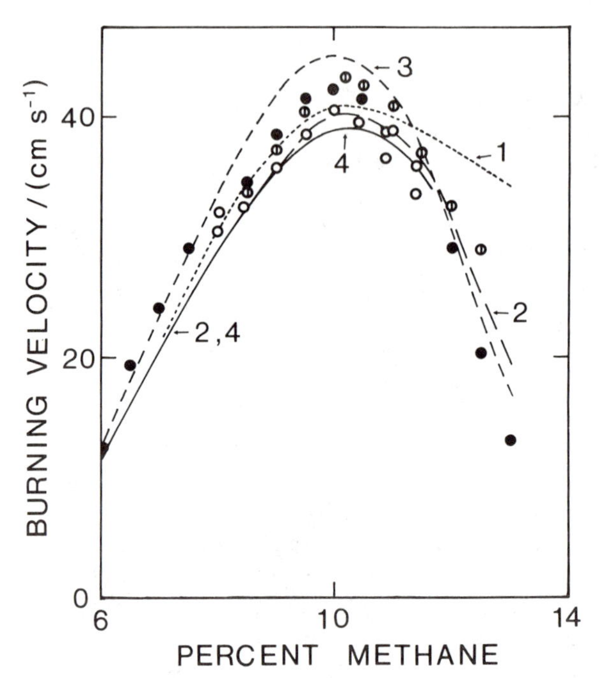

Figure 7. Burning velocities of methane–air flames at atmospheric pressure and $T_u = 298$ K. Comparison of burner measurements (points) with computations by Smoot *et al.* (1976) (curve 1); Tsatsaronis (1978) (curve 2); Warnatz (1981) (curve 3); and Dixon-Lewis (1981) (curve 4). Measurements: ●, Günther and Janisch (1972); ⦶, France and Pritchard (1976); ○, Dixon-Lewis and Islam (1983).

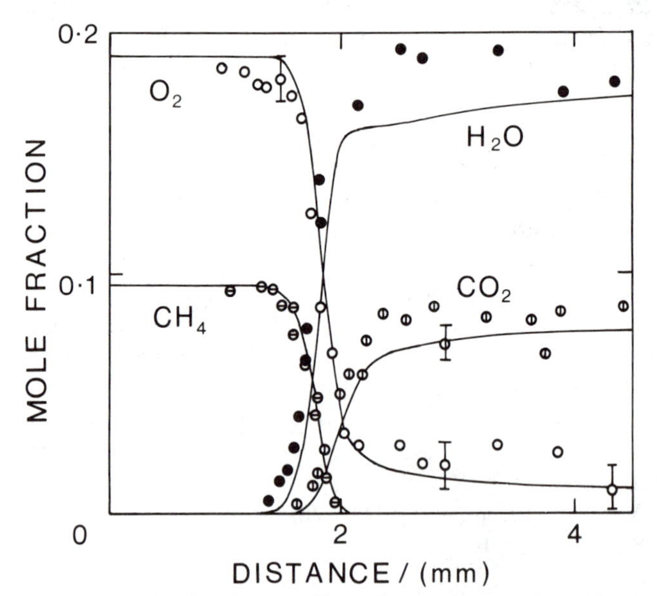

Figure 8. Mole fraction profiles of reactants and products in a stoichiometric methane–air flame at atmospheric pressure and $T_u = 298$ K, comparing measurements of Bechtel *et al.* (1981), represented by points, with line representing computation by the author (Dixon-Lewis, 1981).

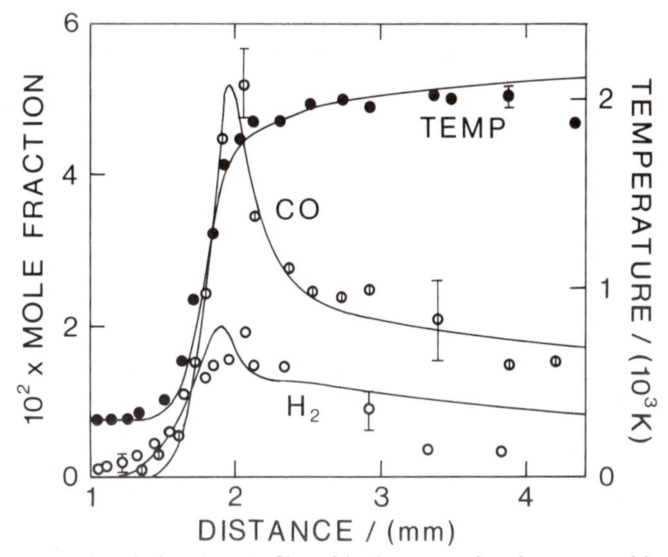

Figure 9. Temperature and mole fraction profiles of hydrogen and carbon monoxide in a stoichiometric methane–air flame at atmospheric pressure and $T_u = 298$ K. Points represent measurements of Bechtel *et al.* (1981) and lines represent computation by the author (Dixon-Lewis, 1981; Dixon-Lewis and Islam, 1983).

6. Two further solution techniques

The foregoing sections describe a generally applicable technique for solving reactive flow problems in one space dimension. The accuracy is first order in both space and time. Two further methods will now be briefly introduced: one of which can give faster convergence though at the same time requiring more accurate initial estimates of the solution before it will converge at all; and one which is capable of higher-order accuracy. A third technique, particularly useful for the determination of radical decay profiles in the burnt gas regions of one-dimensional flames of otherwise known properties, is described in Section 9.

6.1. Newton-type iteration around stationary flame equations

The first approach starts from the stationary flame equations in Eulerian coordinates. In their second-order form these equations are obtained by omitting the time derivative from the equations of type (4.19) (cf. Lagrangian representations where the *distance* derivative on the left-hand side is omitted). The further development, as illustrated by Wilde (1972), is essentially the same as that for the time-dependent equations, except that now the successive time intervals are replaced by successive Newton-type iterations towards a solution. Convergence is rapid when it occurs, but the technique of straightforward undamped Newton iteration lacks the stabilizing influence which the time derivative exerts when δt is small (or P in Eq. (4.23) and subsequent finite-difference algebraic equations is large). It is of course possible to employ a damping technique such as one of those used by Marquardt (1963) or Deuflhard (1974) to reduce the sensitivity of the iteration to the initial estimates. However, and more importantly, it appears that the domain of convergence for the stationary flame technique is larger for coarser grid spacings and that it is possible with experience to use this feature in combination with a damping strategy to provide a very efficient alternative route to an Eulerian solution by successive refinements of an initially coarsely spaced grid (Smooke, 1982).

The linearization as applied to R_i in Eq. (4.39) may also be applied to the right-hand sides of the first-order form of the stationary flame equations, and can again lead to a solution by similar finite-difference techniques (Wilde, 1972). The initial first-order differential equations are obtained by the insertion of appropriate expressions for the j_i and q_i, and for Q_T and q_E, into Eqs. (2.11), (2.12), and (2.21) or (2.22). Convergence is rapid, but again close initial estimates are required.

6.2. Finite-element collocation method

The second approach, applied by Margolis (1978) and by Heimerl and Coffee (1980) to ozone decomposition flames, employs a *method-of-lines* technique. In combination with finite-element collocation methods this technique provides a general approach to the numerical solution of partial differential equations. Taking Eqs. (4.12) and (4.13) as the working examples

$$\partial\sigma_n/\partial t + (a + b\omega)\,\partial\sigma_n/\partial\omega = -\{1/(\psi_u - \psi_b)\}\,\partial/\partial\omega\,(Aj_n/m_n) + R_n/\rho \qquad (n = 1, 2, \ldots, N) \quad (4.12)$$

$$\partial h/\partial t + (a + b\omega)\,\partial h/\partial\omega = -\{1/(\psi_u - \psi_b)\}\,\partial/\partial\omega\,\{A(Q_D + Q_T)\} \quad (4.13)$$

it is assumed that the solution can be written as

$$\sigma_n \cong \sum_{i=1}^{m} C_n^{(i)}(t)F_i(\omega) \qquad (n = 1, 2, \ldots, N) \quad (6.1)$$

$$h \cong \sum_{i=1}^{m} C_{N+1}^{(i)}(t)F_i(\omega) \quad (6.2)$$

where the basis functions $F_i(\omega)$, $i = 1, 2, \ldots, m$ span the solution space at any fixed time t to within a small error tolerance. The time-dependent coefficients $C_n^{(i)}(t)$ are uniquely determined by requiring that the expansions (6.1) and (6.2) satisfy the boundary conditions of the problem and also the differential equations (4.12) and (4.13) exactly at m-2 interior (collocation) points $\omega_1, \omega_2, \ldots, \omega_{m-2}$. At these collocation values of ω, substitution of expansions (6.1) and (6.2) into the initial partial differential equations leads to a system of coupled nonlinear ordinary differential equations which express the development of the system with time. After uncoupling by Gaussian elimination similar to that described in Section 4.3, these equations may be integrated numerically by any standard technique, for example, some variant of Gear's method for stiff equations. The uncoupling will be discussed further below. The initial values for the coefficients are obtained by requiring Eqs. (6.1) and (6.2) to satisfy the initial conditions and profiles [cf. Section 4.2(a)] at all m collocation points.

An important feature of the method of lines is selection of the basis functions $F_i(\omega)$, which determines the precision of (spatial) curve fitting. The piecewise polynomials known as B splines meet the requirements. Curve fitting by means of spline functions entails division of the solution space into subintervals by means of a series of points called "knots." Knots may be either single or multiple, a multiple knot being formed by the coincidence of two or more such points. They are numbered in nondecreasing order of location $s_1, s_2, \ldots, s_i, \ldots$. A normalized B spline of order k takes nonzero values only over a range of k subintervals between knots, and, for example, $B_{i,k,\mathbf{s}}(\omega)$, the *ith* normalized B spline of order k for the knot sequence $\mathbf{s}$, is zero outside the interval $s_i \leqslant \omega \leqslant s_{i+k}$, nonnegative at $\omega = s_i$ and $\omega = s_{i+k}$, and strictly positive for $s_i < \omega < s_{i+k}$. In any subinterval between knots, such that $s_i \leqslant s_j < \omega < s_{j+1} \leqslant s_{i+k}$, $B_{i,k,\mathbf{s}}(\omega)$ is a polynomial of order k (degree $k - 1$). If s_l, $s_i \leqslant s_l \leqslant s_{i+k}$, is a knot of multiplicity $(k - v)$, then $d^v(B_{i,k,\mathbf{s}})/d\omega^v$ is a discontinuity and all lower derivatives are continuous at $\omega = s_l$. For the special case of equally spaced knots with $\delta\omega \neq 0$, $B_{i,k,\mathbf{s}}(\omega)$ is a symmetric bell-shaped function about $\omega = s_i + (1/2)k\delta\omega$.

A spline function of order k with knot sequence $\mathbf{s}$ is then defined as any linear combination $\Sigma_i \alpha_i B_{i,k,\mathbf{s}}$ of B splines of order k for the particular knot sequence.

Further detailed discussion of splines is found in an exposition by de Boer (1978). It follows from the B-spline representation theorem of Curry and Schoenberg (1966) that, given a set of $l + 1$ strictly increasing locations ξ_i at which the polynomials are to be joined (that is, a set of $l + 1$ breakpoints), together with the number of continuity conditions $v_i (v_i \leqslant k)$ to be satisfied at each interior breakpoint, a representation of order k requires the specification of a set of $r + k = k(l + 1) - \Sigma_{i=2}^{l} v_i$ knots. The interior knots $s_{k+1} \ldots s_r$ are identified with the interior breakpoints $\xi_2 \ldots \xi_l$, each with multiplicity $(k - v_i)$. The initial and final breakpoints, located at the ends of the range of interest, both become knots of multiplicity k. The ends of the range thus become discontinuities, and the individual splines extend across the ranges s_i to s_{i+k}, $i = 1, 2, \ldots, r + 1$. Over the whole interval $s_k \leqslant \omega \leqslant s_{r+1}$ we have $\Sigma_i B_i(\omega) = 1$, and for the particular case where $v_i = k - 1$, $i = 2 \ldots l$, the interior knots are all single.

The jth derivative of a B spline function of order k, with coefficients α_i, is given (de Boer, 1978) by

$$D^j\left[\sum_i \sigma_i B_{i,k}\right] = \sum_i \alpha_i^{(j)} B_{i,k-j} \tag{6.3}$$

where

$$\alpha_i^{(j)} \begin{cases} = \alpha_i & \text{for } j = 0 \\ = \dfrac{\alpha_i^{(j-1)} - \alpha_{i-1}^{(j-1)}}{(s_{i+k-j} - s_i)/(k-j)} & \text{for } j > 0 \end{cases} \tag{6.4}$$

We now return to Eqs. (6.1) and (6.2). If the right-hand sides of these equations are taken to be B spline functions of order k, then it can be shown (Margolis, 1978) that the spatial truncation error, that is, the amount by which the approximate solution fails to solve the partial differential equations (4.12) and (4.13) at time t, is $0(\delta\xi^{k-2})$. Next, if the number of continuity conditions to be satisfied at each interior ($i = 2 \ldots l$) breakpoint is v, then only the first $m = kl - v(l-1)$ knots of the sequence form the *origins* of fresh polynomial pieces. It follows that the number of *pp* coefficients $C^{(i)}(t)$, and hence the number of collocation points required, takes this same value m. The collocation points must be distributed at $k - v$ per subinterval between breakpoints, the same in each interval, together with a further v appropriately interspersed points or other conditions. In a typical one-dimensional flame problem three of the extra constraints would normally be provided by the boundary conditions—one for the first space derivative of each of the dependent variables at each end of the range, and one for the input composition. In this context the zero-gradient conditions at flame boundaries are conveniently introduced (Margolis, 1978) by way of additional differential equations at the boundary collocation points

$$d/dt\left[D\left\{\sum_i C_n^{(i)}(t) B_{i,k,\mathbf{s}}(\omega)\right\}\right] = 0$$

$$(\omega = 0, 1; n = 1, 2, \ldots, N+1) \tag{6.5}$$

The space derivative is defined in Eq. (6.3), and its initial value is set at zero.

The system of ordinary differential equations resulting from the collocation procedure may be written in the form

$$\mathbf{A}\dot{\mathbf{c}}_n = \mathbf{g}_n[(\mathbf{AC})_j] = \mathbf{g}_n \qquad (\text{all } j; n = 1, 2, \ldots, N+1) \tag{6.6}$$

where $\mathbf{A} = \{A_{ji}\}$ is the matrix of values of the ith B spline at the jth collocation point, $\mathbf{C} = \{C_{in}\} = \{C_n^{(i)}(t)\}$, and $\dot{\mathbf{c}}_n = \{\dot{c}_i\}_n = \{dC_n^{(i)}/dt\}$ and $\mathbf{g}_n = \{g_j\}_n$ are column vectors relating to the species n. The g_n are the formal expressions which include all the terms other than the time derivatives from Eqs. (4.12) and (4.13), and the extra subscript denotes the position of evaluation. This system is not fully coupled, since (i) there is no explicit coupling of the time derivatives for separate species in any of the equations; and (ii) at most, k of the B splines are nonzero at any collocation point, by definition. Further, the coefficients A_{ji} in Eq. (6.6) are functions only of position and the same for all species. If $\mathbf{c}_n$ are the column vectors of $\mathbf{C}$, so that $\mathbf{c}_n = \{c_i\}_n \equiv \{C_n^{(i)}(t)\}$, then the same coefficient matrix $\mathbf{A}$ relates the $\mathbf{c}_n$ with the $\sigma_{n,j}$ and h_j at the collocation points, so that

$$\mathbf{A}\mathbf{c}_n = \sigma_{n,j} \qquad (\text{all } j;\ n = 1, 2, \ldots, N) \tag{6.7}$$

$$\mathbf{A}\mathbf{c}_{N+1} = h_j \qquad (\text{all } j) \tag{6.8}$$

The required coefficients $\mathbf{c}_n$ and the ordinary differential equations may now be written

$$\mathbf{c}_n = \mathbf{A}^{-1}\sigma_{n,j} \qquad (\text{all } j;\ n = 1, 2, \ldots, N) \tag{6.9}$$

$$\mathbf{c}_{N+1} = \mathbf{A}^{-1}h_j \qquad (\text{all } j) \tag{6.10}$$

$$\dot{\mathbf{c}}_n = \mathbf{A}^{-1}\mathbf{g}_n \qquad (\text{all } j;\ n = 1, 2, \ldots, N+1) \tag{6.11}$$

and could be set up by inversion of the matrix $\mathbf{A}$. However, the matrix is almost block-diagonal, and the most efficient method of solution is by a Gaussian elimination procedure similar to that described in Section 4.3. More specifically, if the order of the B splines is k, with v continuity conditions to be satisfied at the breakpoints, the coefficient matrix has standard sub-blocks of k-v rows (one row per collocation point) and k columns each. There is an overlap of v columns between subintervals. An extra collocation point within a subinterval introduces one more equation there, that is, one more row into the corresponding sub-block. As an example, if $k = 6$, $v = 4$, and $l = 4$, and if, further, the necessary v side conditions are distributed one in each subinterval between breakpoints, then, showing only the nonzero elements in the coefficient matrix, Eqs. (6.6) appear as

$$\left|\begin{array}{cccccccccccc}
\times & \times & \times & \times & \times & \times & & & & & & \\
\times & \times & \times & \times & \times & \times & & & & & & \\
\times & \times & \times & \times & \times & \times & & & & & & \\
& & \times & \times & \times & \times & \times & \times & & & & \\
& & \times & \times & \times & \times & \times & \times & & & & \\
& & \times & \times & \times & \times & \times & \times & & & & \\
& & & & \times & \times & \times & \times & \times & \times & & \\
& & & & \times & \times & \times & \times & \times & \times & & \\
& & & & \times & \times & \times & \times & \times & \times & & \\
& & & & & & \times & \times & \times & \times & \times & \times \\
& & & & & & \times & \times & \times & \times & \times & \times \\
& & & & & & \times & \times & \times & \times & \times & \times
\end{array}\right|
\left|\begin{array}{c}
\dot{c}_1 \\ \dot{c}_2 \\ \dot{c}_3 \\ \dot{c}_4 \\ \dot{c}_5 \\ \dot{c}_6 \\ \dot{c}_7 \\ \dot{c}_8 \\ \dot{c}_9 \\ \dot{c}_{10} \\ \dot{c}_{11} \\ \dot{c}_{12}
\end{array}\right|
=
\left|\begin{array}{c}
g_1 \\ g_2 \\ g_3 \\ g_4 \\ g_5 \\ g_6 \\ g_7 \\ g_8 \\ g_9 \\ g_{10} \\ g_{11} \\ g_{12}
\end{array}\right| \tag{6.12}$$

A general method of LU decomposition in such cases involves partitioning the coefficient matrix as follows:

(i) Commencing at the top left-hand corner of the main diagonal, a number l ($= 4$ in the example) of $(k - v) \times (k - v)$ submatrices are formed along the diagonal. These correspond with the submatrices $\mathbf{C}$ in Eq. (4.65).
(ii) Immediately to the right and immediately below each diagonal submatrix must be set up respectively a $v \times (k - v)$ and a $(k - v) \times v$ submatrix. These correspond to the submatrices $\mathbf{D}$ and $\mathbf{B}$ in Eq. (4.65). A $v \times v$ square submatrix is left in the bottom right-hand corner.
(iii) The column vector $\mathbf{g}$ is partitioned from the top to give l subvectors of length $k - v$, with a final subvector of length v. These subvectors correspond to the $\mathbf{W}$ of Eq. (4.65).

Following the partitioning, factorization and the first stage of solution by the algorithm (4.69) leads to

$$\left|\begin{array}{cccccccccccc}
1 & & \times & \times & \times & \times & & & & & & \\
 & 1 & \times & \times & \times & \times & & & & & & \\
 & & 1 & & \times & \times & \times & \times & & & & \\
 & & & 1 & \times & \times & \times & \times & & & & \\
 & & & & 1 & & \times & \times & \times & \times & & \\
 & & & & & 1 & \times & \times & \times & \times & & \\
 & & & & & & 1 & & \times & \times & \times & \times \\
 & & & & & & & 1 & \times & \times & \times & \times \\
 & & & & & & & & 1 & & & \\
 & & & & & & & & & 1 & & \\
 & & & & & & & & & & 1 & \\
 & & & & & & & & & & & 1
\end{array}\right| \left|\begin{array}{c} \dot{c}_1 \\ \dot{c}_2 \\ \dot{c}_3 \\ \dot{c}_4 \\ \dot{c}_5 \\ \dot{c}_6 \\ \dot{c}_7 \\ \dot{c}_8 \\ \dot{c}_9 \\ \dot{c}_{10} \\ \dot{c}_{11} \\ \dot{c}_{12} \end{array}\right| = \left[\begin{array}{c} Y_1 \\ Y_2 \\ Y_3 \\ Y_4 \\ Y_5 \\ Y_6 \\ Y_7 \\ Y_8 \\ Y_9 \\ Y_{10} \\ Y_{11} \\ Y_{12} \end{array}\right] \tag{6.13}$$

The coefficient matrix now consists of a number l of $(k - v) \times (k - v)$ unit submatrices and a single $v \times v$ unit submatrix along the main diagonal. Immediately to the right of each of the first l of these is a $v \times (k - v)$ submatrix. These last are the matrices $\mathbf{F}$ of the algorithm (4.69).

We now solve Eq. (6.13) by back-substitution. Taking the last four equations, we have

$$\dot{c}_i = Y_i \qquad (i = (k - v)l + 1, \ldots, (k - v)l + v)$$

Let the remaining $\dot{c}_i$ and Y_i, $i \leq (k - v)l$, be divided successively into groups of $k - v$ each, and further, let the v values of $\dot{c}_i$ immediately following the jth group be formed into the subvector Z_{j+1}. The complete solution, working upwards, then becomes

$$\begin{aligned} \dot{c}_i &= Y_i && (i = (k - v)l + 1, \ldots, (k - v)l + v) \\ \dot{\mathbf{c}}_j &= \mathbf{Y}_j - \mathbf{F}_j \mathbf{Z}_{j+1} && (j < l) \end{aligned} \tag{6.14}$$

After the uncoupling, the ordinary differential equations have the form

$$dC_n^{(i)}/dt = \sum_j \beta_{ij} g_n[(\mathbf{AC})_j] \tag{6.11a}$$

where the β_{ij} determined by the elimination procedure include only the nonzero elements of $\mathbf{A}^{-1}$. For any i, at most k of the β_{ij} are nonzero. Given the expressions for the g_n and initial values of the $C_n^{(i)}$, the equations may be integrated forward in time. The initial values are obtained from the assumed initial profiles of σ_n and h at time $t = 0$ by

$$\mathbf{c}_n = \sum_j \beta_{ij} \sigma_{n,j} \qquad (n = 1, 2, \ldots, N) \tag{6.9a}$$

$$\mathbf{c}_{N+1} = \sum_j \beta_{ij} h_j \tag{6.10a}$$

The principles outlined in this section have been incorporated by Madsen and Sincovec (1977) into a general subroutine package, PDECOL, for solving partial differential equations. This package combines the B-spline subroutines of de Boor with a stiff ordinary differential equation integrator due to Hindmarsh (1976). Given (i) the order of the B-spline representation required, (ii) a set of $l + 1$ breakpoints as defined above, and (iii) a number v of continuity conditions to be satisfied at each breakpoint, PDECOL itself generates the set of $\{kl - v(l - 1)\}$ basis functions and collocation points. The ordinary differential equations at the collocation points are then set up from user-supplied information about the initial values of σ_n $(n = 1, 2, \ldots, N)$ and h [cf. Section 4.2(a)]. Again, since only N-1 of the σ_n are independent, one of Eqs. (4.12) is normally discarded. For Lagrangian formulations of flame problems, the temperature, the density, and the remaining σ_N may then be coupled into the system of equations by

$$c_p(dT/dt) = dh/dt - \sum_{n=1}^{N} H_n(d\sigma_n/dt) \tag{6.15}$$

$$d\rho/dt = -(\rho/T)\, dT/dt - \sum_{n=1}^{N} (\rho/\sum_n \sigma_n)\, d\sigma_n/dt \tag{6.16}$$

$$m_N(d\sigma_N/dt) = 1 - \sum_{n=1}^{N} m_n\sigma_n - \sum_{n=1}^{N-1} m_n(d\sigma_n/dt) \tag{6.17}$$

all of which follow directly from the auxiliary equations (2.25) to (2.27) in a similar manner to Eqs. (4.44) to (4.46). For pure Lagrangian calculations the convection term on the left-hand side of each of Eqs. (4.12) and (4.13) becomes zero. Both c_p and the H_n in Eq. (6.15) are weak functions of temperature, with the H_n being given by expressions of the form (4.17).

The remaining matter requiring attention is the placement of the breakpoints. Similar considerations will apply here as outlined in Section 4.2(a) in connection with the finite-difference grid point distribution. That is, it is important to concentrate the breakpoints at the position where most reaction occurs, with progressively lower interior breakpoint concentrations towards the flame boundaries. For (presumably) an Eulerian calculation of an ozone decomposition flame structure, Heimerl and Coffee (1980) used between 60 and 70 breakpoints. While at the time of writing the author has not had first-hand experience with the method of lines, the most promising procedure for Eulerian calculations would seem to involve: (i) transformation of Eqs. (4.1) and (4.2) into the von Mises system of coordinates, followed by non-dimensionalization of the distance coordinate by Eq. (4.9) to give

$$j_n = -\frac{\rho}{(\psi_u - \psi_b)}\left\{\beta_n^h(\partial h/\partial\omega) + \sum_{j=1}^{N}\beta_{nj}m_j(\partial\sigma_j/\partial\omega)\right\} \tag{6.18}$$

$$Q_D + Q_T = -\frac{\rho}{(\psi_u - \psi_b)}\left\{\gamma^h(\partial h/\partial\omega) + \sum_{j=1}^{N}\gamma_j m_j(\partial\sigma_j/\partial\omega)\right\} \tag{6.19}$$

and then (ii) a sequence of integrations parallel to the sequence described in Section 4.4(b), either block implicit or by an arrangement in which the separate species are uncoupled and dealt with implicitly in rotation. In either case the temperature, density, and convection terms would be updated between complete cycles. In this context, and in the general context of the need for computation of reaction rates and space integral rates, there may also be some virtue in computing σ_n and h between cycles by means of Eqs. (6.7) and (6.8). In the Eulerian steady state the time derivatives of all dependent variables, including the $C_n^{(i)}(t)$ of Eqs. (6.1) and (6.2), become zero.

The formal transformations of Eqs. (4.1) and (4.2) into Eqs. (6.18) and (6.19) are essential for both the Eulerian and Lagrangian approaches by the collocation method.

In terms of *B*-spline representation, finite-difference procedures, corresponding to piecewise linear approximations, have $k = 2$, with discontinuous first derivatives at the single knots. This was the order of approximation used by Bledjian (1973) for the spatial discretization associated with his early method-of-lines solution for the properties of an ozone decomposition flame. Bledjian's procedure involved a straightforward finite-difference representation of the spatial second derivatives which arise on the right-hand sides of equations of the type of Eqs. (4.12) and (4.13) when the expressions for the transport fluxes j_n and $Q_D + Q_T$ are inserted. His integration of the ordinary differential equations at the collocation points was carried out explicitly.

Margolis (1978) used a *B*-spline method of order 6 together with an implicit integrator to examine the ozone decomposition flame. As a consequence of certain points of difference between his results and those of Bledjian, he concluded that the higher-order method is essential to the theoretical prediction of flame properties. On the other hand, Heimerl and Coffee (1980), again using *B*-spline representation and implicit integration on ozone decomposition flames, found that varying the order of the *B*-spline function between 3 and 8 and the number of continuity conditions at the breakpoints between 2 and 7 made no significant difference to the computed results. Heimerl and Coffee (1982) also used the same method to predict the properties of a stoichiometric hydrogen-air flame at atmospheric pressure. Their results are not significantly different from those obtained by finite-difference solution (Dixon-Lewis, 1982; Warnatz, 1982) with the use of identical reaction-rate parameters. In the general one-dimensional context it is as yet uncertain whether the increased precision obtainable by the higher-order finite-element approaches gives sufficiently different results or better insight to merit the greater computational cost.

7. Implicit methods and general reactive flow problems

7.1. Boundary layer flows

It is convenient next to consider the matter of geometric complexity. The differential equations (2.18) and (2.20b) that we have been considering are of the type known as "parabolic." They can be integrated forward with time as the marching coordinate. In fact, the finite-difference method of solution described for the one-dimensional flame problem was developed initially for the treatment of steady, two-dimensional, boundary layer flows. These are flows in which the changes in the dependent variables such as velocity, temperature, or concentration occur predominantly in a cross-stream direction, with the stream-wise direction itself behaving as the time-wise or marching coordinate. Because of the very small gradients in the stream-wise direction, transport fluxes in that direction are neglected. In their simplest form such situations may occur, for example, when a flowing gas stream reaches the leading edge of a stationary flat plate which lies parallel with the flow direction or when a plane or round jet discharges into a surrounding atmosphere.

Figure 10 illustrates the development of the boundary layer at the edge of a jet flow which results from the discharge of a uniform velocity gaseous stream from a nozzle. Only one half of the symmetrical system is shown. The two coordinate directions to be considered may be the two cylindrical polar directions [cf. Eq. (2.2b)], with z as the stream-wise coordinate; or they may be two of the three Cartesian coordinate directions [cf. Eq. (2.2a)]. In the latter case, taking x as the stream-wise coordinate, Eqs. (2.5), (2.7), (2.18), and (2.20a) become, for the steady, two-dimensional, plane boundary layer,

$$\partial/\partial x(\rho v_x) + \partial/\partial y(\rho v_y) = 0 \tag{7.1}$$

$$\rho v_x(\partial v_x/\partial x) + \rho v_y(\partial v_x/\partial y) = -dP/dx - \partial Q_V/\partial y \tag{7.2}$$

$$\rho v_x(\partial \sigma_i/\partial x) + \rho v_y(\partial \sigma_i/\partial y) = -\partial/\partial y(j_i/m_i) + R_i \qquad (i = 1, 2, \ldots, N) \tag{7.3}$$

$$\rho v_x(\partial h/\partial x) + \rho v_y(\partial h/\partial y) = q_E - \partial/\partial y(Q_D + Q_T) + v_x(dP/dx) - Q_V(\partial v_x/\partial y) \tag{7.4}$$

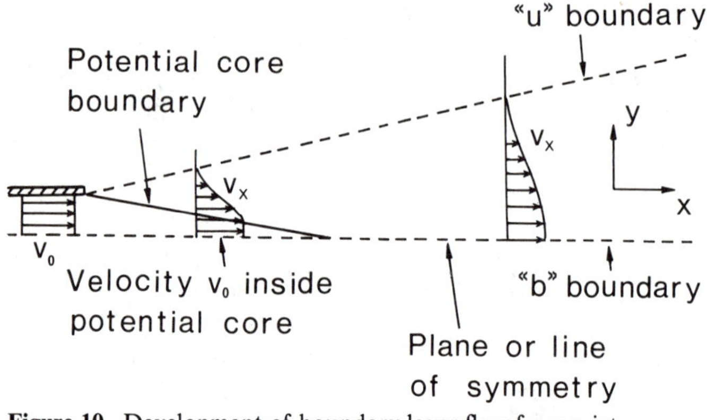

Figure 10. Development of boundary layer flow from a jet.

where Q_V represents the viscous stress due to the y-direction gradient of v_x, and is given in laminar flow by $Q_V = -\eta(\partial v_x/\partial y)$.

To convert these equations to von Mises coordinates we replace Eqs. (4.8) by

$$\begin{aligned} d\psi/dy &= \rho v_x \\ d\psi/dx &= -\rho v_y \end{aligned} \tag{7.5}$$

The continuity equation (7.1) is thus satisfied automatically. Then, putting $\rho v_y = M_y = a + b\omega$ as before, we find the result to have the same form as Eqs. (4.12) and (4.13). The convective fluxes at the "u" and "b" boundaries of the flow become rates of entrainment into the growing boundary layer, and because of the nondimensionalization of ψ by Eq. (4.9), the number of grid points across the region of significant concentration change does not vary with stream-wise distance x. Implicit finite-difference solutions of the boundary layer equations have been described by Blottner (1964, 1970, 1975), and at length by Patankar and Spalding (1970).

The entrainment rates at the "u" and "b" surfaces of the boundary layer require further discussion, since they determine the size of the region to be considered. They must therefore be chosen so that all significant changes in the dependent variables occur within the range $0 \leq \omega \leq 1$. Frequently, as in Fig. 10, one of the surfaces is at a plane or line of symmetry in the flow. At such a boundary both the mass entrainment rate and the gradients of the dependent variables in the normal direction are zero [cf. Section 4.2(c)]. Other types of surface include (i) those at walls, where the boundary conditions will be defined at specific locations, and (ii) those at free boundaries, at which the boundary layer fluid properties are supposed to merge gradually into those of the surrounding medium. The free boundary is more frequently encountered, since it occurs at the interface of every jet flow with its surrounding medium. The position of this type of boundary is less clearly defined than the others, and the most satisfactory approach is to use an equation similar to Eqs. (4.72) and (4.73) to define an entrainment rate. Thus, for a "u" boundary associated at the other end of the grid with a "b" boundary at a wall or symmetry plane, the entrainment rate might be given by

$$(M_y)_u = \{\Gamma_\phi/(y_u - y_{uu})\}M_{n'} \tag{7.6}$$

where

$$\left.\begin{aligned} M_{n'} &= \frac{\phi_{i,uu} - \phi_{i,u}}{\phi_{i,b} - \phi_{i,u}} F_u \qquad \text{at } t = n'\,\delta t,\ n' = 0 \\ M_{n'+1} &= M_{n'}\left[\left|\frac{\phi_{i,uu} - \phi_{i,u}}{\phi_{i,b} - \phi_{i,u}}\right| F_u\right]^{0.1} \end{aligned}\right\} \tag{7.7}$$

Here uu denotes a point near the u boundary, F_u is an adjustable parameter which is aimed towards securing a specific fraction $|(\phi_{i,uu} - \phi_{i,u})/(\phi_{i,b} - \phi_{i,u})|$ for property ϕ_i at the point uu, y is the physical cross-stream distance, and Γ_ϕ is the transport coefficient relating to the quantity ϕ near the boundary. It is well known, for example, that the thermal boundary layer at a hot plate may develop in the cross-stream direction at a different rate from the velocity boundary layer, depending on the relative values of the thermal diffusivity and the viscosity of the fluid. It is important that the transport coefficient employed in Eq. (7.6) be properly related to the phenomenon under investigation so as to give maximum spatial resolution while at the same time permitting correct development of the thickest boundary layer of interest.

It may also be useful in practice to provide an override condition so that if it is found during a computation that too high a rate of entrainment is being obtained, so that $\phi_{i,uu}$ is approaching the free outside stream condition $\phi_{i,u}$ too closely, a zero rate of entrainment is enforced over one or more iterations.

7.2. Counterflow flame geometries. Stretched one-dimensional flames

Counterflow flame configurations, such as may be established in the stagnation region between two opposed gaseous streams (Tsuji 1982, 1983), are of particular interest for the study of flame extinction phenomena and of flame behavior near flammability limits. The flow through such flames is characterized by a velocity gradient across the flame front, and the flames themselves are therefore suitable for the study of aerodynamic quenching or "flame stretch" phenomena. Figures 11 and 12 illustrate possible counterflow premixed and diffusion flame arrangements respectively. In Fig. 11 two axisymmetric premixed gas streams of identical composition, and with uniform and equal velocities, impinge normally on the stagnation plane between them, and premixed flames are established back to back on either side of the stagnation region. Figure 12 represents a counterflow diffusion flame

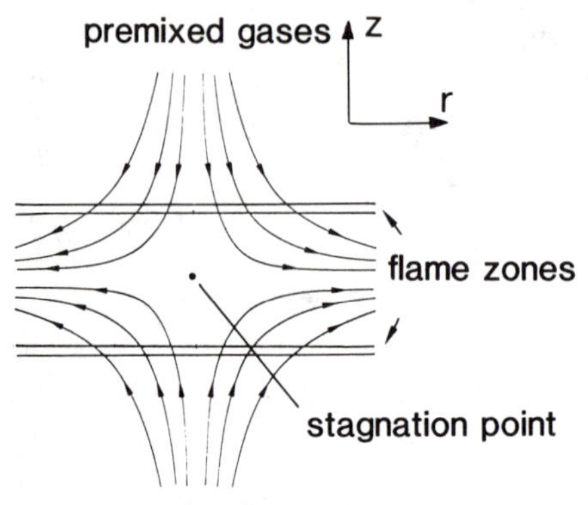

Figure 11. Counterflow premixed flame geometry.

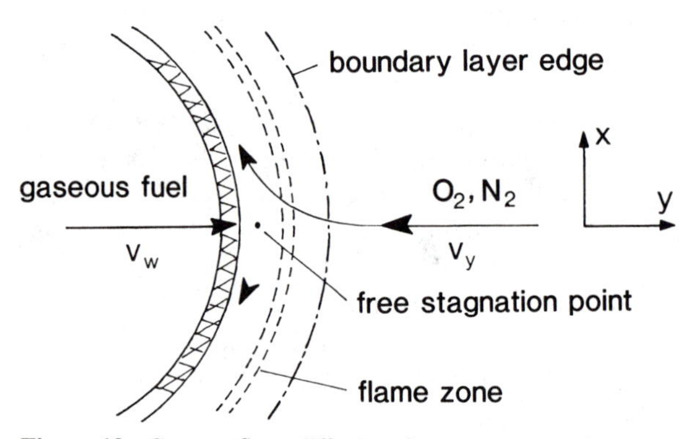

Figure 12. Counterflow diffusion flame geometry in forward stagnation region of a porous cylinder.

geometry in the forward stagnation region of a porous cylinder, as studied experimentally by Tsuji and Yamaoka (1967, 1969, 1971). Physically, a fuel gas such as methane is blown from a porous cylinder into an oncoming air stream so that a free stagnation line parallel with the cylinder axis is obtained in front of the porous surface. The line is in the symmetry plane of a stagnation point boundary layer, and combustion takes place within a thin flame zone around the location of the stoichiometric mixture. The location of the flame zone is within the boundary layer but outside the stagnation point, and there is a convective backflow of combustion products into the fuel-rich region of the flame (Tsuji and Yamaoka 1971).

For axisymmetric, inviscid, steady state potential flow of an incompressible fluid, the velocity distribution in the neighborhood of the stagnation point

$r = z = 0$ in Fig. 11 is given (Schlichting 1960) by

$$v_r = (1/2)\,ar$$
$$v_z = -az \tag{7.8}$$

where v_r and v_z are the radial and axial velocity components, and a is a velocity gradient. For the plane potential flow of Fig. 12 the corresponding expressions are (stagnation point at $x = 0$, $y = y_0$)

$$v_x = ax$$
$$v_y = -a(y - y_0) \tag{7.9}$$

The two configurations will now be considered in turn.

(a) The stretched one-dimensional premixed flame

As remarked in Section 2.2, most flames are essentially constant pressure systems, and if the flow is inviscid it is then not necessary to consider the momentum equation at all. Inviscid flow will be assumed for the purpose of the following discussion. Since the flame lies in the radial direction (Fig. 11), the gradients of all dependent variables except the radial velocity are zero in this direction, and the radial diffusive fluxes also vanish. By use of Eqs. (2.1) and (2.2b), conservation of mass in the steady state axisymmetric flow gives

$$\partial/\partial z\,(\rho v_z) + (\rho/r)\,\partial/\partial r\,(r v_r) = 0 \tag{7.10}$$

If the velocity gradient $(1/r)\,\partial/\partial r\,(r v_r)$ is given the constant value a (in s^{-1} units), Eq. (7.10) reduces for steady incompressible flow to Eqs. (7.8). Increasing the modulus of a increases the degree of stretch on the flame. The sign of a may be either positive or negative depending on whether there is expansion or contraction of the cross-section of the flow on moving downstream. However, a stable stationary flame solution is only possible when a is positive.

Treating the variable density flow in the flame by a procedure parallel to that of Section 4.1, we introduce the transformation [cf. Eq. (4.8)]

$$d\psi/dz = \rho \tag{7.11}$$

Putting $\rho v_z = M_z$, Eq. (7.10) then leads to

$$dM_z/d\psi + a = 0 \tag{7.12}$$

and

$$M_z = -a(\psi - \psi_0) \tag{7.13}$$

where ψ_0 is the value of ψ at the stagnation plane ($z = 0$).

By methods similar to those outlined in Sections 2.3 and 2.4, the species and energy continuity equations for the steady state system become

$$M_z(d\sigma_i/d\psi) = -d/d\psi\,(j_i/m_i) + R_i/\rho \qquad (i = 1, 2, \ldots, N) \tag{7.14}$$

$$M_z(dh/d\psi) = -d/d\psi\,(Q_D + Q_T) \tag{7.15}$$

or, in terms of the finite total mass interval $(\psi_e - \psi_0)$ and the dimensionless stream function $\omega = (\psi - \psi_0)/(\psi_e - \psi_0)$ [cf. Eq. (4.9)], and substituting for M_z from Eq. (7.13)

$$-a\omega\,(d\sigma_i/d\omega) = -\{1/(\psi_e - \psi_0)\}\,d/d\omega\,(j_i/m_i) + R_i/\rho$$
$$(i = 1, 2, \ldots, N) \tag{7.14a}$$

$$-a\omega\,(dh/d\omega) = -\{1/(\psi_e - \psi_0)\}\,d/d\omega\,(Q_D + Q_T) \tag{7.15a}$$

The subscript "e" here refers to the free stream side of the flame. The boundary conditions when a is positive are

$$\begin{aligned} &\omega = 0, 1 \qquad d\sigma_i/d\omega = 0, \qquad dh/d\omega = 0 \\ &\omega = 1, \qquad \sigma_i = \sigma_{ie}, \qquad h = h_e(T_e, \sigma_{ie}) \end{aligned} \tag{7.16}$$

In order to obtain a solution it is convenient to retain the steady state flow velocity as in Eqs. (7.14a) and (7.15a), but otherwise to convert to the time-dependent forms

$$\partial\sigma_i/\partial t - a\omega\,(\partial\sigma_i/\partial\omega) = -\{1/(\psi_e - \psi_0)\}\,\partial/\partial\omega\,(j_i/m_i) + R_i/\rho$$
$$(i = 1, 2, \ldots, N) \tag{7.17}$$

$$\partial h/\partial t - a\omega\,(\partial h/\partial\omega) = -\{1/(\psi_e - \psi_0)\}\,\partial/\partial\omega\,(Q_D + Q_T) \tag{7.18}$$

For any fixed value of the velocity gradient a, the Eulerian steady state solution is then obtained by a parallel approach to that outlined in Section 4. The entrainment formulas (4.72) and (4.73) are no longer required. Since the flame now moves itself into a grid location appropriate to the velocity gradient, the interval $(\psi_e - \psi_0)$ must be chosen so as to allow this to occur. It also becomes necessary during the computation to employ adaptive gridding so as to retain spatial resolution in the flame regions of high reactivity [cf. Section 4.4(a)].

(b) Counterflow diffusion flames

Compared with the situation outlined in Section 7.2(a), the experimental arrangement of Fig. 12 has the added feature of a viscous boundary layer in the mixing region near the cylinder wall. Although the first of Eqs. (7.9) is still valid in the free stream region outside the boundary layer (subscript e), that is

$$v_{x,e} = ax \tag{7.9a}$$

the second of these equations requires modification everywhere to allow for the viscous effects and the variable density. The thermodynamic properties are constant in the free stream and at the cylinder wall.

The continuity equations for mass, x-direction momentum, chemical species and energy in the plane, stationary, laminar boundary layer flow have already been given as Eqs. (7.1) to (7.4). The stream function ψ, by means of which the mass continuity equation is automatically satisfied, is defined by Eqs. (7.5). Following the approaches of Lees (1956), Fay and Riddell (1958), and Chung (1965), self-similar solutions in the stagnation region are obtained via transformations from (x, y) co-ordinates to the two new variables

$$\xi = \{v_{x,e}/(2s)^{1/2}\} \int_0^y \rho dy \tag{7.19}$$

$$s = \int_0^x (\rho\eta)_e v_{x,e} dx \tag{7.20}$$

Then, defining a non-dimensional stream function

$$f(\xi, s) = \psi/(2s)^{1/2} \tag{7.21}$$

Eqs. (7.5) lead to

$$f_\xi = v_x/v_{x,e} \tag{7.22}$$

$$\rho v_y = -[(2s)^{1/2} f_\xi(\partial\xi/\partial x) + \{(2s)^{1/2} f_s + (2s)^{-1/2} f\}(ds/dx)] \tag{7.23}$$

where the subscripts ξ and s denote partial differentiation with respect to these variables. Since also $\partial v_{x,e}/\partial y = 0$, then from Eq. (7.2)

$$\begin{aligned} dP/dx &= -\rho_e v_{x,e}(\partial v_{x,e}/\partial x) \\ &= -\rho_e v_{x,e}(\partial v_{x,e}/\partial s)(ds/dx) \end{aligned} \tag{7.24}$$

and Eqs. (7.2) to (7.4) transform from (x, y) to (ξ, s) co-ordinates as follows

$$(C f_{\xi\xi})_\xi + f f_{\xi\xi} + (2s/v_{x,e})(dv_{x,e}/ds)\{\rho_e/\rho - (f_\xi)^2\} = 2s(f_\xi f_{\xi s} - f_s f_{\xi\xi}) \tag{7.2a}$$

$$\begin{aligned} f(\sigma_i)_\xi - \{(2s)^{1/2}/(ds/dx)\}(j_i/m_i)_\xi + 2sR_i/\{\rho v_{x,e}(ds/dx)\} \\ = 2s\{f_\xi(\sigma_i)_s - f_s(\sigma_i)_\xi\} \end{aligned} \tag{7.3a}$$

$$f h_\xi - \{(2s)^{1/2}/(ds/dx)\}(Q_D + Q_T)_\xi = 2s(f_\xi h_s - f_s h_\xi) \tag{7.4a}$$

where $C = (\rho\eta)/(\rho\eta)_e$, radiative losses are neglected ($q_E = 0$), and the contributions of the kinetic energy (or pressure gradient) and viscous terms in Eq. (7.4) are also neglected. The terms which contain s-direction derivatives other than $dv_{x,e}/ds$ have been collected on the right-hand sides of the equations. These right-hand sides all become zero at the stagnation point where $s = 0$. By use of Eqs. (7.9a) and (7.20), it can also be shown that in the stagnation region

$$
\begin{aligned}
(2s/v_{x,e})(dv_{x,e}/ds) &= 1 \\
(2s)^{1/2}/(ds/dx) &= \{(\rho\eta)_e a\}^{-1/2} \\
2s/\{v_{x,e}(ds/dx)\} &= 1/a
\end{aligned}
\tag{7.25}
$$

Exact ordinary differential equations may therefore be set up for the variables σ_i, h and $df/d\xi$ (or f' in the equations below) in the neighborhood of the stagnation point, independently of the magnitudes of the chemical reaction rates R_i. Putting $V = -f$ and $\omega = (\xi - \xi_w)/(\xi_e - \xi_w)$ (where the subscript w refers to conditions at the wall), and remembering also the relations (7.25), the complete set of equations becomes

$$dV/d\omega + (\xi_e - \xi_w)f' = 0 \tag{7.26}$$

$$V(df'/d\omega) = \{1/(\xi_e - \xi_w)\}\, d/d\omega\, \{C(df'/d\omega)\} + (\xi_e - \xi_w)\{\rho_e/\rho - (f')^2\} \tag{7.2b}$$

$$V(d\sigma_i/d\omega) = -\{(\rho\eta)_e a\}^{-1/2}\, d/d\omega\,(j_i/m_i) + (\xi_e - \xi_w)R_i/(a\rho) \quad (i = 1, 2, \ldots, N) \tag{7.3b}$$

$$V(dh/d\omega) = -\{(\rho\eta)_e a\}^{-1/2}\, d/d\omega\,(Q_D + Q_T) \tag{7.4b}$$

It further follows from Eq. (7.23) that

$$V = \rho v_y/\{(\rho\eta)_e a\}^{1/2} \tag{7.27}$$

The transport fluxes j_i, Q_D and Q_T in Eqs. (7.2b) to (7.4b) are entirely in the y-direction of coordinates. The differential equations are to be solved subject to the boundary conditions (7.28) at the cylinder wall, and (7.29) in the free stream

$$\omega = 0, f' = 0, V = V_w, T = T_w, V_w(G_{i,w}/m_i - \sigma_i) = \{(\rho\eta)_e a\}^{-1/2}(j_i/m_i)_w \tag{7.28}$$

$$\omega = 1, f' = 1, T = T_e, \sigma_i = \sigma_{i,e}, \quad \text{zero gradients of all dependent variables except } V \tag{7.29}$$

Here $G_{i,w}$ denotes the mass fraction of the ith component in the input stream from the cylinder.

Since neither C nor ρ_e/ρ varies strongly with mixture composition, there is no strong direct coupling between the momentum equation (7.2b) and the species equations. Consequently, for purposes of practical solutions, it has been found convenient to uncouple the momentum from the species and energy equations and to solve it by a damped Newton method (Section 6.1) running in parallel with the time-dependent approach for the remaining conservation equations. For a fixed value of the velocity gradient a, convergence of f' and V is rapid, to profiles consistent with the prevailing temperatures and densities. However, application of the Newton method to the solution of the species and energy equations is not straightforward, and at present the time-dependent approach is probably to be preferred. The appropriate time-dependent relations to replace Eqs. (7.3b) and (7.4b) are

$$(1/a)\,\partial\sigma_i/\partial t + V(\partial\sigma_i/\partial\omega) = -\{(\rho\eta)_e a\}^{-1/2}\,\partial/\partial\omega\,(j_i/m_i) + (\xi_e - \xi_w)\,R_i/(a\rho) \qquad (i = 1, 2, \ldots, N) \quad (7.30)$$

$$(1/a)\,\partial h/\partial t + V(\partial h/\partial\omega) = -\{(\rho\eta)_e a\}^{-1/2}\,\partial/\partial\omega\,(Q_D + Q_T) \quad (7.31)$$

The Eulerian steady state solution to these equations may be obtained by a parallel approach to that outlined in Sections 4 and 7.2(a), with the velocities constrained by Eqs. (7.2b) and (7.26) as above. Equation (7.26) must be integrated on the assumption of linear variation of f' with ω between nodes so as to evaluate the reduced velocities V at the cell boundaries.

7.3. Multidimensional flows

The still more complex two- and three-dimensional flows in practical combustion systems can only be properly described by the full Navier-Stokes and other conservation equations, including all the distance derivatives of Eqs. (2.2) in the flux divergence term of the continuity equation (2.1). One problem in solving them is the rapid escalation of computing costs as the number of dimensions is increased, caused by the increased number of grid points, the correspondingly increased size of the set of Eqs. (4.65), and the increased complexity of the matrix elimination procedure required to treat the multidimensional spatial coupling. In Lagrangian reference frames there is also the problem of possible grid distortion, while when recirculating flow is present yet another problem arises since the system of equations becomes "elliptical" in nature instead of parabolic. It is not then possible to identify a marching coordinate as can be done with parabolic systems. A review of the inherent problems and of the techniques available for the solution has been given by McDonald (1979). However, because of the difficulties outlined, implicit methods have not yet been used on multidimensional flows with detailed chemistry.

Grid distortion problems are probably best treated by the use of triangular instead of rectangular surface elements, and of tetrahedral instead of cubic elements in three dimensions. Following Crowley (1971), Fritts, Oran, and Boris (1981) developed a Lagrangian finite-difference method based on a variable triangular mesh that can be adjusted to suit local flow conditions. While keeping the numbers of both triangles and nodes (or triangle vertices) in the mesh constant, the code allows connections between nodes to be adjusted as the calculation proceeds so that each node remains connected only to nearest neighbors. A procedure is also available by which the spatial resolution in any part of the flow field may be altered if necessary.

The treatment of multidimensional reactive flows by finite-element methods, as described in Section 6.2 but using surface piecewise polynomial approximations over triangular, quadrilateral, or rectangular elements, may also show promise, but it is still in its infancy. For the background principles of the finite-element method the interested reader is referred to the monograph by Strang and Fix (1973), and, for surface fitting, to articles by Powell (1974), Schumaker (1976), Barnhill (1977), and Lawson (1977).

8. Operator splitting techniques in multidimensional systems

Because of the computational complexity of the full block implicit approach outlined in Section 4, it is natural that there have been attempts at simplification either (i) by uncoupling the equations governing the separate species in a chain-reaction mechanism or (ii) by avoiding simultaneous implicit treatment of both the fluid mechanics and the chemistry. We have already in Section 4.4(b) encountered the former approach to uncoupling the Eulerian system of equations. However, it is not possible to uncouple the Lagrangian system in a similar manner, and here the approach of directly uncoupling the implicit treatments may offer an alternative. This is the basis of the operator-splitting approach of Yanenko (1971). If, for a one-dimensional Cartesian system, appropriate expressions for the diffusive fluxes are inserted into the Lagrangian equations (2.28) to (2.31), they assume the general form

$$D\phi_i/Dt = \sum_{k=1}^{N+1} a_{ik}(y, t)(\partial^2 \phi_k/\partial y^2) + R_i(y, t, \mathbf{\Phi}) \tag{8.1}$$

The first term on the right-hand side takes care of the transport processes, and the second term is the chemical source term. The operator-splitting approach treats these two terms separately and implicitly, and then combines the results explicitly to give, for the time interval $n'\delta t \rightarrow (n' + 1)\delta t$

$$\begin{aligned} (\phi_i^* - \phi_i^{n'})/\delta t &= \sum_{k=1}^{N+1} a_{ik}(\partial^2 \phi_k^*/\partial y^2) \\ (\phi_i^{n'+1} - \phi_i^*)/\delta t &= R_i(\mathbf{\Phi}^{n'+1}) \end{aligned} \tag{8.2}$$

The approach has the advantage that the separate solutions are simpler than for the coupled system. The final step requires solution as for a homogeneous system, with the chemical source terms being solved by a standard implicit package such as Gear's method.

The satisfactory functioning of the operator-splitting technique depends on the coupling between the transport and the chemical terms not being too strong, since it neglects cross-perturbation effects. It must be of particular concern that the overall change from any $\phi_i^{n'}$ to $\phi_i^{n'+1}$ is not a small difference which results from summing two large effects of opposite sign. In that case it is necessary to reduce the time step so as to keep the change produced by each separate operator within bounds. The method may well meet difficulty in handling flamelike problems where the overall effect depends specifically on the interplay between the two operators.

The operator-splitting approach is likely to be most useful in the treatment of systems having more than one spatial dimension, for example, two-dimensional, time-dependent, boundary layer flow, where the governing equations are of the form

$$D\phi_i/Dt = \sum_{k=1}^{N+1} a_{ik}(x, y, t)\{(\partial^2\phi_k/\partial x^2) + (\partial^2\phi_k/\partial y^2)\} + R_i(x, y, t, \mathbf{\Phi}) \quad (8.3)$$

The "alternating direction" methods, reviewed by Douglas and Gunn (1964), solve this by an operator-splitting technique as follows

$$\frac{(\phi_i^* - \phi_i^{n'})}{(\delta t/2)} = \sum_{k=1}^{N+1} a_{ik}\{(\partial^2\phi_k^*/\partial x^2) + (\partial^2\phi_k^{n'}/\partial y^2)\} + R_i(x, y, t, \mathbf{\Phi}^*)$$

$$\frac{(\phi_i^{n'+1} - \phi_i^*)}{(\delta t/2)} = \sum_{k=1}^{N+1} a_{ik}\{(\partial^2\phi_k^*/\partial x^2) + (\partial^2\phi_k^{n'+1}/\partial y^2)\} + R_i(x, v, t, \mathbf{\Phi}^{n'+1}) \quad (8.4)$$

where the implicit solutions are reduced to the tridiagonal elimination procedure characteristic of one-dimensional systems.

9. Chemical quasi-steady-state and partial equilibrium assumptions in reactive flow modeling

An alternative to the implicit method for the solution of "stiff" chemical kinetic systems is the introduction of the quasi-steady-state or partial equilibrium conditions in place of one or more of the radical continuity equations. Both conditions imply rapid rates of at least one of the elementary reactions forming and removing the intermediate product concerned, the situations which lead to stiffness. In numerical terms the procedure is essentially a zero-order asymptotic approximation for the concentration of a labile species present in small quantities. The particular condition to be used depends on the time scale of the kinetic phenomenon to be investigated. The quasi-steady-state condition is used effectively to approximate the *distribution* of radicals within a reacting system. It may be used at times which are long compared with the time scales responsible for this distribution. The overall *size* of the radical pool is determined by kinetic factors. In most of the more common flames of the hydrocarbon-hydrogen-carbon monoxide-oxygen system, the quasi-steady-state conditions relax into the partial equilibrium conditions at times which are long compared with the time scale for the initial removal of fuel or molecular oxygen. This relaxation occurs early in the radical recombination regions of the flames, where it has been found both theoretically and experimentally that H, O, OH, and O_2 are effectively equilibrated among themselves to provide a "partial equilibrium" by reactions (i), (ii), (iii), and (xviii)

$$OH + H_2 \rightleftharpoons H_2O + H \tag{i}$$

$$H + O_2 \rightleftharpoons OH + O \tag{ii}$$

$$O + H_2 \rightleftharpoons OH + H \tag{iii}$$

$$OH + OH \rightleftharpoons O + H_2O \tag{xviii}$$

(The numbering of these reactions follows a scheme used frequently in the literature; see Dixon-Lewis and Williams, 1977.)

In the main reaction zones of flames, the radical concentrations are above their values at full equilibrium. In hydrogen-oxygen-supported flames the decay of the pool of radicals towards full equilibrium occurs by slower recombination steps

$$H + O_2 + M \rightleftharpoons HO_2 + M \tag{iv}$$

$$H + H + M \rightleftharpoons H_2 + M \tag{xv}$$

$$H + OH + M \rightleftharpoons H_2O + M \tag{xvi}$$

$$H + O + M \rightleftharpoons OH + M \tag{xvii}$$

together with additional bimolecular reactions

$$H + HO_2 \rightleftharpoons OH + OH \tag{vii}$$

$$H + HO_2 \rightleftharpoons O + H_2O \tag{viia}$$

$$H + HO_2 \rightleftharpoons H_2 + O_2 \tag{xii}$$

$$OH + HO_2 \rightleftharpoons H_2O + O_2 \tag{xiii}$$

$$O + HO_2 \rightleftharpoons OH + O_2 \tag{xiv}$$

It is these steps which, subject to a distribution of radicals by the partial equilibria (i), (ii), (iii), and (xviii), exercise kinetic control over the rate of final equilibration. In flames which contain carbon monoxide and carbon dioxide in the burnt gas, reaction (xxi) may also be equilibrated, and reactions (xxii) and (xxiii) may contribute kinetically

$$OH + CO \rightleftharpoons CO_2 + H \tag{xxi}$$

$$O + CO + M \rightleftharpoons CO_2 + M \tag{xxii}$$

$$H + CO + M \rightleftharpoons HCO + M \tag{xxiii}$$

The incorporation of partial equilibrium (or quasi-steady-state) conditions into a reactive flow calculation thus necessitates consideration specifically of the overall size of the radical pool and the distribution of species within it. The overall size is determined kinetically by explicit rate processes, and the species distribution is determined by the new conditions. The approach involves the use of "composite fluxes" in the manner described by Dixon-Lewis, Goldsworthy, and Greenberg (1975). The technique has been used by Dixon-Lewis *et al.* (1979, 1981) for studies of radical recombination in flames and for the study of hydrogen-air flame structures. The one-dimensional flame equations are integrated in the Stefan-Maxwell form which, with detailed transport flux formulation, gives the relations [cf. Eqs. (2.12), (2.13), (3.64), (2.22), and (3.38a)]

$$dG_i/dy = q_i/M_y = m_i R_i/M_y \qquad (i = 1, 2, \ldots, N) \tag{9.1}$$

$$dX_i/dy = \sum_{j=1}^{N} \left\{ \frac{X_i X_j (1 - \Delta_{ij})}{[\mathscr{D}_{ij}]_1} \frac{M_y}{\rho_{\text{molar}}} \left(\frac{G_j}{m_j X_j} - \frac{G_i}{m_i X_i} \right) \right\} - (5/2T)(\mathbf{L}^{01}\mathbf{K}\mathbf{R}_1^1)_i \, (dT/dy) \qquad (i = 1, 2, \ldots, N) \tag{9.2}$$

$$dT/dy = (M_y/\lambda_\infty) \sum_{i=1}^{N} \left\{ (G_i h_i - G_{iu} h_{iu}) + (5/2)RT \sum_{k=1}^{N} \sum_{r,s}^{M}{}' X_i (\mathbf{K}\mathbf{L}^{10})_{ik} \frac{G_k - w_k}{m_k X_k} \right\} \tag{9.3}$$

where $\sum_{r,s}^{M}{}'$ means that terms with $rs = 00$ are omitted,

$$\Delta_{ij} = (25P/16T)([\mathscr{D}_{ij}]_1/X_i X_j)(\mathbf{L}^{01}\mathbf{K}\mathbf{L}^{10})_{ij} \tag{9.4}$$

$$\lambda_\infty = \lambda_{\infty,\,\text{tr}} + \lambda_{\infty,\,\text{int}}$$

$$\lambda_{\infty,\,\text{tr}} = -4 \sum_{i=1}^{N} X_i (\mathbf{K}\mathbf{R}_1^1)_{i10} \tag{9.5}$$

$$\lambda_{\infty,\,\text{int}} = -4 \sum_{i=1}^{N} X_i (\mathbf{K}\mathbf{R}_1^1)_{i01}$$

and, referring to Eqs. (3.49), the matrix $\mathbf{L}$ is partitioned into the four submatrices

$$\begin{array}{c|c:c} \mathbf{L}^{00,00} & \mathbf{L}^{00,10} & \mathbf{L}^{00,01} \\ \hline \mathbf{L}^{10,00} & \mathbf{L}^{10,10} & \mathbf{L}^{10,01} \\ \hdashline \mathbf{L}^{01,00} & \mathbf{L}^{01,10} & \mathbf{L}^{01,01} \end{array} \tag{9.6}$$

comprising an $N \times N$ order matrix $\mathbf{L}^{00}$ with elements $L_{ij}^{00.00}$ only; an $\{(M-1)N - N^*\} \times N$ order matrix $\mathbf{L}^{01}$ with elements $L_{ij}^{00,mn}$; an $N \times \{(M-1)N - N^*\}$ order matrix $\mathbf{L}^{10}$ with elements $L_{ij}^{rs,00}$; and an $\{(M-1)N - N^*\} \times \{(M-1)N - N^*\}$ order matrix $\mathbf{L}^{11}$ with elements $L_{ij}^{rs,mn}$, where neither rs nor mn are 00. Here M is defined immediately following Eq. (3.22) and N^* is the number of monatomic components in the N-component mixture. The absence of both inelastic-collision effects and diffusion of internal energy in the case of monatomic components means that the corresponding terms are omitted from the overall set of relations [cf. Eqs. (3.49b) and (3.50)].

The vector $\mathbf{R}^1$ which represents the right-hand sides of Eqs. (3.25a) may be similarly partitioned. The subvector $\mathbf{R}_1^1$ then contains the elements R_{irs}^1 where $rs \neq 00$. The matrix $\mathbf{K}$ in Eqs. (9.5) is the inverse of $\mathbf{L}^{11}$. The final terms in Eqs. (9.2) and (9.3) are due to thermal diffusion and its reciprocal effect, the Dufour effect.

If inelastic-collision effects are entirely neglected [cf. Section 3.3(a)], then the matrices $\mathbf{L}^{01}$, $\mathbf{L}^{10}$ and $\mathbf{L}^{11}$ may be truncated so that *rs* and *mn* take only the value 10 [cf. Eqs. (3.49b)]. $\lambda_{\infty,\mathrm{int}}$ must then be assigned the value

$$\lambda_{\infty,\mathrm{int}} = \frac{P}{T} \sum_i \frac{c_{i,\mathrm{int}} X_i}{k_B \left(\sum_k X_k / \mathscr{D}_{i,\mathrm{int},k} \right)} \tag{3.51}$$

Equations (9.1), (9.2), and (9.3) are ordinary differential equations in which distance is the independent variable. The technique of integration is to start from a perturbed full equilibrium condition at the hot boundary of the flame and integrate backwards across the flame by an explicit method. Dixon-Lewis *et al.* (1979a,b; 1981) used a fourth-order Runge-Kutta procedure with variable step size for this purpose. We continue here by reviewing briefly the application of the method with both partial equilibrium and quasi-steady-state assumptions.

(a) Partial equilibrium assumptions

The partial equilibrium assumptions by themselves in conjunction with Eqs. (9.1), (9.2), and (9.3), and a reaction mechanism as outlined above, do not permit construction of a complete model from which an eigenvalue burning velocity and full profiles may be computed *ab initio*. On the other hand the assumptions are extremely useful when dealing with H–N–C–O flame systems, since their application to reactions (i), (ii), (iii), and (xviii) above allows us to calculate many of the species profiles, and particularly the free radical profiles, on close approach to full equilibrium. The time-dependent computation does not economically do this directly. The computations require an input mass flux or burning velocity which must be either a measured or a separately calculated value. For composite flux calculation purposes the overall radical pool is chosen so as to represent a total flux of free electron spins, that is, spins belonging to H, O, OH, *and* O_2 (Dixon-Lewis *et al.*, 1975).

Flame profiles based on partial equilibrium assumptions are comparatively easy to compute, with a single integration through the zone being sufficient. The initial values of the dependent variables must represent a partial equilibrium condition which is a perturbation of full equilibrium (Dixon-Lewis and Greenberg, 1975). Such computations were used by Dixon-Lewis (1979) in connection with the analysis of measured radical concentration profiles in both fuel-rich and fuel-lean hydrogen–oxygen–nitrogen flames. In the same paper, partial equilibrium profiles for H, O, and OH in the burnt gas regions of the whole series of hydrogen-air flames at atmospheric pressure are compared with profiles which result from complete flame calculations with the application of the quasi-steady-state assumptions. The comparison allows an assessment of the range of validity of the partial equilibrium assumptions and contributes also to detailed understanding of the flame mechanism and of such matters as radical overshoot. A similar comparison of partial equilibrium and quasi-steady-state profiles in a 60% hydrogen-air flame inhibited by the addition of 4% hydrgoen bromide showed reaction (xxiv) to be more or less equilibrated over virtually the whole flame (Dixon-Lewis, 1979b).

$$H + HBr \rightleftharpoons Br + H_2 \tag{xxiv}$$

Dixon-Lewis and Simpson (1977) also investigated computationally the sensitivity of the inhibitory effect of HBr to the equilibrium constant assumed for reaction (xxiv) by reducing the equilibrium constant so as to conform with the analogous reaction of HCl. The inhibitory effect of the halogen acid was drastically reduced, thus confirming the importance of the thermodynamic (equilibrium) effects.

In order to examine the achievement of partial equilibrium in hydrocarbon combustion products and other carbon-monoxide-containing gases, it is necessary to consider also the equilibration of reaction (xxi). Cherian *et al.* (1981b) examined several hydrogen–carbon monoxide–air flames and found that on the fuel-rich side of stoichiometric the water-gas equilibrium [via reactions (i) and (xxi)] is not achieved in the burnt gas when the flame temperature is below about 1500 K. Figures 13 and 14 show the profiles of the stable species mole fractions in two hydrogen–carbon monoxide–air flames, one of which has a final adiabatic temperature of about 1650 K and the other about 1350 K. The existence of water overshoots in the two flames and the failure to achieve the water-gas equilibrium in the second case are clearly illustrated.

As a final example of the utility of the partial equilibrium approach, it is interesting to consider the formation of nitric oxide by the fixation of atmospheric nitrogen in the burnt gas region of a one-dimensional flame. The fixation occurs by the three reactions

$$O + N_2 \rightleftharpoons NO + N \tag{xxv}$$

$$N + O_2 \rightleftharpoons NO + O \tag{xxvi}$$

$$N + OH \rightleftharpoons NO + H \tag{xxvii}$$

which form the extended Zeldovich mechanism. The rate-controlling step is reaction (xxv), which has an activation energy near 315 kJ/mol. The nitric oxide formation is therefore quite slow even at 2000 K or so, and can conveniently be studied in the burnt gas region of a hot flame, with

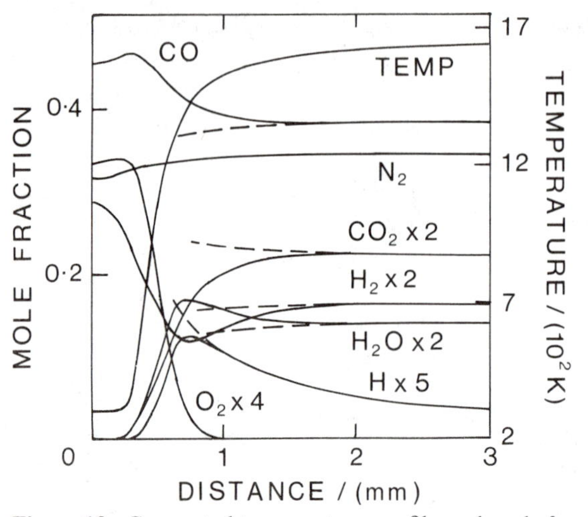

Figure 13. Computed temperature profile and mole fraction profiles of stable species and hydrogen atoms in 60% (24.1% hydrogen + 75.9% carbon monoxide) + 40% air flame at atmospheric pressure and $T_u = 298$ K (from Cherian *et al.*, 1981b). Solid lines: time-dependent flame calculation; broken lines: calculation with partial equilibrium assumptions on reactions (i), (ii), (iii), and (xxi). (By courtesy of The Royal Society.)

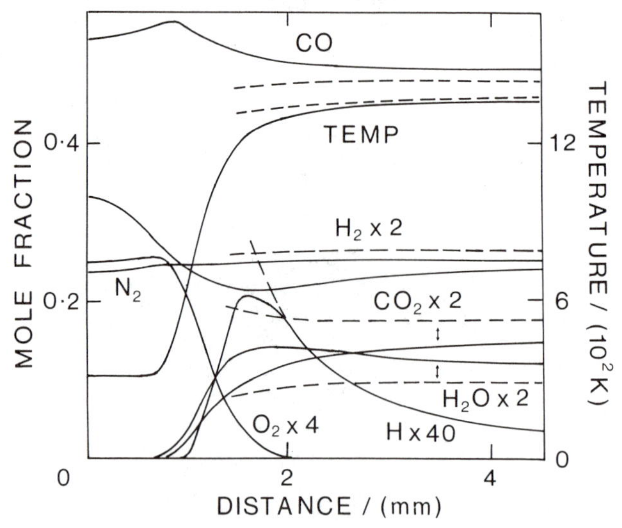

Figure 14. Computed temperature profile and mole fraction profiles of stable species and hydrogen atoms in 70% (24.1% hydrogen + 75.9% carbon monoxide) + 30% air flame at atmospheric pressure and $T_u = 298$ K (Cherian *et al.*, 1981b). Solid lines: time-dependent flame calculation; broken lines: calculation with partial equilibrium assumptions on reactions (i), (ii), (iii), and (xxi). (By courtesy of The Royal Society.)

computational analysis by means of an oxygen atom profile calculated with the aid of partial equilibrium assumptions (Dixon-Lewis, to be published). Because of the high sensitivity of the reaction rate to the temperature, allowance must be made in the computation for heat losses to the surroundings, achieved by means of the term q_E in Eqs. (2.19) and (2.21).

(b) Quasi-steady-state assumptions

In contrast to partial equilibrium computations, assumption of quasi-steady-state conditions allows one to calculate complete flame properties. In the case of a hydrogen-oxygen or hydrocarbon-oxygen reaction mechanism the technique involves separating molecular oxygen from the radical pool consisting of H, O, and OH; and the start of an integration through a flame now involves a quasi-steady-state perturbation of a partial equilibrium condition near the hot side of the reaction zone. The size of the perturbation (of the molecular oxygen concentration) must be related to the size of the remaining "pool" in such a way that the boundary conditions at the cold side of the flame are satisfied at the end of the integration. Because of mathematical instability of the equations themselves, the solution has to be constrained precisely at all stages of the integration. Additionally, the final solution requires trial and error adjustment of the input mass flux M_y in order to find its eigenvalue, which is determined by the constraint that the space integral rates of both reactants and products at the end of the integration must be consistent with the input value of M_y. Finding the final solution thus becomes a complex boundary value problem; and because of computational cost, it is necessary to define the problem so that there is not more than one missing condition at the starting hot boundary. This confines the method to hydrogen–oxygen and other very simple kinetic systems.

Even for hydrogen–oxygen flames the solution of the stationary flame equations with quasi-steady-state assumptions to relate the concentrations of H, O, OH and HO_2 is tedious. Before the development of the approach described in Section 6.2 there was, however, still a compensating advantage in using the quasi-steady-state method, namely, the fourth-order accuracy of the Runge-Kutta integration technique. The higher precision of the integration serves as a check on the properties calculated by the implicit approach described in Section 4. Table 4 compares the results of the computations on several hydrogen–air flames by the two methods. Despite the lower, first-order accuracy of the implicit method, the results obtained by the two approaches agree closely.

The composite flux method with quasi-steady-state assumptions has been used in studies of hydrogen–oxygen–nitrogen flames (Dixon-Lewis, 1979a) and of a hydrogen–air flame inhibited by hydrogen bromide (Dixon-Lewis, 1979b). However, the same order of accuracy is now obtainable by the finite-element technique described in Section 6.2.

Table 4. Computed properties of atmospheric pressure hydrogen–air flames having $T_u = 298$ K, comparing results of (a) stationary flame Runge-Kutta integration with quasi-steady-state assumptions and (b) time-dependent finite-difference solution without quasi-steady-state assumptions (from Dixon-Lewis, 1979a)

	Stationary flame solution				Time-dependent solution			
$(si)_H$/nm =	0.35		0.30		0.35		0.30	
$X_{H_2,u}$	S_u (cm/s)	$10^2 X_{H,max}$	S_u (cm/s)	$10^2 X_{H,max}$	S_u (cm/s)	$10^2 X_{H,max}$	S_u (cm/s)	$10^2 X_{H,max}$
0.70	78	1.25	81	1.22	77	1.26	80	1.23
0.50	229	6.18	—	—	—	—	240	6.09
0.41	254	7.52	265	7.45	—	—	267	7.46
0.30	200	4.63	209	4.66	200	4.69	211	4.65
0.20	87	1.02	90	1.01	—	—	89	1.07

10. Concluding remarks

This chapter has been written primarily as an introduction to the present state of the art of reactive flow modeling. It should guide the reader, after some effort, to appreciate the relevant literature, use or improve existing modeling codes, or write new ones. The basic hydrodynamic conservation equations and the expressions for the transport fluxes were presented and the finite-difference solution of the time-dependent equations for the one-dimensional laminar flame described in some detail. The most general transport flux model has been retained on the basis that it is easier to simplify than to construct. Conversion to more approximate formulations of transport fluxes, for example, those of Section 3.4, presents no difficulty where these are felt to be more appropriate.

Other methods of solution and computations with more complex geometries have been described only briefly. An important topic which has not been included is turbulent reactive flows. Because of the uncertainities and complexities in the detailed representation of both the aerodynamics of such flows *and* the chemistry, their treatment is very much more speculative or "global" in approach than that of the simpler flow systems dealt with here. They fall into the general category of recirculating flows (see Section 7.3), and they must be solved by Gauss–Seidel or similar iterative techniques. These flows had to remain outside the scope of the present development.

11. Nomenclature

Notation confined to small sections of the text is occasionally omitted. Also, when appropriate, limitations of usage of defined symbols are given in parentheses.

Symbol	Meaning	Equation of first mention
a	defined in Eq. (4.14)	(4.12)
a	velocity gradient (Section 7.2)	(7.8)
A, A_+, A_-	stream tube area	(2.4)
A^*_{ij}	ratio of reduced collision integrals (Section 3)	(3.46)
b	defined by Eq. (4.15)	(4.12)
B^*_{ij}	ratio of reduced collision integrals (Section 3)	(3.46)
$\mathbf{B}_{j'}$	coefficient matrix for $\delta\mathbf{\Phi}_{j'-1}$ in finite-difference equations	(4.55)
$\mathscr{B}_{iI}$	tensor function	(3.9)
B_{ik}	element of $\mathbf{B}_{j'}$	(4.56)

Symbol	Meaning	Equation of first mention
$B_{i,k,s}$	ith normalized B spline of order k for knot sequence **s** (Section 6.2)	(6.3)
$c_{i,\text{int}}$	internal specific heat per molecule of ith species (Section 3.1)	(3.22a)
$c_{i,\text{rot}}$	rotational specific heat per molecule of ith species (Section 3.1)	(3.49)
c_p, c_v	specific heat per unit mass	(4.4)
C_{ij}^*	ratio of reduced collision integrals (Section 3)	(3.46)
C_p, C_v	molar specific heat	(3.58)
$C_{i,\text{int}}$	molar internal specific heat of ith species	
$C_{i,\text{rot}}$	molar rotational specific heat of ith species	
$\mathbf{C}_{j'}$	coefficient matrix for $\delta\mathbf{\Phi}_{j'}$ in finite-difference equations	(4.55)
C_{ik}	element of $\mathbf{C}_{j'}$	(4.56)
$C_n^{(i)}(t)$	time-dependent coefficient in B-spline function relating to nth species	(6.1)
d	defined in Eq. (3.10); at constant pressure it is the mole fraction gradient vector (Section 3.1)	(3.9)
d	rate terms in generalized conservation equation (Section 4 *et seq.*)	(4.19)
D_{ij}	multicomponent diffusion coefficient [Section 3 and Eq. (4.3)]	(3.32)
D_{im}	diffusion coefficient of trace species i in mixture, defined by Eq. (3.69) (Section 3.4)	(3.68)
D_{ik}	element of $\mathbf{D}_{j'}$ (Section 4.2(b) *et seq.*)	(4.56)
$\mathbf{D}_{j'}$	coefficient matrix for $\delta\mathbf{\Phi}_{j'+1}$ in finite-difference equations	(4.55)
D_i^T	multicomponent thermal diffusion coefficient	(3.32)
$(eps)_{ij}$	Lennard-Jones (12 : 6) potential parameter; single suffix denotes pure component	(3.44)
$(eps)_0$	Stockmayer potential parameter for polar molecule (Section 3)	(3.54)

Symbol	Meaning	Equation of first mention
E	molecular internal energy	(3.2)
f	distribution function for molecular velocity (Section 3.1)	(3.1)
$f^{(0)}$	equilibrium distribution function (Section 3.1)	(3.6)
f	non-dimensional stream function defined by Eq. (7.21) (Section 7.2)	(7.21)
$\mathbf{F}$	flux, with subscript to denote particular conserved quantity	(2.1)
F_x, F_y, F_z, F_r	flux component; additional nondirectional subscript denotes a particular conserved quantity	(2.2)
F_u, F_b	factors controlling entrainment rates at u and b boundaries	(4.72)
$\mathbf{g}$, $\mathbf{g}'$	relative molecular velocities before and after a binary collision, $\mathbf{g} \equiv \mathbf{v} - \mathbf{v}_1$ (Section 3.1)	(3.5)
G_i	mass flux fraction of ith species	(2.11a)
G	defined by Eq. (4.24) (Section 4)	(4.24)
h	specific enthalpy; with subscript for pure component	(2.19a)
H	molar enthalpy; with subscript for pure component	(4.17)
$H_{ij}^{rs,mn}$	element of coefficient matrix for viscosity calculation, computed according to Eq. (3.52)	(3.52)
I	differential scattering cross section (Section 3.1)	(3.5)
$\mathbf{j}$	diffusive mass flux vector	(3.35)
j_i	diffusive mass flux of ith species in y direction	(2.11)
j'	integer defining spatial position in finite difference (ω, t) grid	(4.28)
$J_{\text{tot}\,i,b}$, $J_{\text{tot}\,i,u}$	total flux of ith species across boundary	(4.62)
$\delta J_{i,b}$, $\delta J_{i,u}$	correction to $J_{\text{tot}\,i}$ to allow for catalytic formation of ith species at boundary surface	(4.62)
k	reaction-rate coefficient (Section 2.3)	(2.14)
k	order of B spline or B-spline function (Section 6.2)	(6.3)

Symbol	Meaning	Equation of first mention
k_B	Boltzmann constant	(3.6)
K	equilibrium constant (Section 2.3)	(2.14)
L_+	convective flux across (+) boundary, defined by Eq. (4.25)	(4.25)
L_-	convective flux across (−) boundary, defined by Eq. (4.26)	(4.26)
L_0	convective flux $(M_yA)_b$ across b boundary of finite-difference grid ($\omega = 0$)	(4.57)
L_{M+2}	convective flux $(M_yA)_u$ across u boundary of finite-difference grid ($\omega = 1$)	(4.59a)
$L_{ij}^{rs,mn}$	element of coefficient matrix used in computation of diffusion, thermal conduction, and thermal diffusion coefficents, defined by Eqs. (3.49)	(3.23)
m_i	molar mass of ith species	(2.8)
m_G	average molar mass	(4.3)
m_i'	molecular mass of ith species	(3.2)
m	number of collocation points (Section 6.2)	
M	number of finite-difference intervals in grid, giving $M + 3$ nodes numbered $0, 1, \ldots, M + 2$	
M_y	overall mass flux in y direction	(2.5)
n	molecular number density (Section 3.1)	(3.1)
n'	integer defining temporal position in finite-difference (ω, t) grid	
N	total number of chemical components in system	(2.8)
N^*	number of monatomic components (Section 9)	
P, P'	scalar pressure; tensor notation $\mathscr{P}$ is only shown following Eq. (2.7) and in Eqs. (3.27) to (3.29). In Section 4, $P \not\equiv$ pressure	(2.7)
P	defined in Eq. (4.23) (Section 4)	(4.23)
q_k	rate of formation of quantity k, per unit volume	(2.1)
q_E	rate of receipt of energy from surroundings, per unit volume	(2.19)
q'	heat flux vector	(3.2)

Symbol	Meaning	Equation of first mention
Q	viscous pressure (Section 2)	(2.7)
Q	transport flux in y direction (Section 4 *et seq.*)	(4.19)
Q_D	heat flux in y direction due to diffusion	(2.19)
Q_T	heat flux in y direction due to thermal conduction [including Dufour effect, see Eq. (3.38a)]	(2.19)
Q_V	viscous stress	(7.2)
r	radial coordinate	
r	distance between centers of molecules (Section 3.2)	(3.54)
$\mathbf{r}$	position vector (Section 3.1)	(3.1)
R	molar gas constant	(2.25)
R_i	molar rate of formation of ith species, per unit volume	(2.13)
s	transformation variable defined by Eq. (7.20) (Section 7.2)	(7.20)
$\mathbf{s}$	knot sequence in B-spline representation (Section 6.2)	
$(si)_{ij}$	Lennard-Jones (12 : 6) potential parameter; single suffix indicates pure component	(3.44)
$(si)_0$	Stockmayer potential parameter for polar molecule (Section 3)	(3.54)
S_u, S_b	linear flow velocity normal to stationary one-dimensional flame front, referred to "u" or equilibrium "b" boundary	(2.6)
S_i	space integral rate	(4.38)
$S_{D,ik}$	element $(\partial R_i/\partial \phi_k)_B$ of Jacobian matrix	(4.40)
t	time	(2.1)
T	temperature	(2.15)
T_{ij}^*	reduced temperature $(= k_B T/(eps)_{ij})$	(3.45)
T_{ik+}, T_{ik-}	coefficients controlling transport fluxes at finite-difference cell boundaries	(4.32)
$\hat{U}$	specific internal energy	(2.19)
$\hat{U}^{(0)}$	specific internal energy under Boltzmann equilibrium conditions (Section 3.1)	(3.7c)
$\mathcal{U}$	unit tensor	(3.13)

Symbol	Meaning	Equation of first mention
$\mathbf{v}$	molecular velocity (Section 3.1)	(3.1)
$\mathbf{v}_0$	mass average velocity (Section 3.1)	(3.4)
v_x, v_y	components of mass average velocity	(2.7)
$\mathbf{V}_i$	peculiar velocity of ith species	(3.1)
$\mathbf{V}_i$	diffusion velocity of ith species	(3.1)
V	reduced velocity, defined by Eq. (7.27) (Section 7.2)	(7.26)
w_i	mass fraction of ith species	(2.8)
$\mathbf{W}_i$	reduced velocity $(m_i'/2k_BT)^{1/2}\,\mathbf{V}_i$ (Section 3.1)	(3.6)
$\mathbf{W}_{j'}$	vector of right-hand sides of finite-difference equations (Section 4.2 *et seq.*)	(4.55)
x	Cartesian coordinate	
X_i	mole fraction of ith species	(2.8)
y	Cartesian coordinate	
$\delta y_{j'}$	physical distance between finite-difference grid points j' and $j' + 1$	(4.34)
$\Delta y_{j'}$	physical distance between finite-difference grid points $j' - 1$ and $j' + 1$ $(= \delta y_{j'} + \delta y_{j'-1})$	(4.41a)
z	Cartesian coordinate; axial cylindrical polar coordinate	
α_n^*	reduced polarizability of nonpolar molecule (Section 3)	(3.56)
β_{ij}, β_i^h, β_{ik}'	effective transport coefficients in conservation equation for ith species when composition and energy variables are σ and h, respectively	(4.1)
γ	reduced relative velocity between molecules $(\mu_{ij}/2kT)^{1/2}g_{ij}$ (Section 3.1)	(3.39)
γ_j, γ^h	effective transport coefficients in energy conservation equation when composition and energy variables are σ and h, respectively	(4.2)
Γ_ϕ	transport coefficients for quantity ϕ	(7.6)
δ_{ij}	Kronecker delta ($=0$ for $i \neq j$; $=1$ for $i = j$)	(3.14)
ε	reduced molecular internal energy E/k_BT (Section 3.1)	(3.6)

Symbol	Meaning	Equation of first mention
$\Delta\varepsilon$	reduced internal energy change on collision (Section 3.1)	(3.42)
ζ	coefficient of volume viscosity (Section 2.2)	(2.7a)
ζ_{ij}	collision number for rotational relaxation (Section 3)	(3.48)
η	coefficient of shear viscosity, with subscript if pure component. η_i computed by Eq. (3.53)	(2.7a)
λ_i	thermal conductivity of ith component	(3.72)
$\lambda_0, \lambda_{\text{mix}}, \lambda_\infty$	thermal conductivity of mixture	(3.36)
μ	reduced mass, defined in Eq. (3.40) (Section 3.1)	(3.39)
$\bar{\mu}$	dipole moment of polar molecule (Section 3.2)	(3.54)
$\bar{\mu}^*$	reduced dipole moment (Section 3)	(3.56)
v	number of continuity conditions to be satisfied at breakpoints (Section 6.2)	
ξ_t	factor controlling maximum time step in finite-difference computation	(4.78)
ξ	transformation variable defined by Eq. (7.19) (Section 7.2)	(7.19)
$\boldsymbol{\xi}$	sequence of breakpoints (Section 6.2)	
ρ	density	(2.1)
σ	composition variable ($\sigma_i = w_i/m_i$)	(2.14)
ϕ	Stockmayer molecular interaction potential (Section 3.2)	(3.54)
ϕ	element of $\boldsymbol{\Phi}$ (see below)	
ϕ_{iI}	perturbation parameter in Eq. (3.8) (Section 3.1)	(3.8)
ϕ_{ik}	parameter used in "approximate" formulations of transport coefficients (Section 3.4)	(3.62)
$\boldsymbol{\Phi}_{j'}$	vector of dependent variables at fixed value of ω	(4.19)
$\delta\boldsymbol{\Phi}_{j'}$	corrections to $\boldsymbol{\Phi}_{j'}$	(4.55)
ψ	von Mises coordinate, defined by Eq. (4.8) or (locally in Section 7) by Eq. (7.5) or (7.11)	(4.8)

Symbol	Meaning	Equation of first mention
ω	dimensionless stream function, defined by Eq. (4.9) or by similar local definitions in Section 7.2	(4.9)
$\Omega_{ij}^{(l,s)}$	collision integral	(3.43)
$\Omega_{ij}^{(l,s)*}$	reduced collision integral	(3.45)
Ω	$\omega_{j'+1} - \omega_{j'-1}$	(4.28)
Ω_+	$\omega_{j'+1} - \omega_{j'}$	(4.29)
Ω_-	$\omega_{j'} - \omega_{j'-1}$	(4.30)
$\mathscr{D}_{ij}$	binary diffusion coefficient, evaluated by Eq. (3.45)	(3.45)
$\langle\ \rangle$	denotes integration defined by Eq. (3.39)	(3.39)
Subscripts		
b	hot boundary of flame; boundary of finite-difference grid at $\omega = 0$	(2.6)
B	term to be evaluated before updating	(4.39)
e	free stream (Section 7.2)	(7.14a)
$h, i, j, \ldots$	hth, ith, jth chemical species $(1, 2, \ldots, N)$	(2.8)
i	identifier of B spline (Section 6.2)	(6.3)
int	internal	(3.2)
$I, J, \ldots$	Ith, Jth ... quantum state (Section 3.1)	(3.5)
j	identifier of collocation point (Section 6.2)	(6.6)
j'	defines spatial position in finite-difference (ω, t) grid, $j' = 0, 1, \ldots, M + 2$	(4.28)
k	order of B spline or B spline function (Section 6.2)	(6.3)
m, n	identifiers of terms in Sonine polynomial expansion (Sections 3 and 9)	(3.20)
n'	defines temporal position in finite-difference (ω, t) grid	
$N + 1$	specific enthalpy (h)	(4.20)
$N + 2$	density (ρ)	(4.40)
$N + 3$	temperature (T)	(4.40)
$N + 4$	physical distance interval (δy)	(4.40)
r	radial direction	(2.2)
$\mathbf{s}$	knot sequence (Section 6.2)	

Symbol	Meaning	Equation of first mention
tr	translational	(3.2)
u	cold boundary of flame; boundary of finite-difference grid at $\omega = 1$	(2.6)
U	term to be evaluated at beginning of time step	(4.22)
w	wall boundary (Section 7.2)	(7.28)
x, y, z	Cartesian coordinate directions	(2.2)
$(+)$	defines finite-difference cell boundary at $(j' + \frac{1}{2})(\delta\omega)$	(4.22)
$(-)$	defines finite-difference cell boundary at $(j' - \frac{1}{2})(\delta\omega)$	(4.22)
Superscripts		
h	enthalpy (Section 4 *et seq.*)	(4.1)
$h, i, j, \ldots$	hth, ith, jth ... chemical component (Section 3)	(3.9)
$I, J, \ldots$	Ith, Jth ... quantum state (Section 3.1)	(3.5)
(i)	coefficient of ith B spline (Section 6.2)	(6.2)
(j)	coefficients for jth derivative of B spline (Eqs. (6.3) and (6.4) only)	(6.3)
rs, mn	identifiers of terms in Sonine polynomial expansion (Sections 3 and 9)	(3.22a)
T	thermal	
(0)	values under Boltzmann equilibrium conditions (Section 3.1)	(3.6)
0	to be evaluated at zeroth or "b" boundary point on finite-difference grid	(4.57)
$M + 2$	to be evaluated at $(M + 2)$th or "u" boundary point on finite-difference grid	(4.59a)

12. References

Barnhill, R. E. (1977). "Representation and approximation of surfaces," in *Mathematical Software III*, J. R. Rice, Ed., Academic Press, New York, p. 69.

Bechtel, J. H., Blint, R. J., Dasch, C. J., & Weinberger, D. A. (1981). Combustion and Flame, **42,** 197.

Bledjian, L. (1973). Combustion and Flame, **20,** 5.

Blottner, F. G. (1964). A.I.A.A. Journal, **2,** 1921.

Blottner, F. G. (1970). A.I.A.A. Journal, **8,** 193.

Blottner, F. G. (1975). Comp. Methods Appl. Mech. Eng., **6,** 1.

Bonilla, C. F., Wang, S. J., & Weiner, H. (1956). Trans. Am. Soc. Mech. Engrs., **78,** 1285.

Boris, J. P. & Book, D. L. (1973). J. Comp. Phys., **11,** 38.

Buddenberg, J. W. & Wilke, C. R. (1949). Ind. Eng. Chem., **41,** 1345.

Chapman, S. & Cowling, T. G. (1970). *The mathematical theory of non-uniform gases*, 3rd ed., The University Press, Cambridge.

Cherian, M. A., Rhodes, P., Simpson, R. J., & Dixon-Lewis, G. (1981a). *Eighteenth Symposium (International) on Combustion*, The Combustion Institute, Pittsburgh, p. 385.

Cherian, M. A., Rhodes, P., Simpson, R. J., & Dixon-Lewis, G. (1981b). Phil. Trans. Roy. Soc. Lond., A **303,** 181.

Cheung, R., Bromley, L. A. & Wilke, C. R. (1962). Am. Inst. Chem. Eng. Journal, **8,** 221.

Chung, P. M. (1965). *Advances in Heat Transfer*, **2,** 109.

Clifford, A. A., Gray, P., Mason, R. S., & Waddicor, J. I. (1982). Proc. Roy. Soc. Lond., A **380,** 241.

Coffee, T. P. & Heimerl, J. M. (1981). Combustion and Flame, **43,** 273.

Crowley, W. P. (1971). "FLAG: A Free Lagrange method for numerically simulating hydrodynamic flows in two dimensions," in *Proc. 2nd. International Conference on Numerical Methods in Fluid Dynamics*, Springer-Verlag, New York.

Curtiss, C. F. & Hirschfelder, J. O. (1949). J. Chem. Phys., **17,** 550.

de Boer, C. (1978). *A practical guide to splines*, Springer-Verlag, New York.

Deuflhard, P. (1974). Numer. Math., **22,** 289.

Dixon-Lewis, G. (1968). Proc. Roy. Soc. Lond., A **307,** 111.

Dixon-Lewis, G. (1979a). Phil. Trans. Roy. Soc. Lond., A **292,** 45.

Dixon-Lewis, G. (1979b). Combustion and Flame, **36,** 1.

Dixon-Lewis, G. (1981). *First Specialists' Meeting (International) of The Combustion Institute*, Section Francaise du "Combustion Institute," p. 284.

Dixon-Lewis, G. (1982). Communication to GAMM Workshop (Aachen, 1981) on "Numerical methods in laminar flame propagation," N. Peters and J. Warnatz, Eds., Vieweg, Wiesbaden.

Dixon-Lewis, G. (1983). Combustion Science and Technology, **34,** 1.

Dixon-Lewis, G., Goldsworthy, F. A., & Greenberg, J. B. (1975). Proc. Roy. Soc. Lond., A **346**, 261.

Dixon-Lewis, G. & Greenberg, J. B. (1975). J. Inst. Fuel, 132.

Dixon-Lewis, G. & Islam, S. M. (1983). *Nineteenth Symposium (International) on Combustion*, The Combustion Institute, Pittsburgh, p. 283.

Dixon-Lewis, G. & Simpson, R. J. (1977). *Sixteenth Symposium (International) on Combustion*, The Combustion Institute, Pittsburgh, p. 1111.

Dixon-Lewis, G. & Williams, D. J. (1977). "The oxidation of hydrogen and carbon monoxide", in *Comprehensive Chemical Kinetics*, C. H. Bamford and C. F. H. Tipper, Eds., vol. 17, Elsevier, Amsterdam. p. 1.

Douglas, J. & Gunn, J. (1964). Numer. Math., **6,** 428.

Enskog, D. (1917). "Kinetische Theorie der Vorgänge in massig verdünnten Gasen," Dissertation, Uppsala.

Fay, J. A. & Riddell, F. R. (1958). *J. Aeronaut. Sci.*, **25,** 73.

France, D. H. & Pritchard, R. (1976). J. Inst. Fuel, **49,** 79.

Fritts, M. J., Oran, E. S., & Boris, J. P. (1981). *Lagrangian fluid dynamics for combustion modelling*, NRL Memorandum Report 4570, Naval Research Laboratory, Washington, D.C.

Günther, R. & Janisch, G. (1972). Combustion and Flame, **19,** 49.

Heimerl, J. M. & Coffee, T. P. (1980). Combustion and Flame, **39,** 301.

Heimerl, J. M. & Coffee, T. P. (1982). Communication to GAMM Workshop (Aachen, 1981) on "Numerical methods in laminar flame propagation," N. Peters and J. Warnatz, Eds., Vieweg, Wiesbaden.

Hellund, E. J. (1940). Phys. Rev., **57,** 319, 328.

Hellund, E. J. & Uehling, E. A. (1939). Phys. Rev., **56,** 818.

Herzfeld, K. F. & Litowitz, T. A. (1959). *Absorption and dispersion of ultrasonic waves*, Academic Press, New York.

Hilsenrath, J. *et al.* (1960). *Tables of thermodynamic and transport properties*, Pergamon Press, New York.

Hindmarsh, A. C. (1976). Lawrence Livermore Laboratory Report UCID-30150.

Hirschfelder, J. O. & Curtiss, C. F. (1949). *Third Symposium (International) on Combustion*, Williams and Wilkins Co., Baltimore, p. 121.

Hirschfelder, J. O., Curtiss, C. F., & Bird, R. B. (1954). *Molecular theory of gases and liquids*, John Wiley & Sons, New York.

Kestin, J. & Wang, H. E. (1960). Physica, **26,** 575.

Landau, L. D. & Liftshitz, E. M. (1959). *Fluid mechanics* (trans. from Russian by J. B. Sykes and W. H. Reid), Pergamon Press, New York.

Lawson, C. L. (1977). "Software for C^1 surface absorption," in *Mathematical Software III*, J. R. Rice, Ed., Academic Press, New York, p. 161.

Lees, L. (1956). *Jet Propulsion*, **26,** 259.

Lund, C. M. (1978). Report UCRL-52504, Univ. of California, Lawrence Livermore Laboratory.

Madsen, N. K. & Sincovec, R. F. (1977). *PDECOL: General collocation software for partial differential equations*, Lawrence Livermore Laboratory Preprint UCRL-78263 (Rev 1).

Margolis, S. B. (1978). J. Comput. Phys., **27,** 410.

Marquardt, D. W. (1963). J. Soc. Industrial and Applied Math., **11,** 431.

Mason, E. A. & Monchick, L. (1962). J. Chem. Phys., **36,** 1622.

Mason, E. A. & Saxena, S. C. (1958). Phys. Fluids, **1,** 361.

Mason, E. A., Vanderslice, J. T., & Yos, J. M. (1959). Phys. Fluids, **2,** 688.

McDonald, H. (1979). Progr. in Energy and Combustion Science, **5,** 97.

Monchick, L. & Mason, E. A. (1961). J. Chem. Phys., **35,** 1676.

Monchick, L., Munn, R. J., & Mason, E. A. (1966). J. Chem. Phys., **45,** 3051.

Monchick, L., Pereira, A. N. G., & Mason, E. A. (1965). J. Chem. Phys., **42,** 3241.

Monchick, L., Yun, K. S., & Mason, E. A. (1963). J. Chem. Phys., **39,** 654.

Muckenfuss, C. & Curtiss, C. F. (1958). J. Chem. Phys., **29,** 1273.

Oran, E. S. & Boris, J. P. (1981). Progr. in Energy and Combustion Science, **7,** 1.

Patanker, S. V. & Spalding, D. B. (1970). *Heat and mass transfer in boundary layers*, Intertext Books, London.

Picone, J. M. & Oran, E. S. (1980). *Approximate equations for transport coefficients of multicomponent mixtures of real gases*, NRL Memorandum Report 4384, Naval Research Laboratory, Washington, D.C.

Powell, M. J. D. (1974). "Piecewise quadratic surface fitting for contour plotting," in *Software for Numerical Mathematics*, D. J. Evans, Ed., Academic Press, London, p. 253.

Richtmyer, R. D. & Morton, K. W. (1967). *Difference methods for initial value problems*, 2nd. ed., Interscience Publishers, New York.

Roache, P. J. (1972). *Computational fluid dynamics*, Hermosa Publishers, Albuquerque, N. M.

Schlichting, H. (1960) *Boundary layer theory*, 4th. ed. McGraw-Hill, New York.

Schumaker, L. L. (1976). "Fitting surfaces to scattered data," in *Approximation Theory II*, G. G. Lorentz, C. K. Chui, and L. L. Schumaker, Eds., Academic Press, New York, p. 203.

Shifrin, A. S. (1959). Teploenergetika, **6,** 22.

Smooke, M. D. (1982). J. Comput. Phys., **48,** 72.

Smoot, L. D., Hecker, W. C., & Williams, G. A. (1976). Combustion and Flame, **26,** 323.

Spalding, D. B. & Stephenson, P. L. (1971). Proc. Roy. Soc. Lond., A **324,** 315.

Spalding, D. B., Stephenson, P. L., & Taylor, R. G. (1971). Combustion and Flame, **17,** 55.

Stephenson, P. L. & Taylor, R. G. (1973). Combustion and Flame, **20,** 231.

Strang, G. & Fix, G. J. (1973). *An analysis of the finite element method*, Prentice-Hall, Englewood Cliffs, N. J.

Svehla, R. A. (1962). Technical Report R-132, NASA, Washington, D.C.

Taxman, N. (1958). Phys. Rev., **110,** 1235.

Tsatsaronis, G. (1978). Combustion and Flame, **33,** 217.

Tsuji, H. (1982). *Progr. in Energy and Combustion Science*, **8,** 93.

Tsuji, H. (1983). *ASME-JSME Thermal Engineering Conference*, Honolulu, Hawaii, **4,** 11.

Tsuji, H. and Yamaoka, I. (1967). *Eleventh Symposium (International) on Combustion*, The Combustion Institute, Pittsburgh. p. 979.

Tsuji, H. and Yamaoka, I. (1969). *Twelfth Symposium (International) on Combustion*, The Combustion Institute, Pittsburgh. p. 997.

Tsuji, H. and Yamaoka, I. (1971). *Thirteenth Symposium (International) on Combustion*, The Combustion Institute, Pittsburgh. p. 723.

Waldmann, L. (1947). Z. Phys., **124,** 175.

Waldmann, L. (1958). "Transporterscheinungen in Gasen von mittlerem Druck," in *Handbuch der Physik*, S. Flügge, Ed., Vol. 12, Springer-Verlag, Berlin.

Wang Chang, C. S. & Uhlenbeck, G. E. (1951). *Transport phenomena in polyatomic gases*, University of Michigan Engineering Research Report No. CM-681.

Wang Chang, C. S., Uhlenbeck, G. E., & de Boer, J. (1964). *Studies in statistical mechanics*, J. de Boer and G. E. Uhlenbeck, Eds., Vol. 2, John Wiley & Sons, New York.

Warnatz, J. (1978a). Ber. Bunsenges. phys. Chem., **82,** 193.

Warnatz, J. (1978b). Ber. Bunsenges. phys. Chem., **82,** 643.

Warnatz, J. (1978c). Ber. Bunsenges. phys. Chem., **82,** 834.

Warnatz, J. (1979). Ber. Bunsenges. phys. Chem., **83,** 950.

Warnatz, J. (1981). *Eighteenth Symposium (International) on Combustion*, The Combustion Institute, Pittsburgh, p. 369.

Warnatz, J. (1982). Communication to GAMM Workshop (Aachen, 1981) on "Numerical methods in laminar flame propagation," N. Peters and J. Warnatz, Eds., Vieweg, Wiesbaden.

Westbrook, C. K. (1980). Combustion Science and Technology, **23,** 191.

Westbrook, C. K. & Dryer, F. L. (1980). Combustion and Flame, **37,** 171.

Wilde, K. A. (1972). Combustion and Flame, **18,** 43.

Williams, F. A. (1965). *Combustion theory*, Addison-Wesley, Reading, Mass.

Yanenko, N. N. (1971). *The method of fractional steps*, Springer-Verlag, New York.

Chapter 3

Bimolecular Reaction Rate Coefficients

Reinhard Zellner

1. Introduction

The study of bimolecular gas reaction rate coefficients has been one of the primary subjects of kinetics investigations over the last 20 years. Largely as a result of improved reaction systems (static flash photolysis systems, flow reactors, and shock tubes) and sensitive detection methods for atoms and free radicals (atomic and molecular resonance spectrometry, electron paramagnetic resonance and mass spectrometry, laser-induced fluorescence, and laser magnetic resonance), improvements in both the quality and the quantity of kinetic data have been made. Summarizing accounts of our present knowledge of the rate coefficients for reactions important in combustion chemistry are given in Chapters 5 and 6.

It is a common—and largely justified—practice to express the temperature dependence of the rate coefficients of bimolecular reactions in the so-called Arrhenius form (Arrhenius, 1889):

$$k = Ae^{-E_A/RT} \tag{1.1}$$

where the preexponential factor A and the activation energy E_A are temperature-independent parameters. However, for some bimolecular reactions studied with high accuracy and over a large temperature range it has been found that Eq. (1.1) does not yield an adequate representation (Zellner, 1979a,b) and that more complex temperature functions, usually of the form

$$k = BT^b e^{-E^0/RT} \tag{1.2}$$

have to be invoked. Equation (1.2) and other rate-coefficient functions which differ from Eq. (1.1) are said to describe *non-Arrhenius* behavior.

One of the main objectives of this chapter is to describe the theoretical basis of the temperature dependence of bimolecular rate coefficients. It will be shown that both thermodynamics (van't Hoff, 1884) and bimolecular rate reaction theory (Tolman, 1927; Kassel, 1932; Eyring, 1935; Wigner, 1938; Glasstone *et al.*, 1941; Eliason and Hirschfelder, 1959; Marcus, 1965, 1974) require that A and E_A be functions of temperature. Theoretically, deviations from Eq. (1.1) should be the rule rather than exceptions. Why then can representations of log k versus $(1/T)$ (Arrhenius plots) be used at all?: Because the exponential temperature dependence, represented, for example, by E^0/RT in Eq. (1.2), is usually large compared to the T^b factor. Therefore, curvature in an Arrhenius plot of laboratory data easily gets lost within the error limits of experiments, in particular if these are available over only a limited range of temperature. While interpreting non-Arrhenius behavior of bimolecular reactions may appear not to be a cornerstone to modeling combustion, it is at least one of its important side issues, since extrapolation of data and applications of expressions such as Eq. (1.2) require some knowledge of the physical background. Non-Arrhenius behavior is thus treated in detail here, because conventional presentations of bimolecular kinetics do not explain it adequately if at all.

This chapter is divided into three sections. In the first section we outline fundamental concepts and explain the relationship between microscopic and macroscopic descriptions of reaction kinetics. The second section is devoted to *a priori* estimation of bimolecular reaction rate coefficients and their temperature dependence using classical rate theory (Tolman, 1927; Kassel, 1935; Eliason and Hirschfelder, 1959) and transition state theory (TST) (Eyring, 1935; Wigner, 1938; Glasstone *et al.*, 1941; Marcus, 1965, 1974). In the third section a comparison between theoretical concepts and experimental rate data for some selected reactions is made.

It is intended that the reader have both a user's guide and the conceptual framework of the molecular origin of a macroscopic (i.e., measurable) rate coefficient and its temperature dependence. Because the presentation must be kept short, the reader occasionally will be referred to comprehensive textbooks of gas kinetics (Bunker, 1966; Johnston, 1966; Gardiner, 1969; Pratt, 1969; Mulcahy, 1973; Levine and Bernstein, 1974; Weston and Schwartz, 1976; Smith, 1980).

2. Fundamental concepts

Before considering rate coefficients in terms of theoretical models we introduce the fundamental concepts underlying the rate coefficients of bimolecular gas reactions.

2.1. The rate coefficient and the Arrhenius equation

Measurements of rate coefficients of elementary reactions are usually made in situations where the constituents of the reaction mixture are thermally equilibrated and the temperature is well defined. This situation differs drastically from rate measurements of reactions of atoms or molecules in specific quantum states, where nonequilibrium concentrations of reactive species prevail; these are discussed in Section 3. The observed variation of rate coefficient with temperature can then be described by the Arrhenius equation (1.1), from which the *activation energy* E_A may be obtained by differentiation with respect to $(1/T)$

$$E_A = -R d \ln k/d(1/T) \tag{2.1}$$

Equations (1.1) and (2.1) are used to define A and E_A as the characteristic reaction rate parameters as obtained from the temperature dependence of the measurable quantity k.

It is often found that rate-coefficient measurements can be well represented, within the limits of experimental accuracy, by temperature-independent parameters A and E_A, yielding straight lines on $\log k$-vs-$(1/T)$ plots (see Section 4 and Chapters 5 and 6). Moreover, linear extrapolations of such Arrhenius plots are common methods used to estimate rate coefficients at temperatures where they have not been measured directly. Arrhenius plots may be curved, however, and temperature-dependent parameters $A(T)$ and $E_A(T)$, usually both increasing with increasing temperature, may have to be used. This is not common practice over narrow temperature ranges, but it does not conflict with thermodynamics or with the Arrhenius equation, which was originally derived on a thermodynamic basis.

2.2. Thermodynamic predictions

Rate coefficients and thermodynamic quantities are connected via equilibrium constants. For the elementary reaction

$$A + B \rightleftarrows C + D$$

at equilibrium, while the net change of concentrations is zero, the elementary reactions proceed at equal rates in forward (f) and reverse (r) directions

$$k_f[A]_e[B]_e = k_r[C]_e[D]_e \tag{2.2}$$

Identifying the concentration product ratio $[C]_e[D]_e/[A]_e[B]_e$ with the equilibrium constant K_c expressed in terms of concentrations, we can rearrange Eq. (2.2) to

$$k_f/k_r = K_c \tag{2.3}$$

This fundamental relation is a form of the *principle of detailed balancing.* It allows us to relate the rate-coefficient ratio to thermodynamic quantities.

To do this we make use of the van't Hoff isochore

$$d \ln K_P/d(1/T) = -\Delta H^0/R \tag{2.4}$$

where K_P is the equilibrium constant in terms of partial pressures and ΔH^0 is the standard enthalpy change of reaction. Because no change in the number of molecules results from this reaction, $K_P = K_c$ and thus

$$d \ln K_C/d(1/T) = -\Delta H^0/R \tag{2.5}$$

Forming the logarithm of Eq. (2.3), differentiating with respect to $(1/T)$, and substituting Eqs. (2.1) and (2.5) gives

$$E_{A_f} - E_{A_r} = \Delta H^0 \tag{2.6}$$

(If the number of molecules does change in reaction, the equation changes slightly; thus the difference of activation energies for forward and reverse reactions is more generally equal to ΔU^0, the change of standard internal energy.)

Similarly, the preexponential factors A_f and A_r may be related to the change of entropy. Using the integrated form of the van't Hoff isochore

$$\ln K_P = -\Delta G^0/RT = \Delta S^0/R - \Delta H^0/RT \tag{2.7}$$

and comparing again with Eq. (2.3) and the definition of k in Eq. (1.1) leads to

$$\ln(A_f/A_r) = \Delta S^0/R \tag{2.8}$$

Equations (2.6) and (2.8) can be used to calculate the Arrhenius parameters for the reverse reaction (A_r and E_{Ar}) from measurements of A_f and E_{Af}, and vice versa. This is the reason why accurate thermodynamic data for reactive species are an essential prerequisite for modeling the kinetics of reactions where both forward and reverse directions of the same elementary reaction occur. This is particularly so in high-temperature combustion, where many reactions—even those with large ΔH^0 values—proceed close to equilibrium conditions.

Let us note two predictions that can be derived for the temperature dependence of the Arrhenius parameters A and E_A:

1. Since ΔH^0 is generally a function of temperature, Eq. (2.6) predicts that either E_{Af} or E_{Ar}, or both, are also temperature dependent.
2. The preexponential factors A_f and A_r can be any functions of temperature so long as Eq. (2.8) is fulfilled. Thermodynamics places no restrictions on rate coefficients in one direction. This is an important result; thermodynamics deals with rate-coefficient ratios only and offers no prediction for the absolute magnitude or temperature dependence of the components of these ratios.

2.3. Macroscopic and microscopic kinetics

A thermal rate coefficient is a purely macroscopic quantity; definition of a temperature implies that k is a measure of the average result of a multitude of individual molecular events that differ from one another in many respects such as collision energy, relative orientations, and details of atomic motion. These molecular-level details are hidden by thermal averaging. Insight into k has to start at the molecular level of microscopic kinetics. Since the physical laws governing thermal distributions of translational and internal energy are well known, evaluation of k can be made once the corresponding microscopic rate parameters are known.

The fundamental microscopic molecular quantity to consider in this context is the energy or velocity dependent reactive cross section $\sigma(E)$ or $\sigma(v)$. In essence, the product $v\sigma(v)$ is the cylindrical volume swept out per second by one molecule A moving at velocity v; if the concentration of its reaction partner B is N_B, then the probability of reaction of that A in 1 s is $v\sigma(v)N_B$. The reactive cross section is connected with the thermal rate coefficient via

$$k = \langle v\sigma(v) \rangle \tag{2.9}$$

where the brackets denote the appropriate thermal averaging. The definition of σ and the nature of this averaging are described in the next section.

Experimentally, $\sigma(v)$ can in principle be determined in crossed molecular beam studies (Bernstein, 1966; Pruett *et al.*, 1975; Menzinger and Jokozeki, 1977). Such studies, however, require large cross sections in order to produce detectable signals. This in turn means that only fast reactions or high collision energies (usually beyond the range of interest for thermal kinetics systems) can be studied. Usable experimental $\sigma(v)$ data therefore do not exist for any combustion reaction.

The second approach to $\sigma(v)$ comes from theory. The largest amount of theoretical $\sigma(v)$ information has been obtained in Monte Carlo classical trajectory studies (Polanyi, 1972; Polanyi and Schreiber, 1974; Porter, 1976), which provide an inventory of all microscopic and macroscopic quantities. The necessary prerequisite, however, is knowledge of the potential energy of the interacting species for all atomic configurations. This can be readily obtained by quantum mechanical methods only for three-atom systems. Among combustion reactions, such computations have indeed been done for O—H—H, for the reaction $O + H_2 \rightarrow OH + H$ (Clary *et al.*, 1979; Schinke and Lester, 1979; Walch *et al.*, 1980), and H—O—O for the reaction $H + O_2 \rightarrow OH + O$ (Gauss, 1978) with useful accuracy. The third of the important chain-branching reactions in the H_2/O_2 explosion, $OH + H_2 \rightarrow H_2O + H$, requires the four-atom problem H—O—H—H to be solved. Already at this level the number of atomic arrangements to be considered and the number of electrons present are so large that accurate computations become very expensive (Walch and Dunning, 1980; Schatz and Walch, 1980). A further complication arises when both singlet and triplet interactions must be considered, as is required for $O + C_2H_2$ (Harding, 1981). Larger systems than this can be treated by sophisticated quantum-mechanical methods, but the accuracy of the results is only sufficient to answer major questions about reaction pathways, not to provide interpretation of $\sigma(v)$ or k behavior.

While the relationship between k and $\sigma(v)$ is a simple equation [Eq. (2.9)], it will be more useful for our later purposes to consider k as it is related to the cross section as a function of the relative translational kinetic energy E. We will be able to generate connections then between the activation energy E_A and the *threshold energy* E^0, the latter being defined as the relative translational energy below which the cross section is zero:

$$\sigma(E)\begin{cases} >0, & E > E^0 \\ =0, & E \leq E^0 \end{cases} \tag{2.10}$$

Collisions with $E \leq E^0$ are thus not reactive. This usually applies to the majority of all collisions, since the condition $E > E^0$ is met only by a small fraction of collisions in the high-energy tail of the energy distribution function. A relationship between E_A and E^0 will be derived in Section 3.

The origin of both E_A and E^0 is the barrier height V^* associated with the variation of the electronic potential energy V along the *reaction coordinate q*. This is a reflection of how the electronic energy varies with the position of the nuclei during approach of the reactants and separation of the products (Levine and Bernstein, 1974; Smith, 1980). The reaction coordinate q is used here in a schematic manner to indicate the lowest-energy reaction path. A typical example of $V(q)$ for an exothermic reaction of the type

$$A + BC \rightarrow AB + C \tag{2.11}$$

where A, B, and C are atoms is shown in Fig. 1. The energy of the reactants,

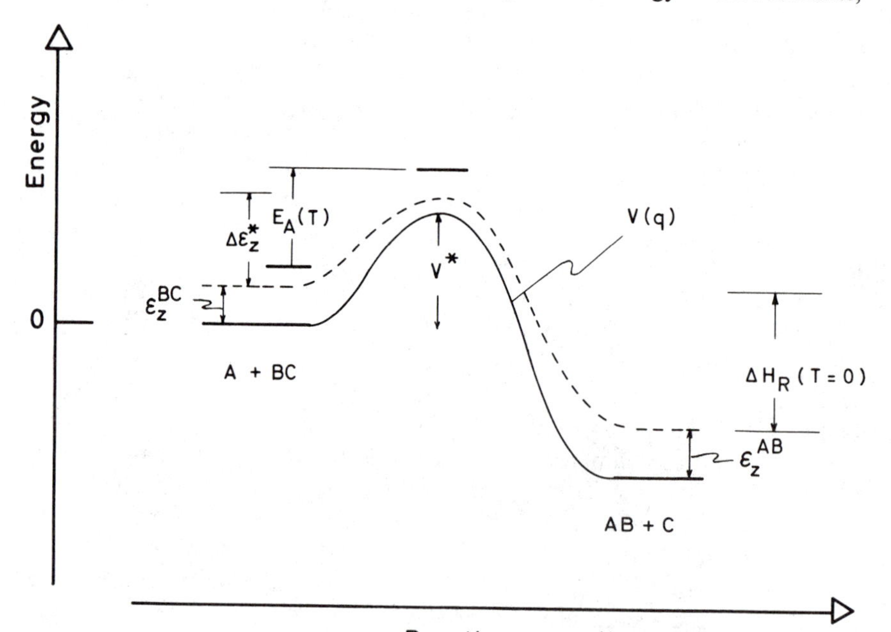

Figure 1. Energies related to an exothermic reaction of the type $A + BC \rightarrow AB + C$. The activation energy E_A is the (temperature-dependent) difference between the average energy of reacting molecules in the barrier configuration and the average total energy of the reactants at the same temperature.

arbitrarily set to zero, corresponds to complete separation of A and BC with no forces acting between them and with the electronic potential energy of BC (minimum of the BC diatomic potential function) also zero. As A approaches BC, it encounters repulsive forces and therefore V increases. This increase in V creates the barrier; the minimum relative translational energy of an $A + BC$ collision that can lead to reaction is the barrier height (V^*). $V(q)$ and hence V^* depend on the relative orientations of the reactants. The most favorable encounters—the ones with the smallest V^*—are usually collinear ones, with A, B, and C lying on a straight line.

So far we have neglected the quantized behavior of atoms moving under the influence of the potential. In fact, an energy of zero is unattainable by the system due to the zero-point vibrational energy of the molecules. Therefore, on the reactant and product side we must add to $V(q)$ the corresponding zero-point energies of BC and AB, respectively. Similar quantum effects are present at all points along the potential curve, but the magnitude of ε_z varies. The zero-point energy difference $\Delta\varepsilon_z^*$ is not incorporated into V^*, for which reason V^* is often termed the classical barrier height.

The different energies discussed above are compared in Fig. 1. The macroscopic nature of the activation energy E_A as a thermally averaged quantity is again stressed.

3. Theoretical prediction of bimolecular reaction rate coefficients

In this section we introduce collision theory and derive expressions for simple cross-section models allowing for variation of reactivity with relative translational energy. We then introduce quantum-state-selected reactivity and discuss the influence of vibrational rate enhancement on the thermal rate coefficient. As an alternative approach we then turn to transition state theory and consider the evaluation of rate coefficients from molecular properties.

Note on units. Rate coefficients computed from the equations developed in this chapter are (except in Section 3.2.4) in cm^3 $molecule^{-1}$ s^{-1} units. They may be converted to the cm^3 mol^{-1} s^{-1} units used elsewhere in this book through multiplication by 6×10^{23}.

3.1. Collision theory

The basis of the collision theory of bimolecular reactions is the idea that molecular encounters that lead to reaction are associated with *reactive cross sections.* Consider a general reaction between species A and B moving with velocities $\mathbf{v}_A$ and $\mathbf{v}_B$

$$A(\mathbf{v}_A) + B(\mathbf{v}_B) \rightarrow \text{products} \tag{3.1}$$

The reaction rate is the product of the A concentration and the probability per second of A undergoing reaction $v\sigma(v)[B]$

$$-d[A(\mathbf{v}_A)]/dt = [A(\mathbf{v}_A)][B(\mathbf{v}_B)]v\sigma(v) \tag{3.2}$$

where v is the magnitude of the relative velocity $|\mathbf{v}_A - \mathbf{v}_B|$ and $\sigma(v)$ is the corresponding reactive cross section. For a detailed derivation of Eq. (3.2) from the kinetic theory of gases the reader is referred to standard textbooks (e.g., Gardiner, 1969). For the rate coefficient at this relative velocity we obtain

$$k(v) = v\sigma(v) \tag{3.3}$$

If the velocities $\mathbf{v}_A$ and $\mathbf{v}_B$ are not considered fixed we have in place of Eq. (3.3)

$$k(f(v)) = \int v\sigma(v)f(v)\,dv \tag{3.4}$$

where $f(v)$ is the normalized distribution function of relative velocities. Equation (3.4) is the basis of collision theory.

In order to derive k as a function of temperature we let the molecular velocities have a Maxwell distribution

$$f_A(v_A, T) = 4\pi v_A^2(m_A/2\pi k_B T)^{3/2}e^{-m_A v_A^2/2k_B T} \tag{3.5}$$

Forming the corresponding expression for the distribution of relative velocities $f(v, T)$ (see Fowler and Guggenheim, 1949) and substituting into Eq. (3.4) we obtain

$$k = 4\pi(\mu/2\pi k_B T)^{3/2}\int v^3\sigma(v)e^{-\mu v^2/2k_B T}\,dv \tag{3.6}$$

where $\mu = m_A m_B/(m_A + m_B)$ is the reduced mass of the reactants. From Eq. (3.6) an analogous expression in terms of the relative translational energy E can be derived by making the substitutions $v^2 = 2E/\mu$ and $v\,dv = dE/\mu$

$$k = (8k_B T/\pi\mu)^{1/2}(k_B T)^{-2}\int_{E^0}^{\infty} E\sigma(E)e^{-E/k_B T}\,dE \tag{3.7}$$

where we have used the threshold energy E^0 as the lower limit of integration [see Eq. (2.10)].

Equation (3.7) shows that k is the convolution of an energy distribution function $E\exp(-E/k_B T)$ with the *excitation function* $\sigma(E)$. The contributions of the factors appearing in Eq. (3.7) are presented graphically in Fig. 2 for two different temperatures. We call the product of the energy distribution function and the excitation function the *reaction function*.

It should be stressed that for reactions with threshold energies E^0 much larger than $k_B T$, only a small fraction of encounters are reactive. The reaction function is shifted by the excitation function to higher energies than the energy distribution function. Since reacting molecules are continuously removed from the system, their energy distribution function may be perturbed by reaction (Prigogine, 1961). However, quantitative investigations of trans-

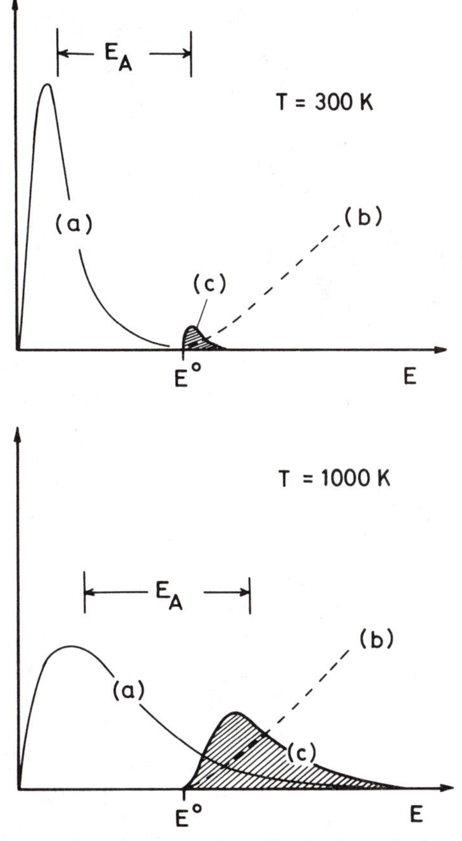

Figure 2. Comparison of energy distribution function $E\exp(-E/k_BT)$ (a); excitation function $\sigma(E)$ (b); and their product, the reaction function (c) at two different temperatures. The integral of (c) is proportional to the rate coefficient k. Also indicated is the difference between the average energy of reactive collisions and the average energy of all collision, the activation energy E_A.

lational nonequilibrium effects lead to the conclusion that even in extreme cases the rate coefficient is smaller by no more than a few percent compared to the full equilibrium condition (Present 1959; Shizgal and Karplus, 1970, 1971).

It is useful to derive from Eqs. (2.1) and (3.7) the general expression for the activation energy E_A

$$E_A = \frac{6 \times 10^{23} \int_{E^0}^{\infty} E^2 \sigma(E) e^{-E/k_BT}\, dE}{\int_{E^0}^{\infty} E\sigma(E) e^{-E/k_BT}\, dE} - 1.5RT \tag{3.8}$$

where Avogadro's number accommodates the translation from molecular-level to molar-level energies. The ratio of the integrals corresponds to the average energy of those collisions that actually lead to reaction, whereas $1.5RT$ is the mean energy (per mole) of all collisions $\langle E \rangle = 3 \times 10^{23}\ \mu\langle v^2 \rangle$. We may therefore write in short notation

$$E_A = \langle E^* \rangle - \langle E \rangle \tag{3.9}$$

Equations (3.8) and (3.9), first derived by Tolman (1927), are compared in Fig. 2. It is seen that E_A increases with increasing temperature due to the fact that the average energy of reactive collision grows faster with temperature than the average energy of all collisions. Equation (3.8) also predicts that $E_A \to E^0$ as $T \to 0$, which may be regarded as a second definition for the threshold energy, after Eq. (2.10).

3.1.1. Simple models for reactive collisions

In the previous section we derived the general relation between $\sigma(E)$ and k without making reference to the functional form of $\sigma(E)$. In this section we consider forms of excitation functions derived from simple collision models. Although these treatments are approximate, they correctly predict the general behavior observed for reactions with and without energy barriers.

A simple starting assumption is to view the colliding particles as structureless hard spheres and to assume that reaction occurs when the total collision energy exceeds a critical value E^0. This *hard-sphere cross section* is the same for all relative energies greater than E^0

$$\sigma(E) = \begin{cases} 0 & \text{for} \quad E \leq E^0 \\ \pi d^2 & \text{for} \quad E > E^0 \end{cases} \tag{3.10}$$

where $d = R_A + R_{BC}$ is a reactive collision diameter (cf. Fig. 3). Substituting Eq. (3.10) into Eq. (3.7) yields

$$k = (8\pi k_B T/\mu)^{1/2} d^2 \left(1 + \frac{E^0}{k_B T}\right) e^{-E^0/k_B T} \tag{3.11}$$

An alternative and more plausible assumption would be that not the entire collision energy but only its component along the *line of centers* can overcome a potential barrier. This takes into account the fact that a head-on collision ought to have a greater chance of leading to reaction than a glancing one of equal collisional energy. From simple geometric considerations (Gardiner, 1969; Smith, 1980) it can be shown that in this case

$$\sigma(E) = \begin{cases} 0 & \text{for} \quad E \leq E^0 \\ \pi d^2(1 - E^0/E) & \text{for} \quad E > E^0 \end{cases} \tag{3.12}$$

This *line-of-centers cross section* increases linearly with E above E^0 and approaches the hard-sphere cross section at high collision energies (Fig. 3). When this function is used in Eq. (3.7) we obtain for the thermal rate coefficient

$$k = (8\pi k_B T/\mu)^{1/2} d^2 e^{-E^0/k_B T} \tag{3.13}$$

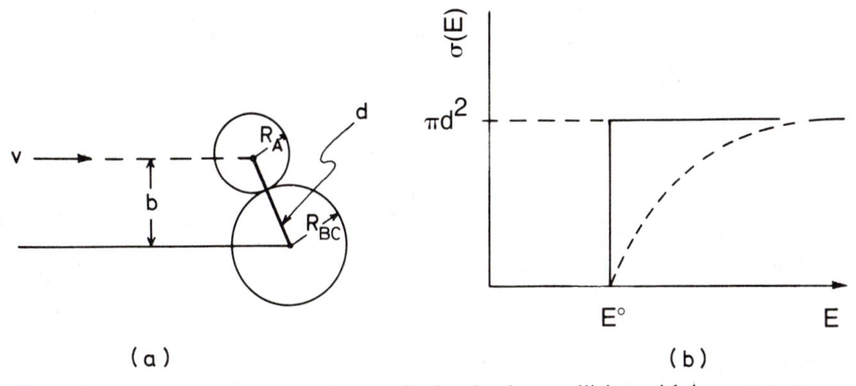

Figure 3. (a) Schematic representation of a hard sphere collision with impact parameter b between an atom A and a molecule BC. (b) Excitation functions for total relative energy (——) and its component along the line of centers (– – –) being effective in overcoming the energy barrier.

This expression has a particular physical meaning. If we set E^0 equal to zero and replace πd^2 with the hard-sphere collision cross section σ_{KT} of the kinetic theory of gases then we obtain the collision frequency Z (Hirschfelder *et al.*, 1959).

$$k = (8k_B T/\pi\mu)^{1/2}\sigma_{KT} = Z \tag{3.14}$$

where $(8k_B T/\pi\mu)^{1/2}$ is the average relative velocity of the reactants [see Eq. (3.3)]. Substituting into Eq. (3.13) yields

$$k = Z(\pi d^2/\sigma_{KT})e^{-E^0/k_B T} \tag{3.15}$$

Thus, we find that in this simple model the preexponential factor A in Eq. (1.1) can in principle be calculated if the reactive collision cross section πd^2 were known. It has become common practice to define a steric *factor p*

$$p = (\pi d^2/\sigma_{KT}) \tag{3.16}$$

to describe, empirically, the deviation of the reactive cross section from the kinetic theory of gases hard-sphere cross section. It can be derived by comparing calculated and measured A factors. It is found (Gardiner, 1969) that p generally decreases with increasing complexity of the reactants. This suggests that collisions in order to be reactive may require special relative orientations of the reactants; the stricter this requirement, the smaller the value of p.

It is of interest to see what the line-of-centers collision theory rate coefficient predicts for the activation energy. Applying $E_A = -Rd \ln k/d(1/T)$ to Eq. (3.13) we obtain

$$E_A = 6 \times 10^{23}(E^0 + k_B T/2) \tag{3.17}$$

Hence, E_A increases with temperature for the line-of-centers model (cf. Fig. 2).

This increase of E_A pertains not only to the line-of-centers cross section but to any excitation function increasing with energy (Zellner, 1979). The general shape of the excitation function (for $E > E^0$)

$$\sigma(E) \propto E^{-1}(E - E^0)^{n+1/2} \tag{3.18}$$

(cf. Fig. 4) corresponds to rate coefficients

$$k \propto T^n e^{-E^0/k_B T} \tag{3.19}$$

and hence activation energies

$$E_A = 6 \times 10^{23}(E^0 + nk_B T) \tag{3.20}$$

Equation (3.18) is only a formal representation for which there are no corresponding mechanical models except when $n = 1/2$.

3.1.2. Refinement of simple collision models

Crossed molecular beam and hot atom experiments suggest that real excitation functions are different from the ones assumed in the foregoing collision models. Two observations are particularly important:

(1) At $E = E^0$ real cross sections rise fairly rapidly to a peak value, after which
(2) At large E they gradually decrease and eventually decay to zero.

Unfortunately, there are not many experimental $\sigma(E)$ functions known. Therefore, these observations can only be explored using $\sigma(E)$ functions that possess the correct general behavior near E^0 and for large E (Menzinger and Wolfgang, 1969; LeRoy, 1969). One such function is

$$\sigma(E) = \begin{cases} 0 & \text{for} \quad E \leq E^0 \\ \alpha[(E - E^0)^n/E]e^{-m(E-E^0)} & \text{for} \quad E > E^0 \end{cases} \tag{3.21}$$

Using this in Eq. (3.7) yields

$$k = \Gamma(n + 1)\alpha(8k_B T/\pi\mu)^{1/2}[(k_B T)^{n-1}/(1 + mk_B T)^{n+1}]e^{-E^0/k_B T} \tag{3.22}$$

and for the activation energy

$$E_A = 6 \times 10^{23}\left[E^0 + (n - 1/2)k_B T - \frac{m(n + 1)(k_B T)^2}{1 + mk_B T}\right] \tag{3.23}$$

The principal temperature dependence is again exponential, just as for simple collision models. Unless the parameters n and m in Eq. (3.21) are known, the rate-coefficient expression (3.22) is indistinguishable from the rate-coefficient functions derived previously. Therefore, although realistic excitation functions are better representations of bimolecular reactivity, their characteristic features are lost in the corresponding k expressions because of thermal averaging. Their value is in refining the physical model behind rate coefficients that vary more strongly with temperature than the line-of-centers rate

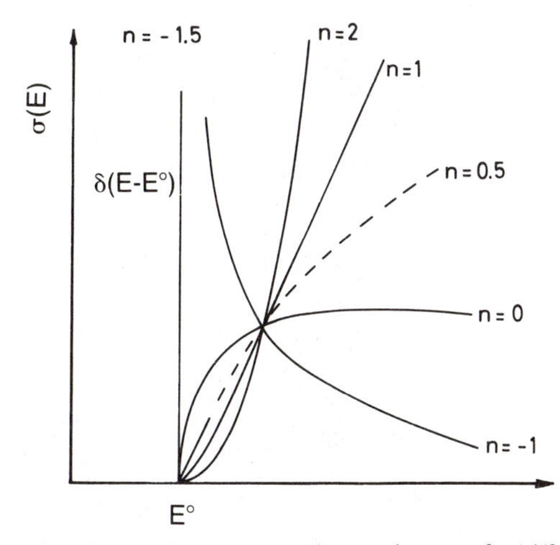

Figure 4. Normalized representation of excitation functions $\sigma(E) \propto E^{-1}(E - E^0)^{n+1/2}$ corresponding to rate-coefficient expressions $T^n \exp(-E^0/k_B T)$. The curve with $n = 0.5$ is the line-of-centers cross section; $n = 0$ corresponds to Arrhenius behavior with no preexponential temperature dependence.

coefficient $T^{1/2} \exp(-E^0/k_B T)$. Non-Arrhenius behavior may be viewed as being associated with excitation functions that rise convexly upwards above threshold (cf. $n = 1$ and $n = 2$ of Fig. 4).

An important refinement of the hard-sphere cross-section formula [Eq. (3.14)] arises when there are long-range attractive interactions. For reaction without barrier this interaction can be described by a spherically symmetric potential of the form

$$V(R) = -C/R^s \tag{3.24}$$

where C is a constant and $s = 3$, 4, or 6, corresponding to the interactions between two permanent dipoles, an ion and a polarizable molecule, and two neutral molecules (Johnston, 1966). This potential is modified by a dynamical term associated with the centrifugal energy ($V_Z = Eb^2/R^2$) so that the effective potential is (Fig. 5)

$$V(R)_{\text{eff}} = -C/R^s + Eb^2/R^2 \tag{3.25}$$

Unlike $V(R)$, the effective potential shows a maximum at $R_{\max}$ for each b, the so-called centrifugal barrier $V_Z = Eb^2/R_{\max}^2$. With the assumption that reaction occurs if E exceeds the centrifugal barrier, the maximum impact parameter $b_{\max}$ for each energy E and hence the reaction cross section $\sigma(E) = \pi b_{\max}^2$ may be calculated. One obtains for this *close-collision* model (Levine and Bernstein, 1974)

$$\sigma(E)_{\text{cc}} = \pi(s/2)[(s\text{-}2)/2]^{-(s-2)/s}(C/E)^{2/s} \tag{3.26}$$

Hence $\sigma(E)$ decreases monotonically with E (cf. Fig. 5). For ion-molecule

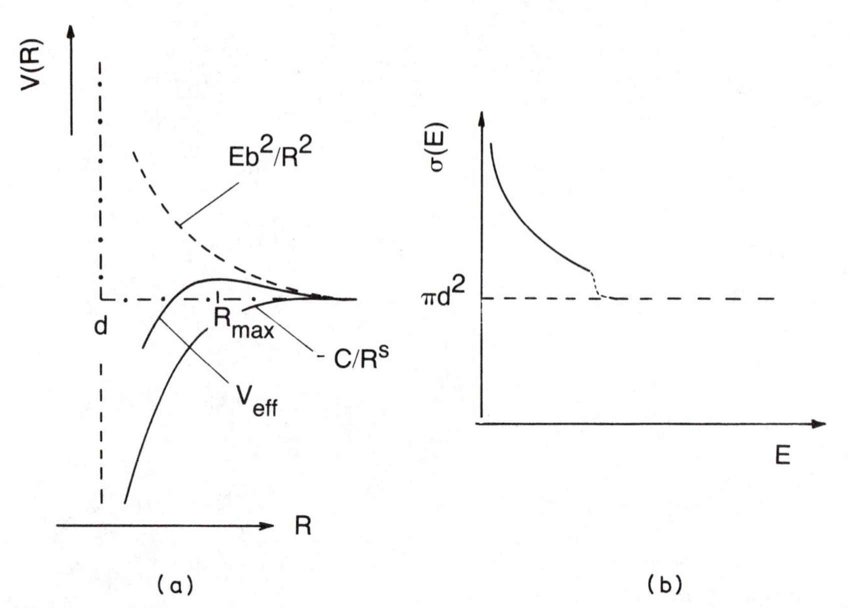

Figure 5. (a) Variation of potential energy in collisions between hard spheres with attraction. (b) Corresponding cross section for the close-collision model. (–·–·–) indicates the interaction appropriate for the hard-sphere, energy invariant cross section πd^2.

reactions in particular, one obtains $\sigma(E) \propto E^{-1/2}$, the Langevin-Gioumousis cross section. This corresponds to a temperature-independent rate coefficient, as is often found for ion-molecule reactions without threshold, e.g., $Ar^+ + D_2$ (Gioumousis and Stevenson, 1958; Dubrin, 1973).

The main effect of the attractive interaction is to enhance the cross section over its corresponding hard-sphere value, giving rate coefficients larger than collision numbers inferred from the kinetic theory of gases. This is indeed found in ion-molecule reactions, but not in reactions between neutral species. Even so, the close-collision concept and cross sections decreasing with E do explain unusual temperature dependences (Menzinger and Jokozeki, 1977).

3.1.3. State-specific collision theory rate coefficients and their averages

So far we have neglected internal quantum states of the reactants and attributed the temperature dependence of the rate coefficient only to the variation of cross section with relative translational energy. In this section we extend the collision theory concept to state-specific rate coefficients and their thermal average behavior.

State-specific and thermal rate coefficients The general collision theory rate-coefficient expression [Eq. (3.7)] may be extended to a reaction between molecules in defined internal quantum states i and j

$$A(i, v_A) + B(j, v_B) \rightarrow C + D \tag{3.27}$$

In this case we have for the specific rate coefficient k_{ij}

$$k_{ij} = (8k_BT/\pi\mu)^{1/2}(k_BT)^{-2}\int_0^\infty E\sigma_{ij}(E)e^{-E/k_BT}\,dE \tag{3.28}$$

where σ_{ij} is the state-specific excitation function. The corresponding expression for the overall thermal average rate coefficient is then obtained by allowing the internal states to have a Boltzmann distribution, i.e.,

$$[A(i)] = [A]e^{-\varepsilon_{Ai}/k_BT}/Q_A \tag{3.29}$$

where the brackets denote concentrations. Then

$$k = (Q_AQ_B)^{-1}\sum_{ij} k_{ij}e^{-(\varepsilon_{Ai}+\varepsilon_{Bj})/k_BT} \tag{3.30}$$

where Q_A and Q_B are the internal partition functions. Substitution of Eq. (3.28) leads to

$$k = (8k_BT/\pi\mu)^{1/2}(k_BT)^{-2}(Q_AQ_B)^{-1}\sum_{ij} e^{-(\varepsilon_{Ai}+\varepsilon_{Bj})/k_BT}\int_0^\infty E\sigma_{ij}(E)e^{-E/k_BT}\,dE \tag{3.31}$$

Hence the thermal rate coefficient is the sum of the state-specific rate coefficients weighted by their corresponding Boltzmann factors (Eliason and Hirschfelder, 1959).

The significance of Eq. (3.31) can be demonstrated by considering a reaction between A and a diatomic molecule BC, which we allow for simplicity to have only two internal states of energy difference ε_{BC}

$$A + BC(i = 0, 1) \rightarrow AB + C$$

The temperature-dependent part of the thermal rate-coefficient expression is then

$$k \propto (T^{3/2}Q_{BC})^{-1}\left\{\int_0^\infty E\sigma_0(E)e^{-E/k_BT}\,dE + e^{-\varepsilon_{BC}/k_BT}\int_0^\infty E\sigma_1(E)e^{-E/k_BT}\,dE\right\} \tag{3.32}$$

or in terms of specific reaction functions F_0 and F_1

$$k(T) \propto (T^{3/2}Q_{BC})^{-1}\int_0^\infty [F_0 + e^{-\varepsilon_{BC}/k_BT}F_1]\,dE \tag{3.33}$$

The integrand in Eq. (3.33) may be considered to be an effective, temperature-dependent reaction function, F_{eff} (cf. Fig. 6). Unlike F_0 and F_1, F_{eff} spreads to higher *and* lower energies with increasing temperatures; the energy spectrum of reacting molecules strongly increases with temperature.

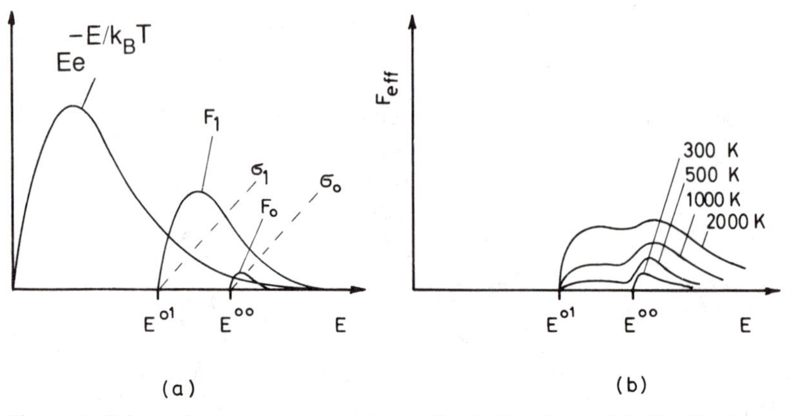

Figure 6. Schematic representation of specific F_1, F_0 (a) and effective F_{eff} (b) reaction functions for $A + BC(i = 0, 1) \rightarrow AB + C$. The excitation functions σ_i were assumed to have the same shape, corresponding to the line-of-centers model, but different threshold energies. The ratio of specific rate coefficients $k_1/k_0 = \int F_1\, dE / \int F_0\, dE$ was chosen to be approximately 20 at 300 K.

Equations (3.28) and (3.31) are rate-coefficient expressions typical for the collision theory concept. The influence of internal excitation of the reagents is introduced via excitation functions $\sigma_{ij}(E)$ that are generally inaccessible to experiments. No prediction is made in them for effectiveness of internal energy in overcoming a reaction barrier. This approach is therefore fundamentally different from statistical concepts in which it is assumed that the internal degrees of freedom of the reactants contribute indiscriminantly to overcoming the threshold energy. One such concept is the classical *activation in many degrees of freedom* assumption in which the rate coefficient is assumed to be the product of the hard-sphere collision frequency and the probability that the total energy in a specified number of degrees of freedom of the colliding molecules exceeds a threshold energy value E^0. Using the methods of statistical mechanics it can be shown that the rate-coefficient expression in this case becomes (Fowler, 1936).

$$k = Z_{KT}\Gamma(s)^{-1}(E^0/k_B T)^{s-1} e^{-E^0/k_B T} \tag{3.34}$$

where s is the number of effective oscillators (Chapter 4). Unlike Eq. (3.31), in Eq. (3.34) the contribution of internal states is explicitly described.

Non-Arrhenius behavior as a result of vibrational rate enhancement In the preceding section we derived the general relation between thermal and state-specific rate coefficients. We now use this concept to predict the temperature dependence of reactions for which measurements of the state-specific rate coefficients are available.

Investigations of the influence of internal energy on the rate of reaction have been concentrated mainly on vibrational excitation. Usually it is found that the reaction rate is enhanced, sometimes by several orders of magnitude, with the most dramatic effects observed in endothermic reactions (Moore and Smith, 1979; Birely and Lyman, 1975; Kneba and Wolfrum, 1980; Glass and Chaturvedi, 1981). Corresponding investigations of the effect of rotational excitation are comparatively scarce (Douglas and Polanyi, 1976; Stolte *et al.*, 1976). Experimental results as well as trajectory calculations indicate that in this case both rate increases and rate decreases are observed, the total effect, however, being small compared to vibrational excitation (Truhlar and Wyatt, 1976; Schinke and Lester, 1979). In the following we therefore focus attention on vibrational rate enhancement only.

We again consider an $A + BC$ reaction where BC $(i = 0, 1)$ now corresponds to BC molecules in vibrational ground and first excited states. Upon application of Eq. (3.30) we obtain for the thermal rate coefficient

$$k = Q_{BC}^{-1}[k_{v=0} + e^{-E_{\mathrm{vib},BC}/RT}k_{v=1}] \tag{3.35}$$

The overall temperature dependence depends on both $k_{v=0}$ and $k_{v=1}$. Generally $k_{v=0}$ can be identified with k at lower temperatures where the contribution of the excited-state reaction is negligible. On the other hand $k_{v=1}$ must be measured separately and, due to the experimental difficulties associated with such measurements, it is usually only known at one temperature or over a very limited temperature range. The interpretation of such measurements led to the conclusion that $k_{v=1}$ is best represented in the form (Birely and Lyman, 1975)

$$k_{v=1} = A_{v=1}e^{-(E^0 - \alpha \cdot E_{\mathrm{vib},BC})/RT} \tag{3.36}$$

Here α is the fraction of the BC vibrational energy which is effectively utilized to reduce the reaction barrier E^0. Equation (3.36) implies that the total rate enhancement is not an effect of a single reaction parameter but jointly caused by an enhanced preexponential factor $(A_{v=1})$ and a reduced reaction barrier. Such a separation is equivalent to saying that vibrational excitation may change both the shape and the threshold of the excitation function.

Equation (3.36) is an Arrhenius form. If we assume, in a first approximation, the same to hold for the rate coefficient of the ground-state reaction, i.e.,

$$k_{v=0} = A_{v=0}e^{-E^0/RT} \tag{3.37}$$

then the rate coefficient of the thermal reaction (3.36) may still show significant departures from simple Arrhenius behavior due to the interaction of the two terms. The necessary condition for this, however, is that a large fraction of the total rate enhancement is due to an increased preexponential factor. Only if $A_{v=1}$ is enhanced to such an extent that the inequality

$$(A_{v=1}/A_{v=0}) > e^{E_{\mathrm{vib},BC}/RT} \tag{3.38}$$

holds at temperatures around 2000 K will non-Arrhenius behavior be detected with rate-coefficient measurements accurate to a factor of 2. Any rate enhancements caused by reduction of the energy barrier ($\alpha > 0$) is compensated by the Boltzmann factor and does not contribute to non-Arrhenius behavior.

In summary, non-Arrhenius temperature dependence of a thermal rate coefficient occurs if a substantial fraction of the total reactive flux at higher temperatures occurs via excited states, provided that the enhancement of the state-specific rate coefficients is largely an effect of enhancement of preexponential factors.

The non-Arrhenius temperature dependence predicted from measurement of $k_{v=0}$ and $k_{v=1}$ can now be demonstrated for the reaction

$$OH + H_2(v = 0, 1) \rightarrow H_2O + H$$

Because of its importance in combustion systems, this reaction has been the subject of numerous experimental studies (Section 4 and Chapter 5). The rate coefficient $k_{v=0}$ may be derived from measurements below 700 K, where

$$k_{v=0} = 5.6 \times 10^{12} e^{-18\,\mathrm{kJ}/RT}\ \mathrm{cm^3/mol\ s}$$

The corresponding rate coefficient for the vibrationally excited reaction was determined to be (Zellner and Steinert, 1981)

$$k_{v=1} = 3.6 \times 10^{13} e^{-11\,\mathrm{kJ}/RT}\ \mathrm{cm^3/mol\ s}$$

Hence we have a rate enhancement of $k_{v=1}/k_{v=0} = 120$ at 300 K, of which a factor of $A_{v=1}/A_{v=0} = 6.5$ is due to increased preexponential factor. The reduction of the energy barrier is 7 kJ, corresponding to $\alpha = 0.14$ of the H_2 vibrational quantum of 50 kJ (see also Glass and Chaturvedi, 1981).

State-specific and thermal rate coefficients for this reaction are shown in Fig. 7. Substantial Arrhenius curvature is predicted above 1500 K, although the curvature predicted on the basis of the vibrational rate enhancement is insufficient to account for all of the experimentally observed behavior (see also Section 4).

Nonequilibrium effects In principle, chemical reaction always perturbs the distribution of states due to the energy dependence, and hence the energy selectivity, of the reactive cross section. Therefore, in a reactive system, a thermal equilibrium distribution can only be maintained if energy transfer in inelastic collisions is sufficiently rapid to more than counterbalance the reactive loss from higher energy states. This is usually fulfilled for translational and rotational degrees of freedom, which tend to be equilibrated within a few collisions (Present, 1959; Prigogine, 1961; Stevens 1967; Shizgal and Karplus, 1970, 1971; Lambert, 1977). A notable exception is the rotational relaxation of H_2, where several hundred collisions are required (Winter and Hill, 1976). The equilibration of vibrational states, on the other hand, is comparatively slow, requiring typically thousands of collisions. Hence, the probability of vibrational energy transfer may become comparable to or less than a typical reaction probability. As a consequence, an equilibrium distribution of vibrational

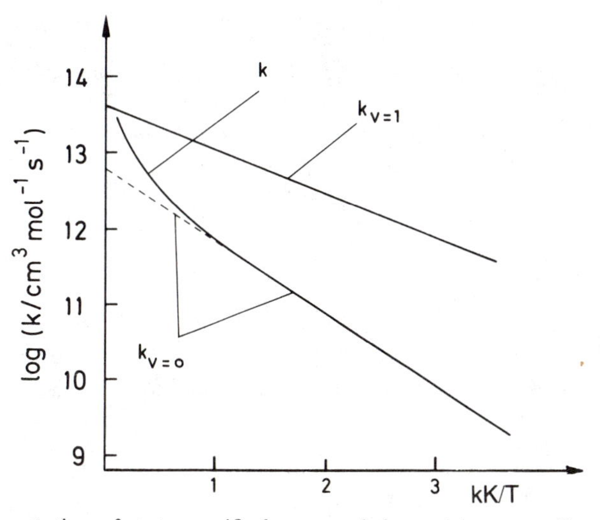

Figure 7. Arrhenius representation of state-specific $k_{v=0,1}$ and thermal k rate coefficients for $OH + H_2$ $(v = 0, 1) \rightarrow H_2O + H$. The thermal rate coefficient k was calculated from $k = Q_{\text{vib}}^{-1} \cdot [k_{v=0} + \exp(-E_{\text{vib}}/RT)k_{v=1}]$.

states may not be maintained if all or part of the reactive flux occurs via excited vibrational states. In this case the rate coefficient becomes time dependent and less than its vibrational equilibrium value (Widom, 1971; Clark *et al.*, 1977). We first outline the general formalism and then discuss whether nonequilibrium effects will be important for typical combustion conditions and for conditions under which rate coefficients of bimolecular reactions are measured.

The time-dependent rate coefficient for a system with comparable time constants for reaction (from different vibrational states) and energy relaxation (within these states) is described by a *master equation* (Troe and Wagner, 1967; Pritchard, 1975). The important aspects are illustrated by a two level system for which the mechanism of reaction and energy transfer is

$$\begin{aligned} &(0) && A + BC(v = 0) \rightarrow AB + C \\ &(1) && A + BC(v = 1) \rightarrow AB + C \qquad (3.39) \\ &(M) && BC(v = 0) + M \rightleftarrows BC(v = 1) + M \end{aligned}$$

In the limit of fast relaxation, where $[BC(v = 1)]$ is given by its equilibrium value

$$[BC(v = 1)]_{\text{eq}} = k_M[BC(v = 0)]/k_{-M} \qquad (3.40)$$

or

$$[BC(v = 1)]_{\text{eq}} = e^{-E_{\text{vib},BC}/RT} \cdot [BC(v = 0)] \qquad (3.41)$$

the thermal rate coefficient is identical to the expression derived previously [Eq. (3.35)]. If, however, relaxation to $BC(v = 1)$ is slow, its concentration becomes time dependent and approaches a steady-state value

$$[BC(v = 1)]_{ss} = k_M[M][BC(v = 0)]/(k_{-M}[M] + k_1[A]) \qquad (3.42)$$

The approach to steady state depends on whether $[BC(v = 1)]$ was at equilibrium at the onset of reaction. The former might apply to stationary reaction systems at high temperatures where reaction is initiated instantaneously, i.e., by flash photolysis. In this case we obtain for the $BC(v = 1)$ concentration

$$[BC(v = 1)]_t = \{[BC(v = 1)]_{eq} - [BC(v = 1)]_{ss}\}e^{-t/\tau} \qquad (3.43)$$

where $\tau = (k_{-M}[M] + k_1[A])^{-1}$ is the time constant for the approach to steady state. If, on the other hand, reaction is induced thermally (e.g., by shock heating a gas mixture) then $[BC(v = 1)]$ rises from zero to steady state, showing a typical induction behavior

$$[BC(v = 1)]_t = k_M[M][BC(v = 0)] \cdot \tau \cdot (1 - e^{-t/\tau}) \qquad (3.44)$$

Equations (3.43) and (3.44) are represented graphically in Fig. 8. In the derivation of these equations we have assumed $[A]$ and $[BC(v = 0)]$ to be time independent. Generally, however, at least one of them (say, the atom concentration A) varies with time. In this case the steady-state concentration of $BC(v = 1)$ also changes.

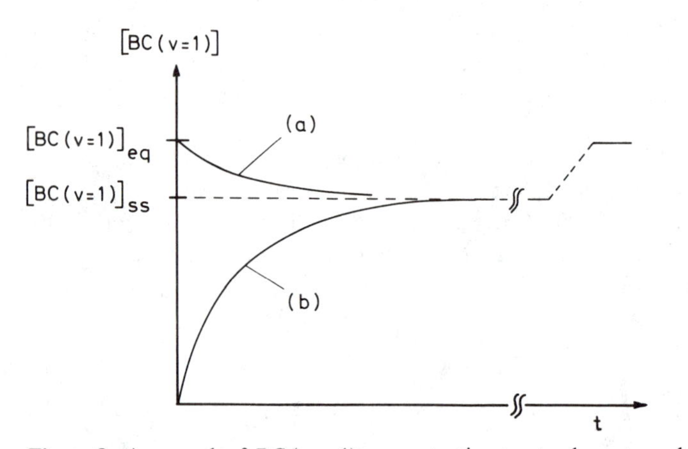

Figure 8. Approach of BC $(v = 1)$ concentration to steady state under nonequilibrium conditions. The curves refer to different experimental conditions: (a) instantaneous induction (e.g., flash photolysis) [Eq. (3.43)]; (b) thermal induction [Eq. (3.44)]. The deviation of $[BC(v = 1)]_{ss}$ from its equilibrium value depends on the reactant concentration A. Therefore, strictly speaking, $[BC(v = 1)]_{ss}$ becomes time dependent as the concentration of A changes and approaches its equilibrium value once A is completely consumed.

To investigate the effects of nonequilibrium concentrations on the thermal rate coefficient we write instead of Eq. (3.35)

$$k_{n\cdot \mathrm{eq}} = Q_{BC}^{-1} \cdot (k_{v=0} + e^{-E_{\mathrm{vib},BC}/RT} k_{v=1}^{\mathrm{eff}}) \tag{3.45}$$

where the effective rate coefficient $k_{v=1}^{\mathrm{eff}}$ describes the fractional deviation of $[BC(v = 1)]$ from its equilibrium value. At steady state

$$\begin{aligned} k_{v=1}^{\mathrm{eff}} &= k_{v=1}[BC(v = 1)]_{\mathrm{ss}}/[BC(v = 1)]_{\mathrm{eq}} \\ &= k_{v=1}/(1 + k_{v=1}[A]/k_{-M}[M]) \end{aligned} \tag{3.46}$$

Hence the thermal rate coefficient for nonequilibrium kinetics is always smaller than the one appropriate for equilibrium conditions, the deviation depending on the relative rates of relaxation and reaction ($k_{-M}[M]/k_1[A]$) for $BC(v = 1)$ molecules.

We have demonstrated that vibrational rate enhancement can cause positive departures from a simple Arrhenius relation. We now investigate how this is modified under nonequilibrium conditions. To do this we again use $OH + H_2$ $(v = 0, 1) \rightarrow H_2O + H$ as a reference reaction and consider first a situation in which the reactants are highly diluted in an inert gas such as Ar.

Figure 9 is a joint Arrhenius representation of $k_{v=1}$ for $OH + H_2$ and of k_{-M} for the vibrational relaxation of H_2 by Ar, $H_2(v = 1)$

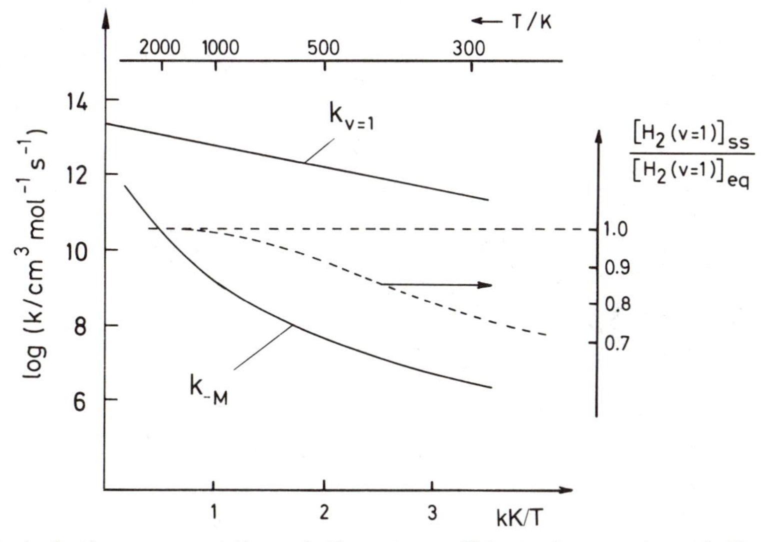

Figure 9. Arrhenius representation of the rate coefficients for reaction of H_2 $(v = 1)$ with OH ($k_{v=1}$) and for vibrational relaxation of H_2 $(v = 1)$ by Ar (k_{-M}). Also shown is the relative deviation of the steady-state H_2 $(v = 1)$ concentration from its equilibrium value as calculated from $[H_2(v = 1)]_{\mathrm{ss}}/[H_2(v = 1)]_{\mathrm{eq}} = [1 + (k_{v=1}[OH]/k_{-M}[Ar])]^{-1}$. The [Ar]/[OH] ratio was taken to be 3.6×10^5.

$+ \text{Ar} \rightarrow \text{H}_2(v = 0) + \text{Ar}$. The latter may be described by the Landau-Teller expression $\log(p\tau/\text{atm s}) = 45.09T^{-1/3} - 8.956$ (Kiefer and Lutz, 1966; Dove and Teitelbaum, 1974), which in terms of rate coefficients corresponds to $\log(k_{-M}/\text{cm}^3\,\text{mol}^{-1}\,\text{s}^{-1}) = 10.90 + \log(T/K) - 45.09T^{-1/3}$. Vibrational relaxation times and rate coefficients are always given for the exothermic [deactivating $v(1 \rightarrow 0)$] pathway. The corresponding quantity for the activating route is $k_M = k_{-M} \exp(-hc\omega_0/k_B T)$. Figure 9 also shows the ratio $[\text{H}_2(v = 1)]_{\text{ss}}/[\text{H}_2(v = 1)]_{\text{eq}}$ as calculated from the relation $[1 + (k_{v=1}[\text{OH}]/k_{-M}[\text{Ar}])]^{-1}$ for typical reaction conditions. Because of the strong increase of the energy transfer rate coefficient with temperature, deviations of $[\text{H}_2(v = 1)]$ from its equilibrium concentration only occur at lower temperatures. Since in this temperature range, however, the contribution of the vibrationally excited-state reaction to the overall thermal rate coefficient is negligible (cf. Fig. 7), no nonequilibrium effect of k is produced.

The situation is different if the total reactant concentration increases relative to the Ar diluent. The effect then predicted is shown in Fig. 10, where $k_{v=1}^{\text{eff}}$ [Eq. (3.46)] is shown for decreasing [Ar]/[OH] ratios. It can be seen that $k_{v=1}^{\text{eff}}$ shows deviations from its corresponding equilibrium value, $k_{v=1}$. This causes a noticeable modification at [Ar]/[OH] ratios less than 400.

In summary, vibrational nonequilibrium effects in bimolecular reactions are important only if a substantial fraction of the total reactive flux occurs via the excited state and if vibrational relaxation is slow. The effects are enhanced during the induction period preceding attainment of the steady-state condition. Hence, it may be surmised that the ignition delay of shock-

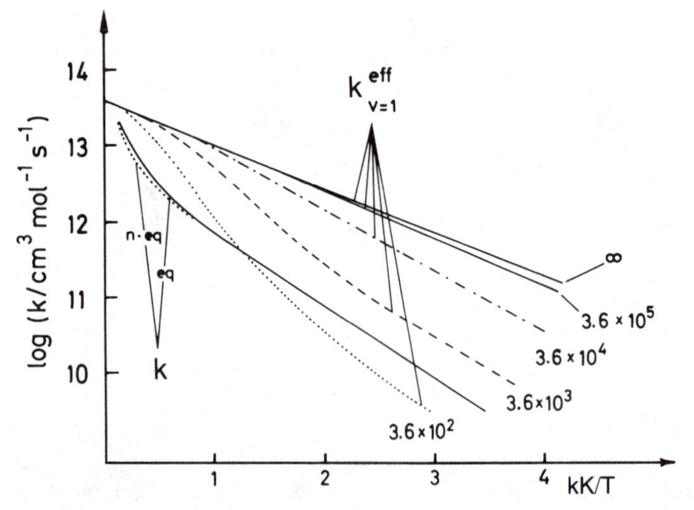

Figure 10. Arrhenius representation of effective rate coefficient $k_{v=1}^{\text{eff}}$ for $\text{OH} + \text{H}_2(v = 1) \rightarrow \text{H}_2\text{O} + \text{H}$ for different [Ar]/[OH] ratios. Also shown is the thermal rate coefficient for equilibrium (——) and nonequilibrium (·····) conditions.

heated $H_2/O_2/Ar$ mixtures may be prolonged by incomplete relaxation of H_2 in the chain reactions $O + H_2$ $(v = 0, 1) \rightarrow OH + H$ and $OH + H_2$ $(v = 0, 1) \rightarrow H_2O + H$.

3.2. Transition-state theory

One of the main drawbacks of collision theory is its need for the energy-dependent reactive cross section. As shown above, while this may be obtained from simple or refined collision models, the exact approach would involve determining the complete potential hypersurface and then carrying out elaborate quantum scattering calculations. In view of this complication and because a theory of chemical kinetics should be capable of predicting a rate coefficient from properties of the reacting molecules, considerable effort has been invested in devising simpler, chemically intuitive theories of reaction rates. The most common such theory is the *transition-state theory* (TST), also referred to as activated complex theory, developed mainly by Henry Eyring (Eyring, 1935; Glasstone *et al.*, 1941). It differs from collision theory in that it requires only limited information about the potential energy surface.

We base the following discussion on conventional TST in its *canonical* version, which applies to an assembly of reactant molecules for which the distribution among states is defined by the Boltzmann distribution laws for a single temperature. It should be mentioned, however, that this is not the only way statistical laws can be applied. Statistical mechanics can also be applied in a microcanonical version to derive rate coefficients $k(\epsilon)$ for molecular events at a specified energy (Wigner, 1938; Marcus, 1965, 1974; Kupperman, 1979; Garrett and Truhlar, 1979; Truhlar, 1979; Garrett *et al.*, 1980). From this the thermal rate coefficient can be obtained by averaging over the microscopically different events. Moreover, since conventional TST provides only an upper bound to the equilibrium rate coefficient, quantum mechanical versions have been formulated in which the transition state is no longer regarded as "fixed"; the rate coefficient is calculated by variationally minimizing the activated complex partition function along the reaction coordinate. This approach is referred to as *variational transition-state theory* (Garrett and Truhlar, 1979; Truhlar, 1979; Garrett *et al.*, 1980). It is equivalent to minimizing the free energy change in forming the transition state.

3.2.1. Fundamental assumptions and basic TST expression

It is a fundamental supposition of TST that one can define a region of the potential surface, identified with a small length δ of the reaction coordinate around the maximum of $V(q)$, as corresponding to a transition structure or "activated complex" (Fig. 11). The reaction is then thought of as proceeding from reactants to products via this transition structure

$$A + BC \rightarrow [ABC^{\ddagger}] \rightarrow AB + C$$

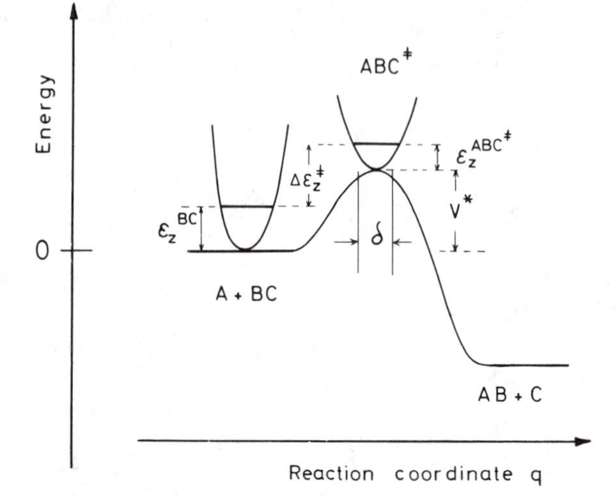

Figure 11. Location of transition-state species within a length δ along reaction coordinate. Also shown are zero-point energy levels of reactants and the activated complex. For the latter they include all vibrational degrees of freedom except the one which is considered a translational motion in the reaction coordinate.

In order to derive a rate-coefficient expression, reactants and activated complexes are assumed to be in equilibrium

$$A + BC \rightleftarrows ABC^{\ddagger} \rightarrow AB + C \tag{3.47}$$

in which case the rate of reaction is the net rate of transformation of activated complexes into product molecules

$$\text{Rate}_f \equiv k_f[A][BC] = [ABC^{\ddagger}]\langle v_{qf}\rangle/2\delta \tag{3.48}$$

where the index f denotes the forward direction and $\langle v_{qf}\rangle$ is the average decomposition velocity of complexes in one direction along the reaction coordinate q. The term $\langle v_{qf}\rangle/2\delta$ is identical with the average frequency at which $ABC^{\ddagger}$ decomposes into products.

Equation (3.48) describes the rate of forward reaction at chemical equilibrium between reactants and products. It is assumed in TST that Eq. (3.48) also describes the rate in the forward direction even if the system is not at equilibrium, i.e., when net forward reaction occurs. One condition for this is that the reactants are in internal thermodynamic equilibrium, so that their states are populated according to the Boltzmann distribution law. If this is true, Eq. (3.48) is unaffected by the extent to which the reverse reaction is occurring (Smith, 1980).

We are thus able to calculate a rate coefficient if we have a means to evaluate the concentration of activated complexes $ABC^{\ddagger}$ and their decomposition frequency. Assuming that $[ABC^{\ddagger}]$ is given by an equilibrium constant $K^{\ddagger}$

$$[ABC^{\ddagger}] = K^{\ddagger}[A][BC] \tag{3.49}$$

we obtain with Eq. (3.48)

$$k_f = K^{\ddagger}(\langle v_{qf} \rangle / 2\delta) \tag{3.50}$$

This rate-coefficient expression can be evaluated from molecular properties by using statistical mechanics formulas. To do this one writes the equilibrium constant $K^{\ddagger}(T)$ in terms of partition functions per unit volume Q

$$K^{\ddagger}(T) = (Q^{\neq}_{ABC}/Q_A \cdot Q_{BC})e^{-\Delta\varepsilon_z^{\ddagger}/k_B T} \tag{3.51}$$

where $\Delta\varepsilon_z^{\ddagger}$ is the difference between the zero-point energies of activated complex and reactants (cf. Figs. 1 and 11). The partition functions have energies referenced to these zero-point levels. The motion of $ABC^{\ddagger}$ along the reaction coordinate is assumed to be separable from other degrees of freedom. Therefore, the total partition function $Q^{\neq}_{ABC}$ can be factored into $Q^{\ddagger}_q$, the translational partition function for a small length δ of the reaction coordinate

$$Q^{\ddagger}_q = (2\pi\mu_q k_B T)^{1/2}\,\delta/h \tag{3.52a}$$

and the partition function $Q^{\ddagger}_{ABC}$ for the remaining $(3N\text{-}4)$ coordinates

$$Q^{\neq}_{ABC} = Q^{\ddagger}_q Q^{\ddagger}_{ABC} \tag{3.52b}$$

Using this separation and the kinetic theory of gases formula for the average velocity in one direction for reduced mass μ_q

$$\langle v_{qf} \rangle = (2k_B T/\pi\mu_q)^{1/2} \tag{3.53}$$

one obtains by combining Eqs. (3.50)–(3.53)

$$k = (k_B T/h)(Q^{\ddagger}_{ABC}/Q_A Q_B)e^{-\Delta\varepsilon_z^{\ddagger}/k_B T} \tag{3.54}$$

This is the general TST expression for the thermal rate coefficient. It contains a universal frequency factor ($k_B T/h = 6.25 \times 10^{12}\ \mathrm{s}^{-1}$ at 300 K) which is independent of the nature of the reactants. Specific molecular properties appear in the ratio of the partition functions and in $\Delta\varepsilon_z^{\ddagger}$. Equation (3.54) differs therefore from the collision theory rate-coefficient expression [Eq. (3.7)] in that all quantities contained in this equation are at least in principle derivable from molecular properties.

In deriving the TST rate-coefficient formula it was assumed that the rate of reaction is identical with the rate of passage of activated complexes in one direction [Eq. (3.48)]. If TST is evaluated in terms of reactive trajectories through a dividing surfaces in phase space (equivalent to a transition state) it can be shown that the theory is exact (Pechukas and Pollak, 1979), provided that all trajectories move into the product region and none of them are reflected. To allow for reflective failures in the free passage assumption, the conventional TST expression is multiplied by a *transmission coefficient* κ

having a value in the range $0 < \kappa < 1$. Instead of Eq. (3.54) we may therefore write

$$k = \kappa(k_B T/h)(Q^{\ddagger}_{ABC}/Q_A Q_B)e^{-\Delta\varepsilon^{\ddagger}_{z}/k_B T} \tag{3.55}$$

In this formulation κ appears as the ratio of the exact TST rate coefficient to the TST expression with $\kappa = 1$. Evaluation of κ requires experimental or computational knowledge of the exact rate coefficient; if this were available, there would be no need for TST expressions. As a consequence it is customary to set $\kappa = 1$ except for situations where quantum-mechanical tunneling is expected to be important (Johnston, 1966).

The calculation of the partition functions to be used in Eq. (3.54) proceeds by factoring Q into a product of translational, electronic, rotational, and vibrational partition functions

$$Q = Q_{\text{tr}} Q_{\text{el}} Q_{\text{rot}} Q_{\text{vib}} \tag{3.56}$$

where Q_{el} is equal to the electronic degeneracy of the molecule and Q_{tr}, Q_{rot}, and Q_{vib} are obtained from the statistical mechanical formulas in the ideal gas, rigid rotor, harmonic oscillator approximation as

$$\begin{aligned} Q_{\text{tr}} &= (2\pi m k_B T/h^2)^{3/2} && && (3.57) \\ Q_{\text{rot}} &= 8\pi^2 I k_B T/h^2 && \text{linear} \\ Q_{\text{rot}} &= \pi^{1/2}(8\pi^2 (I_x I_y I_z)^{1/3} k_B T/h^2)^{3/2} && \text{nonlinear} \\ Q_{\text{vib}} &= \prod_i (1 - e^{-hc\omega_i/k_B T}) \end{aligned}$$

The required input information (moments of inertia I and vibrational wavenumbers ω_i) is always available for the reactant molecules from spectroscopic data. Alternatively, for molecules for which tables of the free energy function $(G^\circ_T - H^\circ_0)/RT$ are available, the partition functions per unit volume can also be calculated from

$$Q = 7.34 \times 10^{-21} T^{-1} e^{-(G^\circ_T - H^\circ_0)/RT} \text{ cm}^{-3} \tag{3.58}$$

Calculating partition functions for the activated complex, however, involves an estimation of its structure (to obtain I_x, I_y, I_z) and vibrational frequencies. How this can be done approximately will be discussed in Section 3.2.3. These estimations are the weakest point of the theory and lead to the sometimes unfavorable reputation of TST as being so flexible that it can explain almost any k expression by making adjustments to the structural parameters assumed for the activated complex.

3.2.2. Arrhenius parameters

TST makes no prediction about the height of the potential barrier, which usually is more important in determining the rate coefficient than the partition functions. TST is therefore more useful for predicting the temperature

Table 1. Loss of rotational degrees of freedom (n) and gain of vibrational degrees of freedom (m) in various types of bimolecular reactions.

Reactants	Activated complex	n	m
atom + linear	linear	0	2
	nonlinear	-1	1
atom + nonlinear	nonlinear	0	2
linear + linear	linear	2	4
	nonlinear	1	3
linear + nonlinear	nonlinear	2	4
nonlinear + nonlinear	nonlinear	3	5

dependence of the activation energy, by way of the temperature dependence of the preexponential factor, than for predicting rate coefficients themselves.

The temperature dependence of the preexponential factor in the TST rate-coefficient expression arises mostly from the conversion of translational and rotational degrees of freedom of the reactant molecules into vibrational degrees of freedom of the activated complex. The temperature dependence of the TST rate-coefficient expression (3.54) can be summarized by

$$k = \text{constant} \cdot T^{-(n+1)/2}[Q^{\ddagger}_{\text{vib}}/(Q_{\text{vib}\,A}Q_{\text{vib}\,B})]e^{-\Delta\varepsilon^{\ddagger}_{z}/k_B T} \tag{3.59}$$

where the factor $T^{-(n+1)/2}$ accounts for the temperature dependence of the $k_B T/h$ factor and the ratios of the translational ($T^{-3/2}$) and rotational partition functions ($T^{-n/2}$). The value of n, the number of rotational degrees of freedom lost in forming the activated complex, depends on molecular structure (Table 1) and leads to the strongest temperature dependence ($n = 3$) for reactions between nonlinear molecules. In most cases n is positive, corresponding to a net loss of rotational degrees of freedom in formation of the activated complex. Only in the case of reaction between an atom and a linear molecule to form a nonlinear complex is there a gain in the number of rotational degrees of freedom, giving $n = -1$. This case is rather unrealistic, however, since for triatomic activated complexes the reaction path of minimum energy usually has a linear structure.

Equation (3.59) may now be used to explore the TST prediction for the temperature dependence of the activation energy $E_A = -Rd \ln k/d(1/T)$. Taking this derivative of Eq. (3.59) gives

$$\begin{aligned} E_A = 6 \times 10^{23}\, \Delta\varepsilon^{\ddagger}_{z} - (n+1)RT/2 - d \ln Q^{\ddagger}_{\text{vib}}/d(1/T) \\ + d \ln Q_{\text{vib}\,A}/d(1/T) + d \ln Q_{\text{vib}\,B}/d(1/T) \end{aligned} \tag{3.60}$$

While exact evaluation of Eq. (3.60) requires knowledge of the vibrational frequencies of reactants and activated complex, the limiting values of E_A at low and high temperatures can be found directly. At low temperatures all Q_{vib}

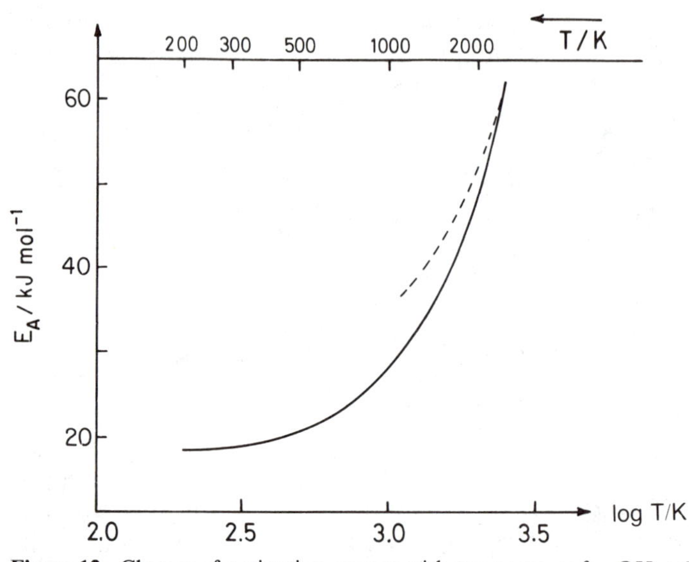

Figure 12. Change of activation energy with temperature for $OH + H_2 \rightarrow H_2O + H$ as calculated from Eq. (3.60). A nonlinear transition state with vibrational frequencies $\omega_i/\text{cm}^{-1} = 3227,\ 2353,\ 939,\ 572,\ 453$, and potential maximum $V^* = 21$ kJ was assumed (Smith and Zellner, 1974). The dashed curve corresponds to the classical limit of fully excited vibrations.

approach unity and Eq. (3.60) becomes

$$E_A\,(\text{low } T) = 6 \times 10^{23}\,\Delta\varepsilon_z^\ddagger - (n + 1)RT/2 \qquad (3.61)$$

At high temperatures each vibrational degree of freedom contributes a factor $k_B T/hc\omega_i$ to Q_{vib}. The high-temperature vibrational contribution to E_A therefore depends upon the number of vibrational degrees of freedom gained in forming the activated complex, given by $m = 2 + n$, as T^m. In the high-temperature limit then

$$E_A\,(\text{high } T) = 6 \times 10^{23}\,\Delta\varepsilon_z^\ddagger + (n + 3)RT/2 \qquad (3.62)$$

We note that although this is a correct high-temperature limit for E_A, the actual value of Q_{vib} remains much smaller than its high-temperature limit $k_B T/hc\omega_i$ at all temperatures where molecules exist.

As an illustration of the change of E_A over a large temperature range we consider a reaction between two diatomic molecules, such as $OH + H_2 = H_2O + H$, having a four-atom nonlinear activated complex and therefore $n = 1$ (Table 1). The high-temperature limit E_A is greater than $6 \times 10^{23}\,\Delta\varepsilon_z^\ddagger$ by $2RT$ [Eq. (3.62)]. At the low-temperature limit E_A is less than $6 \times 10^{23}\,\Delta\varepsilon_z^\ddagger$ by $(n + 1)RT/2 = RT$ [Eq. (3.61)]. The change in E_A between the low- and high-temperature limits is thus $R(2T_{\text{high}} - T_{\text{low}})$. If the limiting temperatures are 300 and 2000 K, then the difference in activation energy is

30 kJ; this amounts to a factor of 2 change in E_A for this particular reaction. In general

$$E_A(\text{high } T) - E_A(\text{low } T) = R[(n + 3)T_{\text{high}}/2 + (n + 1)T_{\text{low}}/2] \quad (3.63)$$

In order to see how Eq. (3.60) approaches its high-temperature limiting value one must assign a set of vibrational frequencies to the activated complex. In Fig. 12 a comparison of an exact calculation using the activated complex vibrational frequencies proposed by Smith and Zellner (1974) to the prediction of Eq. (3.63) is shown; as expected, the approach to the high-temperature limit is slower than the simple formula predicts.

3.2.3. Approximation formulas for structures and vibrational frequencies of activated complexes

In order to find the TST rate-coefficient expression one must have the partition function of the activated complex. While the translational part of this partition function is known exactly, the vibrational and rotational parts can only be found by using estimates of the structures and vibrational frequencies of the activated complex. These depend on the electronic energy of the activated complex as a function of intramolecular coordinates, the so-called *potential energy hypersurface*.

The most rigorous method available for finding potential energy hypersurfaces is the *ab initio* quantum mechanical method. As the computer time requirement for this method increases rapidly with the number of electrons to be considered, virtually all results of "chemical accuracy" have been obtained for three atoms (Eyring and Lin, 1974; Kuntz, 1976). Elementary reactions for which *ab initio* potential energy hypersurfaces have been calculated include $O + OH \rightarrow O_2 + H$ (Gauss, 1978) and $O + H_2 \rightarrow OH + H$ (Schinke and Lester, 1979; Walch *et al.*, 1980). A few four-atom hypersurfaces have also been calculated to chemical accuracy, among them $OH + H_2 \rightarrow H_2O + H$ (Walch and Dunning, 1980; Schatz and Walch, 1980).

In view of the high cost of *ab initio* quantum mechanical calculations, simplified procedures have been used in attempts to capture the essential features of potential energy hypersurfaces by approximate means. We discuss two of these, the BEBO (bond-energy-bond-order) method (Johnston and Parr, 1963; Johnston, 1966) and the LEPS (London-Eyring-Polanyi-Sato) method (Sato, 1955).

The BEBO method involves consideration of a collinear model for the transfer of a hydrogen atom from atom A to atom B, which can be represented as

$$A + HB \rightarrow A(\uparrow)\underset{r_1}{\cdots\cdots}H(\downarrow)\underset{r_2}{\cdots\cdots}B(\uparrow) \rightarrow AH + B$$

where the vertical arrows denote electron spins. The electronic energy is assumed to be the sum of the pair interactions

$$V(r_1, r_2) = -E_1(AH) - E_2(HB) + E_3(AB) \quad (3.64)$$

where the first two terms are binding interactions (antiparallel spins) and the third term is antibinding (parallel spins, triplet repulsion). The course of reaction can be described in terms of a bond-order coordinate n: $n = 0$ corresponds to no bond between A and H; $n = 1$ corresponds to A being bonded to H. There is an empirical relationship between bond order and the bond energy E and length r (Pauling, 1947)

$$E = E_s \cdot n^p$$
$$r = r_s - 0.26 \ln n \tag{3.65}$$

where p is an empirical constant and the subscript s denotes the energy and bond length of a single bond between the atoms concerned. For E_1 and E_2 this relationship gives

$$E_1 = E_{1s} n^p$$
$$E_2 = E_{2s}(1 - n)^{p'} \tag{3.66}$$

An "anti-Morse" function (Hirschfelder and Linnett, 1950; Sato 1955)

$$E_3 = (E_{3s}/4)[e^{-2\beta_{H_2}\Delta r_3} + 2e^{-\beta_{H_2}\Delta r_3}] \tag{3.67}$$

is used to represent the theoretical triplet repulsion energy, where $\beta_{H_2} = 0.194\ \text{nm}^{-1}$ is the Morse parameter of H_2 and Δr_3 is the difference between the $A - B$ distance $r_1 + r_2$ and the equilibrium bond distance of AB. Combining Eqs. (3.64)–(3.67) and adding E_{2s} so that the energy refers to A + HB rather than $A + H + B$ gives

$$V(n) = E_{2s}[1 - (1 - n)^{p'}] - E_{1s} n^p + (E_{3s}/4)[e^{-2\beta_{H_2}\Delta r_3} + 2e^{-\beta_{H_2}\Delta r_3}] \tag{3.68}$$

In Eq. (3.68) n is the only variable, Δr_3 being implied by using Eq. (3.65) to find r_1 and r_2 for any value of n, and all other quantities being molecular properties of AH, BH, and AB or the semiempirical parameters p and p' as estimated by Johnston and Parr (1963). By evaluating Eq. (3.65) for a number of values of n between 0 and 1 it is therefore possible to generate $V(n)$ and the interatomic distances r_1 and r_2 over the course of reaction. The maximum of $V(n)$ is then the energy difference between activated complex and reactants, the corresponding r_1 and r_2 values giving the geometry of the activated complex.

The LEPS approximation to the potential energy hypersurface is also based on the assumption of a linear activated complex. It takes the London equation for the H_2 molecule (Pitzer, 1953) as a starting point. For a three-atom system it extends to

$$V(r_1, r_2, r_3) = [1/(1 + \Delta)]\{Q_{12} + Q_{23} + Q_{13} - (1/2) \cdot [(J_{12} - J_{23})^2 + (J_{23} - J_{13})^2 + (J_{13} - J_{12})^2]\} \tag{3.69}$$

where Q and J are Coulomb and exchange integrals and Δ is an adjustable parameter usually selected by trial and error to make the calculated barrier height equal to the experimental activation energy. Again, borrowing from the

Heitler-London theory of H_2 one has equations for the binding (attractive) and antibinding (repulsive) potentials

$$(Q - J)/(1 - \Delta) = (E_s/4)(e^{-2\beta\Delta r} + 2e^{-\beta\Delta r}) \tag{3.70}$$

where the London forms on the left-hand side are expressed as Morse and anti-Morse functions on the right-hand side (Morse, 1929; Sato, 1955).

Values for the Morse parameters β may be derived from standard reference tables (Herzberg, 1950; Rosen, 1970; Huber and Herzberg, 1979). The quantities tabulated are the fundamental vibration frequencies ω_e and dissociation energies D_e in cm^{-1}. These are related to the Morse parameter β by $\beta = 1.22 \times 10^7\ \omega_e(\mu/D_e)^{1/2}$, where μ is the reduced mass in amu.

Once the potential energy surface has been calculated either by *ab initio* methods or by the semiempirical approximations, the remaining calculations of activated complex structure and vibration are straightforward. The procedure is (Johnston, 1966):

1. The barrier height V^* and the interatomic distances r_1^* and r_2^* in the activated complex are found by locating the maximum of V, either by trial and error or by systematic use of the Newton-Raphson extrapolation.
2. The potential function around the maximum is expanded in a Taylor's series to second order

$$V - V^* = (1/2)f_{11}r_1^2 + f_{12}r_1r_2 + (1/2)f_{22}r_2^2 \tag{3.71}$$

   where the f_{ij} are the force constants for the stretching vibrations defined by $f_{11} = -(\partial^2 V/\partial r_1^2)$, $f_{12} = -(\partial^2 V/\partial r_1 \partial r_2)$, and $f_{22} = -(\partial^2 V/\partial r_2^2)$. A fourth force constant for the bending motion is obtained from $f_{bend} = -(r_1r_2/r_3)(\partial V/\partial r_3)_{r_1r_2}$.
3. Vibration frequencies can be calculated from the force constants by assigning reduced masses to the motions in the appropriate coordinates and using the relationship $\omega\,(cm^{-1}) = \nu/c = (1/2\pi c)(f/\mu)^{1/2}$. For the linear triatomic system the frequencies are a symmetric stretch, two equal bending frequencies, and an asymmetric stretching frequency in the reaction coordinate that is imaginary, corresponding to the negative force constant appearing in this square root.

When the above procedure is done for the first time, the activation energy calculated from Eq. (3.60) will not be the same as the experimental value. One can adjust p and p' in the BEBO method or Δ in the LEPS method to get these to agree, after which the preexponential factor according to TST is fixed.

The semiempirical methods deal with three-particle collinear activated complexes, and so they can really describe only atom + diatomic molecule reactions. The basic ideas can still be used for more complicated cases, however, by considering the three central atoms with the semiempirical formulas and adding structural and vibration frequency information from other sources (see Smith and Zellner, 1974; Ernst *et al.*, 1977, 1978; Zahniser *et al.*, 1978).

Table 2. Calculated transition structure properties for HOHH.

	BEBO[a]	LEPS[a]	*ab initio*[b]
r_{H-O}/Å	0.96	0.96	0.98
$r_{O..H}$/Å	1.24[c]	1.21	1.33
r_{H-H}/Å	0.85[d]	0.85	0.85
H–O–H angle	104.6	104.6	97.6
O–H–H angle	180	180	165
V^*	21	21	31(26)[e]
ω_i/cm^{-1}	3727, 909i, 2354 939, 453, 578	3741, 1244i, 1936 1036, 557, 758	3368, 1655i, 1945 1248, 440, 686
E_A/kJ (300 K)[f]	20	19	23(19)[g]

[a] Smith and Zellner (1974).
[b] Walch and Dunning (1980); Schatz and Walch (1980).
[c] Equilibrium bond length of OH is 0.98 Å.
[d] Equilibrium bond length of H_2 is 0.74 Å.
[e] The smaller barrier was obtained with a larger basis set.
[f] Calculated from Eq. (3.60).
[g] Includes a tunneling correction.

In Table 2 the transition structure properties for the reaction $OH + H_2 \rightarrow H_2O + H$ are shown for semiempirical (Smith and Zellner, 1974) and *ab initio* (Walch and Dunning, 1980; Schatz and Walch, 1980) potential energy functions. The results given for the BEBO and LEPS surfaces apply to the four-atom model; the H—O—H angle was assumed to be 104.6° as in H_2O and the OHH arrangement was assumed to be linear. The *ab initio* calculation suggests a slightly different structure, the most notable difference being a 15° deviation from collinearity in the OHH structure. The properties of importance for the TST rate-coefficient calculation are nonetheless quite similar for all three types of calculation. Since the activation energies for the BEBO and LEPS calculations were made to agree with experiment by varying their adjustable parameters, they cannot be compared with experiment. The *ab initio* value can, however, and here the agreement is also good. A tunneling correction was included to remove a 4 kJ discrepancy (Schatz and Walch, 1980).

While the procedures described above for calculating transition structures do give satisfactory agreement with experimental rate coefficients, one must be reserved about drawing the conclusion that they are accurate representations of the real transition structures. In the first place, the methods used to calculate them are quite approximate, and in the second place quite different transition structures may give identical rate coefficients

In conclusion, we note that transition structures and vibration frequencies can also be found by noting analogies between presumed transition structures and known stable molecules. Useful rate-coefficient calculations can still be made even if substantial errors in geometry and frequencies are made.

3.2.4. Thermodynamic formulation of transition-state theory

The equations derived previously for the TST rate coefficient [Eqs. (3.54) and 3.55)] can be viewed as defining an equilibrium constant for activation, according to which

$$k = (k_B T/h) \cdot K^{\ddagger} \tag{3.72}$$

In place of using statistical mechanics formulas to evaluate $K^{\ddagger}$, one can estimate it by thermochemical arguments using procedures developed by Benson (1976) and co-workers. The value of this method is that with some intuition and practice one can make reasonable guesses of the transition structure, compute its entropy of formation, and thus find preexponential factors. As with TST generally, no prediction is made for the barrier height and so the activation energy must be taken from experiments. We present here an outline of the thermochemical data manipulations required in this formulation and illustrate them with an application.

The equilibrium constant K (and hence $K^{\ddagger}$) is expressed in concentration units. It is related to the equilibrium constant K_p by

$$K = K_p(R'T)^{-\Delta v} \tag{3.73}$$

where $R' = 82.06\ \text{cm}^3\ \text{atm/Kmol}$ and Δv is the change in number of moles due to reaction. For a bimolecular reaction forming an activated complex, $\Delta v = -1$ and hence

$$K^{\ddagger} = K_p^{\ddagger} \cdot R'T \tag{3.74}$$

Expressing K_p by $\exp(-\Delta G^\circ/RT)$ with $\Delta G^\circ = \Delta H^\circ - T\Delta S^\circ$ and inserting into Eq. (3.72), one obtains

$$k = (k_B T/h) R'T e^{\Delta S^{\ddagger}/R - \Delta H^{\ddagger}/RT} \tag{3.75}$$

(In contrast to the other equations in this chapter, the value of k from Eq. (3.75) is in $\text{cm}^3\ \text{mol}^{-1}\ \text{s}^{-1}$ units, as in the value of A from Eq. (3.77) below.) Here $\Delta S^{\ddagger}$ and $\Delta H^{\ddagger}$ denote the entropy and enthalpy changes associated with forming the activated complex. Equation (3.75) is the fundamental equation of the thermodynamic formulation of TST.

Using the definition of activation energy [Eq. (2.1)] we obtain from Eq. (3.75)

$$E_A = \Delta H^{\ddagger} + 2RT \tag{3.76}$$

Substituting Eq. (3.76) into Eq. (3.75) and comparing with the Arrhenius equation (1.1) show that the Arrhenius A factor may be identified with

$$A = e^2 (k_B T/h) R'T e^{\Delta S^{\ddagger}/R}\ \text{cm}^3/\text{mols} \tag{3.77}$$

The enthalpy and entropy of activation generally depend on temperature, such that

$$\begin{aligned} \Delta H^{\ddagger}(T_2) &= \Delta H^{\ddagger}(T_1) + \langle \Delta C_p^{\ddagger} \rangle (T_2 - T_1) \\ \Delta S^{\ddagger}(T_2) &= \Delta S^{\ddagger}(T_1) + \langle \Delta C_p^{\ddagger} \rangle \ln(T_2/T_1) \end{aligned} \tag{3.78}$$

Table 3. Arrhenius parameters for $O + CH_4 \rightarrow OH + CH_3$ assuming a linear transition structure.[a]

T/K	300	500	1000	2000
$\log A^{b}_{calc}$	13.08	13.54	14.14	14.74
$E^{c}_{A calc}$/kJ mol^{-1}	37	40	49	65
$\log k_{calc}$	6.64	9.35	11.61	13.04
$\log k^{d}_{exp}$	6.64	9.32	11.62	13.08

[a] The calculations were done by Golden (1979) using the molecular parameters for CH_3F (ΔS° for formation from CH_3 and F = -115.9 J/Kmol) as a starting point. Correction terms are (J/Kmol units): spin 5.9, external rotation 7.9, translation -0.4, C–F vibration 1100 cm^{-1} loss -0.4, OH stretch vibration 2000 cm^{-1} 0.0, 2 O–H–C bending vibrations 600 cm^{-1} 4.2, 3 F–C–H bending vibrations 1200 cm^{-1} going to 3 (OH)—C—H vibrations 700 cm^{-1} 3.8; the sum is $\Delta S^\ddagger = -94.9$ J/Kmol.

[b] Units are cm^3/mol s. A temperature-independent $\Delta S^\ddagger$ was assumed.

[c] Adjusted to give experimental E_A at 300 K; otherwise calculated from Eq. (3.76).

[d] Experimental results of Roth and Just (1977).

The average heat capacity of activation $\langle \Delta C_p^\ddagger \rangle$ usually proves to be so small that its effect upon the energy of activation is insignificant compared to the $2RT$ term of Eq. (3.76) (Shaw, 1977, 1978; Golden, 1979).

Equations (3.76) and (3.77) show that in the thermodynamic formulation of TST the activation energy and the preexponential factor of the Arrhenius equation (1.1) are related to $\Delta H^\ddagger$ and $\Delta S^\ddagger$, respectively. While $\Delta H^\ddagger$ cannot be estimated in this theory, $\Delta S^\ddagger$ can, and so it does offer the possibility of predicting the absolute value of the preexponential factor as well as the temperature dependence of the activation energy. Because formation of the activated complex involves transformation of translational into vibrational degrees of freedom, $\Delta S^\ddagger$ is always negative. In order to estimate its value one utilizes the fact that the entropies of structurally similar molecules are close to one another (Golden, 1973; Benson, 1975). One assumes a transition structure that seems reasonable based upon the overall structural change of the reaction, the structures of the reactant and product molecules, and perhaps stable molecules similar to the assumed transition structure. From this structure an entropy value is estimated, which is then corrected if necessary for electronic degeneracy, symmetry, and vibrational frequency effects. While this would seem to involve much guesswork, in fact the entropy is relatively insensitive to small changes in geometry or vibration frequencies, and so guesswork about the transition structure has little effect on the computed value of $\Delta S^\ddagger$ (Benson, 1975, 1976).

A particularly simple situation arises for the reaction of a light atom (H or D) with a large polyatomic molecule. The H (or D) atom has little effect upon the entropy of the molecule, and so the changes arising from formation of the activated complex are those due to the different symmetry and ground-state spin degeneracy. For the reaction $H + C_2H_6 \rightarrow C_2H_7^\ddagger$ as an example, $\Delta S^\ddagger$ can be approximated by $-S^\circ(H) + R \ln 12$, where the $R \ln 12$ term comes from the $6 \rightarrow 1$ change in symmetry number and the $1 \rightarrow 2$ change in degeneracy.

For reactions of heavier atoms and radicals, other contributions must be included. This is shown in Table 3 for the reaction $O + CH_4 \rightarrow OCH_4^\ddagger \rightarrow OH + CH_3$ as computed by Golden (1979). The transition structure was assumed to be similar to CH_3F. In Table 3 there is also a comparison to the experimental Arrhenius parameters, the activation energy for the calculation having been chosen to equal the measured value at 300 K. While the near-perfect agreement is impressive, it is also to a degree fortuitous, because if $\langle \Delta C_p^\ddagger \rangle$ were not assumed to be constant and/or if the transition structure were assumed to be bent rather than linear, then the agreement is reduced to being within a factor of 2. This level of accuracy is the best that can be expected of any TST approach.

3.2.5. Limitations of TST

Transition-state theory has some special limitations in addition to its quantitative uncertainty. There are situations in which it becomes ambiguous or even inapplicable.

In the preceding sections we have assumed that the potential surface has the qualitative form shown in Fig. 11, where a single potential maximum at the transition structure separates reactant and product species. A reaction of this type is said to be "direct." There are examples known where in place of the potential maximum there is a potential well along the reaction coordinate (Fig. 13) corresponding to the formation of an intermediate bound complex. One sees that there are then two potential maxima that must be traversed, and both of them may have an effect upon the rate of reaction. Moreover, if the bound complex is stable enough, the rate coefficient may become pressure dependent. This can be seen from the mechanism that then pertains

$$A + BC \underset{-1}{\overset{1}{\rightleftharpoons}} ABC^\ddagger \rightarrow AB + C$$
$$ABC^\ddagger \xrightarrow{3} ABC$$

where the activated complex $ABC^\ddagger$ is also a vibrationally excited ABC molecule. Application of the steady-state approximation to $ABC^\ddagger$ yields

$$k = \frac{k_1(k_2 + k_3[M])}{k_{-1} + k_2 + k_3[M]} = k_1 \varepsilon(T, M) \tag{3.79}$$

Hence the rate coefficient k that would be observed differs from the rate coefficient k_1 appropriate for a direct reaction by the efficiency factor $\varepsilon(T, M)$,

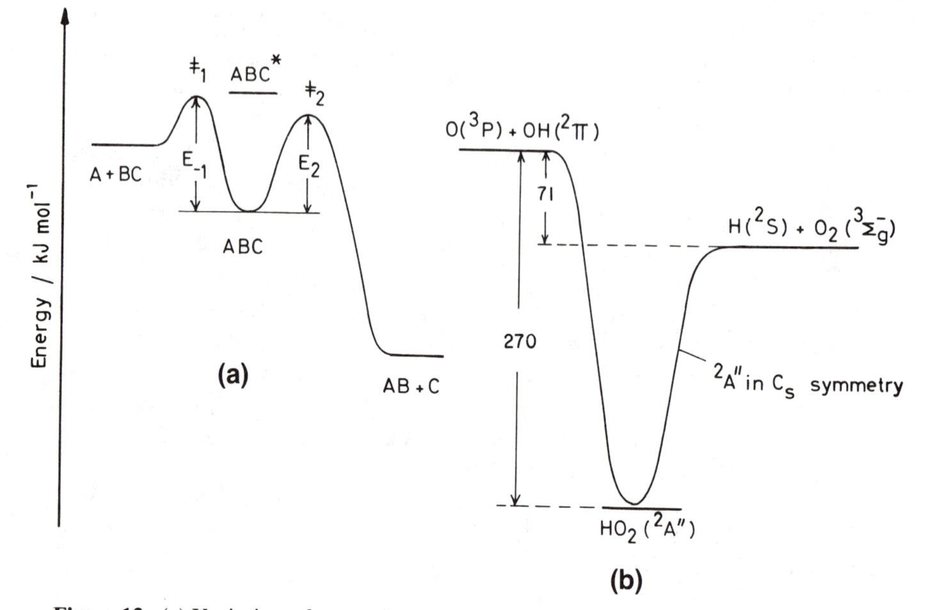

Figure 13. (a) Variation of potential energy for a bimolecular reaction proceeding via formation of a bound complex. If the barriers E_{-1} and E_2 are of comparable height, either of the two transition states may be rate determining. (b) In case the second barrier is very much lower (as in $O + OH \rightarrow O_2 + H$) the reaction does not "notice" the existence of the potential valley.

which can depend strongly on temperature and M concentration (Golden, 1979). The observed k is equal to k_1 only when $\varepsilon(T, M)$ becomes unity, which means practically that k_2 must be much greater than the sum $k_{-1} + k_3[M]$. This is turn implies that the second barrier is much lower than the first. A reaction to which this applies is

$$O + OH \rightarrow O_2H^\ddagger \rightarrow O_2 + H$$

This reaction is described by a potential surface in which there is a 270 kJ well of the stable HO_2 radical (Fig. 13). Because the reaction is very exothermic and there is no barrier for the HO_2 formation reaction from H and O_2, the excited $HO_2^\ddagger$ species formed from O + OH decomposes immediately to H and O_2.

The situation is different when the two barriers are comparable to one another. In the limit when k_{-1} is much greater than k_2 and k_2 is much greater than $k_3[M]$, the observed rate coefficient becomes $k_1 k_2/k_{-1} = K_1 k_2$, corresponding to reaction of an equilibrated intermediate species. The efficiency factor in this case is k_2/k_{-1}, a temperature-dependent function much smaller than unity. The temperature dependence can be either positive or negative, since k_2 and k_{-1} have a ratio that depends on both the entropies of activation

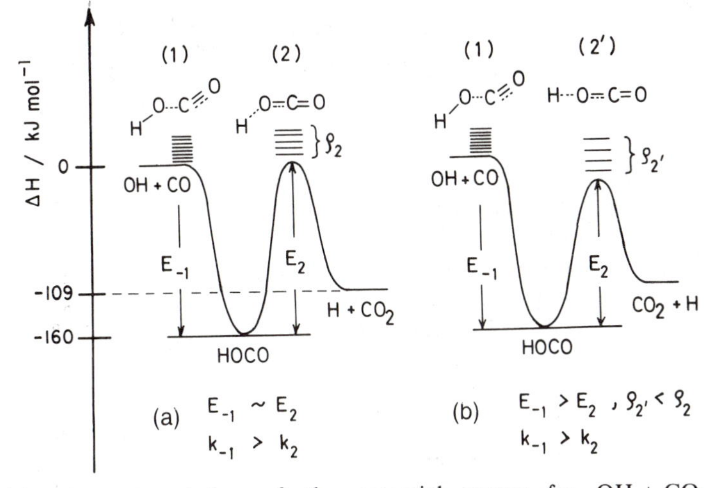

Figure 14. Schematic representation of the potential energy for $OH + CO \rightarrow CO_2 + H$ proceeding via a stable HOCO complex. ΔH_f (HOCO) $= -230$ kJ/mol and hence $E_{-1} = 160$ kJ/mol (O'Neal and Benson, 1973). Also shown are structures of activated complexes associated with the decomposition of $HOCO^*$ to $OH + CO$ (k_{-1}) and $H + CO_2$ (k_2). The density of states ρ (A factor) of the totally bent structure (1) is always large enough to maintain the inequality $k_{-1} > k_2$ even when $E_2 < E_{-1}$. This is because the density of states and hence the A factor is lowered when structure (2) becomes linear as in (2′). The assumed properties of the transition structures were: (2): HOC angle 120°, OCO angle 180°; $\omega_i/\text{cm}^{-1} = 2300$, 1300, 700, 600 (2) (Smith and Zellner, 1974); (2′): HOC angle = OCO angle 180°, $\omega_i/\text{cm}^{-1} = 1833$, 1300, 840 (2), 632 (2) (Golden, 1979).

and the energy barriers. A reaction to which this applies is

$$OH + CO \rightarrow CO_2 + H$$

From the temperature dependence of the observed rate coefficient it was concluded that the potential energy changes along the reaction coordinate as shown in part a of Fig. 14 (Smith and Zellner, 1973; Smith, 1977). The reactants and products are separated by a valley 160 kJ deep corresponding to the HOCO radical (Haney and Franklin, 1969; Milligan and Jacox, 1971; O'Neal and Benson, 1973; Smith, 1977; Gardiner *et al.*, 1978) In order to reconcile the small and nearly temperature-independent values of k at low temperature with the higher values found over 1000 K, it was proposed that the rate-controlling transition structure is the second one corresponding to a linear OCO with the H atom off axis (Smith and Zellner, 1973). The height of this barrier was at first assumed to be the same as that of the reactants. An alternative suggestion is that a lower barrier by even 20 kJ (part b of Fig. 14) could be compensated by a more linear transition structure (Golden, 1979). In the absence of direct information about the barrier heights the transition structure assignment must remain ambiguous.

4. Comparison between experiment and theory for rate coefficients of selected bimolecular gas reactions

In previous sections the theoretical concepts needed to discuss bimolecular gas reactions were presented. In this section we compare theoretical predictions with experimental measurements for some bimolecular combustion reactions that have been studied extensively. Additional comparisons are made in Chapters 5 and 6. The data sources and further references to the original literature for these reactions can be found in Chapter 5.

4.1. $O + H_2 \rightarrow OH + H$

The rate coefficient for this reaction can be represented by

$$k = 1.5 \times 10^7 T^{2.0} e^{-31.6\,\mathrm{kJ}/RT}\ \mathrm{cm^3/mol\ s}$$

over the temperature range 300–2500 K with an uncertainty factor $\log F = \pm 0.15$ (Fig. 15 and Chapter 5). The high-temperature experiments defining the preexponential T^2 dependence were analyzed in terms of a BEBO

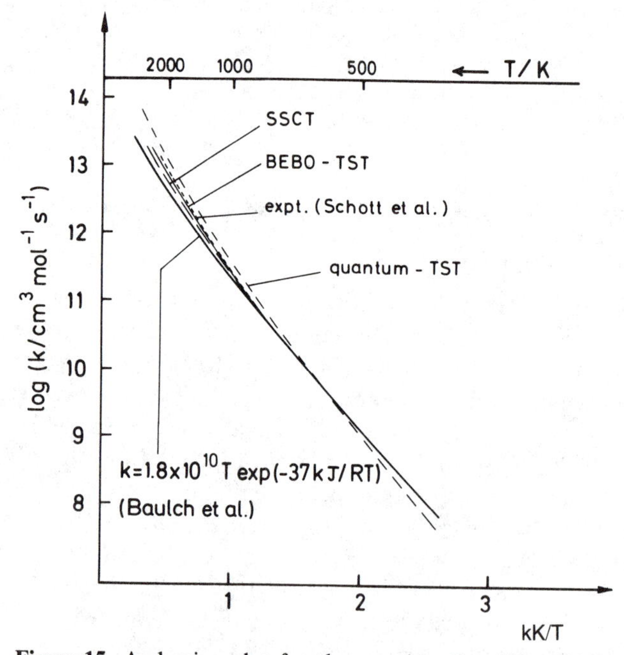

Figure 15. Arrhenius plot for the reaction $O + H_2 \rightarrow OH + H$. The Baulch *et al.* (1972) expression is a recommended empirical fit excluding the experiments of Schott *et al.* (1974). The theoretical results refer to: SSCT, Light (1978); BEBO-TST, Schott *et al.* (1974); quantum-TST, Walch *et al.* (1980).

TST model by Schott *et al.* (1974). After adjusting the magnitude of the rate coefficient and the activation energy to experimental data near 600 K, the rate coefficients over 1000 K were predicted in very good agreement with experiment. A TST calculation with molecular parameters based upon an *ab initio* calculation (Walch *et al.*, 1980) on this reaction also predicted a curved Arrhenius graph in close agreement with experiment. The exponential part of the temperature dependence was somewhat stronger than in the BEBO calculation. Since no adjustments of the barrier height were made in the *ab initio* calculation, its agreement with the experimental data must be regarded as particularly satisfying.

Arrhenius graph curvature for this reaction has also been interpreted in terms of state-specific rate coefficients in the collision theory framework. The ratio of rate coefficients for the $v = 1$ and $v = 0$ states of H_2 was measured to be 2600 at 300 K by Light (1978). By assigning to the $v = 1$ rate coefficient a temperature dependence derived from an LEPS trajectory study (Johnson and Winter, 1977) a major part of the supplementary reaction over 1000 K was predicted to be due to reaction in the $v = 1$ channel. The amount of supplementary reaction, however, was not sufficient to account for the data of Schott *et al.*

There is still experimental as well as theoretical uncertainty about the extent of the Arrhenius graph curvature for this reaction below 400 K. This is discussed by Light (1979).

4.2. $OH + H_2 \rightarrow H_2O + H$

The rate coefficient for this reaction can be represented as

$$k = 1.0 \times 10^8\, T^{1.6} e^{-13.8\,\mathrm{kJ}/RT}\ \mathrm{cm^3/mol\,s}$$

over the temperature range 300–2500 K (Fig. 16 and Chapter 5). A simple BEBO TST calculation gives good agreement with the experimental data (Smith and Zellner, 1974). A TST calculation based upon an *ab initio* potential energy surface also fits the experimental data very well providing that a tunnelling correction is added (Schatz and Walch, 1980). The tunneling correction amounts to an increase of the room-temperature rate coefficient by a factor of 3.

The importance of low-temperature tunneling may shed light upon the failure of the state-specific collision theory approach in computing the rate coefficient of this reaction (Zellner and Steinert, 1981). The assumption was made in this computation that the activation energy measured near room temperature corresponds to reaction through the ground vibrational state only. The occurrence of the tunneling leads to an underestimate of barrier height, and consequently the computed rate coefficients should be increased, in the direction of improving agreement with experiment.

An important feature of the state-specific collision theory study of this reaction was the large effect of vibrational excitation of H_2, which was

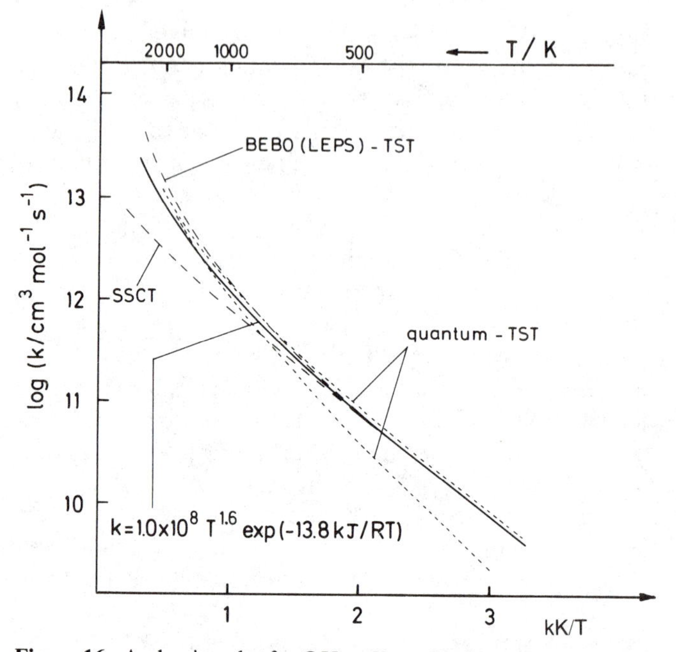

Figure 16. Arrhenius plot for $OH + H_2 \rightarrow H_2O + H$. The heavy curve corresponds to the expression recommended by Zellner (1979). Theoretical results refer to: BEBO(LEPS)-TST, Smith and Zellner, (1974); quantum-TST, Schatz and Walch (1980) (the upper curve includes a tunneling correction); SSCT, Zellner and Steinert (1981).

measured by Zellner and Steinert (1981) and Glass and Chaturvedi (1981) to give a factor of 120 to 150 increase in the rate coefficient for $v = 1$ compared to $v = 0$ at room temperature. This reaction illustrates the general idea that vibrational excitation of a bond that is broken in a reaction is more effective than other excitations in promoting reaction: The factor of 150 derives from addition of 53 kJ/mol into the H_2 vibration, while addition of 45 kJ/mol to the OH vibration increases the rate by only 50% (Spencer *et al.*, 1977).

4.3 $O + CH_4 \rightarrow OH + CH_3$

This reaction has been assigned rather inconsistent rate-coefficient expressions in studies over the temperature range 400–2000 K. Roth and Just (1977) and Felder and Fontijn (1979) favor a preexponential temperature dependence near T^2, while Klemm *et al.* (1981) favor one near $T^{1/2}$ (Fig. 17). As discussed in Section 3.2.4, the T^2 dependence is the one predicted by TST with a temperature-independent $\Delta S^{\ddagger}$. No further theoretical interpretations have been suggested so far.

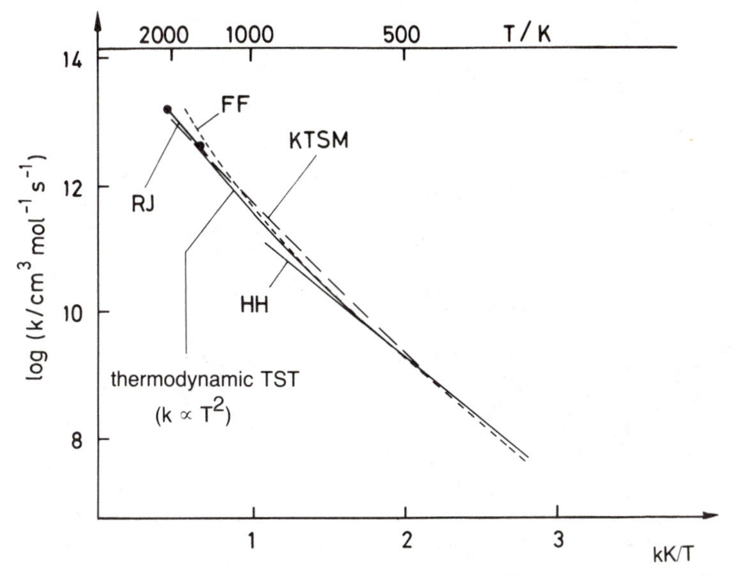

Figure 17. Arrhenius plot for $O + CH_4 \rightarrow OH + CH_3$. The solid curve corresponds to the TST rate expression as derived by Golden (1979) and recommended by Roth and Just (1977). The solid line is an evaluation by Herron and Huie (1973) based on low-temperature data only. Dashed curves refer to experiments by Felder and Fontijn (1979) and Klemm *et al.* (1981).

4.4. $OH + CO \rightarrow CO_2 + H$

This reaction appears to be unique in that its rate coefficient is essentially independent of temperature below 500 K while increasing sharply at higher temperatures (Fig. 18). It has been suggested that the best representation of the experimental data is in the form

$$k = \exp(24.99 + 9.2 \times 10^{-4} T)\, \mathrm{cm^3/mol\, s}$$

(Baulch and Drysdale, 1974). The origin of the temperature dependence was discussed in terms of TST in Section 3.2.5 (see also Dryer *et al.*, 1971, and Smith and Zellner, 1973). No alternative theoretical approaches have been explored for this reaction so far.

Experiments with vibrationally excited OH (Spencer *et al.*, 1977) and CO (Dreier and Wolfrum, 1981) were undertaken in order to see whether the temperature dependence might be interpretable in terms of increased reactivity of vibrationally excited reagents. In both cases it was found that the enhancement is insufficient to account for the observed temperature dependence, perhaps because of the formation of a bound intermediate complex. Independent evidence for complex formation is provided by the observation of pressure dependence of the rate coefficient at room temperature (Chan *et al.*, 1977; Biermann *et al.*, 1978).

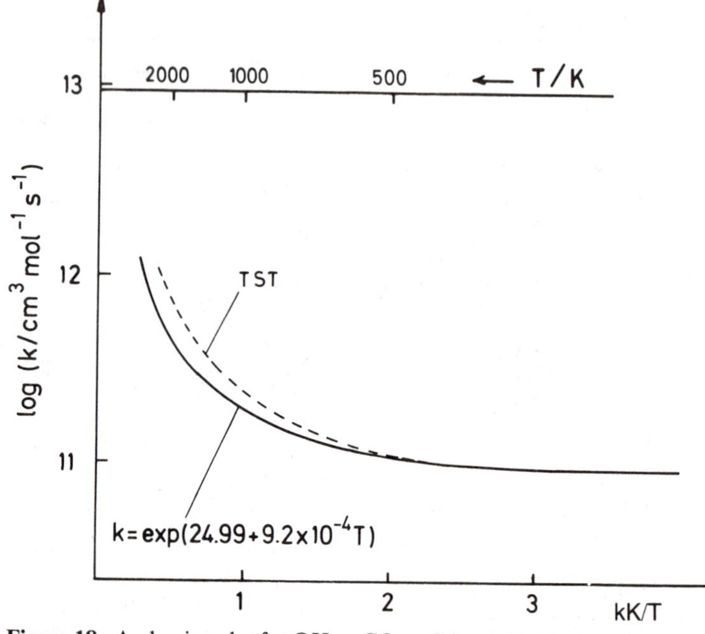

Figure 18. Arrhenius plot for $OH + CO \rightarrow CO_2 + H$. The heavy curve corresponds to the expression recommended by Baulch and Drysdale (1974) and by Steinert and Zellner (1981). The TST curve refers to a calculation by Smith and Zellner (1973).

5. Summary and conclusions

Understanding the magnitudes and temperature dependences of bimolecular gas reactions is important for detailed modeling of combustion reactions. In this chapter the theoretical background required for this was developed and compared to experimental data.

The theoretical interpretation of the temperature dependence of bimolecular rate coefficients has become a topic of fascinating complexity. This has come about in part by the development of increasingly accurate data bases, owing to use of improved experimental techniques, and in part because data have been collected over much wider temperature ranges than was possible for many years. The development may be expected to continue. For some reactions, such as those discussed in the previous section, the rate-coefficient expressions are both well defined and understandable with current theoretical methods. This is not the situation in general with bimolecular gas reactions, for many of the important ones have not only large error bars but also unknown product distributions (see Chapters 5 and 6). Experimental and theoretical advances may be expected to permit at least the most prominent of these reactions to be discussed with a degree of confidence comparable to that we now have in our understanding of the reactions discussed in this chapter.

6. Acknowledgments

The author is indebted to Professor H. Gg. Wagner for stimulation, continuous interest, and discussion of this work. Support by the Deutsche Forschungsgemeinschaft through a Heisenberg Stipendium is gratefully acknowledged. Thanks are also due to my past and present co-workers, including J. Ernst, K. Erler, and W. Steinert. Help, advice, and critical discussion with Professor W. C. Gardiner during the preparation of this chapter is thoroughly appreciated.

7. References

Arrhenius, S. (1889). Z. Physik. Chemie **4,** 226.

Baulch, D. L. & Drysdale, D. D. (1974). Comb. Flame **23,** 215.

Baulch, D. L., Drysdale, D. D., Horne, D. G., & Lloyd, A. C. (1972). *Evaluated Data for High Temperature Reactions*, Vol. 1, Butterworths, London.

Benson, S. W., (1975). J. Chem. Kin. Symp. Vol. 1, 359.

Benson, S. W., (1976). *Thermochemical Kinetics*, 2nd Ed., Wiley, New York.

Bernstein, R. B. (1966). In Adv. Chem. Phys. **10,** p. 75, J. Ross (Ed.), Interscience Publishers.

Biermann, H. W., Zetzsch, C, & Stuhl, F. (1978). Ber. Bunsenges. Phys. Chem. **82,** 633.

Birely, J. H. & Lyman, J. L. (1975). J. Photochem. **4,** 269.

Bunker, D. L., (1966). *Theory of Elementary Gas Reaction Rates*, Pergamon Press, Oxford University Press, New York.

Chan, W. H., Uselman, M., Calvert, J. G., & Shaw, J. H. (1977). Chem. Phys. Lett. **45,** 340.

Clark, T. C., Dove, J. E., & Finkelman, M. (1977). *Proc. 6th Int. Coll. Gasdynamics of Explosions and Reactive Systems*, Stockholm.

Clary, D. C., Connor, J. N. L., & Edge, C. J. (1979). Chem. Phys. Lett. **68,** 154.

Douglas, D. J. & Polanyi, J. C., (1976). Chem. Phys. **16,** 1.

Dove, J. E. & Teitelbaum, H. (1974). Chem. Phys. **6,** 431.

Dreier, T., Wolfrum, J. (1981). 18th Symp. (Int.) on Combustion, The Combustion Institute, Pittsburgh, p. 801.

Dryer, F., Naegeli, D., & Glassman, I. (1971). Comb. Flame **17,** 270.

Dubrin, J. (1973). Ann. Rev. Phys. Chem. **74,** 97.

Eliason, M. & Hirschfelder, J. O. (1959). J. Chem. Phys. **30,** 1426.

Ernst, J., Wagner, H. Gg., & Zellner, R. (1977). Ber. Bunsenges. Phys. Chem. **81,** 1270.

Ernst, J., Wagner, H. Gg., & Zellner, R. (1978). Ber. Bunsenges. Phys. Chem. **82,** 409.

Eyring, H. (1935). J. Chem. Phys. **3,** 107.

Eyring, H. & Lin, S. H. (1974). In *Physical Chemistry: An Advanced Treatise*, Vol. VI-A, H. Eyring, D. Henderson and W. Jost, Eds., Academic Press, New York.

Felder, W. & Fontijn, A. (1979). Chem. Phys. Lett. **67,** 53.

Fowler, R. H. & Guggenheim, E. A. (1949). *Statistical Thermodynamics*, Cambridge University Press, London.

Fowler, R. H. (1936). *Statistical Mechanics*, Cambridge University Press, London.

Gardiner, W. C., Jr. (1969). *Rates and Mechanisms of Chemical Reactions*, Benjamin, New York.

Gardiner, W. C., Jr., Olson, D. B., & White, J. N. (1978). Chem. Phys. Lett. **53,** 134.

Garrett, B. C. & Truhlar, D. G. (1979). J. Phys. Chem. **83,** 200.

Garrett, B. C., Truhlar, D. G., & Grev, R. S. (1980a). J. Chem. Phys. **73,** 231.

Garrett, B. C., Truhlar, D. G., & Grev, R. S. (1980b). J. Phys. Chem. **84,** 805.

Gauss, A., Jr. (1978). Chem. Phys. **68,** 1689.

Gioumousis, G. & Stevenson, D. P. (1958). J. Chem. Phys. **29,** 294.

Glass, G. P. & Chaturvedi, B. K. (1981). J. Chem. Phys. **75,** 2749.

Glasstone, S., Laidler, K., & Eyring, H. (1941). *The Theory of Rate Processes*, McGraw-Hill, New York.

Golden, D. M. (1973). 14th Symp. (Int.) on Combustion, The Combustion Institute, Pittsburgh, p. 121.

Golden, D. M. (1979). J. Phys. Chem. **83,** 108.

Haney, M. H. & Franklin, J. L., (1969). Trans. Far. Soc. **65,** 1794.

Harding, L. B. (1981). J. Phys. Chem. **85,** 10.

Herron, J. T. & Huie, R. E. (1973). J. Phys. Chem. Ref. Data **2,** 467.

Herzberg, G. (1950). *Molecular Spectra and Molecular Structure.* I. *Spectra of Diatomic Molecules*, Van Nostrand, New York.

Hirschfelder, J. O. & Linnett, J. W. (1950). J. Chem. Phys. **18,** 130.

Hirschfelder, J. O., Curtiss, C. F., & Bird, R. B. (1954). *Molecular Theory of Gases and Liquids*, Wiley, New York.

Huber, K. P. & Herzberg, G. (1979). *Molecular Spectra and Molecular Structure. Constants of Diatomic Molecules*, Van Nostrand, New York.

Johnson, B. R. & Winter, N. W. (1977). J. Chem. Phys. **66,** 4116.

Johnston, H. S. & Parr, C. (1963). J. Am. Chem. Soc. **85,** 2544.

Johnston, H. S. (1966). *Gas Phase Reaction Rate Theory*, Ronald Press, New York.

Kassel, L. S. (1932). *The Kinetics of Homogeneous Gas Reactions*, Chemical Catalog Company, New York.

Kiefer, J. H. & Lutz, R. W. (1966). J. Phys. Chem. **44,** 658, 668.

Klemm, R. B., Tanzawa, T., Skolnik, E. G., & Michael, J. V. (1981) 18th Symp. (Int.) on Combustion, The Combustion Institute, Pittsburgh, p. 785.

Kneba, M. & Wolfrum, J. (1980). Ann. Rev. Phys. Chem. **31,** 47.

Kuntz, P. J. (1976). In *Dynamics of Molecular Collisions*, Part B. W. H. Miller, Ed., Plenum, New York.

Kupperman, A. (1979). J. Phys. Chem. **83,** 171.

Lambert, J. D. (1977). *Vibrational and Rotational Relaxation in Gases*, Clarendon Press, Oxford.

LeRoy, R. L. (1969). J. Chem. Phys. **73,** 4338.

Levine, R. D. & Bernstein, R. B. (1974). *Molecular Reaction Dynamics*, Oxford University Press, New York.

Light, G. C. (1978). J. Chem. Phys. **68,** 2831.

Light, G. C. (1979). Presented at 17th Symp. (Int.) on Combustion, The Combustion Institute, Pittsburgh.

Marcus, R. A. (1965). J. Chem. Phys. **43,** 1598.

Marcus, R. A. (1974). In *Techniques of Chemistry*, Vol. 6, E. S. Lewis, Ed., Interscience, New York.

Menzinger, M. & Wolfgang, R. (1969). Angew. Chemie Int. Ed. **8,** 438.

Menzinger, M. & Jokozeki, A. (1977). Chem. Phys. **22,** 273.

Milligan, D. E. & Jacox, M. E. (1971). J. Chem. Phys. **54,** 927.

Moore, C. B. & Smith, I. W. M. (1979). In *Kinetics of State Selective Species*, Faraday Disc. No. 67, Chemical Society, London.

Morse, P. M. (1929). Phys. Rev. **34,** 57.

Mulcahy, M. F. R. (1973). *Gas Kinetics*, Nelson, London.

O'Neal, H. E. & Benson, S. W. (1973). In *Free Radicals*, J. K. Kochi, Ed., Wiley, New York.

Pauling, L. (1947). J. Am. Chem. Soc. **69,** 542.

Pechukas, P. & Pollak, E. (1979). J. Chem. Phys. **71,** 2062.

Pitzer, K. S. (1953). *Quantum Chemistry*, Prentice-Hall, New York.

Polanyi, J. C. (1972). Acc. Chem. Res. **5,** 161.

Polanyi, J. C. & Schreiber, J. L. (1974). In *Physical Chemistry: An Advanced Treatise*, Vol. VI A, H. Eyring, Ed., Academic Press, New York.

Porter, R. N. (1976). In *Dynamics of Molecular Collisions*, Part B, W. H. Miller, Ed., Plenum Press, New York.

Pratt, G. L. (1969). *Gas Kinetics*, Wiley, London.

Present, R. D. (1959). J. Chem. Phys. **31,** 7471.

Prigogine, I. (1961). J. Phys. Coll. Chem. **55,** 765.

Pritchard, H. O. (1975). In *Reaction Kinetics*, Vol. 1, Specialists Periodical Reports, Chem. Soc., London.

Pruett, J. G., Grabiner, F. R., & Brooks, P. R. (1975). J. Chem. Phys. **63,** 1173.

Quack, M. & Troe, J. (1977). Ber. Bunsenges. Phys. Chem. **81,** 329.

Rosen, B. (1970) (Ed.). *Spectroscopic Data Relative to Diatomic Molecules*, Pergamon, Oxford.

Roth, P. & Just, Th. (1977). Ber. Bunsenges. Phys. Chem. **82,** 572.

Sato, S. (1955). J. Chem. Phys. **23,** 592.

Schatz, G. C. & Walch, S. P. (1980). J. Chem. Phys. **72,** 776; J. Chem. Phys. **65,** 4642, 4648.

Schinke, R. & Lester, W. A. Jr. (1979). J. Chem. Phys. **70,** 4893.

Schott, G. L., Getzinger, R. W., & Seitz, W. A. (1974). Int. J. Chem. Kin. **6,** 921.

Shaw, R. (1977). Int. J. Chem. Kin. **9,** 929.

Shaw, R. (1978). J. Phys. Chem. Ref. Data **7,** 1179.

Shizgal, B. & Karplus, M. (1970). J. Chem. Phys. **42,** 4262.

Shizgal, B. & Karplus, M. (1971). J. Chem. Phys. **54,** 4345, 4357.

Siegbahn, P. & Liu, B. (1978). J. Chem. Phys. **68,** 2457.

Smith, I. W. M. (1977) Chem. Phys. Lett. **49,** 112.

Smith, I. W. M. (1980). *Kinetics and Dynamics of Elementary Gas Reactions*, Butterworths, London.

Smith, I. W. M. & Zellner, R. (1973). J. Chem. Soc. Farad. II. **69,** 1617.

Smith, I. W. M. & Zellner, R. (1974). J. Chem. Soc. Farad. II. **70,** 1045.

Spencer, J. E., Endo, H., & Glass, G. P. (1977). 16th Symposium (Int.) on Combustion, The Combustion Institute, Pittsburgh, p. 829.

Stevens, B. (1967). *Collisional Activation in Gases*, Int. Encyclop. Phys. Chem., Vol. 3, Pergamon, London.

Stolte, S., Henry, J. M., & Bernstein, R. B. (1976). J. Chem. Phys. **62,** 2506.

Tolman, R. C. (1927). *Statistical Mechanics with Applications to Physics and Chemistry*, Chapter 27, Chemical Catalog Company, New York.

Troe, J. & Wagner, H. Gg. (1967). Ber. Bunsenges. Phys. Chem. **71,** 937.

Truhlar, D. G. (1979). J. Phys. Chem. **83,** 188.

Truhlar, D. G. & Wyatt, R. E. (1976). Ann. Rev. Phys. Chem. **27,** 1.

Tully, F. B. & Ravishankara, A. R. (1980). J. Phys. Chem. **83,** 46.

van't Hoff, J. H. (1884). *Etudes de Thermodynamique Chimique*, F. Muller, Amsterdam.

Walch, S. P. & Dunning, Th.H. Jr., (1980). J. Chem. Phys. **72,** 1303.

Walch, S. P. *et al.* (1980). J. Chem. Phys. **72,** 2894.

Weston, R. E. Jr. & Schwarz, H. A. (1976). *Chemical Kinetics*, Prentice Hall, New York.

Widom, B. (1971). J. Chem. Phys. **55,** 44.

Wigner, E. (1938). Trans. Faraday Soc. **34,** 29.

Winter, T. G. & Hill, G. C. (1967). J. Accoust. Soc. Amer. **42,** 848.

Zahniser, M. S., Berquist, B. M., & Kaufman, F. (1978). Int. J. Chem. Kin. **10,** 15.

Zellner, R. (1979a). J. Phys. Chem. **83,** 18.

Zellner, R. (1979b). Habilitationsschrift, University of Göttingen.

Zellner, R. & Steinert, W. (1981). Chem. Phys. Lett. **81,** 568.

Chapter 4

Rate Coefficients of Thermal Dissociation, Isomerization, and Recombination Reactions

William C. Gardiner, Jr.
Jürgen Troe

1. Introduction

In this chapter we describe the estimation and interpretation of rate coefficients for thermal dissociation, isomerization, and recombination reactions. While such reactions are only a small fraction of the elementary reactions of combustion mechanisms, some of them do play essential roles. Their kinetic behavior is governed by the competition of unimolecular "chemical" changes in molecular structure with bimolecular "physical" collisional energization and deenergization processes.

Dissociation reactions have the pattern

$$A + M \rightarrow B + C + M$$

Their reverse reactions, usually called recombinations, are then

$$B + C + M \rightarrow A + M$$

Isomerization reactions take the form

$$A + M \rightarrow B + M$$

Here A, B, and C stand for the reacting atoms, radicals, or molecules and M represents "inert" collision partners. Although M appears on both sides of the chemical equations, we will see that it cannot be left out except in special circumstances. Its role as collisional energizer and deenergizer is of crucial importance. For most conditions of combustion reactions, the rates of these reactions depend on the nature and concentration of the "inert" (or "bath-gas") molecules M.

Thermal dissociation may be a relatively rapid process at the high temperatures of combustion reactions. Usually atoms and/or radicals are produced, as in the decompositions of methane and propane

$$CH_4 + M \rightarrow CH_3 + H + M$$
$$C_3H_8 + M \rightarrow CH_3 + C_2H_5 + M$$

Dissociation may also, however, lead to stable products, such as in the "molecular elimination" process of ethylene

$$C_2H_4 + M \rightarrow C_2H_2 + H_2 + M$$

In some cases dissociation proceeds along competing channels. Ethylene, for example, also dissociates via

$$C_2H_4 + M \rightarrow C_2H_3 + H + M$$

Structural isomerization reactions, such as the interconversion of normal and isopropyl radicals

$$CH_3CH_2CH_2 + M \rightarrow CH_3CHCH_3 + M$$

are often so much faster under combustion conditions than other reactions of the species concerned that for modeling purposes one can ignore the existence of such isomers or assign more or less arbitrary high values to the interconversion rate coefficients.

The reverses of simple bond fission reactions, e.g.,

$$H + CH_3 + M \rightarrow CH_4 + M$$
$$H + C_2H_2 + M \rightarrow C_2H_3 + M$$

are called recombination, combination, or radical association reactions. As far as one knows today, for all experimental conditions under which combustion reactions take place the rate coefficients in the forward and reverse directions, k_{diss} and k_{rec}, respectively, are related to one another by the corresponding equilibrium constant K_c

$$k_{diss}/k_{rec} = K_c$$

It must be emphasized that this relationship only applies when both directions of reaction are considered under identical conditions, i.e., identical temperature, pressure, and composition. Because of the equilibrium constant relationship, with the analysis of the rate coefficient in one direction we simultaneously cover both directions. In writing a reaction, we can choose the direction in which a reaction actually proceeds during a process of interest, or the direction in which its rate coefficient has been measured, or the direction which gives the more convenient theoretical formulas.

Dissociation reactions have their first important roles in the preinduction chemistry of explosive and thermal detonations, where chain centers are generated by thermal dissociation of normally stable molecules. In an H_2–O_2 detonation, for example, chain centers are produced at early times predominantly by the reaction

$$H_2 + M \rightarrow H + H + M$$

The second important role played by dissociation reactions arises in induction zones of explosions or flames, where larger intermediate species dissociate to produce smaller ones. Some well-recognized examples are

$$C_2H_5 + M \rightarrow C_2H_4 + H + M$$
$$CH_3O + M \rightarrow CH_2O + H + M$$
$$CH_2O + M \rightarrow CHO + H + M$$

Recombination reactions dominate the postflame chemistry of hydrogen, carbon monoxide, and acetylene flames, and also are generally important pathways in high-pressure combustion. Their essential role in a CO flame, for example, can be seen from the "overall" equation

$$2CO + O_2 \rightarrow 2CO_2$$

where three molecules of reactants are transformed into two product molecules. No mechanism composed of a combination of bimolecular reactions alone could lead to such a reduction in the number of molecules. Instead, in CO flames the fastest recombination processes combine with the fastest bimolecular processes to reduce the number of species and give the overall chemistry observed. In the CO case, traces of hydrogen-containing species permit the recombination reaction

$$H + O_2 + M \rightarrow HO_2 + M$$

to occur in concert with

$$OH + CO \rightarrow CO_2 + H$$

and

$$OH + HO_2 \rightarrow H_2O + O_2$$

The sum of these three reactions

$$CO + 2OH \rightarrow H_2O + CO_2$$

shows the net effect of the one recombination reaction and the two bimolecular reactions in permitting the 3 : 2 reduction in the number of molecules present. Other bimolecular reactions of the H_2–O_2 reaction also participate similarly in reducing the number of molecules of the $CO/H_2/O_2$ reaction system. (In total absence of H-containing species, the only possibility for recombination is $CO + O + M \rightarrow CO_2 + M$, a relatively slow "spin-forbidden" elementary reaction.)

So far we have not specified the chemical identity of the "inert," "bath-gas," or "third-body" species M. In the simplest view, M would represent collectively all the atoms and molecules of the reacting gas. From the ideal gas law, the bath-gas concentration $[M]$ in this view is then given by

$$[M] = P/RT$$

However, different bath-gas molecules may have very different "collision efficiencies" in promoting a particular dissociation or recombination reaction. For example, the reactions

$$H + H + N_2 \rightarrow H_2 + N_2$$
$$H + H + H_2O \rightarrow H_2 + H_2O$$
$$H + H + H \rightarrow H_2 + H$$

have such different rate coefficients (approximately in the ratio 0.4:1.0:7.0 for $M = O_2, H_2$, and H_2O, respectively; see Chapter 5, Section 1.2.) that the rate of the overall reaction

$$H + H + M \rightarrow H_2 + M$$

varies markedly with changing composition of the gas. If these effects are to be taken into account in a computer model, one includes three independent recombination reactions in the mechanism, each one with a specific identification of M. [This concept of a "linear mixture rule" does have limitations, which, however, prove to be of very minor importance in practice, cf. Troe (1980).]

The chemical reaction types considered in the present chapter are the subject of *unimolecular rate theory*, so called because unimolecular elementary processes are involved. Nevertheless, one must not forget that these unimolecular processes also interact with bimolecular collision processes involving the bath-gas molecules M. Unimolecular rate theory has been described in several monographs and reviews (Robinson and Holbrook, 1972; Forst, 1973; Troe, 1975a; Benson, 1976). In the following sections we extract from the literature a compact formalism that gives a realistic representation of the rate coefficients in order to arrive at a set of formulas simple enough to be incorporated into computer models of combustion or pyrolysis reactions. Simplified versions of unimolecular rate theory have already been presented in earlier articles by one of the authors (Troe, 1977a, 1977b, 1978, 1979, 1981, 1983).

Three areas of unimolecular rate theory are important for combustion chemistry:

(a) If experimental information is available on a reaction of interest, unimolecular rate theory enables one to expand it so as to provide rate-coefficient expressions covering all temperature, pressure, and composition conditions of practical interest. In this application, experimental rate coefficients have to be converted into forms that are suitable for use in computer modeling.
(b) Kinetic data on bond fission reactions are sometimes used to deduce thermochemical quantities, such as heats of formation of species which are difficult to access otherwise. Unimolecular rate theory provides relationships between measured activation energies and thermochemical parameters.

(c) If no experimental rate data at all are available for a reaction, unimolecular rate theory is able to provide *ab initio* estimates of the rate coefficients. These can be based either on purely theoretical arguments or on comparisons and cross-checks with related reactions for which experimental information is available.

The formulas presented in the following sections are addressed to these three applications. We do not discuss the derivation of the equations or details that are not of practical importance. Literature references are given to guide interested readers to additional information.

2. General mechanism of thermal dissociation and recombination reactions

The basic kinetic properties of unimolecular reactions can be understood in terms of simple reaction mechanisms that characterize the types of elementary processes involved. A dissociation reaction comprises three steps: collisional activation to form "excited molecules" AB^*(1), collisional deactivation (-1), and unimolecular fragmentation (2)

$$AB + M \rightarrow AB^* + M \quad (1)$$

$$AB^* + M \rightarrow AB + M \quad (-1)$$

$$AB^* \rightarrow A + B \quad (2)$$

The reverse reaction, recombination, comprises an association step that forms AB^*(-2), and redissociation (2) and collisional deactivation (-1) steps

$$A + B \rightarrow AB^* \quad (-2)$$

$$AB^* \rightarrow A + B \quad (2)$$

$$AB^* + M \rightarrow AB + M \quad (-1)$$

It can readily be seen why an energy-rich intermediate species must participate. In the dissociation direction, reaction can only occur after the molecule AB has accumulated, through collisions, enough energy to rupture the chemical bond holding A and B together; in the recombination direction, the first species formed in an A–B encounter retains all of the energy of the newly formed bond until later events, i.e., subsequent collisions, cause it to be dissipated to the environment.

The intermediate, energetically excited species AB^* does have different energy distributions among its internal quantum states in the dissociation and recombination directions. Nevertheless, it has been shown that (under "steady-state" conditions, which may always, as far as is known, be presumed to pertain for combustion) it is meaningful to treat the reaction scheme for dissociation and recombination reactions with the same formal rate coefficients k_1, k_{-1}, k_2, and k_{-2} for the steps (1), (-1), (2), and (-2), respectively.

We now define an $[M]$-dependent first-order rate coefficient k_{diss} for the dissociation reaction and an $[M]$-dependent second-order rate coefficient k_{rec} for the recombination reaction through the equations

$$k_{\text{diss}} \equiv -\frac{1}{[AB]}\frac{d[AB]}{dt} \tag{2.1}$$

and

$$k_{\text{rec}} \equiv +\frac{1}{[A][B]}\frac{d[AB]}{dt} \tag{2.2}$$

Note that there is *no* stoichiometry factor 2 in Eq. (2.1) when M is identical with AB; see the corresponding remark below in connection with Eq. (3.22).

By applying the steady-state assumption $d[AB^*]/dt = 0$ to $[AB^*]$ in the reaction mechanisms given above one quickly derives the formulas

$$k_{\text{diss}} = k_1[M]\left(\frac{k_2}{k_2 + k_{-1}[M]}\right) \tag{2.3}$$

$$k_{\text{rec}} = k_{-2}\left(\frac{k_{-1}[M]}{k_2 + k_{-1}[M]}\right) \tag{2.4}$$

Equations (2.3) and (2.4) are written such that the first factors are the rate coefficients for the starting processes ($k_1[M]$ and k_{-2}, respectively) and the second factors (in brackets) are the probabilities that after initiation the reactions are completed. The probabilities become unity for dissociation at low pressures and for recombination at high pressures; they tend to zero for dissociations at high pressures and recombinations at low pressures. (Note, however, that the probability for dissociation goes to zero as $[M]^{-1}$, which exactly cancels $[M]$ in front of the bracket so that k_{diss} approaches $k_1 k_2/k_{-1}$. Likewise, k_{rec} at low $[M]$ approaches $k_{-2}k_{-1}[M]/k_2$.)

It is easily verified that at this level of description k_{diss} and k_{rec} follow the equilibrium relationship

$$\frac{k_{\text{diss}}}{k_{\text{rec}}} = \frac{k_1}{k_{-1}}\frac{k_2}{k_{-2}} = \left(\frac{[A][B]}{[AB]}\right)_{\text{eq}} = K_c \tag{2.5}$$

where the equilibrium constant K_c is the product of the equilibrium constants k_1/k_{-1} and k_2/k_{-2} for the two elementary steps, i.e., collisional activation/deactivation and dissociation/association, as is required by the principle of detailed balancing.

Equation (2.3) describes at the first level of approximation the $[M]$ dependence (often called the "pressure" or "density" dependence although concentration is always meant) of k_{diss}. Because of the equilibrium relationship (2.5), it also describes the $[M]$ dependence of k_{rec}. Figure 1 shows Eq. (2.3) in a plot of log k as a function of log$[M]$. (Here, and throughout this paper, when the indices diss and rec are suppressed in k, k_0, or k_∞, we refer to dissociation or recombination rate coefficients indiscriminately.) This repre-

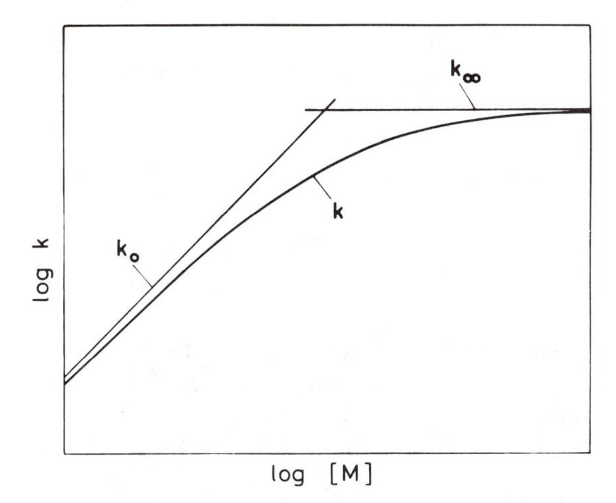

Figure 1. Fall-off curve of a thermal unimolecular dissociation or recombination reaction; from Eq. (2.3).

sentation came to be called the "fall-off curve" of the unimolecular reaction. It shows the transition of the rate coefficient k from the "low-pressure range," where it is proportional to $[M]$, to the "high-pressure range," where it approaches the constant value k_∞. The limiting rate coefficients are clearly the key quantities characterizing the general position of fall-off curves. Equations (2.3) and (2.4) give for these limiting rate coefficients

$$\lim k_{\text{diss}}([M] \to 0) \equiv k_{0,\text{diss}} = k_1[M] \tag{2.6}$$

$$\lim k_{\text{rec}}([M] \to 0) \equiv k_{0,\text{rec}} = \frac{k_{-2}}{k_2} k_{-1}[M] \tag{2.7}$$

$$\lim k_{\text{diss}}([M] \to \infty) \equiv k_{\infty,\text{diss}} = \frac{k_1}{k_{-1}} k_2 \tag{2.8}$$

$$\lim k_{\text{rec}}([M] \to \infty) \equiv k_{\infty,\text{rec}} = k_{-2} \tag{2.9}$$

It is useful to interpret Eqs. (2.6)–(2.9). The dissociation rate at low pressure is equal to $k_1[M]$, the rate of collisional activation [Eq. (2.6)]. At high pressures, the collisional activation/deactivation processes establish an equilibrium ratio of AB and AB^*, described by the rate-coefficient ratio k_1/k_{-1}, and the unimolecular dissociation process of AB^*, reaction (2), becomes rate determining [Eq. (2.8)]. In recombination at low pressures, association and redissociation of AB^* are much more frequent than collisional stabilization, such that an equilibrium between A, B, and AB^* is established, as described by the rate-coefficient ratio k_{-2}/k_2. Collisional stabilization of AB^*, reaction (-1), is then rate determining [Eq. (2.7)]. At high pressures, collisional stabilization is so frequent that the rate of association of A and B, reaction (2), determines the recombination rate [Eq. (2.9)].

A question of considerable practical importance concerns finding a compact general representation of fall-off curves. For this purpose we first define as the "center of the fall-off curve" the concentration $[M]_c$ at which the limiting k_0 ($= k_1[M]$) and k_∞ lines indicated in Fig. 1 intersect. This allows one to express the dependence of k upon $[M]$ in a unifying way by introducing a reduced $[M]$ scale x

$$\frac{[M]}{[M]_c} = \frac{k_0}{k_\infty} = x \tag{2.10}$$

(Equation (2.10) is derived by setting $k_1[M]_c = k_\infty$ and, from Eq. (2.6), $k_{0,\text{diss}} = k_1[M]$.) Second, we express k relative to its limiting high-pressure rate coefficient k_∞. With this procedure, Eqs. (2.3) and (2.4) both lead after some algebra to "reduced fall-off curves" expressed as

$$\frac{k}{k_\infty} = \frac{k_0/k_\infty}{1 + k_0/k_\infty} = \frac{[M]/[M]_c}{1 + [M]/[M]_c} = \frac{x}{1 + x} = F_{LH}(x) \tag{2.11}$$

where $x = [M]/[M]_c$.

Equation (2.11) can be considered as a "switching function" describing the transition of k from k_0 to k_∞ with a value of $k = k_\infty/2$ at the center of the fall-off curve. We call Eq. (2.11), which is based on the model for unimolecular reactions that was first given 50 years ago by Lindemann and Hinshelwood, the Lindemann-Hinshelwood expression. It is shown in Fig. 2.

More detailed theories, as described in the following sections, take into account that in reality the rate coefficients of the Lindemann-Hinshelwood mechanism are averages over the quantum states of AB, AB^*, A, and B. For

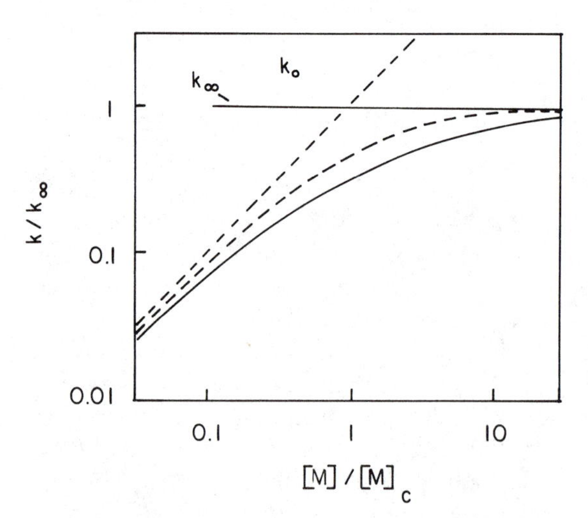

Figure 2. Reduced fall-off curve of a dissociation or recombination reaction. The dashed curve is given by Eq. (2.11); the solid one corresponds to observed fall-off behavior.

each of these states a different rate coefficient for each elementary processes applies. Nonetheless, the general properties of the kinetics turn out to remain qualitatively the same. The detailed theories permit explicit calculation of the two key rate coefficients k_0 and k_∞ from molecular parameters and provide switching functions for the fall-off range that are in closer agreement with experiments. As a general rule, fall-off curves are broader than the Lindemann-Hinshelwood expression of Eq. (2.11) and they are below it at the center, as indicated by the lower curve in Fig. 2. An analysis of recent theories (Troe, 1979, 1983) shows that realistic fall-off curves can be represented well by adding a broadening factor $F(x)$ to Eq. (2.11)

$$\frac{k}{k_\infty} = F_{LH}(x)F(x) \tag{2.12}$$

which in first approximation has the logarithmic form

$$\log F(x) = \left(\frac{1}{1 + \{\log(x)\}^2}\right) \log F_{\text{cent}} \tag{2.13}$$

Here $\log F_{\text{cent}}$ gives the depression of the fall-off curve at the center relative to the Lindemann-Hinshelwood expression in a $\log(k/k_\infty)$-vs-$\log(x)$ plot like Fig. 2. It turns out that F_{cent} is a weak function of T and the nature of M that can be estimated by theory. Experimental fall-off curves can be fitted to the form of Eq. (2.13) and characterized by the three quantities k_0, k_∞, and F_{cent}. In essence, Eq. (2.13) is a first step beyond the Lindemann-Hinshelwood expression. Still more realistic, but more complex, expressions are given in Section 5.

3. Low-pressure rate coefficients

In this section we provide procedures for predicting and analyzing low-pressure rate coefficients k_0 for dissociation or recombination reactions. As discussed in Section 2, the low-pressure rate coefficient for dissociation $k_{0,\text{diss}}$ is given by the rate coefficient for collisional activation k_1 according to $k_{0,\text{diss}} = k_1[M]$. It is convenient to express k_1 by k_{-1} via the principle of detailed balancing

$$\frac{k_1}{k_{-1}} = \left(\frac{[AB^*]}{[AB]}\right)_{\text{eq}} \equiv f^* \tag{3.1}$$

and to express k_{-1} as the product of the gas kinetic collision frequency at unit concentration Z and an effective "collision efficiency" β_c, such that

$$k_{0,\text{diss}} = k_1[M] = \beta_c Z[M]f^* \tag{3.2}$$

If nearly all $M - AB^*$ collisions result in deactivation, $\beta_c \simeq 1$. This situation does pertain for large bath-gas molecules at low temperatures and is called the "strong collision limit." At combustion temperatures, however, β_c is much smaller than unity. Modern theories discuss the relationship between β_c and the process of collisional energy transfer between M and AB^* (Troe, 1977a, 1977b, 1979).

The basic significance of f^* is that it gives the equilibrium fraction of molecules that can dissociate. The simplest, not quite correct, assumption is that all molecules having enough energy to dissociate would be in the AB^* category. In fact, molecules in different rotational quantum states have different minimum total energies for dissociation. Neglecting this distinction, we can compute the equilibrium fraction of molecules having energy equal to or greater than an amount E_0 by integrating over all energies above E_0

$$f^* = \int_{E_0}^{\infty} f(E)\, dE \tag{3.3}$$

where $f(E)$ represents the Boltzmann distribution in those states of AB that have energy above E_0, the minimum energy at which dissociation can occur. E_0, relative to the ground-state energy of AB, is variously called the critical energy or the dissociation threshold energy. For a simple bond fission reaction where the reverse recombination reaction does not involve an energy barrier (such as the dissociation of H_2O into OH and H or the dissociation of a hydrocarbon into two alkyl radicals), E_0 is equal to the enthalpy change of reaction at 0 K ΔH_0°. For complex elimination reactions, such as dissociation of C_2H_4 to C_2H_2 and H_2, where the reverse reaction has an energy barrier, E_0 is equal to ΔH_0° plus this barrier height. At present, such barriers can only be derived from rate measurements. The meanings of E_0 for these two cases are contrasted in Fig. 3.

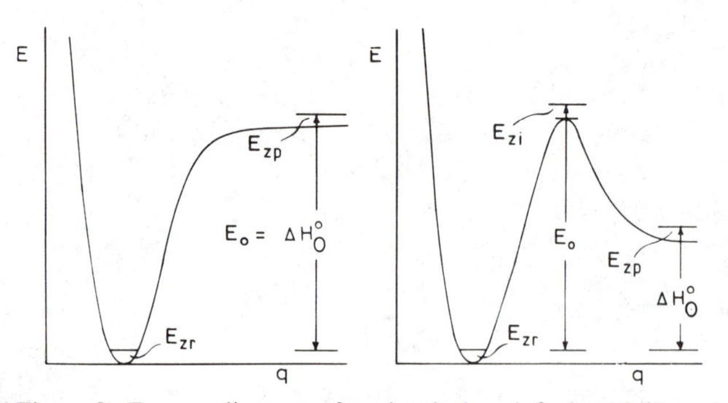

Figure 3. Energy diagrams for simple bond fission (left) and complex elimination (right) reactions. The molecular internal energy is plotted as a function of "reaction coordinate" q, i.e., of a suitable bond length changing markedly during the reaction. The zero-point energies of the reactant, product, and intermediate "activated complex" structures are denoted by E_{zr}, E_{zp}, and E_{zi}, respectively. ΔH_0° and E_0 are the enthalpy of reaction at 0 K and the critical energy for dissociation.

We now describe how to calculate k_0. Statistical mechanics provides expressions for f^*. The simplest of them is based on the approximation that the internal energy of a molecule is contained as vibrational energy of s harmonic oscillators, in equilibrium with a heat bath, and that classical statistical mechanics applies. For a molecule containing N atoms, the total number of internal "degrees of freedom" is given by $3N - 6$ if the molecule is nonlinear and $3N - 5$ if it is linear. For molecules with internal rotations, the number of oscillators plus the number of internal rotors equals the total number of internal degrees of freedom. The fraction of molecules whose total internal energy exceeds E_0 is given in this simplest model by the approximate formula

$$f^* \cong \frac{1}{(s-1)!}\left(\frac{E_0}{k_B T}\right)^{s-1} \exp(-E_0/k_B T) \tag{3.4}$$

With this approximation Eq. (3.2) becomes

$$k_{0,\,\mathrm{diss}} \cong \frac{\beta_c Z[M]}{(s-1)!}\left(\frac{E_0}{k_B T}\right)^{s-1} \exp(-E_0/k_B T) \tag{3.5}$$

which is known as the low-pressure limiting form of the classical RRK rate coefficient (Kassel, 1932). At high temperatures the factor

$$\sum_{i=0}^{s-1} \frac{(s-1)!}{(s-1-i)!}\left(\frac{k_B T}{E_0}\right)^{i}$$

should multiply Eqs. (3.4) and (3.5), as Eq. (3.4) pertains only as a limit for $E_0 \gg k_B T$.

Equation (3.5) incorrectly neglects the quantum nature of molecular vibrations and is found to give rate coefficients that are too large for typical molecules under combustion conditions. Since a lower f^* can be obtained from Eq. (3.4) by taking s to be less than the total number of internal degrees of freedom, the notion of "effective number of oscillators" was introduced, allowing s to be adjusted to agree with experiments. For reasons that will be made clear below, this procedure is not the best one can follow today, and we recommend against its use.

By quantum statistical mechanics one evaluates expressions for f^* which are more appropriate than Eqs. (3.4) and (3.5). Then, to a good approximation, the contributions to f^* can be factored such that Eq. (3.2.2) becomes (Troe, 1977b)

$$k_{0,\,\mathrm{diss}} \cong [M] Z_{LJ} \beta_c \frac{\rho_{\mathrm{vib},h}(E_0) k_B T}{Q_{\mathrm{vib}}} \exp(-E_0/k_B T) F_E F_{\mathrm{anh}} F_{\mathrm{rot}} F_{\mathrm{rot\,int}} \tag{3.6}$$

The significance of the various factors is as follows.

The Boltzmann-factor $\exp(-E_0/k_B T)$ generally gives the main temperature dependence of k_0. However, the other factors also depend on T, in particular the vibrational partition function Q_{vib}. To calculate the other factors of f^*, explicit reference to the molecular vibrations is required. The molecular vibration frequencies ν_i can be taken from tabulations or estimated (see Benson, 1976). These frequencies enter into the vibrational partition function

$$Q_{\text{vib}} = \prod_{i=1}^{s} \left[1 - \exp\left(\frac{h\nu_i}{k_B T}\right)\right]^{-1} \tag{3.7}$$

and contribute to the harmonic oscillator density of states at the threshold energy as

$$\rho_{\text{vib},h}(E_0) = \frac{(E_0 + a(E_0)E_z)^{s-1}}{(s-1)! \prod_{i=1}^{s} (h\nu_i)} \tag{3.8}$$

where the zero-point energy E_z is given by

$$E_z = \tfrac{1}{2} \sum_{i=1}^{s} (h\nu_i) \tag{3.9}$$

and $a(E_0)$ is a correction factor with (Whitten and Rabinovitch, 1963)

$$a(E_0) = 1 - \beta w \tag{3.10}$$

$$\log w = -1.0506(E_0/E_z)^{0.25} \text{ at } E_0 > E_z \tag{3.11}$$

$$w^{-1} = 5(E_0/E_z) + 2.73(E_0/E_z)^{1/2} + 3.51 \text{ at } E_0 < E_z \tag{3.12}$$

$$\beta = (s-1)\left(\sum_{i=1}^{s} \nu_i^2\right) \Big/ \left(\sum_{i=1}^{s} \nu_i\right)^2 \tag{3.13}$$

Values of β in the range 1.2 ± 0.3 are typical. Equations (3.10)–(3.13) are empirical approximations which are accurate enough for nearly all applications. If $E_0 \ll E_z$, however, it is not sufficient. Then the density of states has to be calculated exactly by numerical counting (Stein and Rabinovitch, 1973).

The factor F_E accounts for the energy dependence of the density of states. It is given by

$$F_E = \frac{\int_{E_0}^{\infty} f(E)\,dE}{f(E_0)\,kT} \simeq \sum_{i=0}^{s-1} \frac{(s-1)!}{(s-1-i)!} \left(\frac{kT}{E_0 + a(E_0)E_z}\right)^i \tag{3.14}$$

The factor F_{anh} accounts for anharmonicity of the molecular vibrations. It can be taken approximately as unity for complex elimination reactions; for simple bond fissions it is less approximately given by

$$F_{\text{anh}} \simeq \left(\frac{s-1}{s-3/2}\right)^m \tag{3.15}$$

where m denotes the number of oscillators which disappear during the reaction. The factor F_{rot} takes into account the effect of molecular rotation. It can be approximated by

$$F_{\text{rot}} \simeq F_{\text{rot max}} \left(\frac{I^+/I}{I^+/I - 1 + F_{\text{rot max}}}\right) F_{E_z} \tag{3.16}$$

where for a linear molecule

$$F_{\text{rot max}} \simeq \frac{E_0 + a(E_0)E_z}{sk_B T} \tag{3.17}$$

and for a nonlinear molecule

$$F_{\text{rot max}} \simeq \left(\frac{E_0 + a(E_0)E_z}{k_B T}\right)\frac{(s-1)!}{(s+\frac{1}{2})!} \tag{3.18}$$

For a simple bond fission reaction, I^+/I is given to a first approximation by (Troe, 1977b)

$$I^+/I \simeq 2.15(E_0/k_B T)^{1/3} \tag{3.19}$$

and for molecular elimination reaction one must simply guess the ratio of the principal moment of inertia of the molecule at the reaction barrier to the principal moment of inertia of the molecule at the equilibrium configuration. F_{rot}, as approximated by Eqs. (3.16)–(3.19), generally agrees well with detailed calculations (Troe, 1977a, 1977b, 1979). The factor

$$F_{E_z} = \exp\left(-\frac{\Delta E_{0z}}{k_B T}\right) \tag{3.20}$$

accounts for the appearance of "adiabatic zero-point barriers" in simple bond fission reactions. Only at low temperatures ($\leq 300\ K$) and for very rigid potential surfaces (e.g., O_3) can it be much smaller than unity (Troe, 1979). At combustion temperatures it can always be taken as unity.

Internal rotors are accounted for by the factor $F_{\text{rot int}}$. It is sufficient to treat these degrees of freedom as free rotors. If there are r internal free rotors, one has

$$F_{\text{rot int}} \simeq \frac{(s-1)!}{(s-1+r/2)!}\left(\frac{E_0 + a(E_0)E_z}{k_B T}\right)^{r/2} \tag{3.21}$$

(Note that $s + r$ must be $3N - 6$ for a linear or $3N - 5$ for a nonlinear molecule.)

The Lennard-Jones collision frequency Z_{LJ} can be used as the overall rate coefficient for collisional energy transfer. It is given by

$$Z_{LJ} = N_A \pi \sigma_{AB-M}^2 \sqrt{8\pi k_B T/\mu_{AB-M}}\, \Omega_{AB-M}^{(2,2)*} \tag{3.22}$$

with Avogadro's number N_A, the collision diameter σ_{AB-M}, the reduced mass $\mu_{AB-M} = m_{AB}m_M/(m_{AB} + m_M)$ with the molecular weights m_{AB} and m_M, and the reduced collision integral $\Omega_{AB-M}^{(2,2)*}$.[1] The latter quantity can be approximated by

$$\Omega_{AB-M}^{(2,2)*} \simeq [0.70 + 0.52 \log(k_B T/\varepsilon_{AB-M})]^{-1} \tag{3.23}$$

---

[1] The factor π in Eq. (3.22) was inadvertently omitted from the corresponding equations in two earlier publications, but not in the computations reported therein. (Troe 1977b, 1979)

It can be seen that Eq. (3.23) is a weakly temperature-dependent function of value near unity. If $\Omega_{AB-M}^{(2,2)*}$ is set to unity, then Eq. (3.22) is the hard-sphere collision frequency. The difference between the hard-sphere and the Lennard-Jones collision frequencies is small compared to the uncertainty in β_c, as discussed below. (Note that, consistent with the definition of the rate coefficient of Eq. (2.1), Eq. (3.22) also applies to situations where $M = AB$.) The collision diameters $\sigma_{AB-M} = \sigma_{AB-AB} + \sigma_{M-M}$ and interaction energies $\varepsilon_{AB-M} = \sqrt{\varepsilon_{AB-AB}\varepsilon_{M-M}}$ are determined either from tabulated (Hirschfelder, Curtiss, and Bird, 1963; Reid and Sherwood, 1966) Lennard-Jones parameters or, if the desired values are not tabulated, from the empirical formulas (Reid and Sherwood, 1966)

$$2/3\pi N_A \sigma^3 = 2.3 V_m \tag{3.24}$$

and

$$\varepsilon/k_B \cong 1.18 T_b \tag{3.25}$$

where V_m is the molar volume of the substance at its melting point and T_b is the boiling (or sublimation) temperature.

A major problem is estimation of the weak collision factor (Troe, 1977b, 1979; Quack and Troe, 1977a; Tardy and Rabinovitch, 1977; Endo, Glänzer, and Troe, 1979). This quantity can be related to the average energy transferred per collision $\langle \Delta E \rangle$ by (Troe, 1977a)

$$\frac{\beta_c}{1 - \sqrt{\beta_c}} \cong \frac{-\langle \Delta E \rangle}{F_E k_B T} \tag{3.26}$$

(Equation 3.26 holds for $1 \leq F_E \lesssim 3$. For very large molecules at very high temperatures, F_E may exceed 3; for this case β_c has been calculated by Gilbert, Luther, and Troe, 1983.) For temperatures near room temperature β_c was found to vary with molecular size from 0.1 to 0.8 for a series of bath gases M from monatomic to polyatomic species. Monatomic and diatomic bath gases are characterized by values ranging from 0.1 to 0.5. At 1000–2000 K, dissociation reactions of triatomic molecules generally have been found to give β_c values which increase from 0.02 to 0.2 from monatomic toward polyatomic bath gases (Endo, Glänzer, and Troe, 1979). For these molecules, the temperature dependence of $\langle \Delta E \rangle$ was found to be weak; $\langle \Delta E \rangle$ proportional to $T^{-0.5 \pm 0.5}$ did reproduce the $H_2O + Ar \rightleftarrows H + OH + Ar$ system quite well over the temperature range 200–6000 K (Troe, 1979). For larger reactant molecules (Klein and Rabinovitch, 1978; Klein, Rabinovitch, and Jung, 1977) a much more pronounced decrease of β_c with increasing T was observed near 1000 K. The consequences of this high-temperature behavior have not been fully explored (see Gilbert, Luther, and Troe, 1983); however, it is clear that very low values of β_c are possible. The reason why there is so little knowledge about collision efficiencies in dissociation reactions of polyatomic molecules at high temperatures is easily understood. Under the conditions accessible to experiments, these reactions are almost invariably in the fall-off range, and an analysis to provide k_0 is difficult to carry out. On the other hand, uncertainties in the collision efficiencies for the same reason do not strongly influence the rate-coefficient predictions. For the time being, values for β_c in the range 0.1–0.01 can be used at combustion temperatures, i.e., 1000–3000 K, for estimating k_0 values for the dissociation of polyatomic molecules in simple (monatomic or small polyatomic) bath gases. Whenever experiments are available, a reasonable procedure would be to fit β_c at the most reliable experimental point, derive a value for $\langle \Delta E \rangle$, and use for extrapolation to other conditions the relationship described above assuming a temperature-independent $\langle \Delta E \rangle$. Otherwise $\langle \Delta E \rangle$ can be estimated from typical values for the various bath gases M such as tabulated in Table 1.

Apart from the analysis of the energy transfer properties as expressed by Z_{LJ} and β_c, the foregoing discussion of k_0 represents an elaboration of the low-pressure limit of RRKM theory (Robinson and Holbrook, 1972). Since calculation of the various factors of Eq. (3.6) is time consuming, it is often avoided in favor of the simpler RRK theory, in which all of the factors of Eq. (3.6) are given the form prescribed by classical statistical mechanics, i.e., Eq. (3.5). This simplification can give quite unsatisfactory results. Inspection

Table 1. Average energies transferred per collision for various bath gases M (in kJ/mol).

	(a)	(b)	(c)
He	1.8	2.0	11
Ar	3.7	1.2	3.0
Xe	6.4	0.6	4.4
H_2	2.6		
N_2	2.9	2.0	5.5
O_2	4.4		3.6
CO_2	10	2.6	17
N_2O	7.5		
CH_4	8.8		
CF_4	13	1.5	8
SF_6	9.6	4.6	
C_2H_6	17		
C_3H_8	18		
C_4H_{10}	30		
C_5H_{12}	18		

[a] Troe and Wieters, 1979; from photoisomerization of C_7H_8 at 300 K.
[b] Endo, Glänzer, and Troe, 1979; from the thermal decomposition of ClNO at 1000 K.
[c] Endo Glänzer, and Troe, 1979; from the thermal decomposition of NO_2 at 1800 K.

Table 2. Low-pressure rate coefficients for thermal decomposition of N_2O_5 in N_2.

T/K	$k_0 \exp(+E_0/k_BT)/[N_2]$		
	Exact[a]	RRK $s = 18$[b]	RRK $s = 9$[c]
295	1.4×10^{21}	7.7×10^{25}	4.6×10^{21}
380	3.6×10^{20}	9.3×10^{23}	5.4×10^{20}
520	4.5×10^{19}	3.3×10^{21}	3.2×10^{19}

[a] Equation (3.6), from Malko and Troe, 1981.
[b] Equation (3.5), with $s = 18$, $\beta_c Z$ as from Malko and Troe, 1981.
[c] Same, $s = 9$.

of the various factors in k_0 [Eq. (3.6)] indicates that the simplifications may introduce large errors and should not be made. This is demonstrated in Table 2 where, for the $N_2O_5 + M \rightleftarrows NO_2 + NO_3 + M$ reaction, Eqs. (3.5) and (3.6) are compared.

Equation (3.6) and the elaboration of its factors satisfactorily explain the typical features of k_0:

(i) The preexponential factor of $k_{0,\text{diss}}$ far exceeds the collision frequency $Z_{LJ}[M]$, increasing with molecular complexity to values many orders of magnitude larger than Z_{LJ}.
(ii) The activation energy $E_{a,0} = -R\, d \ln(k_{0,\text{diss}}/[M]d(1/T))$ of $k_{\text{diss},0}$ is less than E_0. It decreases with increasing temperature, more strongly for larger molecules.

The first observation (i) is mainly due to the large increase of $\rho_{\text{vib},h}(E_0)$ with increasing molecular complexity. The second observation (ii) follows mainly from the strong temperature dependence of Q_{vib}. As a rough estimate

$$E_{a,0}(T) \simeq E_0 - U_{\text{int}}(T) \tag{3.29}$$

where $U_{\text{int}}(T)$ is the internal (rotational plus vibrational) part of the thermal energy of the reactant molecule. One can illustrate these properties either by showing the preexponential factors of $k_{0,\text{diss}}$ for various reactions and temperatures or, alternatively, by showing the corresponding third-order recombination rate constants $k_{0,\text{rec}}/[M]$ (using $k_{0,\text{diss}}/k_{0,\text{rec}} = K_c$). Such a representation (Troe, 1978) is shown in Fig. 4 for dissociation–recombination reactions of various molecules in the bath-gas Ar.

A final remark concerns the influence of uncertainties in k_0 on the rate coefficient in the fall-off range. This can be understood from Fig. 2. When a reaction is studied at M concentrations much greater than $[M]_c$, i.e., near the high-pressure limit, an uncertain value of k_0 would have only a very small influence on the predicted rate coefficient, since the fall-off effects are small and only weakly dependent upon k_0 there. Closer to the low-pressure limit, of course, this uncertainty becomes important. Near the center of the fall-off curve the extent of the effect of k_0 uncertainty can be estimated by noting the effects of vertical and horizontal shifts of the relevant fall-off curves.

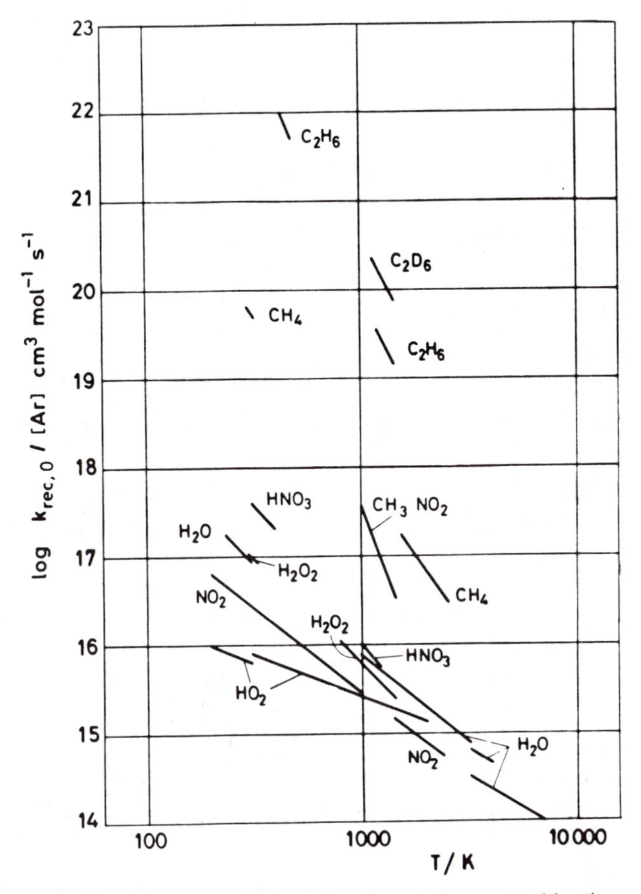

Figure 4. Low-pressure recombination rate coefficients for dissociation-recombination systems of polyatomic molecules (bath gas M = Ar; for references see Troe, 1978).

4. High-pressure rate coefficients

According to Eqs. (2.8) and (3.1), the high-pressure limiting rate coefficient $k_{\infty,\mathrm{diss}}$ is given by

$$k_{\infty,\,\mathrm{diss}} = \frac{k_1}{k_{-1}} k_2 = f^* k_2 \tag{4.1}$$

In this pressure range the actual populations of excited molecular states are close to the equilibrium populations given by f^*. The rate coefficient k_2 is therefore the thermal equilibrium average of the specific rate coefficient $k(E)$ of the unimolecular reaction. The rate coefficient for the reverse recombination of a simple bond fission reaction

$$k_{\infty,\,\mathrm{rec}} = k_{-2} \tag{4.2}$$

is equal to the rate coefficient of association of the two fragments.

Equation (4.1) can be expressed in the form of transition-state theory (Chapter 3)

$$k_{\infty,\,\mathrm{diss}} = \frac{kT}{h}\frac{Q^{\ddagger}}{Q}\exp\left(-\frac{E_0}{k_B T}\right) \tag{4.3}$$

where $Q^{\ddagger}$ is the "activated complex" partition function and Q is the reactant molecule partition function. Equation (4.3) allows one to classify two cases: (i) For rigid activated complex systems, which are characterized by having activation barriers for the reverse reaction, the ratio $Q^{\ddagger}/Q$ is smaller than or close to unity and the reverse reaction has a positive activation energy; (ii) For loose activated complex systems such as simple bond fission reactions, characterized by having no activation barrier for the reverse recombination, the ratio $Q^{\ddagger}/Q$ is of the order 10^1–10^4 and the reverse reaction has no or only weak temperature dependence.

There are various methods for estimating $Q^{\ddagger}/Q$ and hence k_{∞}: (i) In traditional transition-state theory a set of activated complex frequencies is chosen by analogy to related experimental examples and $Q^{\ddagger}$ is calculated by Eq. (3.5) (Chapter 3 and Benson, 1976). (ii) In the statistical adiabatic channel model $Q^{\ddagger}$ is derived by interpolation between reactant and product partition functions or reactant and product vibration frequencies (Troe, 1981; Quack and Troe, 1974, 1977b). The general problem of these approaches is the fact that empirical guesses are required. Therefore, a single good experiment for the reaction under consideration helps to fix the missing unknown parameters.

We do not intend to discuss these approaches but propose instead a simpler procedure which is not much more empirical than they are. Here we have to distinguish various cases.

(i) Simple bond fission reactions: Here it is most convenient to start from $k_{\infty,\,\mathrm{rec}}$ and to obtain $k_{\infty,\,\mathrm{diss}}$ by conversion via the equilibrium constant K_c. Typical values of $k_{\infty,\mathrm{rec}}$ are of the order 10^{13} cm^3 mol^{-1} s^{-1} independent of temperature over the range 200–2000 K (Table 3). For H + radical recombination, $k_{\infty,\,\mathrm{rec}}$ can increase up to about 10^{14} cm^3 mol^{-1} s^{-1}. For sterically unfavorable cases, $k_{\infty,\,\mathrm{rec}}$ can be as low as 10^{12} cm^3 mol^{-1} s^{-1}. A reasonable prediction can be made by comparison with other reactions such as those listed in Table 3. If experiments on the reaction of interest are available, $k_{\infty,\,\mathrm{rec}}$ should be found at one temperature, using the reduced fall-off curves if necessary to extrapolate to the high-pressure limit, and assumed to be nearly temperature independent.

(ii) Bond fission reactions with small (<about 20 kJ) barriers for the reverse recombination: In a number of cases, such as the addition of atoms to unsaturated molecules, which in reverse direction are radical dissociation reactions like $C_2H_5 \rightarrow H + C_2H_4$, one has a so-called rigid potential and an activation barrier in both directions. In this case, $k_{\infty,\,\mathrm{rec}} \simeq (10^{12}$–$10^{13})$ $\exp(-E_{0,\mathrm{rec}}/k_B T)$ cm^{-1} mol^{-1} s^{-1}, where the barrier $E_{0,\mathrm{rec}}$ has to be determined experimentally. Some spin-forbidden reactions, like

Table 3. High-pressure recombination rate coefficients.

Reaction	T/K	$k_{rec,\infty}/cm^3\ mol^{-1}\ s^{-1}$	Comments
Reactions without energy barriers			
$H + CH_3 \rightarrow CH_4$	308	2.3×10^{14}	(a)
$O + NO \rightarrow NO_2$	300	1.7×10^{13}	(b)
$Cl + NO \rightarrow ClNO$	300	5.8×10^{13}	(b)
$2CH_3 \rightarrow C_2H_6$	300	2.9×10^{13}	(b)
$2C_2H_5 \rightarrow C_7H_{10}$	300	8.4×10^{12}	(a)
$2C_3H_7 \rightarrow C_6H_{14}$	300	5.0×10^{12}	(a)
$H + C_7H_7 \rightarrow C_7H_8$	300	2×10^{14}	(c)
Reactions with small energy barriers			
$H + C_2H_2 \rightarrow C_2H_3$		$5.5 \times 10^{12} \exp(-1210/T)$	(d)
$H + C_2H_4 \rightarrow C_2H_5$		$2.2 \times 10^{13} \exp(-1040/T)$	(e)
$H + C_6H_6 \rightarrow C_6H_7$		$5.3 \times 10^{12} \exp(-1360/T)$	(f)
Spin-forbidden reactions			
$O + CO \rightarrow CO_2$		$1.8 \times 10^{10} \exp(-1200/T)$	(g)

[a] References in Frey and Walsh, 1979a.
[b] References in Quack and Troe, 1977a.
[c] Pagsberg and Troe, unpublished results.
[d] Payne and Stief, 1976 (Chapter 5).
[e] Lee *et al.* 1978 (Chapter 5).
[f] Kim *et al.* 1973 (Chapter 5).
[g] Troe, 1975; not valid over 1500 K (Chapter 5).

$CO_2 \rightarrow CO + O$, also fall into this class; these reactions, however, typically have a preexponential factor of $k_{\infty,rec}$ which is 10–100 times smaller. One example is included in Table 3.

(iii) Complex bond fission, elimination and isomerization reactions: These reactions have high barriers in both the forward and reverse directions. Their energy barriers E_0 have to be derived from experiments; the preexponential factors of k_∞ are of the order k_BT/h or smaller. For this class of reactions one really has to rely on rate measurements. If they are simply not available one can only guess by comparison with related reactions. Extensive tables for this type of reaction are available. Examples are given in Table 4 (Benson and O'Neal, 1970; Frey and Walsh, 1979).

Table 4. High-pressure rate coefficients for elimination reactions[a].

Reaction	k_∞/s^{-1}
$C_2H_5Cl \rightarrow C_2H_4 + HCl$	$10^{13.3} \exp(-28400/T)$
$CH_3CF_3 \rightarrow CH_2CF_2 + HCl$	$10^{14.6} \exp(-36600/T)$
cis-2-butene → butadiene + HCl	$10^{13.0} \exp(-33000/T)$
ethyl allyl ether → propane + acetaldehyde	$10^{11.8} \exp(-21900/T)$

[a] Refs. in Quack and Troe, 1977a.

5. Rate coefficients in the intermediate fall-off range

From the limiting low- and high-pressure rate constants k_0 and k_∞, respectively, one immediately derives the pressure at which fall-off behavior is to be expected. The center of the fall-off curve, i.e., that bath-gas concentration $[M]_c$ at which the extrapolated limiting rate coefficients intersect, is given by rearranging Eq. (2.10) to

$$[M]_c = \frac{k_0/[M]}{k_\infty} \tag{5.1}$$

The intermediate rate coefficient k switches from k_0 to k_∞ when $[M]$ increases from values much below $[M]_c$ to values above $[M]_c$. As mentioned in Section 2, this switching function is very complicated and in fact has never been analyzed in full detail. Weak collision effects, rotational effects, and other factors do influence the switching function to some extent. Fortunately, however, according to our present knowledge, all these effects are small, and simple empirical expressions like Eq. (2.13) are useful in practice.

We concentrate attention first on the broadening factor F_{cent} defined in Eq. (2.13), which defines the depression of the fall-off curve at the center relative to the Lindemann-Hinshelwood function. F_{cent} can be expressed to good approximation by the three parameters S_K, B_K, and β_c. S_K is defined by the internal energy $U_{\text{int}}^{\ddagger}$ divided by $k_B T$

$$S_K = 1 + \frac{U_{\text{int}}^{\ddagger}}{k_B T} = 1 - \frac{1}{T}\frac{\partial \ln Q_{\text{int}}^{\ddagger}}{\partial\, 1/T} \tag{5.2}$$

where $Q^{\ddagger}$ is the rotation–vibration partition function of the activated complex. To a good approximation, S_K can be related to the corresponding ratio $U_{\text{int}}/k_B T$ of the parent molecule (Troe, 1983) by

$$S_K \cong \frac{U_{\text{int}}}{k_B T} + 2(\pm 1) \tag{5.3}$$

and hence one can use tabulated molar enthalpies $H_T^\circ - \text{H}_0^\circ$

$$S_K \cong \frac{H_T^\circ - H_0^\circ}{RT} - \tfrac{1}{2}(\pm 1) \tag{5.4}$$

If $H_T^\circ - H_0^\circ$ is not available for the molecule of interest, one has to use Eq. (3.7) for Q_{vib}, adding an internal rotational contribution to S_K of $r/2$ if the molecule has internal rotations.

Using this S_K one defines B_K as

$$B_K = \left(\frac{S_K - 1}{s - 1 + r/2}\right)\left(\frac{E_0 + a(E_0)E_z}{k_B T}\right) \tag{5.5}$$

with E_0, $a(E_0)$, and E_z defined as in Section 3. To a good approximation, F_{cent} is then given by (Troe, 1979)

$$F_{\text{cent}} \simeq \beta_c^{0.14}\{F_1 + F_2 \exp(-B_K/19.5) + [1 - F_1 - F_2] \times \exp(-2.3[B_K/F_3]^{1.5})\} \tag{5.6}$$

$$F_1 = 1.32 \exp(-S_K/4.2) - 0.32 \exp(-S_K/1.4) \tag{5.7}$$

$$F_2 = 1 - \exp(-S_K/30) \tag{5.8}$$

$$F_3 = 7.5 + 0.43 S_K \tag{5.9}$$

Alternative expressions to Eqs. (5.6)–(5.9) have been proposed (Troe, 1983).

Although the expressions in Eqs. (5.3)–(5.8) look somewhat complicated, they are in fact easy to evaluate. Using this procedure to evaluate F_{cent}, the full fall-off curve is given to a first approximation by Eqs. (2.12) and (2.13). To a second approximation, needed particularly at high temperatures, one uses Eq. (2.12) with $F(x)$ containing a scaling factor N defined by

$$\log F(x) = \left(\frac{1}{1 + \left(\frac{\log([M]/[M]_c)}{N}\right)^2}\right) \log F_{\text{cent}} \tag{5.10}$$

$$N \simeq 0.75 - 1.27 \log F_{\text{cent}} \tag{5.11}$$

An example of the broadening effect at combustion temperatures is shown in Fig. 5. At such high temperatures, additional approximations can be added, introducing asymmetry into the fall-off curves. These higher approximations

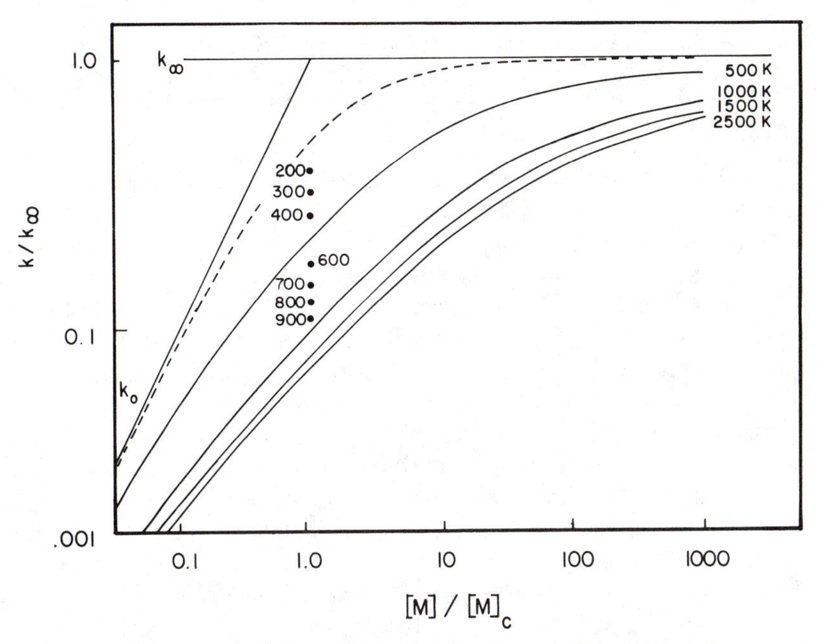

Figure 5. Fall-off curves at high temperatures for thermal dissociation of propane $C_3H_8 \rightarrow CH_3 + C_2H_5$, from Eqs. (5.3)–(5.11). The rapid fall in k/k_∞ at moderate temperatures (dots at $M/M_c = 1$) becomes much slower at combustion temperatures.

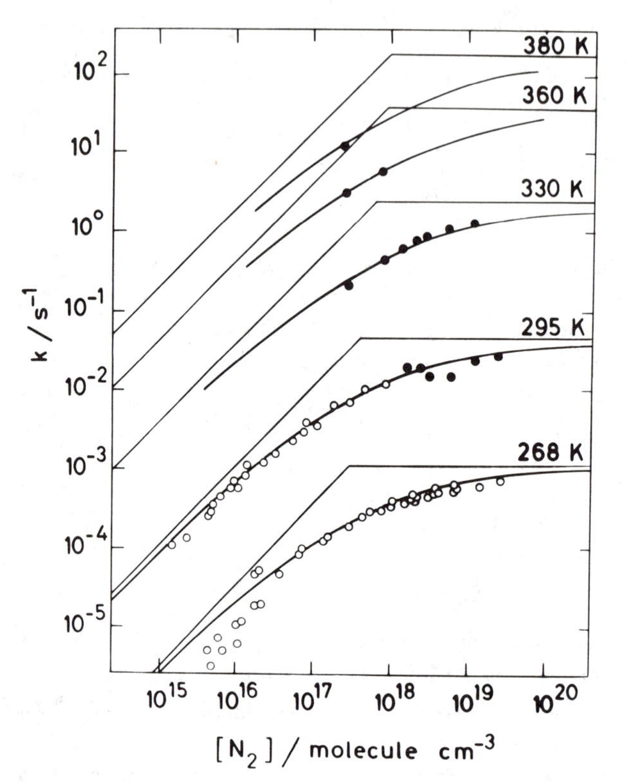

Figure 6. Fall-off curves for the thermal dissociation reaction $N_2O_5 + N_2 \rightarrow NO_2 + NO_3 + N_2$ (experiments ○ = Connell and Johnston, 1979; ● = Viggiano *et al.* 1981; theoretical analysis = full lines, from Malko and Troe, 1981).

are given by Troe (1979, 1983). The required expressions are slightly more complicated than Eqs. (5.10) and (5.11).

The empirical equations (2.12), (2.13), and (5.3)–(5.11) have considerable advantages for use in modeling combustion reactions over the complicated formalisms of conventional unimolecular rate theories. They are both simple enough for modeling purposes and realistic, since they have been derived from the best available theories by empirical fitting. Further improvements in the basic theory will probably lead only to minor modifications in the fitting formulas. Their use has been tested already in many experimental systems, in particular in atmospheric radical recombination reactions (Baulch *et al.*, 1972, 1980). We illustrate with a recent example where the formalism helped to reconcile various experiments, identify experimental errors, locate the fall-off curves and their limits, and represent the rate coefficients in compact form. Figure 6 shows a set of fall-off curves for the thermal dissociation of N_2O_5 to $NO_2 + NO_3$ that was constructed with the methods described above (Malko and Troe, 1981). Extrapolation to shock-wave conditions gave good agreement with the experiments. Conversion to the recombination direction allowed prediction of the rate constant for the reaction $NO_2 + NO_3 \rightarrow N_2O_5$.

6. Conclusions

The theory of unimolecular dissociation and recombination reactions described here provides a compact format for analysis and extrapolation of experimental data as well as for estimation of unknown unimolecular rate coefficients. It has been applied in practice to some selected reactions and should find wider use in the future. The quality of the analysis depends on the availability of experimental data at least for a limited range of conditions. By the use of these experiments, uncertain theoretical parameters can be fitted and a full set of $[M]$- and T-dependent rate coefficients constructed. If no such experimental data are available, the quality of *ab initio* predictions will depend primarily on the accuracy of the thermochemical parameters ΔH_0° and on estimates of weak collision efficiencies β_c and of activated complex structural parameters.

7. References

Baulch, D. L. *et al.* (1980). J. Phys. Chem. Ref. Data **9,** 295.

Baulch, D. L. *et al.* (1972). *Evaluated Kinetic Data for High Temperature Reactions. Vol.* 1, Butterworths, London.

Benson, S. W. & O'Neal, H. E. (1970). In "Kinetic Data on Gas Phase Unimolecular Reactions," NGRDS–NBS 21, Washington.

Benson, S. W. (1976). *Thermochemical Kinetics*, 2nd Ed., Wiley, New York.

Connell, P. & Johnston, H. S. (1979). Geophys. Res. Lett **6,** 553.

Endo, H., Glänzer, K., & Troe, J. (1979). J. Phys. Chem. **83,** 2083.

Forst, W. (1973). *Theory of Unimolecular Reactions*, Academic Press, New York.

Frey, H. M. & Walsh, R. (1979). In "Gas Kinetics and Energy Transfer" Vol. 3, P. G. Ashmore and R. J. Donovan, Eds., Chemical Society, London.

Gilbert, R. G., Luther, K., & Troe, J. (1983). Ber. Bunsenges. Phys. Chem. **87,** 169.

Hirschfelder, J. O., Curtiss, C. F., & Bird, R. B. (1963). *Molecular Theory of Gases and Liquids*, Wiley, London.

Kassel, L. S. (1932). *Kinetics of Homogeneous Reactions*, Chemical Catalog Co., New York.

Kim, P. *et al.* (1973). J. Chem. Phys. **59,** 4593.

Klein, I. E., Rabinovitch, B. S., & Jung, K. H. (1977). J. Chem. Phys. **67,** 3833.

Klein, I. E. & Rabinovitch, B. S. (1978). Chem. Phys. **35,** 439.

Lee, J. H. *et al.* (1978). J. Chem. Phys. **68,** 1817.

Malko, M. W. & Troe, J. (1982). Int. J. Chem. Kin. **14,** 399.

Payne, W. A. & Stief, L. J. (1976), J. Chem. Phys. **64,** 1150.

Quack, M. & Troe, J. (1974). Ber. Bunsenges. Physik. Chem. **78,** 240.

Quack, M. & Troe, J. (1977a). In "Gas Kinetics and Energy Transfer," Vol. 2, P. G. Ashmore and R. J. Donovan, Eds., Chemical Society, London.

Quack, M. & Troe, J. (1977b). Ber. Bunsenges, Physik. Chem. **81,** 329.

Reid, R. C. & Sherwood, T. K. (1966). *The Properties of Gases and Liquids*, 2nd Ed., McGraw-Hill, New York.

Robinson, P. J. & Holbrook, K. A. (1972). *Unimolecular Reactions*, Wiley, London.

Stein, S. E. & Rabinovitch, B. S. (1973). J. Chem. Phys. **58,** 2348.

Tardy, D. C. & Rabinovitch, B. S. (1977). Chem. Rev. **77,** 3691.

Troe, J. (1975a). "Unimolecular Reactions Experiments and Theories" in *Physical Chemistry: An Advanced Treatise*, Vol. VI B, W. Jost, Ed., Academic Press, New York.

Troe, J. (1975b). *15th Symp. (Int.) on Combustion*, The Combustion Institute, Pittsburgh p. 667.

Troe, J. (1977a). J. Chem. Phys. **66,** 4745.

Troe, J. (1977b). J. Chem. Phys. **66,** 4758.

Troe, J. (1978). Ann. Rev. Phys. Chem. **29,** 223.

Troe, J. & Wieters, W. (1979). J. Chem. Phys. **71,** 3931.

Troe, J. (1979). J. Phys. Chem. **83,** 114.

Troe, J. (1980). Ber. Bunsenges. Physik. Chem. **84,** 829.

Troe, J. (1981). J. Chem. Phys. **75,** 226.

Troe, J. (1982). J. Chem. Phys. **77,** 3485.

Troe, J. (1983). Ber. Bunsenges. Phys. Chem. **87,** 161.

Viggiano, A. A. *et al.* (1981). J. Chem. Phys. **74,** 6113.

Whitten, G. Z. & Rabinovitch, B. S. (1963). J. Chem. Phys. **38,** 2466.

Chapter 5

Rate Coefficients in the C/H/O System

Jürgen Warnatz

1. Introduction

1.1. Principles

This chapter is a critical survey of reaction rate coefficient data important in describing high-temperature combustion of H_2, CO, and small hydrocarbons up to C_4. A recommended reaction mechanism and rate coefficient set is presented. The approximate temperature range for this mechanism is from 1200 to 2500 K, which therefore excludes detailed consideration of cool flames, low-temperature ignition, or reactions of organic peroxides or peroxy radicals. Low-temperature rate-coefficient data are presented, however, when they contribute to defining or understanding high-temperature rate coefficients. Because our current knowledge of reaction kinetics is incomplete, this mechanism is inadequate for very fuel-rich conditions (see Warnatz *et al.*, 1982). For the most part, reactions are considered only when their rates may be important for modeling combustion processes. This criterion eliminates considering many reactions among minor species present at concentrations so low that reactions of these species cannot play an essential part in combustion processes. The philosophy in evaluating the rate-coefficient data was to be selective rather than exhaustive: Recent results obtained with experimental methods capable of measuring isolated elementary reaction rate parameters directly were preferred, while results obtained using computer simulations of complex reacting systems were considered only when sensitivity to a particular elementary reaction was demonstrated or when direct measurements are not available. Theoretical results were not considered.

Rate coefficients are given separately for reverse reactions when experimental measurements exist. Reaction products are specified when known;

there are several important cases where the unknown product identities have large influences on modeling studies. When the products are not known, possibilities are suggested and literature references listed.

Rate-coefficient data were rejected, i.e., excluded from consideration for selection of a recommended value, for several possible reasons, including use of insensitive or questionable experimental techniques, excessive scatter, or a large difference from several other results considered more reliable.

It should be emphasized that the recommended rate-coefficient parameters listed are no more than pragmatic representations of experimental data; no physical interpretation is made of the significance of the preexponential factor, temperature dependence, or activation energy. The error limits given include allowance for systematic errors, in spite of the difficulty of estimating these, and are therefore larger than the usually stated error limits that only reflect statistical precision of measurements and not true uncertainty of results.

1.2. Organization

Rate-coefficient data are presented first for the H_2/O_2 system (Tables 2 to 4), next for reactions of CO and CO_2 (Table 5), and then for hydrocarbon reactions in order of increasing size from C_1 to C_4 species with species equal or smaller in size (Tables 6 to 32). Finally, Table 33 gives a list of rate-coefficient recommendations.

Rate coefficients are given in the three-parameter form

$$k = A(T/K)^b \exp(-E/RT)$$

The temperature-independent part A of the preexponential factor is given in units of $(\text{cm}^3/\text{mol})^{n-1}\,\text{s}^{-1}$, which depend on the reaction order n. The exponent b is dimensionless, and E is given in units of kJ/mol ($R = 8.314$ J/K mol, 1 cal = 4.184 J). Two valid digits are given for A and b. The first digit behind the decimal point is given for E. Where measurements are reported at only one temperature, $A = k$.

Estimated errors are given as logarithms of the presumed maximum possible relative deviation from the rate coefficient recommended. For example, an uncertainty amounting to a factor of 2 is given as $\log F = \pm 0.30$, with $F = k_{max}/k_{rec}$ and k_{min}/k_{rec}, respectively.

For pressure-dependent reactions, both low- and high-pressure-limit rate coefficients k_1 and k_∞ are presented where available. For the low-pressure case, collision efficiencies have been reported for several of the third-order reactions in the $H_2/O_2/CO$ system. From these data it can be concluded that the temperature dependences of third-body efficiencies in this system are roughly equal for the different reactions, taking the limits of experimental errors into consideration. Exceptions are collision efficiencies in reactions where the third body interacts chemically with the collison complex, e.g.,

$H + H + H_2$. Here the temperature dependence of the rate coefficient is different from that for the case without chemical interaction (e.g., for noble gases). Recommended collision efficiencies relative to H_2 for k_1 values in the $H_2/O_2/CO$ system are (see references on pressure-dependent reactions in Tables 2 to 5)

H_2	O_2	N_2	H_2O	CO	CO_2	CH_4	Ar	He
1.0	0.4	0.4	6.5	0.75	1.5	6.5	0.35	0.35

No corresponding data for hydrocarbons (besides CH_4) can be given at present.

Experimental fall-off data and theoretical extrapolations to high temperature are given for pressure-dependent reactions. These extrapolations were done with the aid of a quantum mechanical analog of the simple Kassel formula (Robinson and Holbrook, 1972), which was fitted to low-temperature experimental results by variation of the deactivation collision efficiency.

In the last column of the tables presenting detailed rate data are short descriptions of the experimental method used: (1) reactor type, (2) source of reactive species, and (3) method of analysis of products and/or reaction rates. Though these are necessarily very rough descriptions, they do give some feeling for the nature of the data referenced. The abbreviations used are explained in Table 1.

Arrhenius plots (graphs of $\log k$ versus $1/T$) showing the temperature dependence of rate coefficients are given for reactions if the data are too numerous or too scattered to be understood from the representations in the tables alone. In these graphs the data are marked by the initials of the authors,

Table 1. Abbreviations used in the tables.

REACTOR TYPE		RADICAL PRODUCTION	
SR	static reactor	TH	thermal
ST	shock tube	DI	discharge
FR	flow reactor	PH	photolysis
FL	flame	FO	flash photolysis
		HG	Hg sensitization
		PR	pulse radiolysis
ANALYSIS			
GC	gas chromatography	UA	uv absorption
UE	uv emission	IE	ir emission
RA	resonance absorption	P	pressure measurement
RF	resonance fluorescence	MS	mass spectrometry
LF	laser-induced fluorescence	LS	laser schlieren
ES	electron spin resonance	CP	catalytic probe
LM	laser magnetic resonance	CL	chemiluminiscence
SC	saturation current		

who can then be identified from the information given in the tables and in the list of references. The recommended rate-coefficient expression is marked by "rec," rate coefficients determined from equilibrium constants and the reverse-reaction-rate coefficient are denoted by "rev."

Thermochemical data mentioned in the text and used for the calculation of reverse-reaction-rate parameters were taken from the JANAF (1971) tables and addenda. Thermochemical data for species not listed there were taken from Benson (1976).

1.3. Earlier reviews of rate data on hydrocarbon combustion

In the last decade a number of surveys of elementary reactions involved in hydrocarbon combustion have been published. All became outdated within a relatively short time because of the fast progress in experimental methods for obtaining rate data.

One series of publications initiated at the University of Leeds gives complete reviews on selected groups of reactions. They are referred to in this chapter as "Leeds reports."

Evaluated Kinetic Data for High Temperature Reactions, Vol. 1. Homogeneous Gas Phase Reactions of the H_2/O_2 System (Baulch *et al.*, 1972)

Evaluated Kinetic Data for High Temperature Reactions, Vol. 2. Homogeneous Gas Phase Reactions of the $H_2/N_2/O_2$ System (Baulch *et al.*, 1973)

Evaluated Kinetic Data for High Temperature Reactions, Vol. 3. Homogeneous Gas Phase Reactions in the O_2/O_3 System, the $CO/O_2/H_2$ System, and of the Sulfur-containing Species (Baulch *et al.*, 1976)

Evaluated Kinetic Data on Gas Phase Addition Reactions: Reactions of Atoms and Radicals with Alkenes, Alkynes, and Aromatic Compounds (Kerr and Parsonage, 1972)

Evaluated Kinetic Data on Gas Phase Hydrogen Transfer: Reactions of Methyl Radicals (Kerr and Parsonage, 1976)

Because of the availability of these reviews, for many reactions it is not necessary to once again list the rate data given there. Instead, for reactions discussed in these reviews only new measurements are referenced, together with the recommendations given in the evaluations listed above.

Older reviews on important elementary reactions of hydrocarbons are necessarily obsolete since key experimental work on them has been done in the last decade. Nevertheless, three further publications are useful in summarizing older literature:

Kinetic Data on Gas Phase Unimolecular Reactions (Benson and O'Neal, 1970)

Rate Constants of Gas Phase Reactions, Reference Book (Kondratiev, 1972)

Chemical Kinetics of the Gas Phase Combustion of Fuels—A Bibliography on the Rates and Mechanisms of Oxidation of Aliphatic C_1 to C_{10} Hydrocarbons and of their Oxygenated Derivatives (Westley, 1976)

2. General features of high-temperature hydrocarbon combustion

High-temperature hydrocarbon combustion is a complicated chemical process. Nevertheless, it is possible to discern some features common to all fuels. To discuss them, it is convenient to separate the treatment of *radical-poor* and *radical-rich* situations. The former are characteristic of ignition processes, studied in the laboratory mainly by shock-wave initiation, while the latter are found in fully developed flames, studied in the laboratory in burner-stabilized flames. The elementary reactions important in these two situations are in part the same, but in part different.

2.1. Radical-poor situation: ignition and induction periods

In this case, the chemistry is determined by the interaction of chain-branching and chain-terminating processes, while radical-radical reactions are unimportant. This can be seen in combustion modeling by a sensitivity analysis, changing each rate coefficient of a reaction mechanism and looking at the effect upon the computed value of a property of interest, e.g., an induction time, flame velocity, maximum concentration of a species, etc. (Chapter 7). An example is given in Fig. 1, showing the response of the induction time for detectable fuel disappearance to systematic variation of the rate coefficients of a mechanism describing the ignition of lean methane-air mixtures. The mechanism is that of Hidaka *et al.* (1982) and the data those of Eubank *et al.* (1981).

It can be seen that the calculated ignition delay times τ in lean CH_4/O_2 mixtures are sensitive to the choices of rate coefficients for reactions involving CH_4, in particular the rates of the chain-initiating steps $CH_4 + M \rightarrow CH_3 + H + M$ and $CH_4 + O_2 \rightarrow CH_3 + HO_2$. There are also influences of rate coefficients of reactions with chain-terminating character

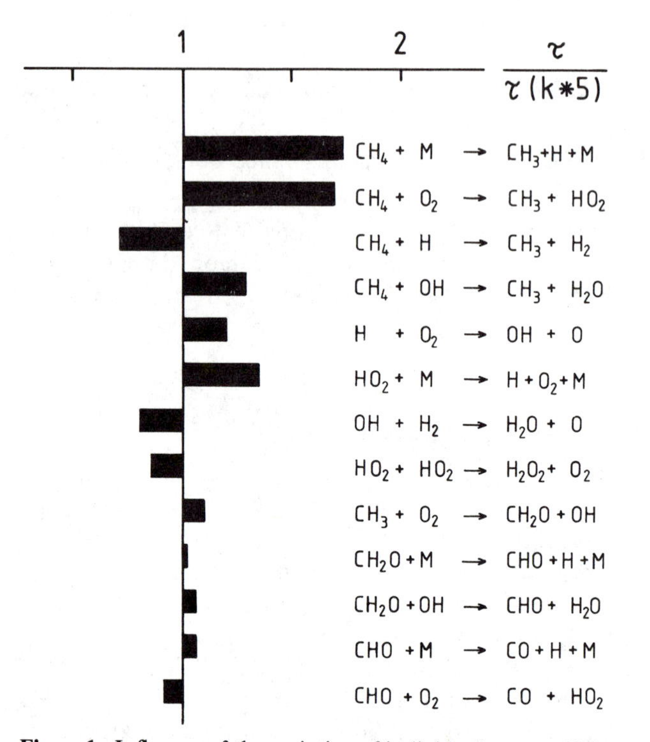

Figure 1. Influence of the variation of individual rate coefficients by a factor of 5 on the calculated ignition delay time in a lean CH_4-air mixture (1% CH_4) at $T = 1400$ K, $P = 2.87$ bar. Only major effects are shown. For references and a complete list of the mechanism used see Gardiner *et al.* (1981).

(e.g., $HO_2 + HO_2 \rightarrow H_2O_2 + O_2$ and $CHO + O_2 \rightarrow CO + HO_2$) and of reactions with chain-branching character (e.g., $H + O_2 \rightarrow OH + O$ and $CHO + M \rightarrow CO + H + M$).

2.2. Radical-rich situation: flame propagation

In this case, radical-radical reactions are important because of the large radical concentrations in a fully developed flame front.

It can be shown (Warnatz, 1981) that the hydrocarbon fuel is attacked by H, O, and OH in the first steps. The alkyl radicals formed in this way decompose rapidly to smaller alkyl radicals and alkenes (Figs. 2 and 3). Only for the smallest alkyl radicals (CH_3 and C_2H_5) does the relatively slow thermal decomposition compete with recombination and with oxidation reactions with O atoms and O_2 (Figs. 2 and 3). This part of the mechanism is rate-controlling in alkane and alkene flames and is responsible for the similarity of all alkane and alkene flame properties.

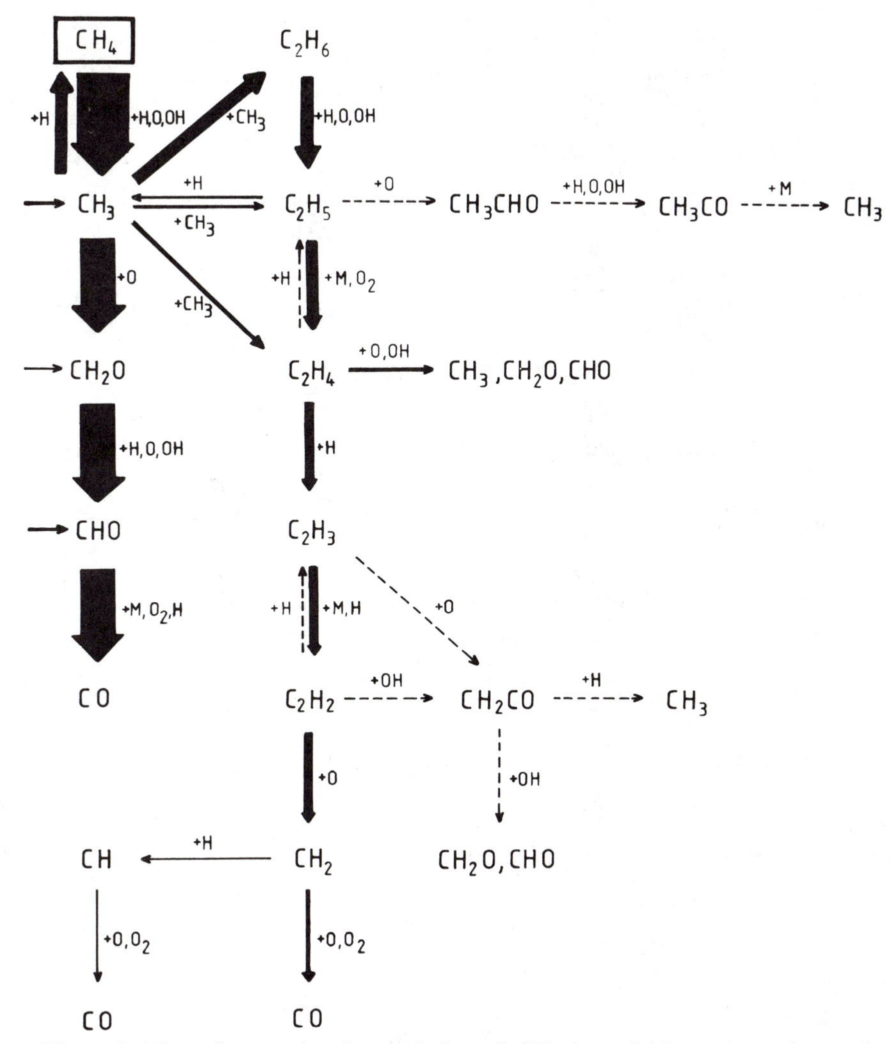

Figure 2. Flow diagram for the oxidation of CH_4 in stoichiometric methane–air flames at $P = 1$ bar, $T_u = 298$ K. The calculations were done with a one-dimensional laminar flat flame model using a mechanism described by Warnatz *et al.* (1982). The thickness of the arrows is proportional to the reaction rates integrated over the whole flame front.

According to this picture, hydrocarbon combustion, at least in lean and moderately rich mixtures, is mainly governed by elementary reactions which are not specific for the fuel considered, e.g., $H + O_2 \rightarrow OH + O$ and $CO + OH \rightarrow CO_2 + H$. This is demonstrated by the sensitivity analysis shown in Fig. 4 for the responses of the flame velocity of freely propagating stoichiometric CH_4-air and C_2H_6-air flames at atmospheric pressure.

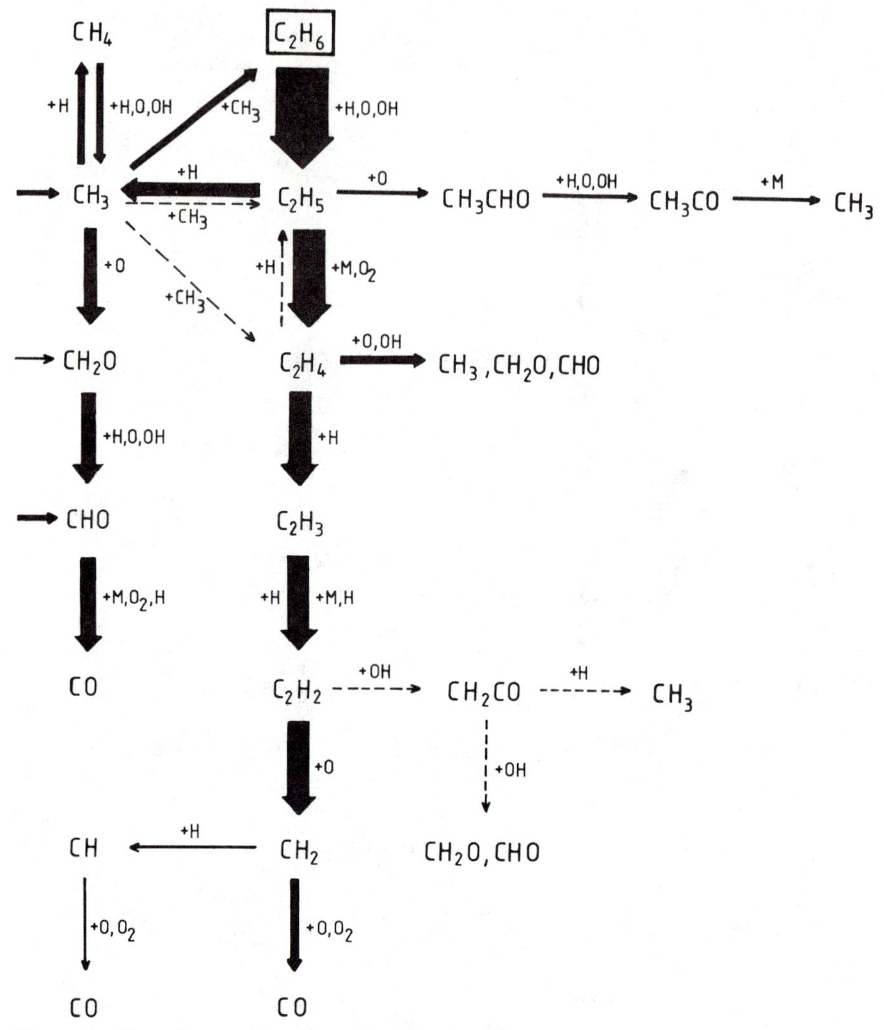

Figure 3. Flow diagram for the oxidation of C_2H_6 in a stoichiometric methane–air flame at $P = 1$ bar, $T_u = 298$ K. Explanations as in Fig. 2.

3. Reactions in the H_2/O_2 system

3.1. Reactions in the H_2/O_2 system not involving HO_2 or H_2O_2

Reactions in the H_2/O_2 system are part of the main chain-branching processes of ignition or maintaining high-temperature combustion of hydrocarbon fuels. Due to their pervasive importance, these reactions have been investigated in detail by many workers using a variety of techniques. This leads to the unique situation that in some cases (e.g., determination of the

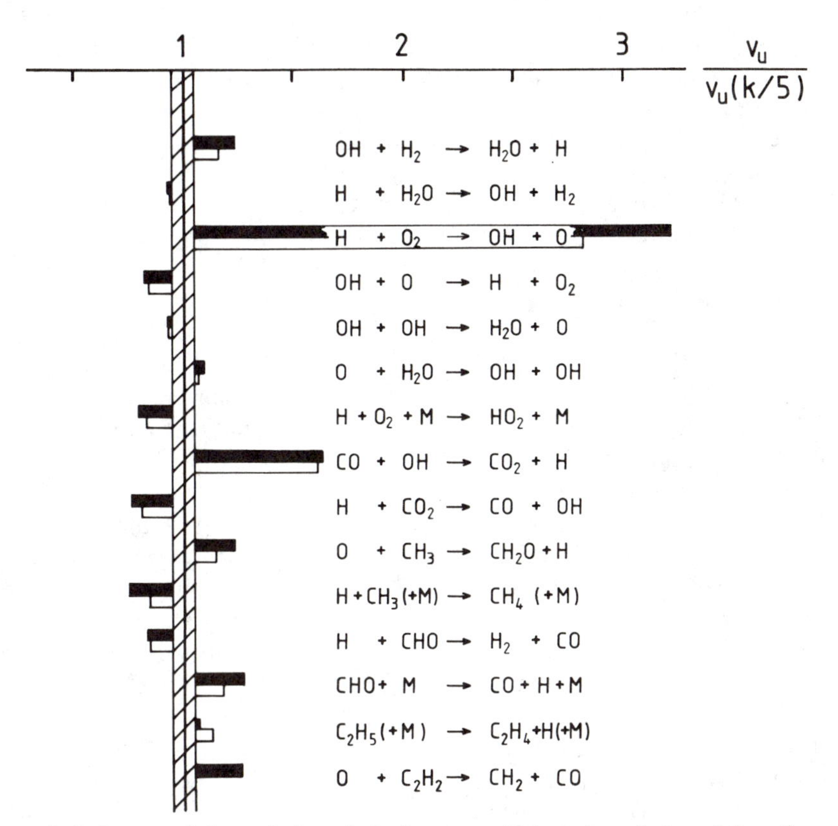

Figure 4. Influence of the variation of single rate coefficients by a factor of 5 on the calculated flame velocity in stoichiometric CH_4-air (dark columns) and stoichiometric C_2H_6-air mixtures (light columns) at $P = 1$ bar, $T_u = 298$ K. Only changes larger than 5% (indicated by the hatched area) are considered. For references and a complete list of the mechanism used see Warnatz (1981).

velocity of free flame propagation, see Peters and Warnatz, 1982) calculations using detailed H_2/O_2 reaction kinetics seem to be more accurate than many carefully planned combustion experiments.

Because of the relative simplicity of this system (8 species, at most about 40 relevant elementary steps), it is possible to give a complete picture of the reactions involved. Since complete reviews have been given by Baulch *et al.* (1972), in this chapter only results published after 1972 are discussed, together with the recommendations given in this Leeds report.

$H + O_2 \rightarrow OH + O$

The reaction of H atoms with molecular oxygen is the basic chain-branching process in high-temperature combustion. About 80% of the O_2 is consumed by this step in typical hydrocarbon-air stoichiometric flames at atmospheric pressure. Due to its relatively large activation energy, caused by the endothermicity of $\Delta H^\circ_{298} = 70.7$ kJ, it is one of the rate-controlling elementary reactions. Therefore, flame propagation and ignition are most sensitive to the value of this rate coefficient (see Figs. 1 and 4).

Because of the parallel occurrence of the reaction $H + O_2 + M \rightarrow HO_2 + M$ there is a pressure- and temperature-dependent competition with $H + O_2 \rightarrow OH + O$ that has consequences for the combustion process. At high temperature and low pressure the reaction $H + O_2 \rightarrow OH + O$ leads to O atoms as a further chain-branching agent. Formation of O and OH is followed by H-atom regeneration in H_2/O_2 flames by $O + H_2 \rightarrow OH + H$ and $OH + H_2 \rightarrow H_2O + H$. On the other hand, for low temperature and high pressure, the reaction $H + O_2 + M \rightarrow HO_2 + M$ (small activation energy, high reaction order) produces HO_2 as a second (relatively unreactive) chain carrier, leading back to H atoms less efficiently via $H + HO_2 \rightarrow OH + OH$ and $OH + H_2 \rightarrow H_2O + H$.

The scatter of the data for $H + O_2 \rightarrow OH + O$ given in the Leeds report (Baulch *et al.*, 1972) is considerable, corresponding to a factor of about 6 at flame temperatures. Because of this scatter, it was impossible to decide whether there is non-Arrhenius behavior of the temperature dependence of the reaction under consideration (as is clearly the case for $O + H_2$ and $OH + H_2$, see below and Chapter 3). New measurements done after 1972 improve this situation (Fig. 5). Especially, the expression given by Schott (1973) results from direct and sensitive measurements. Extrapolation of his expression to lower temperatures, however, leads to incompatibilities with reliable data. The recommendation given by Dixon-Lewis (1983) provides a close match to these data. The higher values given by the Leeds expression at flame temperatures are supported by some indirect but reliable flame modeling studies.

$OH + O \rightarrow H + O_2$

Since this reaction is the reverse of $H + O_2 \rightarrow OH + O$, it distinctly inhibits high-temperature combustion, and there is a moderate sensitivity of flame propagation to its rate coefficient (Fig. 4). Measurements are available only below 425 K. However, it is reasonable to assume zero activation energy for this exothermic radical-radical reaction. This leads to a temperature-independent rate coefficient of 1.8×10^{13} cm^3/mol s (Fig. 6), in agreement with the Leeds report low-temperature recommendation (2×10^{13} cm^3/mol s) and, roughly, with the rate coefficient determined from the reverse reaction and the equilibrium constant.

$O + H_2 \rightarrow OH + H$

After $H + O_2 \rightarrow OH + O$, this slightly endothermic reaction ($\Delta H^\circ_{298} = 8.0$ kJ) is the second important chain-branching step in the H_2/O_2 system

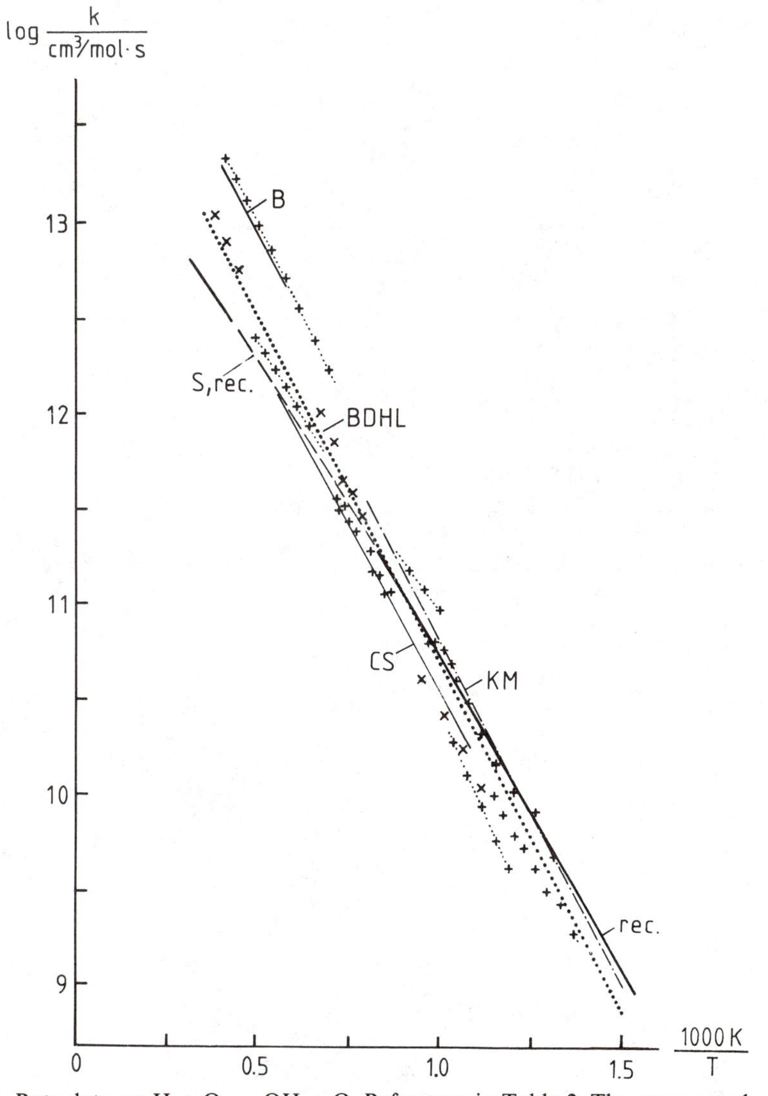

Figure 5. Rate data on $H + O_2 \rightarrow OH + O$. References in Table 2. The crosses and broken lines denote representative experimental results given by Baulch *et al.* (1972).

governing high-temperature combustion chemistry. The sensitivity of flame propagation and ignition to its rate coefficient is small (Figs. 1 and 4) because of its coupling to the rate-determining step $H + O_2$.

The scatter of the data given in the Leeds report is small enough to detect non-Arrhenius behavior of the temperature dependence of this rate coefficient (Chapter 3). The recommended expression differs slightly from the Leeds report recommendation (Fig. 7).

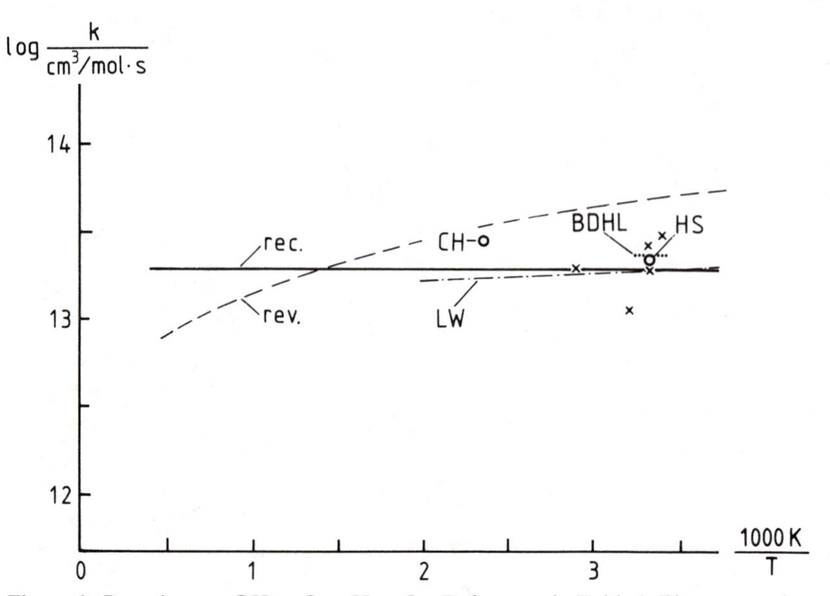

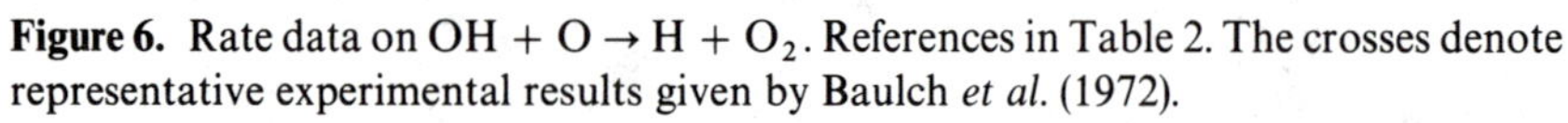

Figure 6. Rate data on $OH + O \rightarrow H + O_2$. References in Table 2. The crosses denote representative experimental results given by Baulch *et al.* (1972).

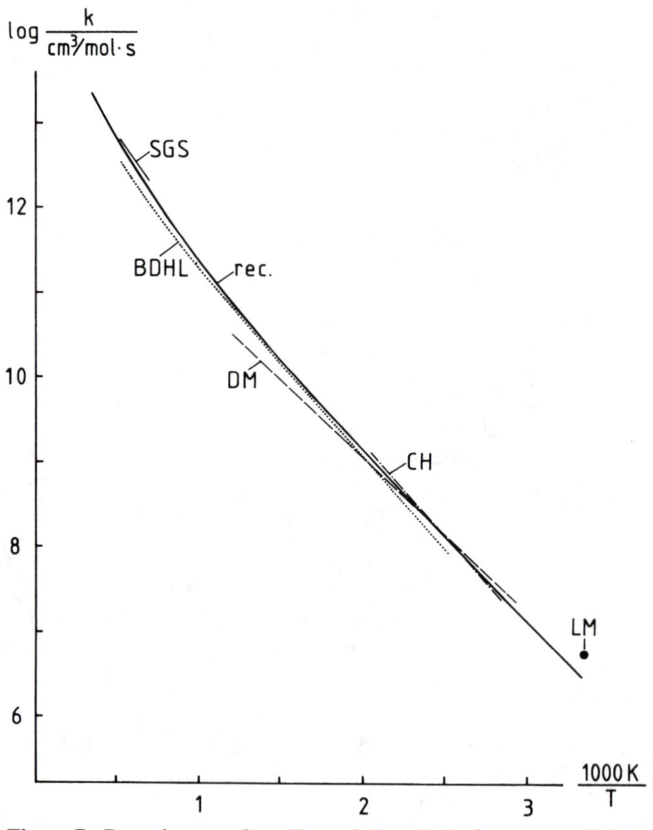

Figure 7. Rate data on $O + H_2 \rightarrow OH + H$. References in Table 2.

$$H + OH \rightarrow O + H_2$$

Like its reverse, this exothermic reaction shows small sensitivity of flame propagation and ignition to its rate coefficient. There are no experimental data in the literature.

$$OH + H_2 \rightarrow H_2O + H$$

Flame propagation shows moderate sensitivity to the rate coefficient of this exothermic ($\Delta H^\circ_{298} = -63.2$ kJ) chain-propagation reaction (Fig. 4). Together with $OH + OH \rightarrow H_2O + O$ and $OH + CH_2O \rightarrow H_2O + CHO$ (see below), this step is the main source of water in typical hydrocarbon–air flames at atmospheric pressure. Together with its reverse reaction and $CO + OH \rightarrow CO_2 + H$ it takes part in establishing the water–gas equilibrium (see below).

The scatter of the rate data in the Leeds report was too large to allow definite identification of non-Arrhenius behavior of the temperature dependence of the rate coefficient of this reaction. However, this is now possible taking into consideration measurements published after 1972 (Fig. 8 and Chapter 3).

$$H + H_2O \rightarrow OH + H_2$$

Because of the sensitivity of flame propagation to the rate coefficient of its reverse reaction (Fig. 4), this reaction has an inhibiting effect on combustion. Together with $OH + H_2 \rightarrow H_2O + H$ and $CO + OH \rightarrow CO_2 + H$, it is

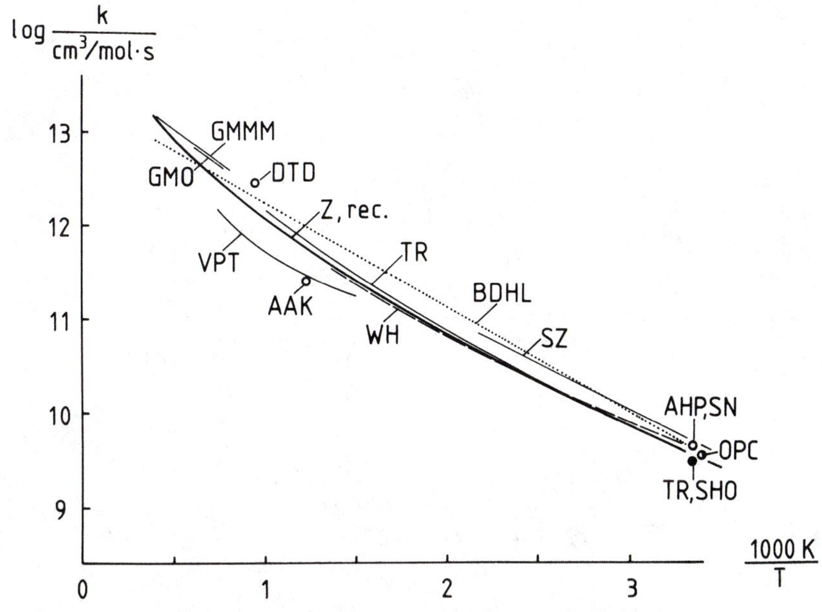

Figure 8. Rate data on $OH + H_2 \rightarrow H_2O + H$. References in Table 2.

mainly responsible for the water–gas equilibration. Because of its relatively large activation energy (and that of $H + CO_2 \rightarrow CO + OH$), water–gas equilibration is rapid only at temperatures above 1800 K (Warnatz, 1979). In rich flames of aliphatic hydrocarbons this reaction generates the decrease of H_2O concentration profiles at long reaction times, because the OH produced is scavenged by the hydrocarbon to prevent the occurrence of the reverse reaction.

There exist some reasonably direct measurements of the rate coefficient of this reaction at high temperature. Nevertheless, the rate coefficient calculated from the reverse reaction seems to be more reliable.

$$OH + OH \rightarrow H_2O + O$$

This exothermic reaction ($\Delta H^\circ_{298} = -71.1$ kJ) is a minor source of H_2O in hydrocarbon flames. Because of its chain-terminating character, it inhibits flame propagation (Fig. 4). The measurements show distinct non-Arrhenius behavior of the temperature dependence of its rate coefficient (Fig. 9). A review is given by Zellner and Wagner (1980) and in Chapter 3.

$$O + H_2O \rightarrow OH + OH$$

This reaction has a moderately accelerating effect on flame propagation (Fig. 4). There exist several fairly direct determinations of the rate coefficient at high temperature, but the expression calculated from the reverse reaction seems to be more reliable.

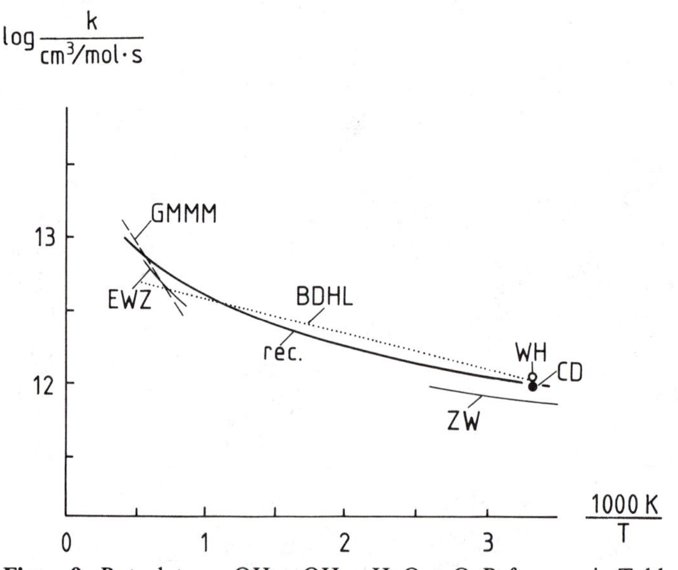

Figure 9. Rate data on $OH + OH \rightarrow H_2O + O$. References in Table 2.

Recombination reactions of H, O, and OH

Because of their third-order behavior, recombination reactions of H, O, and OH (Figs. 10 to 12) are of minor importance in flame propagation at normal or reduced pressure (Fig. 4) but can be relevant at elevated pressure or in ignition or quenching processes because of their chain-terminating character. (The recombination of OH to form H_2O_2 is discussed below.) Relative collision efficiencies are discussed in Section 1.2.

Dissociation reactions of H_2, O_2, and H_2O

On account of their large endothermicity, dissociation reactions of H_2, O_2, or H_2O can play a part only in ignition processes at very high temperature; in flame propagation they are completely unimportant. Relative collision efficiencies are discussed in Section 1.2.

3.2. Formation and consumption of HO_2

$$H + O_2 + M \rightarrow HO_2 + M$$

As mentioned before (see $H + O_2 \rightarrow OH + O$), this highly exothermic reaction ($\Delta H^\circ_{298} = -197$ kJ) is a chain-propagating step competing with the chain-branching reaction $H + O_2 \rightarrow OH + O$. However, because of the low

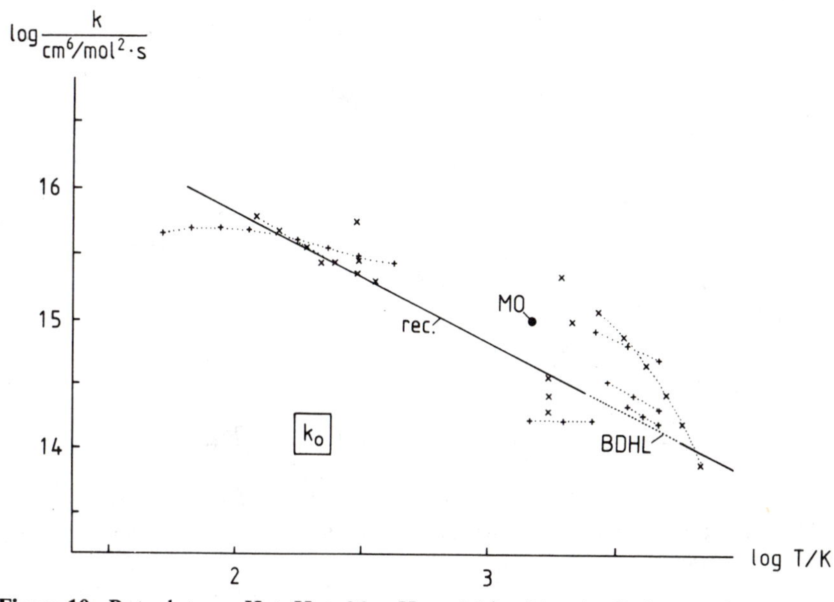

Figure 10. Rate data on $H + H + M \rightarrow H_2 + M$ for M = Ar. References in Table 2. The crosses and broken lines denote representative experimental results given by Baulch *et al.* (1972).

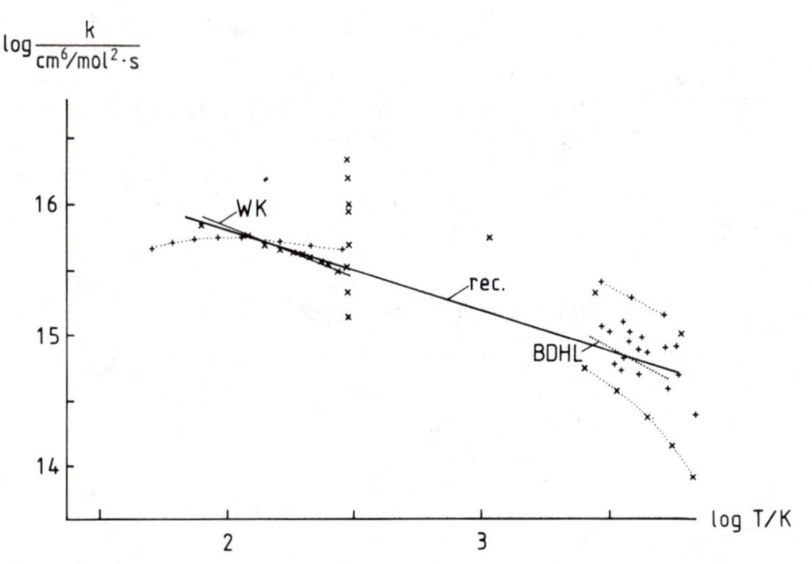

Figure 11. Rate data on $H + H + M \rightarrow H_2 + M$ for $M = H_2$. References in Table 2. The crosses and broken lines denote representative experimental results given by Baulch *et al.* (1972).

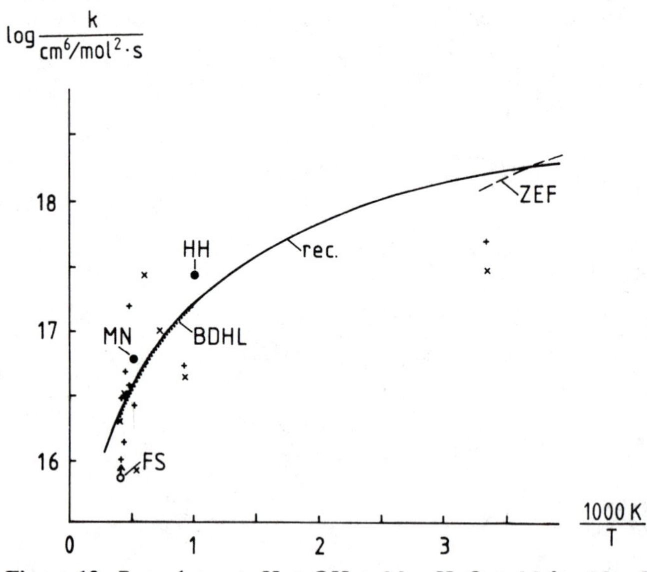

Figure 12. Rate data on $H + OH + M \rightarrow H_2O + M$ for $M = H_2O$. References in Table 2. The crosses and broken lines denote representative experimental results by Baulch *et al.* (1972).

Table 2. Rate data on H_2/O_2 reactions (HO_2/H_2O_2 reactions excluded).

Reference	$A/(cm^3/mol)^{n-1}\ s^{-1}$	b	E/kJ	T/K	Method
		$H + O_2 \rightarrow OH + O$			
Baulch *et al.* (1972).	2.2×10^{14}	0	70.3	700–2500	Review
Kochubei and Moin (1973)	2.7×10^{14}	0	69.5	640–1200	SR,TH,GC
Schott (1973)	1.2×10^{17}	−0.91	69.3	1250–2500	ST,TH,CL
Bowman (1975a)	6.0×10^{14}	0	70.3	1875–2240	ST,TH,UA,UE
Chiang and Skinner (1979)	1.1×10^{14}	0	67.4	925–1825	ST,TH,RA
Recommended ($\log F = \pm 0.15$)	1.2×10^{17}	−0.91	69.1	300–2500	...
		$OH + O \rightarrow O_2 + H$			
Baulch *et al.* (1972)	2.3×10^{13}	...	...	300	Review
Campbell and Handy (1977)	2.6×10^{13}	...	...	425	FR,DI,CL
Lewis and Watson (1980)	1.2×10^{13}	0	−0.9	221–499	FR,DI,RF
Howard and Smith (1980)	2.3×10^{13}	...	...	298	FR,DI,RF
Recommended ($\log F = \pm 0.15$)	1.8×10^{13}	0	0	300–2500	...
		$O + H_2 \rightarrow OH + H$			
Baulch *et al.* (1972)	1.8×10^{10}	1.0	37.2	400–2000	Review
Schott *et al.* (1974)	2.2×10^{14}	0	57.4	1400–1900	ST,TH,CL
Campbell and Handy (1975)	3.1×10^{13}	0	41.2	363–490	FR,DI,CL
Dubinsky and McKenney (1975)	5.3×10^{12}	0	34.9	347–832	FR,DI,RF,CL
Light and Matsumoto (1980)	5.5×10^{6}	...	...	298	FR,DI,LF
Recommended ($\log F = \pm 0.15$)	1.5×10^{7}	2.0	31.6	300–2500	...

Table 2 (*continued*).

Reference	$A/(cm^3/mol)^{n-1}\ s^{-1}$	b	E/kJ	T/K	Method
		$OH + H_2 \rightarrow H_2O + H$			
Baulch *et al.* (1972)	2.2×10^{13}	0	21.5	300–2500	Review
Azatyan *et al.* (1971)	2.5×10^{11}	...	...	810	SR,TH,P
Stuhl and Niki (1972a)	4.3×10^{9}	...	...	300	SR,PH,RF
Gardiner *et al.* (1973)	5.2×10^{13}	0	27.2	1200–2500	ST,TH,UA,IE
Day *et al.* (1973)	2.7×10^{12}	...	...	1050	FL,TH,MS
Westenberg and deHaas (1973a)	1.0×10^{8}	1.6	13.8	298–745	FR,DI,ES
Smith and Zellner (1974)	1.1×10^{13}	0	19.4	210–460	SR,FP,RA
Gardiner *et al.* (1974)	5.2×10^{13}	0	27.0	1350–1600	ST,TH,RA
Trainor and von Rosenberg (1975)	3.2×10^{9}	...	...	300	SR,FP,UA
Atkinson *et al.* (1975)	4.2×10^{9}	...	...	298	SR,FP,RF
Overend *et al.* (1975)	3.5×10^{9}	...	...	295	SR,FP,RA
Vandooren *et al.* (1975)	8.3×10^{-5}	5.0	−15.5	700–1300	FL,TH,MS
Zellner (1979)	1.0×10^{8}	1.6	13.8	300–2000	Review
Sworsky *et al.* (1980)	5.1×10^{9}	...	...	296	SR,FP,UA
Tully and Ravishankara (1980)	5.2×10^{6}	2.0	12.0	298–992	SR,FP,RF
Recommended ($\log F = \pm 0.15$)	1.0×10^{8}	1.6	13.8	300–2500	...
		$H + H_2O \rightarrow OH + H_2$			
Baulch *et al.* (1972)	9.3×10^{13}	0	85.2	300–2500	Review
Recommended ($\log F = \pm 0.20$)	4.6×10^{8}	1.6	77.7	300–2500	...
		$OH + OH \rightarrow H_2O + O$			
Baulch *et al.* (1972)	6.3×10^{12}	0	4.6	300–2000	Review
Gardiner *et al.* (1973)	5.5×10^{13}	...	...	1200–2500	ST,TH,UA,IE
Westenberg and de Haas (1973b)	1.4×10^{12}	...	...	298	FR,DI,ES
Clyne and Down (1974)	8.4×10^{11}	...	...	300	FR,DI,RF
Ernst *et al.* (1977)	3.4×10^{13}	0	21.0	1200–1800	ST,PH,RA

		$O + H_2O \rightarrow OH + OH$			
Baulch *et al.* (1972)	6.8×10^{13}	0	76.8	300–2000	Review
Recommended ($\log F = \pm 0.20$)	1.5×10^{10}	1.14	72.2	300–2500	···
		$H_2 + O_2 \rightarrow$ products			
Baulch *et al.* (1972)	5.5×10^{13}	0	242	···	Review
No recommendation					
		$H + H + M \rightarrow H_2 + M$ (k_1, $M = Ar$)			
Baulch *et al.* (1972)	6.4×10^{17}	−1.0	0	2500–5000	Review
Mallard and Owen (1974)	1.0×10^{15}	···	···	1500	ST,TH,UA
Recommended ($\log F = \pm 0.30$)	6.4×10^{17}	−1.0	0	300–5000	···
		$H + H + M \rightarrow H_2 + M$ (k_1, $M = H_2$)			
Baulch *et al.* (1972)	2.6×10^{18}	−1.0	0	2500–5000	Review
Baulch *et al.* (1972)	3.0×10^{15}	···	···	300	Review
Walkauskas and Kaufman (1975)	8.9×10^{16}	−0.6	0	77–295	FR,TH,CP
Recommended ($\log F = \pm 0.30$)	9.7×10^{16}	−0.6	0	100–5000	···
		$H_2 + M \rightarrow H + H + M$ (k_1, $M = Ar$)			
Baulch *et al.* (1972)	2.2×10^{14}	0	402	2500–5000	Review
Breshears and Bird (1973)	9.3×10^{13}	0	372	3500–8000	ST,TH,LS
Recommended ($\log F = \pm 0.30$)	2.2×10^{14}	0	402	2500–8000	···
		$H_2 + M \rightarrow H + H + M$ (k_1, $M = H_2$)			
Baulch *et al.* (1972)	8.8×10^{14}	0	402	2500–5000	Review
Breshears and Bird (1973)	3.3×10^{15}	0	440	3500–8000	ST,TH,LS
Recommended ($\log F = \pm 0.30$)	8.8×10^{14}	0	402	2500–8000	···

Table 2 (*continued*).

Reference	$A/(cm^3/mol)^{n-1}\,s^{-1}$	b	E/kJ	T/K	Method
	$H + OH + M \rightarrow H_2O + M\ (k_1, M = H_2O)$				
Baulch *et al.* (1972)	1.4×10^{23}	-2.0	0	1000–3000	Review
Homer and Hurle (1969)	2.4×10^{17}	...	...	1000	ST,TH,UA
Friswell and Sutton (1972)	$>7.7 \times 10^{15}$	...	...	2130	FL,TH,CL
Zellner *et al.* (1977)	4.0×10^{24}	-2.6	0	230–300	FR,DI,RF
Recommended (log $F = \pm 0.20$)	1.4×10^{23}	-2.0	0	1000–3000	...
	$H_2O + M \rightarrow H + OH + M\ (k_1, M = (H_2O)$				
Baulch *et al.* (1972)	2.2×10^{16}	0	440	2000–6000	Review
Recommended (log $F = \pm 0.30$)	1.6×10^{17}	0	478	2000–5000	...
	$O + O + M \rightarrow O_2 + M\ (k_1, M = Ar)$				
Baulch *et al.* (1976)	1.9×10^{13}	0	-7.5	190–4000	Review
Recommended (log $F = \pm 0.30$)	1.0×10^{17}	-1.0	0	300–5000	...
	$O_2 + M \rightarrow O + O + M\ (k_1, M = Ar)$				
Baulch *et al.* (1976)	1.8×10^{18}	-1.0	494	3000–18000	Review
Recommended (log $F = \pm 0.50$)	1.2×10^{14}	0	451	2000–10000	...
Zellner and Wagner (1980)	1.9×10^{12}	0	2.3	250–380	SR,FP,RF
Recommended (log $F = \pm 0.20$)	1.5×10^{9}	1.14	0	300–2500	...

reactivity of HO_2, this reaction behaves effectively as a chain-terminating step, especially at high temperature. Furthermore, this reaction is important because a considerable part of the heat of combustion may be released in this step. Corresponding to these facts, there is considerable sensitivity of ignition and flame propagation with respect to the rate coefficient of this reaction (Figs. 1 and 4). Relative collision efficiencies are given in Section 1.2 (Fig. 13).

Reactions of HO_2 with H atoms

The fast reactions $H + HO_2 \rightarrow OH + OH$ and $H + HO_2 \rightarrow H_2 + O_2$ prevent the coexistence of H atoms and HO_2 radicals. This leads to the typical HO_2 concentration peak in the unburnt gas region of flames if there is no HO_2 source other than $H + O_2 + M \rightarrow HO_2 + M$. Flame propagation is rather insensitive to the rate coefficients of these steps because of the competition of the chain-branching reaction leading to $OH + OH$ and the chain-terminating reaction leading to $H_2 + O_2$ (Figs. 14 and 15).

The reaction $H + HO_2 \rightarrow H_2O + O$ is often postulated to compete with the channels leading to $OH + OH$ and $H_2 + O_2$ (see above). There is no direct evidence of its occurrence in the literature, however; indirect determinations show that its rate coefficient is small if the reaction occurs at all. No recommendation is given here.

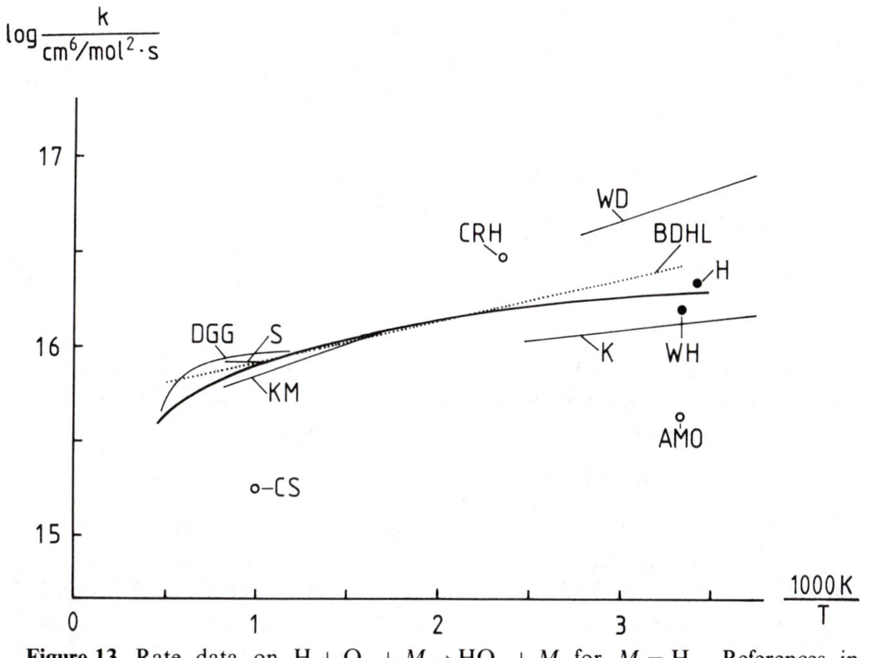

Figure 13. Rate data on $H + O_2 + M \rightarrow HO_2 + M$ for $M = H_2$. References in Table 3.

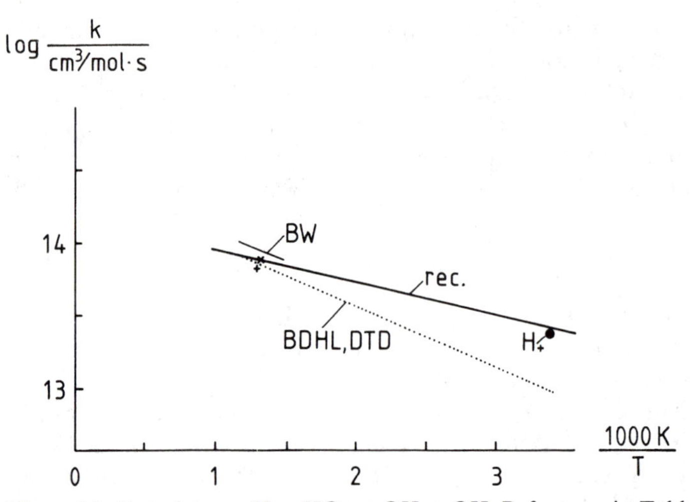

Figure 14. Rate data on $H + HO_2 \rightarrow OH + OH$. References in Table 3. The crosses and broken lines denote representative experimental results given by Baulch *et al.* (1972).

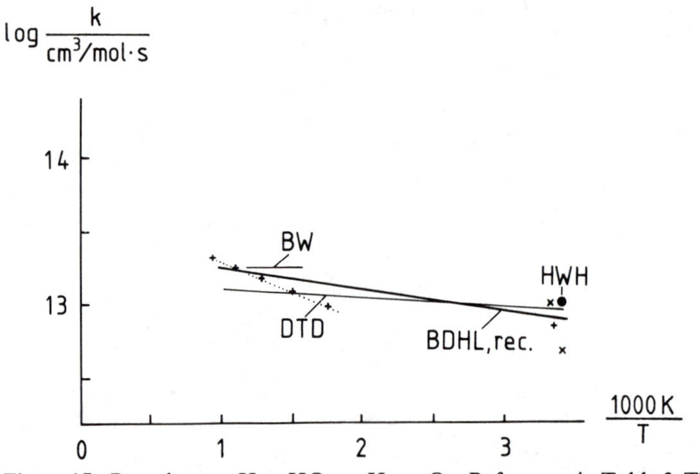

Figure 15. Rate data on $H + HO_2 \rightarrow H_2 + O_2$. References in Table 3. The crosses and broken lines denote representative experimental results given by Baulch *et al.* (1972).

Reactions of HO_2 with O and OH

The reaction of HO_2 with O (Fig. 16) is relatively unimportant for hydrocarbon combustion due to the low O concentration compared with the concentrations of H and OH. More interesting is the reaction of HO_2 with OH to form H_2O and O_2 (Fig. 17), which has attracted much attention in recent years because of its chain-terminating character in the reaction cycles of atmospheric chemistry. Ignition in very lean mixtures may be influenced by

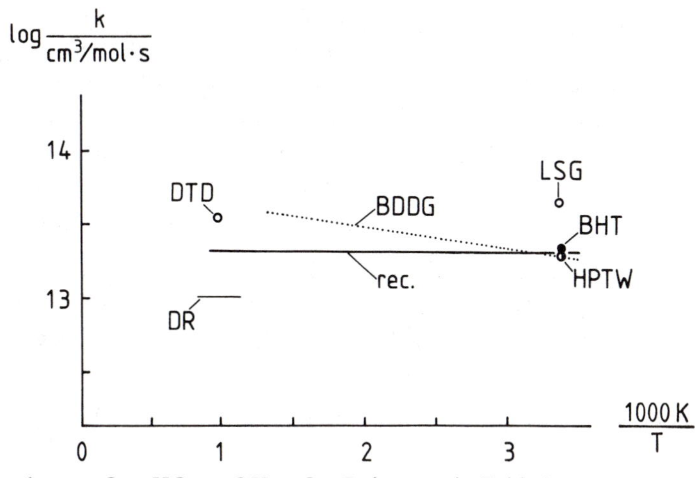

Figure 16. Rate data on $O + HO_2 \rightarrow OH + O_2$. References in Table 3.

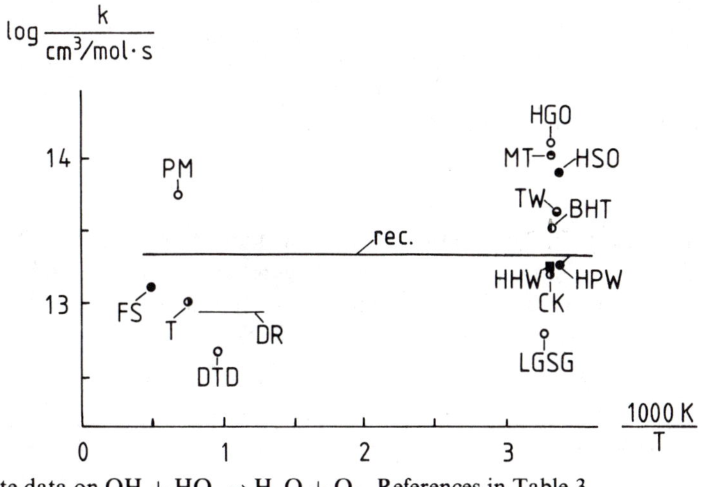

Figure 17. Rate data on $OH + HO_2 \rightarrow H_2O + O_2$. References in Table 3.

this reaction (Fig. 1). Some years ago there was a considerable scatter of the rate data for this reaction, but now the room-temperature rate coefficient seems to be fixed between 2×10^{13} and 6×10^{13} cm^3/mol s. Because of this large value, there cannot be a noticeable activation energy.

$$HO_2 + HO_2 \rightarrow \text{products}$$

This reaction can play a part in ignition at relatively low temperatures, where reaction paths involving HO_2 are preferred (Fig. 1). It is a possible source of by-product H_2O_2, which also may be produced by $OH + OH + M \rightarrow H_2O_2 + M$ and $HO_2 + H_2 \rightarrow H_2O_2 + H$.

As demonstrated by its pressure dependence between 2 and 6 mbar (Thrush and Wilkinson, 1979) and its slightly negative activation energy (Lii *et al.*, 1979, Cox and Burrows, 1979), formation of the reaction products $H_2O_2 + O_2$ (>99.9%, Niki *et al.*, 1980) at low temperature occurs via a complex mechanism, probably via the elementary steps

$$HO_2 + HO_2 \rightarrow H_2O_4$$

$$H_2O_4 \rightarrow H_2O_2 + O_2$$

At high temperature, the tendency to complex formation should decrease, simplifying the reaction mechanism. Investigation of the pressure dependence at high temperature is desirable (Fig. 18).

3.3. Formation and consumption of H_2O_2

$$OH + OH + M \rightarrow H_2O_2 + M$$

Because of the small rate coefficient of $H_2 + HO_2 \rightarrow H_2O_2 + H$ (see below) and the low HO_2 concentrations in high-temperature combustion, hydrogen peroxide is mainly formed by OH recombination (Fig. 19), especially at elevated pressure (see Warnatz, 1982). Relative collision efficiencies for third bodies in this reaction are given in Section 1.2.

$$H_2O_2 + M \rightarrow OH + OH + M$$

This reaction may be important in low-temperature oxidation because of its chain-branching character. The slow H_2O_2 formation rate and the large activation energy of this decomposition reaction mean that it has only minor importance in high-temperature combustion.

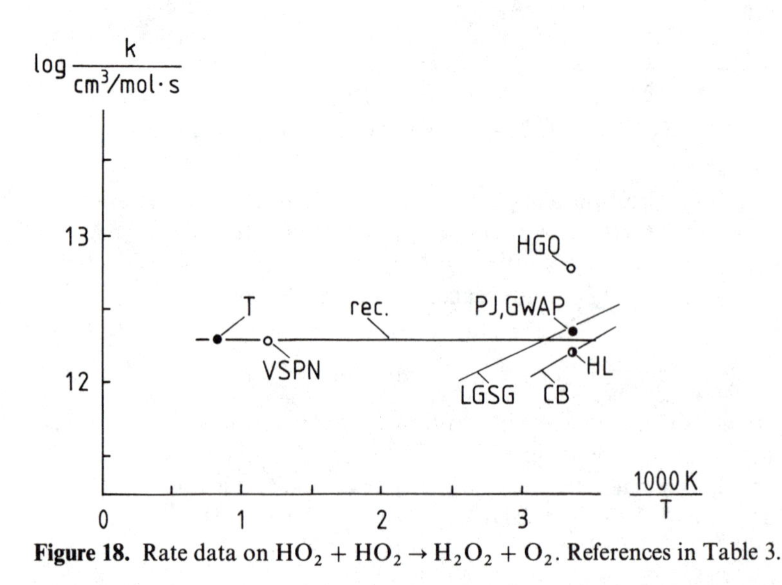

Figure 18. Rate data on $HO_2 + HO_2 \rightarrow H_2O_2 + O_2$. References in Table 3.

Table 3. Rate data on HO_2 reactions (H_2O_2 formation and consumption excluded).

Reference	$A/(cm^3/mol)^{n-1} s^{-1}$	b	E/kJ	T/K	Method
	$H + O_2 + M \rightarrow HO_2 + M$ (k_1, $M = Ar$)				
Baulch *et al.* (1972)	1.5×10^{15}	0	−4.2	300–2000	Review
Ahumada *et al.* (1972)	2.2×10^{15}	…	…	298	SR,HG,RA
Westenberg and deHaas (1972a)	6.8×10^{15}	…	…	300	FR,DI,ES
Kurylo (1972a)	2.4×10^{15}	0	−2.0	203–404	SR,FP,RF
Wong and Davis (1974)	2.4×10^{15}	0	−5.7	220–360	SR,FP,RF
Hack (1977)	9.0×10^{15}	…	…	293	FR,DI,ES
Chiang and Skinner (1979)	7.0×10^{14}	…	…	1000	ST,TH,RA
Recommended ($\log F = \pm 0.20$)	7.0×10^{17}	−0.8	0	300–2000	…
	$H + O_2 + M \rightarrow HO_2 + M$ (k_1, $M = H_2$)				
Baulch *et al.* (1972)	5.0×10^{15}	0	−4.2	300–2000	Review
Ahumada *et al.* (1972)	4.4×10^{15}	…	…	298	SR,HG,RA
Kochubei and Moin (1973)	3.2×10^{15}	0	−6.4	640–1200	SR,TH,P
Wong and Davis (1974)	7.2×10^{15}	0	−5.7	220–360	SR,FP,RF
Dixon-Lewis *et al.* (1975)	4.2×10^{24}	−3.0	0	773–2130	FL,TH,MS
Recommended ($\log F = \pm 0.20$)	2.0×10^{18}	−0.8	0	300–2500	…
	$H + HO_2 \rightarrow OH + OH$				
Baulch *et al.* (1972)	2.5×10^{14}	0	7.9	290–800	Review
Day *et al.* (1973)	2.3×10^{14}	0	7.7	300–770	FL,TH,MS
Hack *et al.* (1978a)	2.5×10^{13}	…	…	293	FR,DI,ES
Baldwin and Walker (1979)	2.8×10^{14}	0	7.2	~800	SR,TH,P
Recommended ($\log F = \pm 0.20$)	1.5×10^{14}	0	4.2	300–1000	…

Table 3 (*continued*).

Reference	$A/(cm^3/mol)^{n-1} s^{-1}$	b	E/kJ	T/K	Method
	$H + HO_2 \rightarrow H_2 + O_2$				
Baulch *et al.* (1972)	2.5×10^{13}	0	2.9	290–800	Review
Day *et al.* (1973)	1.6×10^{13}	0	0.4	300–1050	FL,TH,MS
Hack *et al.* (1978a)	1.0×10^{13}	···	···	293	FR,DI,ES
Baldwin and Walker (1979)	1.7×10^{13}	0	0	~800	SR,TH,P
Recommended ($\log F = \pm 0.20$)	2.5×10^{13}	0	2.9	300–1000	···
	$H + HO_2 \rightarrow H_2O + O$				
Hack *et al.* (1978a)	$<7.0 \times 10^{11}$	···	···	293	FR,DI,ES
Baldwin and Walker (1979)	3.0×10^{13}	0	0	~800	SR,TH,P
No recommendation					
	$O + HO_2 \rightarrow OH + O_2$				
Baulch *et al.* (1976)	6.3×10^{13}	0	2.9	300–800	Review
Day *et al.* (1973)	3.3×10^{13}	···	···	1050	FL,TH,MS
Dixon-Lewis and Rhodes (1975)	1.0×10^{13}	···	···	1000	FL, TH, MS
Burrows *et al.* (1977)	2.0×10^{13}	···	···	293	FR, DI, LM
Hack *et al.* (1979)	2.0×10^{13}	···	···	298	FR,DI,ES,LM
Lii *et al.* (1980b)	4.2×10^{13}	···	···	298	SR, PH, UA
Recommended ($\log F = \pm 0.20$)	2.0×10^{13}	0	0	300–1000	···
	$OH + HO_2 \rightarrow H_2O + O_2$				
Troe (1969)	$<1.0 \times 10^{13}$	···	···	1400	ST,TH,UA
Friswell and Sutton (1972)	1.2×10^{13}	···	···	2130	FL,TH,CL
Hochanadel *et al.* (1972)	1.2×10^{14}	···	···	298	SR,FP,UA

Day *et al.* (1973)	4.0×10^{12}	···	···	1050	FL,TH,MS
Peeters and Mahnen (1973a)	5.0×10^{12}	···	···	1600	FL,TH,MS
deMore and Tschuikow-Roux (1974)	9.6×10^{13}	···	···	298	SR,PH,UA
Hack *et al.* (1975)	$<2.0 \times 10^{13}$	···	···	298	FR,DI,ES
Dixon-Lewis and Rhodes (1975)	8.5×10^{12}	···	···	~ 1000	FL,TH,MS
Burrows *et al.* (1977)	3.0×10^{13}	···	···	293	FR,DI,LM
Hack *et al.* (1978b)	1.8×10^{13}	···	···	293	FR,DI,LM
Chang and Kaufman (1978)	1.5×10^{13}	···	···	295	FR,DI,RF
Lii *et al.* (1980a)	6.0×10^{12}	···	···	308	SR,PH,UA
Hochanadel *et al.* (1980a)	7.0×10^{13}	···	···	296	SR,FP,UA
Temps and Wagner (1982)	4.0×10^{13}	···	···	296	FR, DI, LM
Recommended ($\log F = \pm 0.30$)	2.0×10^{13}	0	0	300–2000	···

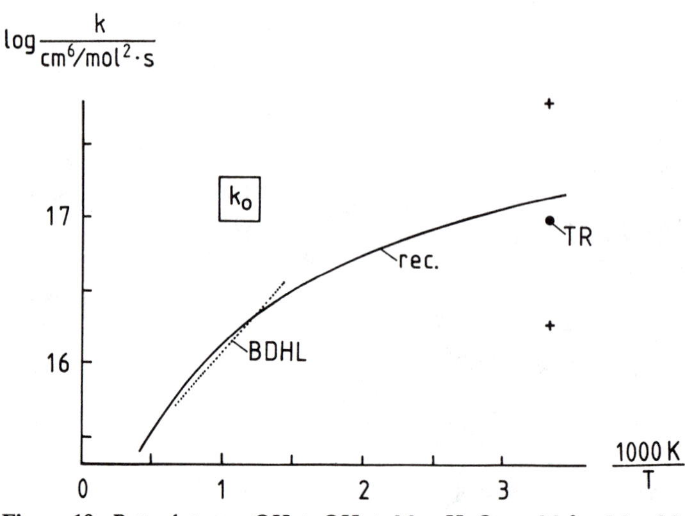

Figure 19. Rate data on $OH + OH + M \rightarrow H_2O_2 + M$ for $M = N_2$. References in Table 4. Crosses and broken lines denote representative experimental results given by Baulch *et al.* (1972).

Reactions of H, O, and OH with H_2O_2

Again, these reactions are relatively unimportant for high-temperature combustion because of the slow H_2O_2 formation rate. Data for $H + H_2O_2$ and $OH + H_2O_2$ are presented in Figs. 20 and 21, since the rate coefficients of the Leeds report cannot be recommended any longer.

$H_2 + HO_2 \rightarrow H + H_2O_2$

No direct measurements are available for this reaction. The rate coefficient recommended in the Leeds report is taken here; it was determined from the rate coefficient of the reverse reaction and the equilibrium constant.

4. Reactions of CO and CO_2

Reactions involving CO and CO_2 play a key part in the hydrocarbon combustion process. CO appears as the first product of hydrocarbon oxidation and is converted subsequently to CO_2 in a slow secondary reaction.

$CO + H + M \rightarrow CHO + M$

This is the reverse of the CHO thermal decomposition reaction discussed below. Because of its third order character, it is too slow to be important in

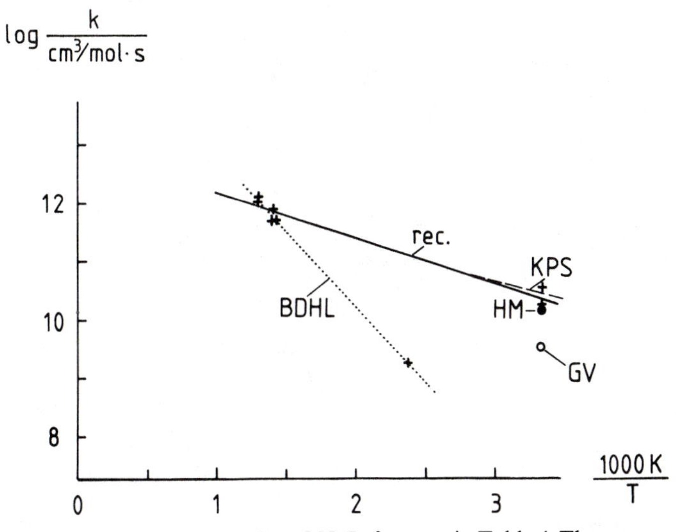

Figure 20. Rate data on $H + H_2O_2 \rightarrow H_2O + OH$. References in Table 4. The crosses and broken lines denote representative experimental results given by Baulch *et al.* (1972).

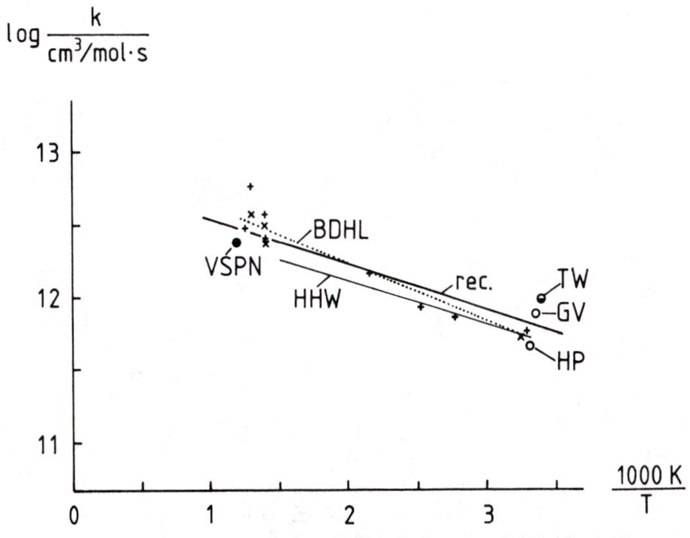

Figure 21. Rate data on $OH + H_2O_2 \rightarrow H_2O + HO_2$. References in Table 4. Crosses and broken lines denote representative experimental results given by Baulch *et al.* (1972).

Table 4. Rate data on H_2O_2 reactions.

Reference	$A/(\mathrm{cm^3/mol})^{n-1}\,\mathrm{s^{-1}}$	b	E/kJ	T/K	Method
	$HO_2 + HO_2 \rightarrow H_2O_2 + O_2$				
Baulch *et al.* (1972)	2.0×10^{12}	...	...	300	Review
Troe (1969)	2.0×10^{12}	...	...	1200	ST,TH,UA
Hochanadel *et al.* (1972)	5.7×10^{12}	...	...	298	SR,FP,UA
Paukert and Johnston (1972)	2.2×10^{12}	...	...	298	SR,PH,UA
Vardanyan *et al.* (1974)	1.8×10^{12}	...	...	850	SR,TH,ES
Hamilton and Lii (1977)	1.5×10^{12}	...	...	298	SR,PR,UA
Graham *et al.* (1979)	2.3×10^{12}	...	...	300	SR,PH,IE
Lii *et al.* (1979)	6.9×10^{10}	0	−8.8	276–400	SR,PR,UA
Cox and Burrows (1979)	2.3×10^{10}	0	−10.4	273–339	SR,PH,UA
Hochanadel *et al.* (1980a)	4.0×10^{12}	...	...	296	SR,FP,UA
Recommended ($\log F = \pm 0.30$)	2.0×10^{12}	0	0	300–1200	...
	$OH + OH + M \rightarrow H_2O_2 + M\ (k_1, M = N_2)$				
Baulch *et al.* (1972)	9.1×10^{14}	0	−21.2	700–1500	Review
Trainor and von Rosenberg (1974)	9.1×10^{16}	...	...	298	SR,FP,UA
Recommended ($\log F = \pm 0.30$)	1.3×10^{22}	−2.0	0	300–1500	...
	$H_2O_2 + M \rightarrow OH + OH + M\ (k_1, M = N_2)$				
Baulch *et al.* (1972)	1.2×10^{17}	0	190	700–1500	Review
Recommended ($\log F = \pm 0.30$)	1.2×10^{17}	0	190	700–1500	...
	$H + H_2O_2 \rightarrow H_2 + HO_2$				
Baulch *et al.* (1972)	1.7×10^{12}	0	15.7	300–800	Review
Gorse and Volman (1974)	1.9×10^{9}	...	...	298	SR,PH,MS,P

Heicklen and Meagher (1974)	1.2×10^{10}	...	...	298	SR,PH,MS,P
Recommended ($\log F = \pm 0.30$)	1.7×10^{12}	0	15.7	300–800	...
	$H_2 + HO_2 \rightarrow H_2O_2 + H$				
Baulch *et al.* (1972)	7.3×10^{11}	0	78.1	300–800	Review
Baldwin and Walker (1979)	1.6×10^{6}	...	...	773	SR,TH,P
Recommended ($\log F = \pm 0.30$)	7.3×10^{11}	0	78.1	300–800	...
	$H + H_2O_2 \rightarrow H_2O + OH$				
Baulch *et al.* (1972)	2.2×10^{15}	0	49.1	400–800	Review
Gorse and Volman (1974)	3.4×10^{9}	...	...	298	SR,PH,MS,P
Heicklen and Meagher (1974)	1.6×10^{10}	...	...	298	SR,PH,MS,P
Klemm *et al.* (1975)	3.1×10^{12}	0	11.6	283–353	SR,FP,RF
Recommended ($\log F = \pm 0.20$)	1.0×10^{13}	0	15.0	300–1000	...
	$O + H_2O_2 \rightarrow OH + HO_2$				
Albers *et al.* (1971)	2.8×10^{-13}	0	26.8	370–800	FR,DI,MS,ES
Davis *et al.* (1974a)	1.6×10^{12}	0	17.7	283–368	SR,PH,RF
Recommended ($\log F = \pm 0.50$)	2.8×10^{13}	0	26.8	300–1000	...
	$OH + H_2O_2 \rightarrow H_2O + HO_2$				
Baulch *et al.* (1972)	1.0×10^{13}	0	7.5	300–800	Review
Gorse and Volman (1972)	7.2×10^{11}	...	...	298	SR,PH,MS,P
Hack *et al.* (1974)	4.8×10^{12}	0	5.6	298–670	FR,DI,ES
Vardanyan *et al.* (1974)	1.9×10^{12}	...	...	850	SR,TH,ES
Harris and Pitts (1979)	4.1×10^{11}	...	...	298	SR,FP,RF
Temps and Wagner (1982)	1.0×10^{12}	...	...	296	FR,DI,LM
Recommended ($\log F = \pm 0.20$)	7.0×10^{12}	0	6.0	300–1000	...

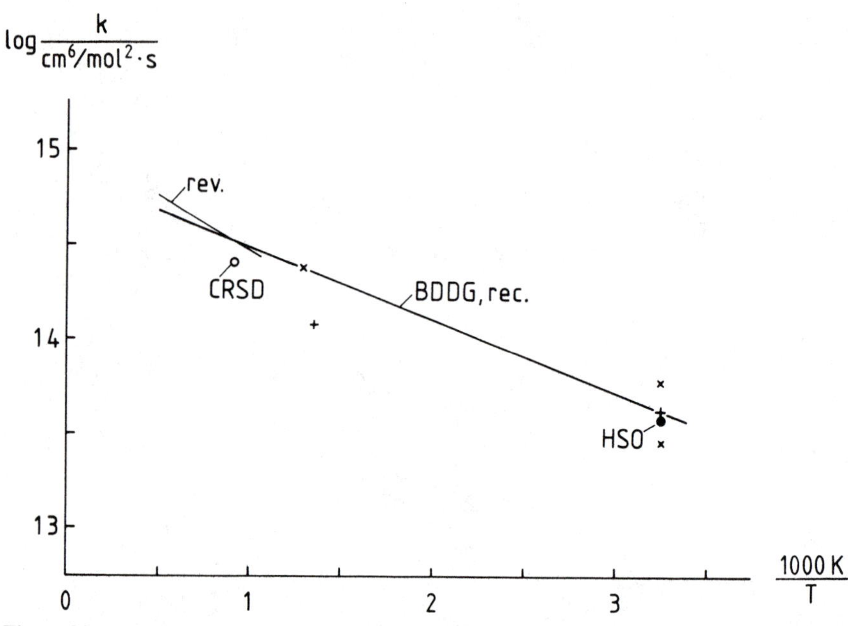

Figure 22. Rate data on $CO + H + M \rightarrow CHO + M$ for $M = H_2$. References in Table 5. The crosses denote representative experimental results given by Baulch *et al.* (1976).

high temperature combustion (Fig. 22). Some influence on the combustion of CO is suggested in the literature (Cherian *et al.*, 1981). Relative collision efficiencies are given in Section 1.2.

$$CO + O + M \rightarrow CO_2 + M$$

This reaction is unimportant for the oxidation of carbon monoxide as long as even trace amounts of hydrogen-containing species are available, e.g., in hydrocarbon combustion. It becomes important only in very dry CO flames (Warnatz, 1978). The rate coefficient (Fig. 23) has an unusual temperature dependence, which can be explained using spin conservation considerations (Baulch *et al.*, 1976). Because of this behavior and the large scatter of the low-temperature data (due to the extreme sensitivity of the relevant experiments to hydrogen-containing impurities), a recommendation is given only for high temperature. For this reaction an extrapolation of the high-temperature Arrhenius expression down to room temperature leads to an error of two orders of magnitude. Relative collision efficiencies are discussed in Section 1.2.

$$CO + OH \rightarrow CO_2 + H \text{ and } H + CO_2 \rightarrow CO + OH$$

These reactions are of basic importance for the oxidation process, because in hydrocarbon combustion CO is oxidized almost exclusively by OH to form

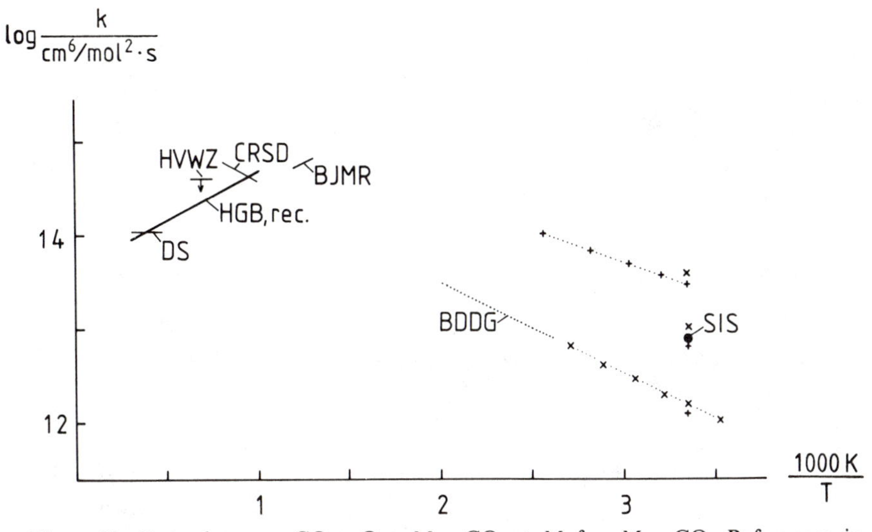

Figure 23. Rate data on $CO + O + M \rightarrow CO_2 + M$ for $M = CO$. References in Table 5. The crosses and broken lines denote representative experimental results given by Baulch *et al.* (1976).

CO_2. Besides, the reverse reaction is necessary to establish the water–gas equilibrium (Warnatz, 1978). Accordingly, variations of the rate coefficients both of $CO + OH \rightarrow CO_2 + H$ and of its reverse reaction strongly influence flame propagation, as demonstrated in Fig. 4. Ignition is not influenced by these steps because of their late occurrence during the combustion process (Fig. 1).

There are many studies of $CO + OH \rightarrow CO_2 + H$ reported in the literature, reviewed by Baulch *et al.* (1976) and in Chapter 3. Figures 24 and 25 present the Leeds report recommendation together with results published after 1976. Most of these studies were done at low temperature, motivated by the importance of this reaction for atmospheric chemistry. There is a definite pressure dependence of the low-temperature rate coefficient (Butler *et al.*, 1978; Chan *et al.*, 1977; Overend and Paraskevopoulos, 1977a), which may be coupled to the presence of trace amounts of oxygen (Biermann *et al.*, 1978). At high temperature this complex formation should be unimportant. Nevertheless, there is need for more precise data over 1000 K, especially in view of the non-Arrhenius behavior.

$$CO + HO_2 \rightarrow CO_2 + OH,\ CO + O_2 \rightarrow CO_2 + O,$$
$$\text{and } CO_2 + O \rightarrow CO + O_2$$

These reactions are too slow to play a considerable part in flame propagation. The recommended expressions are taken from the Leeds report (Baulch *et al.*, 1976).

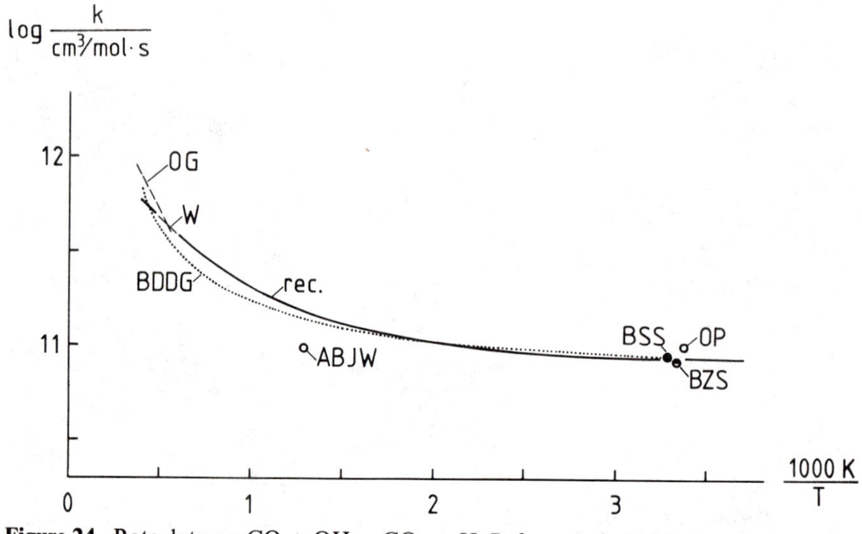

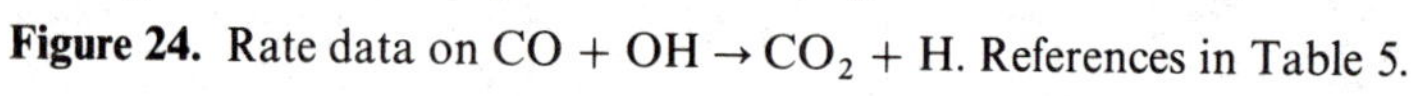

Figure 24. Rate data on $CO + OH \rightarrow CO_2 + H$. References in Table 5.

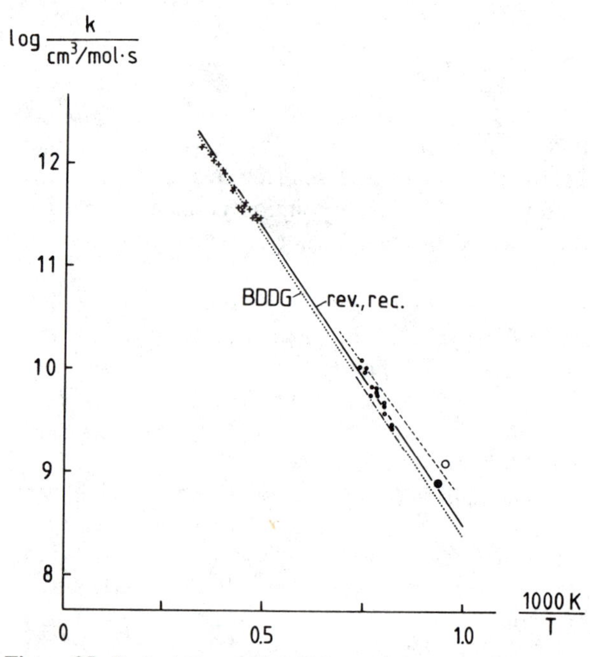

Figure 25. Rate data on $H + CO_2 \rightarrow CO + OH$. References in Table 5. The crosses and broken lines denote representative experimental results given by Baulch *et al.* (1976).

Table 5. Rate data on CO reactions.

Reference	$A/(cm^3/mol)^{n-1}\,s^{-1}$	b	E/kJ	T/K	Method
	$CO + H + M \rightarrow CHO + M\ (k_1,\ M = H_2)$				
Ahumada *et al.* (1972)	2.8×10^{13}	⋯	⋯	298	SR,HG,RA
Baulch *et al.* (1976)	6.9×10^{14}	0	7.0	298–773	Review
Hochanadel *et al.* (1980b)	3.8×10^{13}	⋯	⋯	298	SR,FP,UA
Cherian *et al.* (1981)	5.0×10^{14}	0	6.3	~1100	FL,TH,MS
Recommended ($\log F = \pm 0.20$)	6.9×10^{14}	0	7.0	300–2000	⋯
	$CO + O + M \rightarrow CO_2 + M\ (k_1,\ M = CO)$				
Hardy *et al.* (1974)	4.0×10^{14}	⋯	⋯	~1500	ST,TH,IE
Baulch *et al.* (1976)	2.4×10^{15}	0	18.2	250–500	Review
Dean and Steiner (1977)	1.1×10^{14}	0	0	2100–3200	ST,TH,IE
Hardy *et al.* (1978)	5.3×10^{13}	0	−19.0	1300–2200	ST,TH,IE
Sugawara *et al.* (1980)	6.8×10^{12}	⋯	⋯	296	SR,PR,RA
Cherian *et al.* (1981)	5.4×10^{15}	⋯	19.1	~1100	FL,TH,MS
Recommended ($\log F = \pm 0.50$)	5.3×10^{13}	0	−19.0	1000–3000	⋯
	$CO + OH \rightarrow CO_2 + H$				
Baulch *et al.* (1976)	1.5×10^{7}	1.3	−3.2	250–2000	Review
Cox *et al.* (1976a)	1.6×10^{11}	⋯	⋯	296	SR,PH,CL
Atri *et al.* (1977)	9.6×10^{10}	⋯	⋯	773	SR,TH,P
Overend and Paraskevopoulos (1977a)	1.0×10^{11}	⋯	⋯	296	SR,PH,RF
Olson and Gardiner (1978)	4.0×10^{12}	⋯	33.5	1800–2700	Review
Butler *et al.* (1978)	9.0×10^{10}	⋯	⋯	304	SR,PH,GC
Biermann *et al.* (1978)	8.4×10^{10}	⋯	⋯	300	SR,PH,RA

Table 5 (*continued*).

Reference	$A/(cm^3/mol)^{n-1}\,s^{-1}$	b	E/kJ	T/K	Method
	$CO + OH \rightarrow CO_2 + H$				
Warnatz (1979)	4.4×10^6	1.5	−3.1	300–2000	Review
Recommended ($\log F = \pm 0.15$)	4.4×10^6	1.5	−3.1	300–2000	...
	$CO + HO_2 \rightarrow CO_2 + OH$				
Baulch *et al.* (1976)	1.5×10^{14}	0	98.7	700–1000	Review
Atri *et al.* (1977)	1.9×10^7	...	...	773	SR,TH,P
Colket *et al.* (1977)	5.6×10^9	...	...	1100	FR,TH,GC
Recommended ($\log F = \pm 0.30$)	1.5×10^{14}	0	98.7	700–1100	...
	$CO + O_2 \rightarrow CO_2 + O$				
Baulch *et al.* (1976)	2.5×10^{12}	0	200.0	1500–3000	Review
Recommended ($\log F = \pm 0.50$)	2.5×10^{12}	0	200.0	1500–3000	...
	$CO_2 + H \rightarrow CO + OH$				
Baulch *et al.* (1976)	1.5×10^{14}	0	110	1000–3000	Review
Recommended ($\log F = \pm 0.20$)	1.6×10^{14}	0	110	1000–3000	...
	$CO_2 + O \rightarrow CO + O_2$				
Baulch *et al.* (1976)	1.7×10^{13}	0	220	1500–3000	Review
Recommended ($\log F = \pm 0.50$)	1.7×10^{13}	0	220	1500–3000	...

5. Reactions of C_1-hydrocarbons

As mentioned in Section 2, oxidation of C_1-hydrocarbons (together with C_2-hydrocarbons) plays a central part in hydrocarbon combustion generally because of the ubiquitous formation of methyl radicals.

5.1. Reactions of CH_4

Reactions of H, O, and OH with CH_4

These reactions provide the initial attack upon CH_4 in methane combustion or intermediate CH_4 formed (mainly by combination of CH_3 and H atoms) in the combustion of other fuels. The sensitivity of flame propagation to their rate coefficients is small (Fig. 4), but ignition delay times are sensitive to their rate coefficients (Fig. 1). There are thorough reviews for all of these reactions (Clark and Dove, 1973, and Roth and Just, 1975, for $H + CH_4$; Roth and Just, 1977, and Klemm *et al.*, 1981, for $O + CH_4$; and Ernst *et al.*, 1978, for $OH + CH_4$). The recommendations given in these references are adopted here (Figs. 26 to 28).

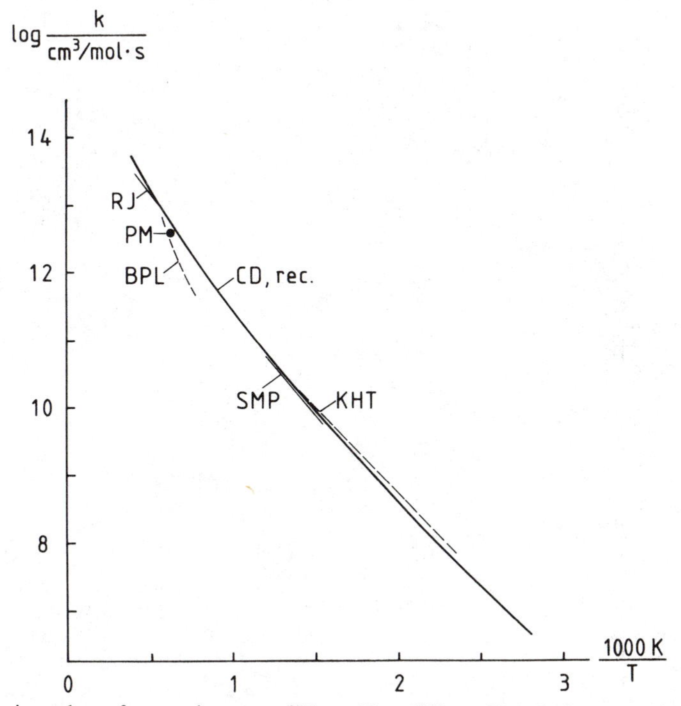

Figure 26. Arrhenius plot of rate data on $CH_4 + H \rightarrow CH_3 + H_2$. References in Table 6.

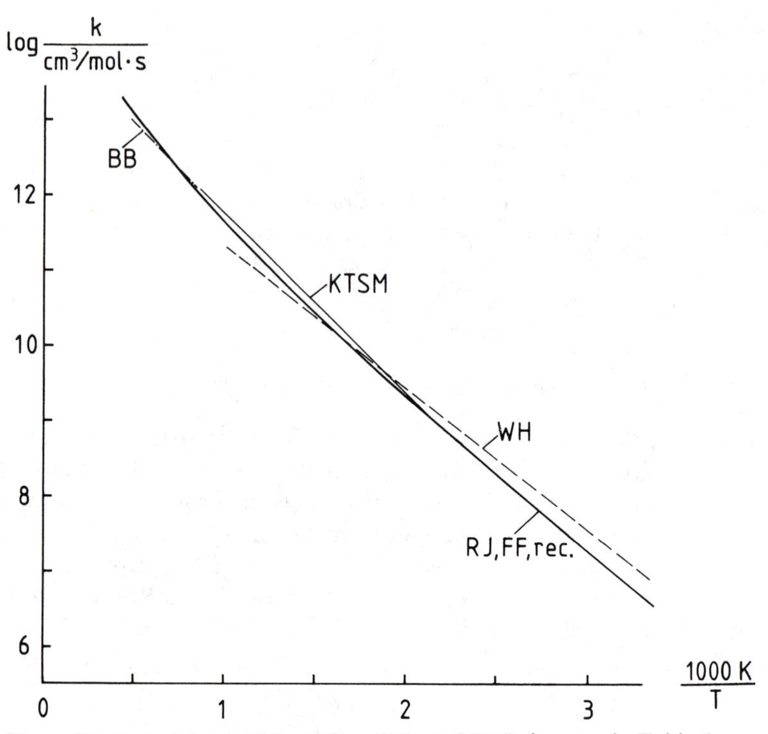

Figure 27. Rate data on $CH_4 + O \rightarrow CH_3 + OH$. References in Table 6.

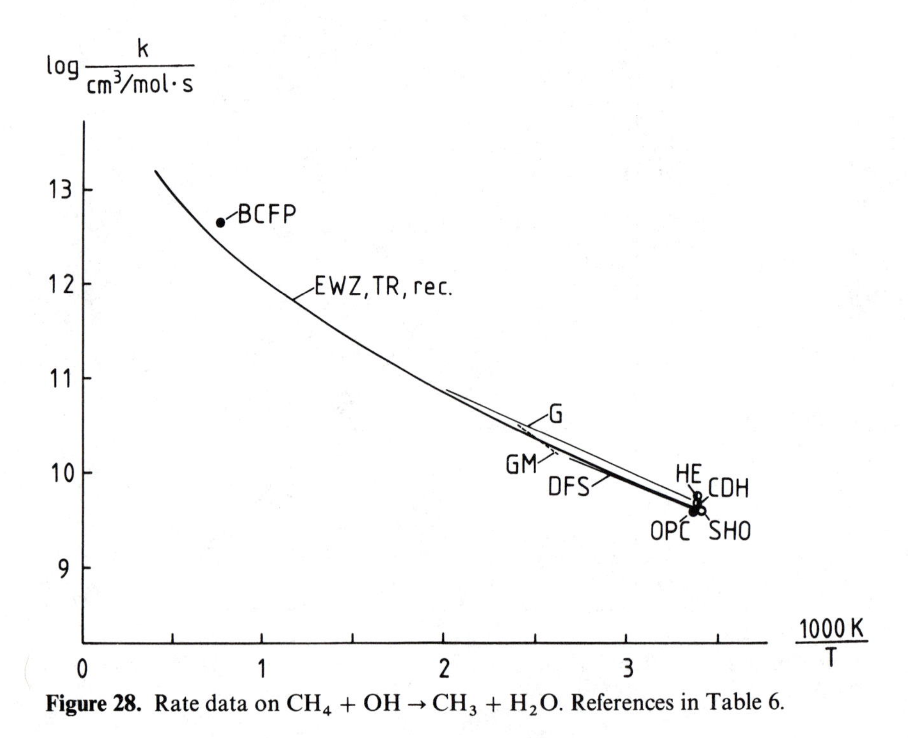

Figure 28. Rate data on $CH_4 + OH \rightarrow CH_3 + H_2O$. References in Table 6.

$$CH_4 + HO_2 \rightarrow CH_3 + H_2O_2 \text{ and } CH_4\ (+M) \rightarrow CH_3 + H\ (+M)$$

These reactions are relatively unimportant for flame propagation because of the small HO_2 concentrations and the small rate coefficient of the HO_2 reaction and because of the large activation energy of the thermal decomposition of methane. On the other hand, CH_4 thermal decomposition strongly influences the ignition of CH_4 (Fig. 1). The pressure dependence of the decomposition reaction has been investigated, enabling the estimation of limiting high- and low-pressure rate coefficients (Fig. 29). Examples of fall-off curves are given in Fig. 30.

5.2. Reactions of CH_3

$$CH_3 + H \rightarrow CH_2 + H_2$$

This endothermic reaction ($\Delta H^\circ_{298} = +21.8$ kJ) is postulated in the interpretation of shock tube work of two experimental groups. No direct measurements are available in the literature. At typical conditions in flame

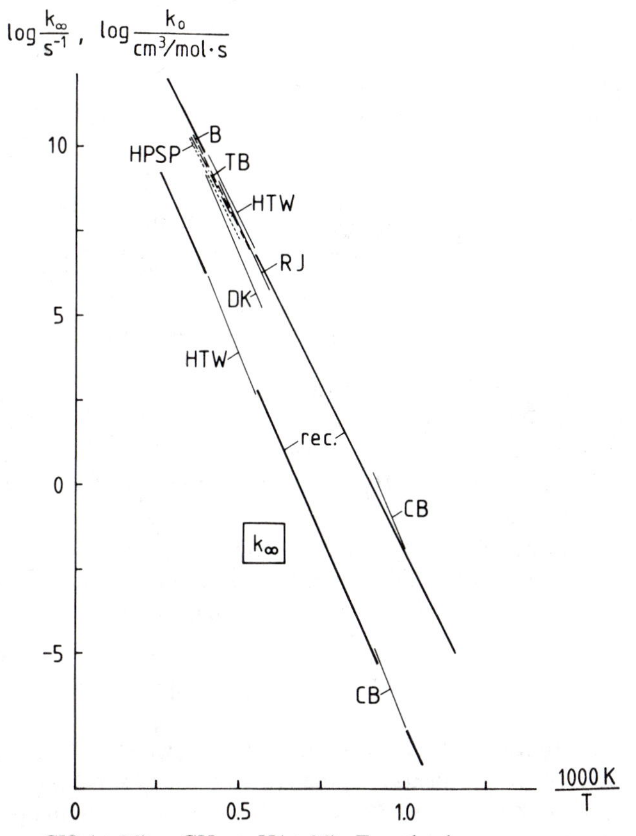

Figure 29. Rate data on $CH_4(+M) \rightarrow CH_3 + H(+M)$. For the low-pressure rate coefficient k_1, M = Ar. References in Table 6. Fall-off curves are given in Fig. 30.

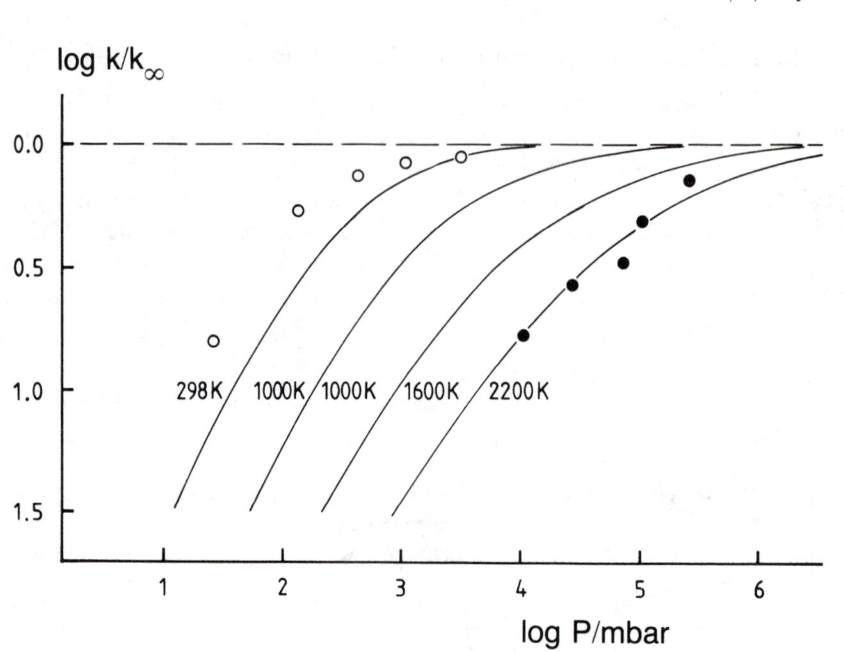

Figure 30. Fall-off curves for $CH_3 + H \rightarrow CH_4$. Lines: calculated as described in Section 1.2. Points: experimental values, ●: Hartig *et al.* (1971) for $T = 2200$ K, ○: Cheng and Yeh (1977) for $T = 308$ K.

propagation this reaction is slow compared with the competing reaction $H + CH_3(+M) \rightarrow CH_4(+M)$. No recommendation is given here.

$$CH_3 + H\ (+M) \rightarrow CH_4\ (+M)$$

The reverse of this exothermic reaction (Fig. 31) has been discussed before, including a presentation of its fall-off behavior (Fig. 30). Flame propagation is rather sensitive to the rate coefficient of this reaction (Fig. 4) because of the importance of CH_3 as an intermediate in the combustion of aliphatic hydrocarbons (Warnatz, 1981) and the chain-terminating character of this step. As no information on the efficiency of third bodies other than noble gases is available, the collision numbers given in Section 1.2 are recommended for use with the low-pressure rate coefficient.

$$CH_3 + O \rightarrow CH_2O + H$$

This reaction ($\Delta H^\circ_{298} = -203.1$ kJ) is very fast and therefore one of the main sinks of CH_3 radicals in flame propagation in addition to CH_3 recombination and combination with H atoms. Consequently, flame propagation is rather sensitive to the rate coefficients of this step (Fig. 4). Though this is a radical–radical reaction, there is a considerable amount of experimental information available in the literature, enabling a relatively safe recommendation (Fig. 32).

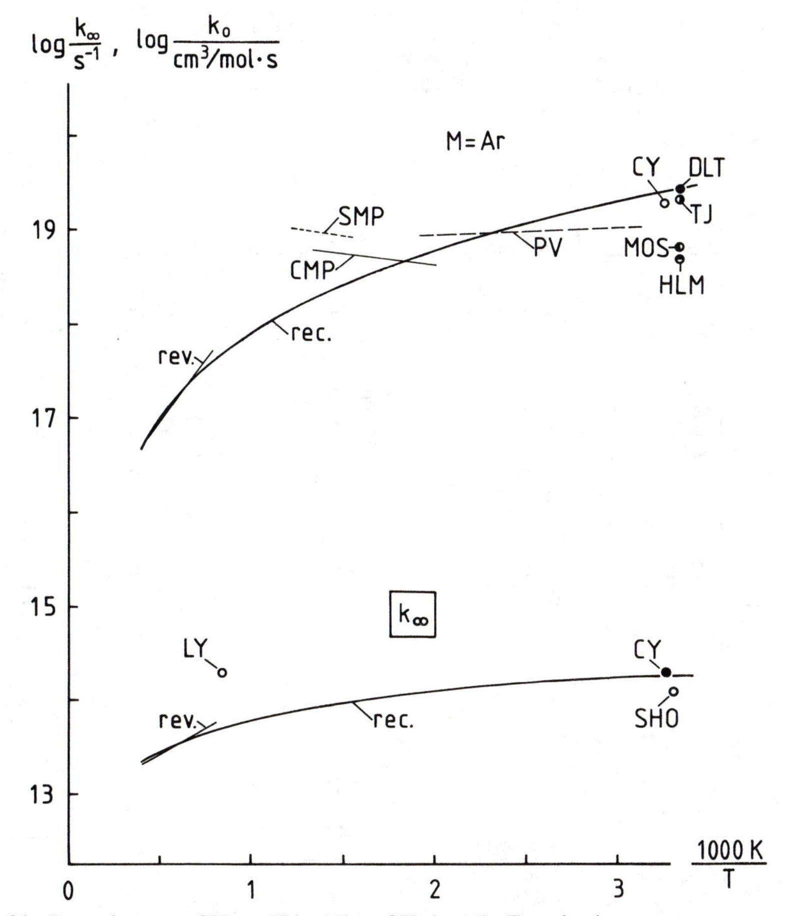

Figure 31. Rate data on $CH_3 + H(+M) \rightarrow CH_4(+M)$. For the low-pressure rate coefficient k_1, M = Ar. References in Table 7. Fall-off curves are given in Fig. 30.

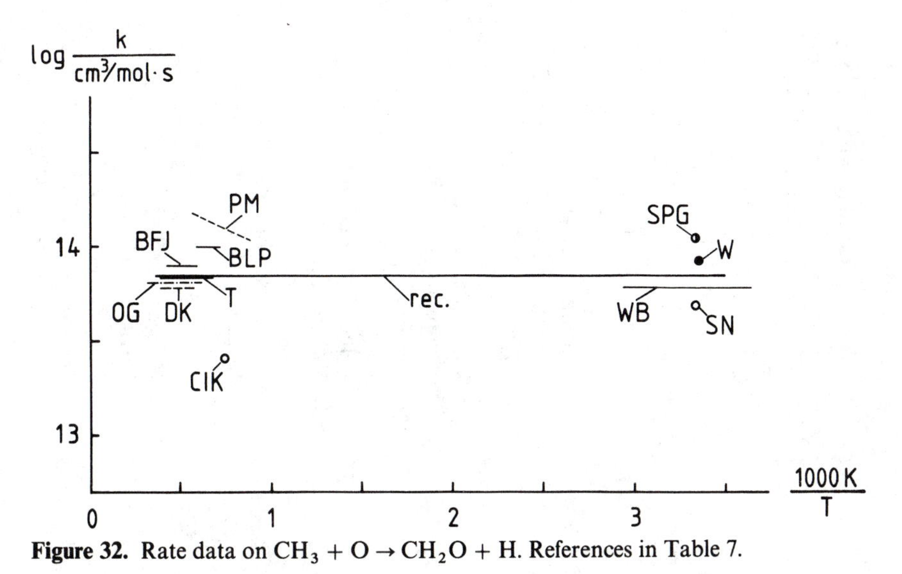

Figure 32. Rate data on $CH_3 + O \rightarrow CH_2O + H$. References in Table 7.

Table 6. Rate data on CH_4 reactions.

Reference	$A/(cm^3/mol)^{n-1} s^{-1}$	b	E/kJ	T/K	Method
	$CH_4 + H \rightarrow CH_3 + H_2$				
Kurylo *et al.* (1970a)	6.2×10^{13}	0	48.6	424–732	FR,DI,ES
Peeters and Mahnen (1973a)	3.2×10^{12}	...	...	1600	FL,TH,MS
Clark and Dove (1973a)	2.2×10^{4}	3.0	36.6	300–1800	Review
Biordi *et al.* (1974)	1.5×10^{5}	3.0	69.0	1300–1700	FL,TH,MS
Roth and Just (1975)	7.2×10^{14}	0	63.0	1700–2300	ST,TH,RA
Sepehrad *et al.* (1979)	1.8×10^{14}	0	55.1	640–818	FR,DI,GC
Recommended ($\log F = \pm 0.15$)	2.2×10^{4}	3.0	36.6	300–2500	...
	$CH_4 + O \rightarrow CH_3 + OH$				
Westenberg and deHaas (1967)	1.7×10^{13}	0	36.4	300–1000	FR,DI,ES
Brabbs and Brokaw (1975)	1.9×10^{14}	0	49.1	1200–2000	ST,TH,UE
Roth and Just (1977)	1.2×10^{7}	2.1	31.9	1500–2250	ST,TH,RA
Roth and Just (1977)	1.2×10^{7}	2.1	31.9	300–2200	Review
Felder and Fontijn (1979)	1.2×10^{7}	2.1	31.9	525–1250	FR,FP,RF
Klemm *et al.* (1981)	1.3×10^{14}	0	45.5	474–1156	FR,DI,RF
Klemm *et al.* (1981)	1.3×10^{4}	0	45.5	474–1156	SR,FP,RF
Felder and Fontijn (1981)	1.6×10^{6}	2.36	31.0	420–1670	FR,FP,RF
Recommended ($\log F = \pm 0.15$)	1.2×10^{7}	2.1	31.9	300–2200	...
	$CH_4 + OH \rightarrow CH_3 + H_2O$				
Greiner (1970b)	3.3×10^{12}	0	15.8	300–500	SR,FP,UA
Davis *et al.* (1974b)	1.4×10^{12}	0	14.2	240–373	SR,FP,RF
Gordon and Mulac (1975)	1.1×10^{14}	0	28.0	381–416	SR,PR,UA
Overend *et al.* (1975)	3.9×10^{9}	...	...	295	SR,FP,RA
Bradley *et al.* (1976)	4.5×10^{12}	...	...	1300	ST,TH,UA

Howard and Evenson (1976b)	5.7×10^9	...	...	296	FR,DI,LM
Zellner and Steinert (1976)	3.5×10^3	3.1	8.4	300–900	SR, FP, RA
Cox *et al.* (1976b)	4.6×10^9	...	...	296	SR,PH,CL
Ernst *et al.* (1978)	2.5×10^{12}	...	...	1300	ST,FP,RA
Ernst *et al.* (1978)	1.6×10^6	2.1	10.3	300–2200	Review
Sworsky *et al.* (1980)	4.2×10^9	...	...	296	SR,FP,UA
Tully and Ravishankara (1980)	1.6×10^6	2.1	10.3	298–1020	SR,FP,RF
Recommended ($\log F = \pm 0.20$)	1.6×10^6	2.1	10.3	300–2200	...
$CH_4 + M \rightarrow CH_3 + H + M$ (k_1, $M = Ar$)					
Hartig *et al.* (1971)	2.0×10^{17}	0	368	1850–2500	ST,TH,IE
Dean and Kistiakowsky (1971)	1.6×10^{18}	0	431	1750–2575	ST,TH,IE
Roth and Just (1975)	4.7×10^{17}	0	390	1700–2300	ST,TH,RA
Bowman (1975a)	2.0×10^{17}	0	370	1875–2240	ST,TH,UA,IE
Heffington *et al.* (1977)	2.2×10^{17}	0	377	2000–2800	ST,TH,IE,UA
Tabayashi and Bauer (1979)	1.0×10^{17}	0	359	1950–2770	ST,TH,LS
Recommended ($\log F = \pm 0.50$)	2.0×10^{17}	0	370	1500–3000	...
$CH_4 \rightarrow CH_3 + H$ (k_∞)					
Hartig *et al.* (1971)	1.3×10^{15}	0	435	1850–2500	ST,TH,IE
Chen and Back (1975)	2.8×10^{16}	0	451	995–1103	SR,TH,GC
Recommended ($\log F = \pm 0.50$)	1.0×10^{15}	0	420	1000–3000	...

$CH_3 + OH \rightarrow$ products

No direct measurements of the rate coefficient of this reaction have been reported, and there is no reliable determination of the reaction products. Because of the doubt about the existence of this reaction and because of the scatter of the data, no recommendation is given. A reaction forming CH_3O (or CH_2OH) and H atoms (analogous to the reaction $CH_3 + O \rightarrow CH_2O + H$) would be endothermic both for CH_2OH ($\Delta H^\circ_{298} = 17$ kJ) and CH_3O ($\Delta H^\circ_{298} = 48$ kJ), preventing a study of this reaction at low temperature. A (relatively fast) recombination (analogous to the reaction $CH_3 + H \rightarrow CH_4$) would seem to be possible also.

$CH_3 + H_2 \rightarrow CH_4 + H$

The reverse of this step has been discussed before; direct determinations of its rate coefficient agree with values calculated from the reverse reaction. A review of this reaction is given by Clark and Dove (1973b).

$CH_3 + O_2 \rightarrow CH_3O + O$

There is much literature concerning a reaction of CH_3 with O_2 assuming rearrangement of the intermediate complex to yield CH_2O and OH (Clark *et al.*, 1971b; Peeters and Mahnen, 1973a; Jachimowski, 1974; Bowman, 1975a; Olson and Gardiner, 1978; Tabayashi and Bauer, 1979). The activation energies given in these references vary from 38 to 88 kJ/mol and the scatter of the absolute values for flame conditions amounts to more than two orders of magnitude. On the other hand, the newer measurements lead to a consistent picture of this reaction if CH_3O and O atoms are postulated to be the primary products, the CH_3O decomposing rapidly to form CH_2O and H atoms. Because of its large activation energy (Fig. 33) the reaction is unimportant for flame propagation (Fig. 4) but extremely important for ignition processes due to its chain-branching character (by fast decomposition of the CH_3O to form CH_2O and H) and the early formation of CH_3 in igniting gas mixtures (Fig. 1).

$CH_3 + CH_3\ (+M) \rightarrow C_2H_6\ (+M)$

This is a very important reaction both for flame propagation and ignition, since it competes with the oxidation reactions of CH_3. Furthermore, it is an important source of C_2-hydrocarbons in methane combustion, leading eventually to the formation of soot and NO in rich flames of CH_4. Values of the high-pressure rate coefficient k_∞ are given in Fig. 34. The pressure dependence of this reaction is well known; examples of fall-off curves are given in Fig. 35. No recommendation of the low-pressure rate coefficient k_1 can be derived from experimental data, but values calculated from the reverse reaction rate coefficient should be reliable at high temperature.

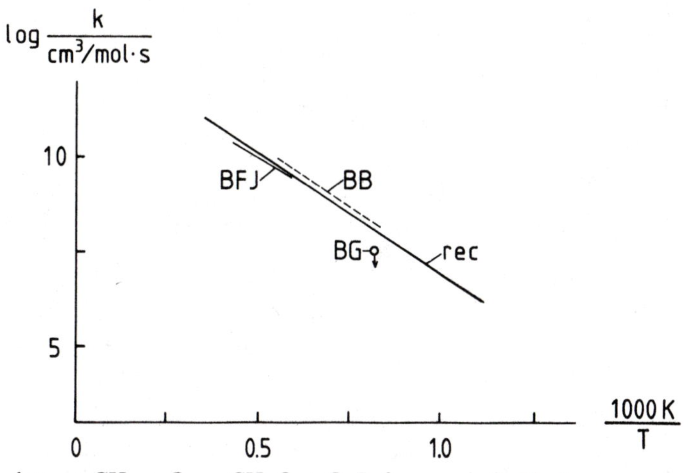

Figure 33. Rate data on $CH_3 + O_2 \rightarrow CH_3O + O$. References in Table 7.

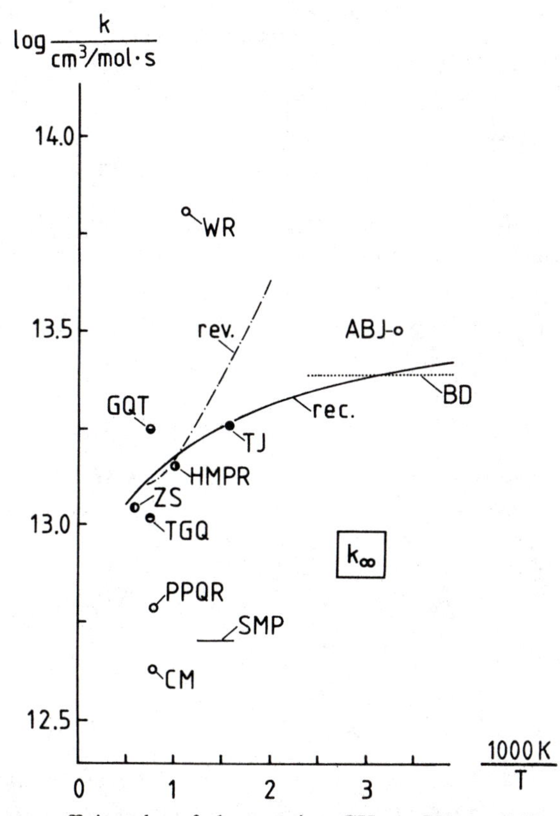

Figure 34. High-pressure rate coefficient k_∞ of the reaction $CH_3 + CH_3 \rightarrow C_2H_6$. References in Table 7 and in the review of Baulch and Duxbury (1980). Fall-off curves are given in Fig. 35.

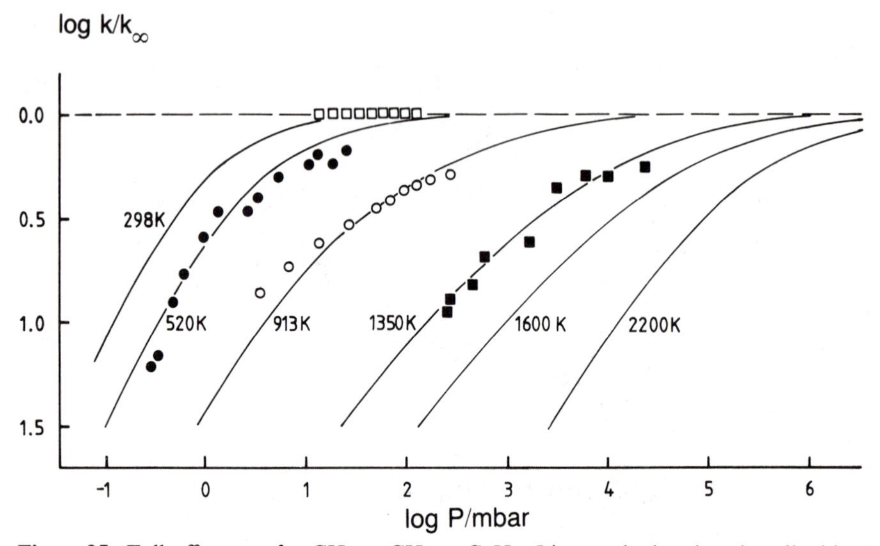

Figure 35. Fall-off curves for $CH_3 + CH_3 \rightarrow C_2H_6$. Lines: calculated as described in Section 1.2. Points: experimental values, ■: Glänzer *et al.* (1977) for $T = 1350$ K, ○: Lin and Back (1966) for $T = 913$ K, ●: Steacie for $T = 520$ K, □: various workers at room temperature. For literature citations see Baulch and Duxbury (1980).

$$CH_3 + CH_3 \rightarrow C_2H_5 + H$$

Relatively direct evidence of this second channel of the reaction $CH_3 + CH_3$ can be derived from H resonance absorption experiments showing hydrogen atoms to be formed in the first step (Roth and Just, 1979). Because of the pressure dependence of the channel forming C_2H_6, C_2H_5 should be formed preferably in low-pressure flames.

$$CH_3 + CH_3 \rightarrow C_2H_4 + H_2$$

There is indirect but compelling evidence of this reaction channel, based on the clear discrepancy between the rate coefficient of the H-atom forming channel (see above) and that part of the $CH_3 + CH_3$ reaction which does not lead to C_2H_6 (Tsuboi, 1978).

5.3. Reactions of CH_2O

Reactions of H, O, and OH with CH_2O

These reactions provide for formaldehyde consumption. Since formaldehyde is a major intermediate produced via $CH_3 + O \rightarrow CH_2O + H$ and $CH_3 + O_2 \rightarrow CH_3O + O$, $CH_3O \rightarrow CH_2O + H$. Both ignition and flame propagation are moderately sensitive to the values of the rate coefficients of

Table 7. Rate data on CH_3 reactions.

Reference	$A/(cm^3/mol)^{n-1}\ s^{-1}$	b	E/kJ	T/K	Method
	$CH_3 + H + M \rightarrow CH_4 + M\ (k_1, M = Ar)$				
Dodonov *et al.* (1969)	2.8×10^{19}	...	...	293	FR,DI,MS
Halstead *et al.* (1970)	5.4×10^{18}	...	...	290	FR,DI,GC
Cowfer *et al.* (1971)	8.5×10^{18}	...	...	298	FR,DI,MS
Teng and Jones (1972)	1.7×10^{19}	...	...	303	FR,DI,GC
Camilleri *et al.* (1974)	1.4×10^{19}	0	5.0	503–753	FR,DI,GC
Pratt and Veltman (1976)	6.9×10^{19}	−0.33	0	321–521	FR,DI,MS
Cheng and Yeh (1977)	2.0×10^{19}	...	...	308	SR,HG,GC
Sepehrad *et al.* (1979)	2.7×10^{19}	0	6.3	640–818	FR,DI,GC
Recommended ($\log F = \pm 0.50$)	8.0×10^{26}	−3.0	0	300–2500	...
	$CH_3 + H \rightarrow CH_4\ (k_\infty)$				
Cheng and Yeh (1977)	2.0×10^{12}	...	...	308	SR,HG,GC
Lee and Yeh (1979)	2.0×10^{14}	...	...	1200	SR,TH,GC
Sworsky *et al.* (1980)	1.2×10^{14}	...	...	296	SR,FP,UA
Recommended ($\log F = \pm 0.30$)	6.0×10^{16}	−1.0	0	300–2500	...
	$CH_3 + H \rightarrow CH_2 + H_2$				
Olson and Gardiner (1978)	7.2×10^{14}	0	63.2	1800–2700	ST,TH,IA
Bhaskaran *et al.* (1979)	1.8×10^{14}	0	63	1700–2300	ST,TH,RA
No recommendation					
	$CH_3 + O \rightarrow CH_2O + H$				
Dean and Kistiakowsky (1971)	6.0×10^{13}	0	0	1750–2575	ST,TH,IE
Clark *et al.* (1971a)	2.5×10^{13}	...	...	1350	ST,TH,MS
Stuhl and Niki (1972b)	4.8×10^{13}	...	...	298	SR,PH,CL

Table 7 (*continued*).

Reference	$A/(cm^3/mol)^{n-1}\,s^{-1}$	b	E/kJ	T/K	Method
	$CH_3 + O \rightarrow CH_2O + H$				
Peeters and Mahnen (1973a)	1.3×10^{14}	0	8.4	1150–1750	FL,TH,MS
Slagle *et al.* (1974)	1.1×10^{14}	...	...	300	FR,DI,MS
Tsuboi (1976)	6.0×10^{13}	0	0	1500–2000	ST,TH,UA
Washida and Bayes (1976)	6.0×10^{13}	0	0	259–341	FR,DI,MS
Biordi *et al.* (1976)	8.2×10^{13}	0	0	1550–1750	FL,TH,MS
Olson and Gardiner (1978)	6.8×10^{13}	0	0	1800–2700	ST,TH,IA
Bhaskaran *et al.* (1979)	8.0×10^{13}	0	0	1700–2300	ST,TH,RA
Washida (1980)	8.3×10^{13}	...	...	298	FR,DI,MS
Recommended ($\log F = \pm 0.20$)	7.0×10^{13}	0	0	300–2500	...
	$CH_3 + OH \rightarrow$ products				
Fenimore (1969)	4.0×10^{12}	0	0	1970–2190	FL,TH,MS
Peeters and Vinckier (1975)	7.4×10^{12}	...	...	1500	FL,TH,MS
Bhaskaran *et al.* (1979)	2.0×10^{16}	0	115	1700–2300	ST,TH,RA
Sworsky *et al.* (1980)	5.6×10^{13}	...	...	296	SR,FP,UA
No recommendation					
	$CH_3 + H_2 \rightarrow CH_4 + H$				
Clark and Dove (1973b)	4.6×10^{10}	...	...	1340	ST,TH,MS
Clark and Dove (1973b)	1.5×10^{13}	0	64.9	1200–2000	Review
Kerr and Parsonage (1976)	8.5×10^{11}	0	45.6	370–700	Review
Recommended ($\log F = \pm 0.20$)	6.6×10^{2}	3.0	32.4	300–2500	...

	$CH_3 + O_2 \rightarrow CH_3O + O$				
Brabbs and Brokaw (1975)	2.4×10^{13}	0	120	1200–1800	ST,TH,CL
Baldwin and Golden (1978)	$<3.0 \times 10^{7}$	...	...	1220	SR,TH,MS
Bhaskaran *et al.* (1979)	7.0×10^{12}	0	107	1700–2300	ST,TH,RA
Recommended ($\log F = \pm 0.50$)	1.5×10^{13}	0	120	1000–2300	...
	$CH_3 + CH_3 \rightarrow C_2H_6\ (k_\infty)$				
Held *et al.* (1977)	1.4×10^{13}	...	...	1005	FR,TH,UA,GC
Sepehrad *et al.* (1979)	5.0×10^{12}	0	0	640–818	FR,DI,GC
Zaslonko and Smirnov (1979)	1.1×10^{13}	...	...	1750	ST,TH,UA
Adachi *et al.* (1980)	3.2×10^{13}	...	...	298	SR,FP,UA
Baulch and Duxbury (1980)	2.4×10^{13}	0	0	250–420	Review
Recommended ($\log F = \pm 0.20$)	2.4×10^{14}	−0.4	0	300–2000	...
	$CH_3 + CH_3 \rightarrow C_2H_5 + H$				
Gardiner *et al.* (1975)	4.0×10^{14}	0	75	2000–2700	ST,TH,MS,IA
Roth and Just (1979)	8.0×10^{14}	0	111	1650–2100	ST,TH,RA
Recommended ($\log F = \pm 0.50$)	8.0×10^{14}	0	111	1500–3000	...
	$CH_3 + CH_3 \rightarrow C_2H_4 + H_2$				
Roth and Just (1979)	1.0×10^{16}	0	134	1650–2100	ST,TH,RA
Recommended ($\log F = \pm 0.50$)	1.0×10^{16}	0	134	1500–2500	...
	$CH_3 + M \rightarrow CH_2 + H + M\ (k_1, M = Ar)$				
Bhaskaran *et al.* (1979)	6.1×10^{15}	0	374	1700–2300	ST,TH,RA
Roth *et al.* (1979)	1.9×10^{16}	0	384	2150–2850	ST,TH,RA
Recommended ($\log F = \pm 0.50$)	1.0×10^{16}	0	379	1500–3000	...

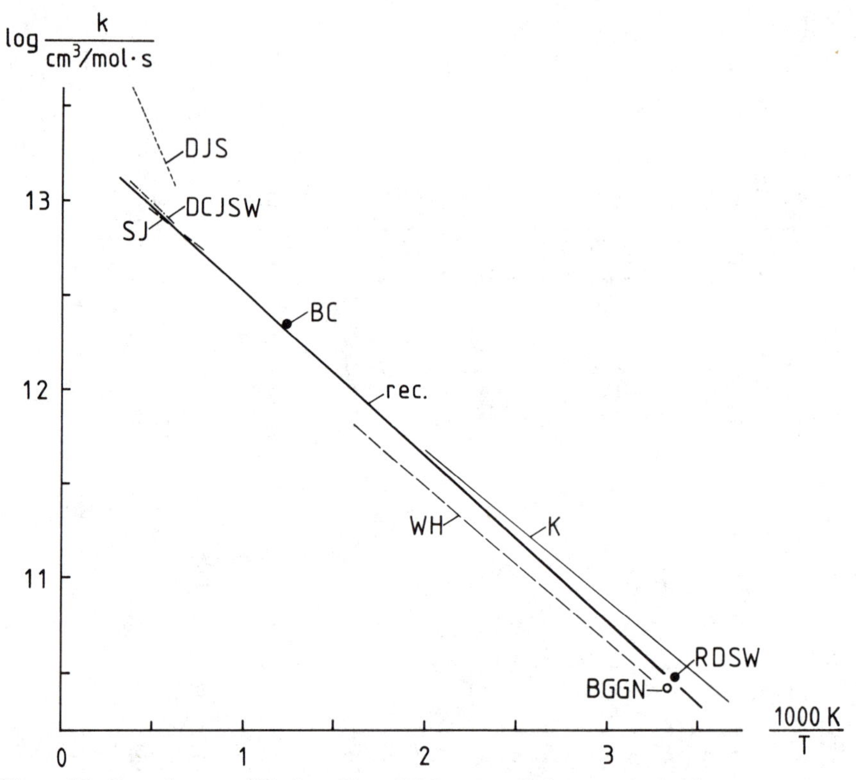

Figure 36. Rate data on $CH_2O + H \rightarrow CHO + H_2$. References in Table 8.

these steps (Figs. 1 and 4). The reactions of H and O are well investigated both at low and high temperature, whereas there is lack of information on the high-temperature rate coefficient of the OH reaction (Figs. 36 to 38). The O reaction may be complicated by the formation of an intermediate H_2CO_2 complex (Chang and Barker, 1979; see discussion of $CH_2 + O_2$ below).

$$CH_2O + HO_2 \rightarrow CHO + H_2O_2,\ CH_2O + CH_3 \rightarrow CHO + CH_4$$

These reactions are too slow to play an important part in flame propagation. The reaction with CH_3 may be important in ignition processes (Hidaka *et al.*, 1982).

Thermal decomposition of CH_2O

There is a persistent discussion in the literature about the nature of the products of this reaction. Independent of the products formed, this reaction is unimportant for flame propagation because of its large activation energy (Fig. 4). Ignition processes may be sensitive to the rate coefficient of this step if CHO and H are the products. Only shock tube work seems to give results direct enough to lead to reliable rate coefficients for this reaction (Fig. 39). The

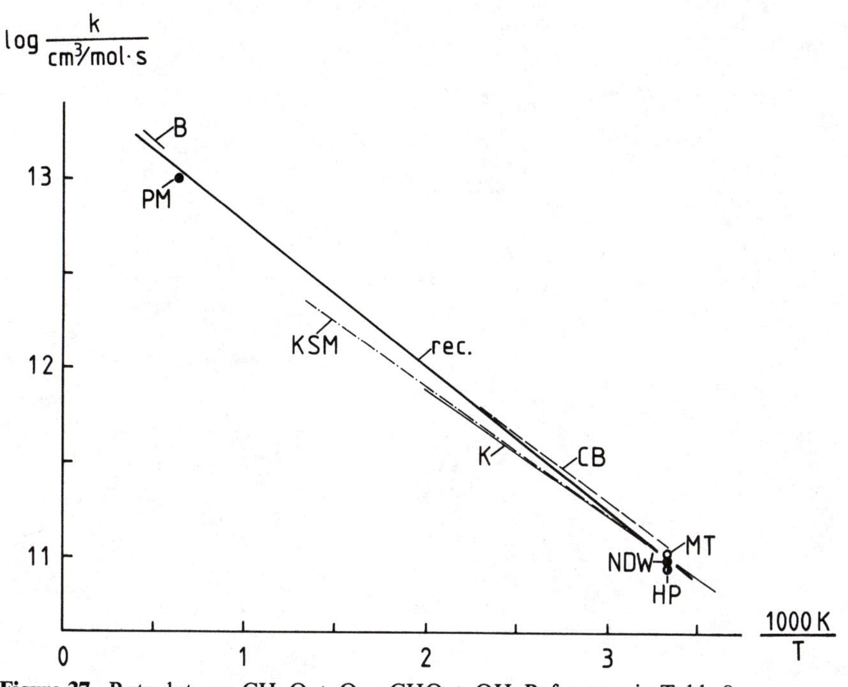

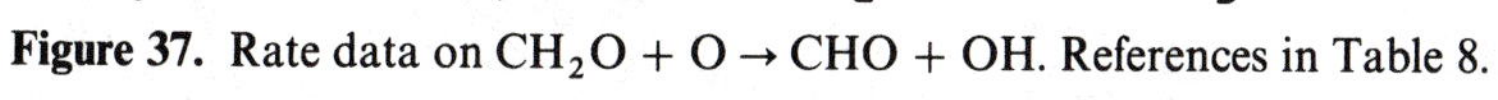

Figure 37. Rate data on $CH_2O + O \rightarrow CHO + OH$. References in Table 8.

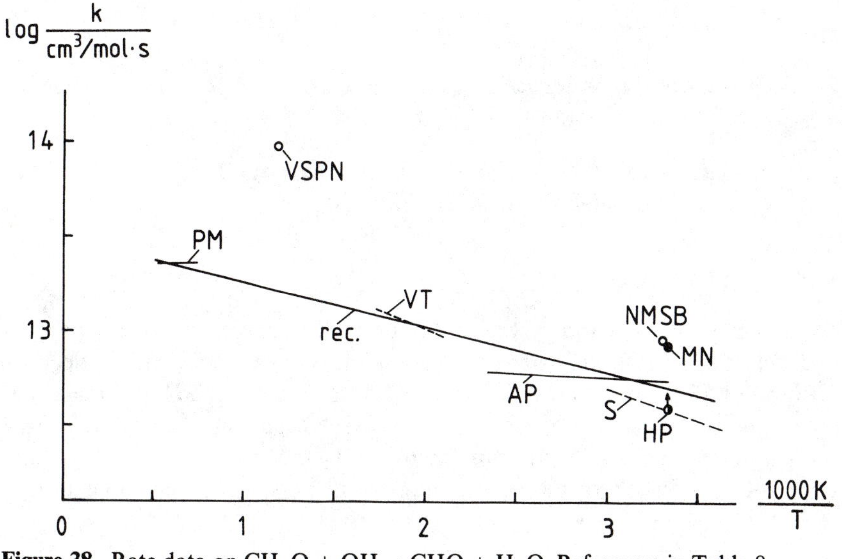

Figure 38. Rate data on $CH_2O + OH \rightarrow CHO + H_2O$. References in Table 8.

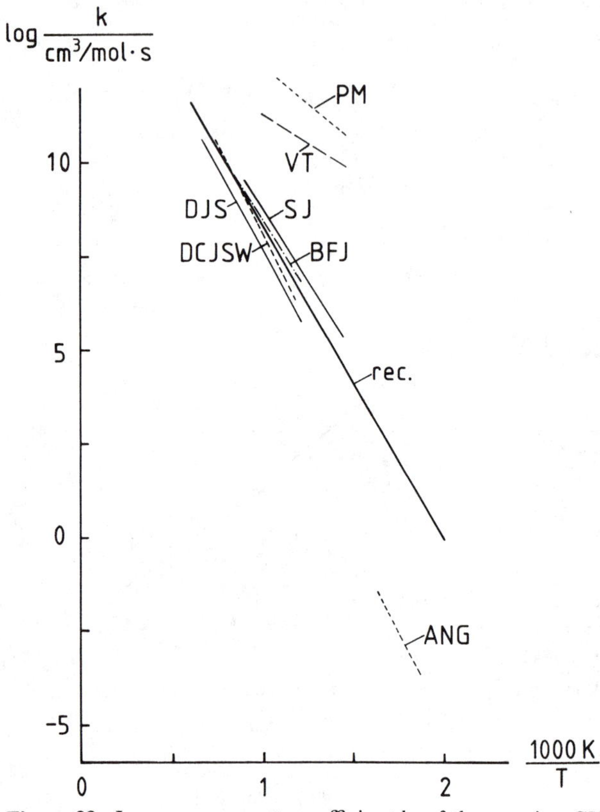

Figure 39. Low-pressure rate coefficient k_1 of the reaction $CH_2O + M \rightarrow$ products for $M = Ar$. References in Table 8.

reaction path leading to CHO + H is favored by these measurements (Schecker and Jost, 1969; Dean *et al.*, 1978, 1980).

5.4. Reactions of CHO

As shown in Figs. 1 and 4, the reactions of CHO play a central part in hydrocarbon oxidation. There is a competition between thermal decomposition of CHO and its reactions with H, OH, and O_2. The latter reactions have chain-terminating character, whereas CHO decomposition has chain-branching character and (in addition to C_2H_5 thermal decomposition and $CH_2 + O_2$) provides for the particle number increase in flame propagation.

Reactions of H, O, and OH with CHO

The rate coefficients of all these reactions (Figs. 40 and 41) are very large; the reaction with H is the fastest bimolecular reaction known at present (if reactions with extremely large cross sections due to ionic effects are

Table 8. Rate data on CH_2O reactions.

Reference	$A/(cm^3/mol)^{n-1}\ s^{-1}$	b	E/kJ	T/K	Method
	$CH_2O + H \rightarrow CHO + H_2$				
Baldwin and Cowe (1962)	2.2×10^{12}	...	...	813	SR,TH,P
Brennen *et al.* (1965)	2.6×10^{10}	...	...	300	FR,DI,MS
Schecker and Jost (1969)	2.0×10^{13}	0	13.8	1400–2200	ST,TH,IA,UA
Westenberg and deHaas (1972b)	1.3×10^{13}	0	15.7	297–652	FR,DI,ES
Ridley *et al.* (1972)	3.2×10^{10}	...	...	297	SR,PH,RF
Dean *et al.* (1979)	2.5×10^{13}	0	16.0	1700–2710	ST,TH,IE
Klemm (1979)	2.0×10^{13}	0	15.4	250–500	SR,FP,RF
Dean *et al.* (1980)	3.3×10^{14}	0	44.0	1600–3000	ST,TH,IE
Recommended ($\log F = \pm 0.20$)	2.5×10^{13}	0	16.7	300–2500	...
	$CH_2O + O \rightarrow (CHO + OH\ ?)$				
Herron and Penzhorn (1969)	9.0×10^{10}	...	...	300	FR,DI,MS
Niki *et al.* (1969a)	9.9×10^{10}	...	...	298	FR,DI,MS
Izod *et al.* (1971)	6.0×10^{13}	0	0	1400–2200	ST,TH,IE
Peeters and Mahnen (1973a)	1.0×10^{13}	...	...	1600	FL,TH,MS
Mack and Thrush (1973)	9.0×10^{10}	...	...	300	FR,DI,ES,CL
Bowman (1975a)	5.0×10^{13}	0	19.1	1875–2240	ST,TH,UA,IE
Klemm (1979)	1.7×10^{13}	0	12.7	250–500	SR,FP,RF
Chang and Barker (1979)	2.3×10^{13}	0	13.2	296–437	FR,DI,MS
Klemm *et al.* (1980)	1.8×10^{13}	0	12.8	250–750	FR,DI,RF
Recommended ($\log F = \pm 0.20$)	3.5×10^{13}	0	14.7	300–2500	...
	$CH_2O + OH \rightarrow CHO + H_2O$				
Herron and Penzhorn (1969)	4.0×10^{12}	...	...	300	FR,DI,MS
Morris and Niki (1971a)	8.4×10^{12}	...	...	300	FR,DI,MS
Peeters and Mahnen (1973a)	2.3×10^{13}	...	...	1400–1800	FR,TH,MS

Table 8 (*continued*).

Reference	$A/(\mathrm{cm^3/mol})^{n-1}\,\mathrm{s^{-1}}$	b	E/kJ	T/K	Method
	$CH_2O + OH \rightarrow CHO + H_2O$				
Vardanyan *et al.* (1974)	9.6×10^{13}	...	...	850	SR,TH,ES
Vandooren and van Tiggelen (1977)	3.9×10^{13}	0	5.9	485–570	FL,TH,MS
Atkinson and Pitts (1978)	7.5×10^{12}	0	0.7	299–426	SR,FP,RF
Niki *et al.* (1978)	9.0×10^{12}	...	...	298	SR,PH,IE
Smith (1978)	4.4×10^{13}	0	6.0	268–334	FR,DI,MS
Recommended ($\log F = \pm 0.30$)	3.0×10^{13}	0	5.0	300–2500	...
	$CH_2O + HO_2 \rightarrow CHO + H_2O_2$				
Baldwin and Walker (1979)	1.0×10^{9}	...	...	773	SR,TH,P
No recommendation					
	$CH_2O + CH_3 \rightarrow CHO + CH_4$				
Kerr and Parsonage (1976)	1.0×10^{11}	0	25.5	300–500	Review
Held *et al.* (1977)	3.2×10^{10}	...	...	1005	FR,TH,UA,GC
Manthorne and Pacey (1978)	2.0×10^{13}	0	60.0	788–935	FR,TH,GC
Recommended ($\log F = \pm 0.30$)	1.0×10^{11}	0	25.5	300–1000	...
	$CH_2O + M \rightarrow (CHO + H + M?)$ (k_1, $M = Ar$)				
Schecker and Jost (1969)	5.0×10^{16}	0	302.0	1400–2200	ST,TH,IA,UA
Peeters and Mahnen (1973a)	2.1×10^{16}	0	147.0	1350–1900	FL,TH,MS
Aronowitz *et al.* (1977)	1.0×10^{14}	0	364.0	1070–1225	FR,TH,GC
Dean *et al.* (1979)	3.7×10^{17}	0	364.0	1700–2710	ST,TH,IE
Bhaskaran *et al.* (1979)	1.3×10^{16}	0	296.0	1700–2300	ST,TH,RA
Dean *et al.* (1980)	3.3×10^{16}	0	339.0	1600–3000	ST,TH,IE
Vandooren and van Tiggelen (1981)	2.5×10^{14}	0	121.0	1400–2000	FL,TH,MS
Recommended ($\log F = \pm 0.50$)	5.0×10^{16}	0	320.0	1000–3000	...

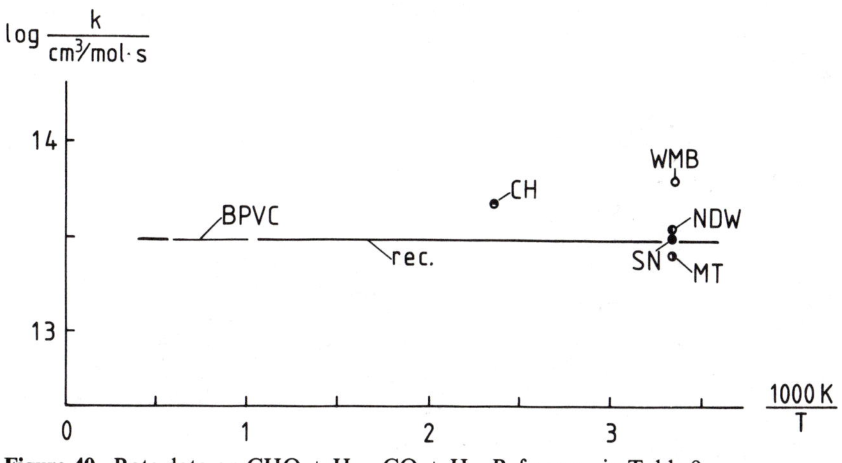

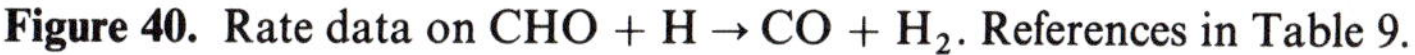
Figure 40. Rate data on $CHO + H \rightarrow CO + H_2$. References in Table 9.

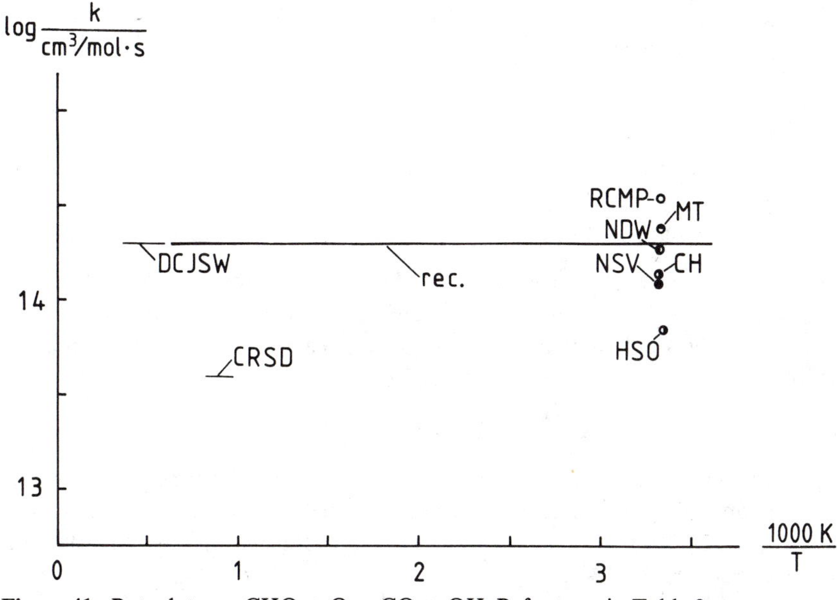

Figure 41. Rate data on $CHO + O \rightarrow CO + OH$. References in Table 9.

disregarded). Most of the rate coefficients for the reactions with H and O are derived from relative measurements, using rate coefficients recommended in this compilation. There is lack of absolute rate data, particularly on the reaction of OH with CHO.

$$CHO + O_2 \rightarrow CO + HO_2$$

This reaction has been of interest because of its importance in atmospheric chemistry. Therefore, the room-temperature rate coefficient is well established,

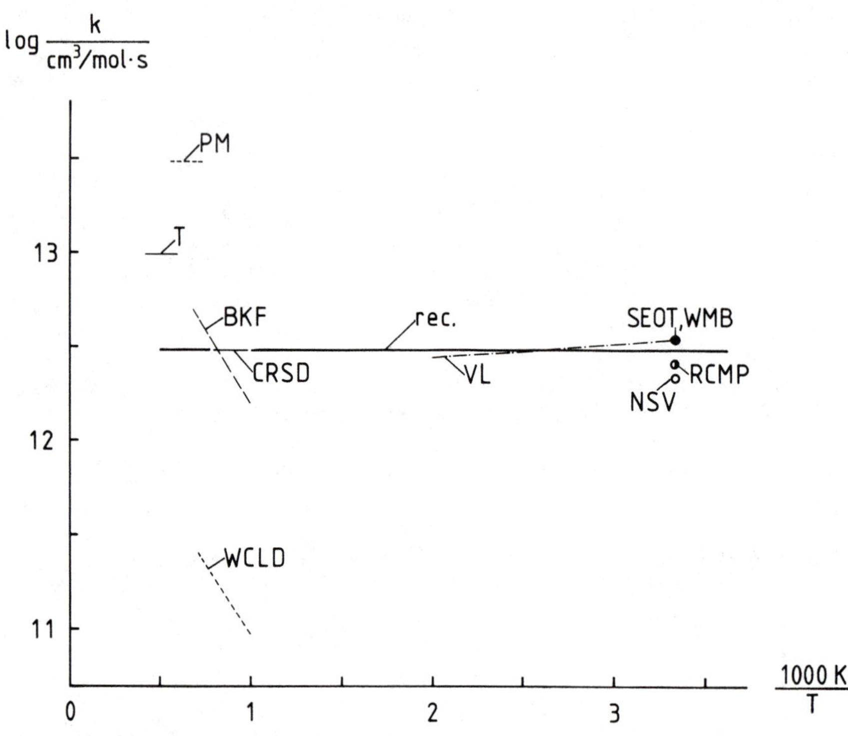

Figure 42. Rate data on $CHO + O_2 \rightarrow CO + HO_2$. References in Table 9.

whereas there is a large scatter of the data at flame temperatures (Fig. 42). Nevertheless, an extrapolation to high temperature seems to be reliable because of the small activation energy of this reaction (Veyret and Lesclaux, 1981).

$$CHO + M \rightarrow CO + H + M$$

More exact rate measurements seem to be desirable for this important reaction (Figs. 1 and 4). The recommended rate coefficient is a mean value of the literature data and is supported by the rate coefficient calculated from the reverse reaction (Fig. 43).

5.5. Reactions of CH_2

CH_2 is mainly formed by the reaction of O atoms with C_2H_2 (see below), which is a very important intermediate in rich hydrocarbon flames as well as an important fuel itself (Warnatz *et al.*, 1982).

$$CH_2 + H \rightarrow CH + H_2,\ CH_2 + O \rightarrow \text{products}$$

The fast reaction of CH_2 with H atoms leads to CH radicals, which may be precursors of non-Zeldovich NO formation in rich hydrocarbon flames. The

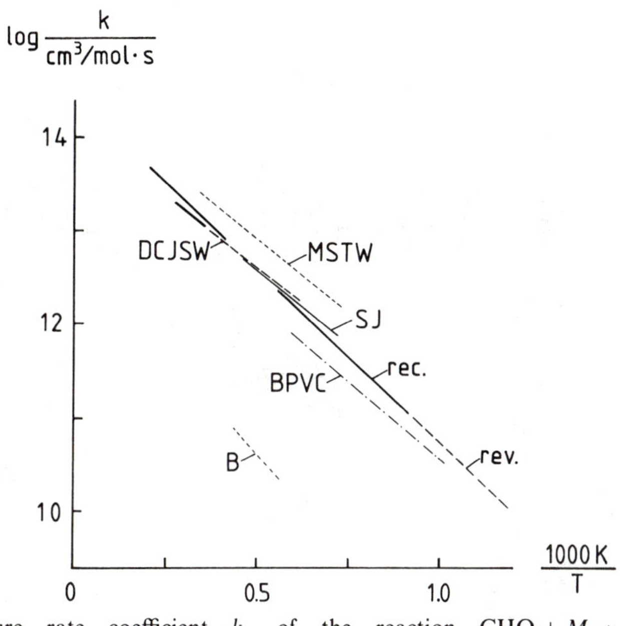

Figure 43. Low-pressure rate coefficient k_1 of the reaction $CHO + M \rightarrow CO + H + M$ for $M = Ar$. References in Table 9.

reaction of CH_2 with O is less important because of the small O concentrations in flames. The products seem to be H and CHO, which decompose very fast to form CO and H (Homann and Schweinfurth, 1981).

$$CH_2 + O_2 \rightarrow \text{products}$$

Flame simulations show this reaction to be a very important chain-branching step. From measurements of lean low-pressure C_2H_2/O_2 flames (Eberius *et al.*, 1973) it has been concluded (Warnatz, 1981) that this reaction leads to the products $CO_2 + H + H$ for two reasons: (1) Only the formation of two H atoms can account for the fast propagation of this flame. Formation, e.g., of $CO + OH + H$, would lead to very slow propagation incompatible with the measurements; (2) CO_2 formation in this flame can be explained only with direct formation in the reaction of CH_2 with O_2. Formation of CO followed by $CO + OH \rightarrow CO_2 + H$ is too slow to reproduce the measured CO_2 flame profiles.

This result on the product distribution is in contrast to room-temperature measurements by Himme (1977) on ketene photolysis in presence of O_2 ($\sim$95% formation of CO). Perhaps this discrepancy can be explained by a temperature-dependent behavior of an intermediate reaction complex H_2CO_2, decomposing at high temperature, and rearranging to form OH at low temperature. The rate coefficient determined by Vinckier and Debruyn (1979b) is adopted here (Fig. 44). Clearly, there is need of measurements both

Table 9. Rate data on CHO reactions.

Reference	$A/(\text{cm}^3/\text{mol})^{n-1}\ \text{s}^{-1}$	b	E/kJ	T/K	Method
	$CHO + H \rightarrow CO + H_2$				
Niki *et al.* (1969b)	1.8×10^{14}	...	...	300	FR,DI,MS
Mack and Thrush (1973)	2.4×10^{14}	...	...	298	FR,DI,ES,GC
Campbell and Handy (1978)	1.2×10^{14}	...	...	425	FR,DI,CL
Reilly *et al.* (1978)	3.3×10^{14}	...	...	298	SR,FP,UA
Dean *et al.* (1979)	2.0×10^{14}	0	0	1700–2710	ST,TH,IE
Nadtochenko *et al.* (1979b)	1.2×10^{14}	...	...	298	SR,FP,UA
Hochanadel *et al.* (1980b)	6.9×10^{13}	...	...	298	SR,FP,UA
Cherian *et al.* (1981)	4.0×10^{13}	...	...	1100	FL,TH,MS
Recommended ($\log F = \pm 0.30$)	2.0×10^{14}	0	0	300–2500	...
	$CHO + O \rightarrow CO + OH$				
Browne *et al.* (1969)	3.0×10^{13}	...	...	1000–1700	FL,TH,GC,UA
Niki *et al.* (1969b)	3.5×10^{13}	...	...	300	FR,DI,MS
Stuhl and Niki (1979b)	3.0×10^{13}	...	...	298	SR,PH,CL
Mack and Thrush (1973)	2.5×10^{13}	...	...	300	FR,DI,ES,GC
Washida *et al.* (1974)	6.5×10^{13}	...	...	297	FR,DI,MS
Campbell and Handy (1978)	5.0×10^{13}	...	...	425	FR,DI,CL
Recommended ($\log F = \pm 0.30$)	3.0×10^{13}	0	0	300–2000	...
	$CHO + O \rightarrow CO_2 + H$				
Niki *et al.* (1969b)	$k(CO_2 + H)/k_{total} = 0.20$			300	FR,DI,MS
Westenberg and deHaas (1972c)	$k(CO_2 + H)/k_{total} = 0.58$			298	FR,DI,MS,ES
Mack and Thrush (1973)	$k(CO_2 + H)/k_{total} = 0.46$			300	FR,DI,GC,ES
Campbell and Handy (1978)	$k(CO_2 + H)/k_{total} = 0.60$			425	FR,DI,CL
Recommended ($\log F = \pm 0.30$)	3.0×10^{13}	0	0	300–2000	...

	$CHO + OH \rightarrow CO + H_2O$				
Browne *et al.* (1969)	3.0×10^{13}	···	···	1000–1700	FL,TH,GC,UA
Bowman (1970)	1.0×10^{14}	···	···	1875–2240	ST,TH,IE,LS
Recommended ($\log F = \pm 0.50$)	5.0×10^{13}	0	0	1000–2500	···
	$CHO + O_2 \rightarrow CO + HO_2$				
Basevich *et al.* (1971)	6.0×10^{13}	0	30.0	1000–1450	FL,TH,?
Peeters and Mahnen (1973a)	3.0×10^{13}	···	···	1400–1800	FL,TH,MS
Vardanyan *et al.* (1974)	6.0×10^{10}	···	···	850	SR,TH,ES
Washida *et al.* (1974)	3.4×10^{12}	···	···	297	FR,DI,MS
Tsuboi (1976)	9.5×10^{12}	···	···	2000	ST,TH,IA
Westbrook *et al.* (1977)	3.3×10^{12}	0	29.3	1000–1350	FR,TH,GC
Shibuya *et al.* (1977)	3.3×10^{12}	···	···	298	SR,FP,UA
Reilly *et al.* (1978)	2.4×10^{12}	···	···	298	SR,FP,UA
Nadtochenko *et al.* (1979a)	2.2×10^{12}	···	···	298	SR,FP,UA
Cherian *et al.* (1981)	3.0×10^{12}	···	···	1100	FL,TH,MS
Veyret and Lesclaux (1981)	3.3×10^{13}	−0.4	0	298–503	SR,FP,RA
Recommended ($\log F = \pm 0.20$)	3.0×10^{12}	0	0	300–2000	···
	$CHO + M \rightarrow CO + H + M$ (k_1, $M = Ar$)				
Browne *et al.* (1969)	7.0×10^{13}	0	63.0	1000–1700	FL,TH,GC,UA
Schecker and Jost (1969)	1.6×10^{14}	0	62.0	1400–2200	ST,TH,IA,UA
Bowman (1975a)	5.0×10^{12}	0	80.0	1800–2300	ST,TH,UA,IE
Dean *et al.* (1979)	1.5×10^{14}	0	61.0	1700–2710	ST,TH,IE
Recommended ($\log F = \pm 0.50$)	2.5×10^{14}	0	70.3	700–2500	···

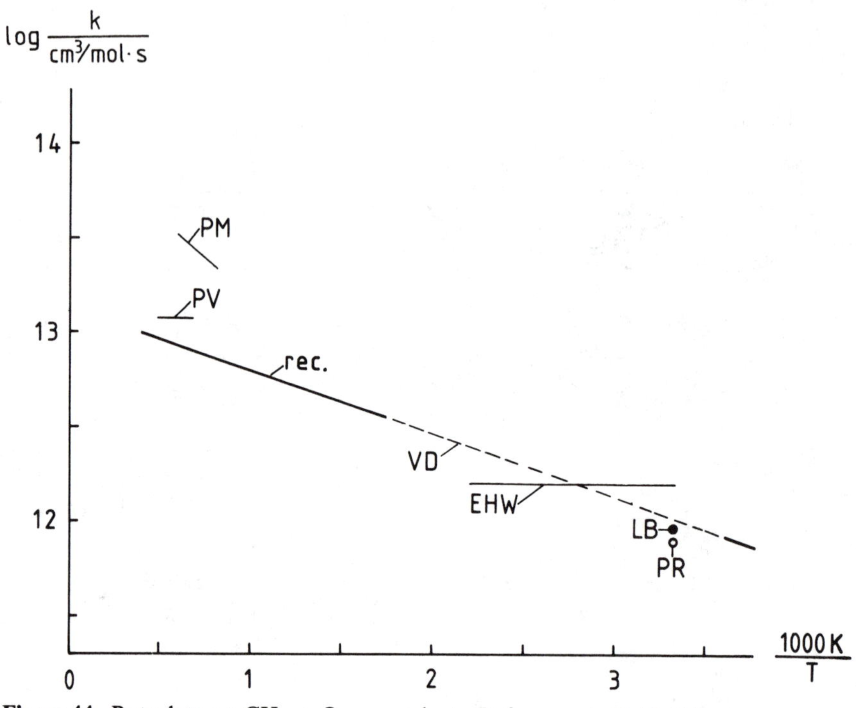

Figure 44. Rate data on $CH_2 + O_2 \rightarrow$ products. References in Table 10.

of the rate coefficient and the product spectrum of this important reaction, especially at high temperatures.

$$CH_2 + CH_2 \rightarrow C_2H_2 + H_2, \; CH_2 + CH_3 \rightarrow C_2H_4 + H$$

There is little information on these reactions, which may be important in rich acetylene flames because of their chain-terminating character (Warnatz *et al.*, 1982).

5.6. Reactions of CH

Since CH is one of the most reactive species present in hydrocarbon combustion, its reactions are not rate-determining for flame propagation or ignition. As mentioned earlier, knowledge of its behavior is important because of its potential role as a precursor of non-Zeldovich NO in rich hydrocarbon flames.

$$CH + O \rightarrow CO + H, \; CHO^+ + e$$

As expected for a radical–radical reaction, consumption of CH by O atoms is very fast. In part (about 0.5% at flame temperatures), the reaction leads to CHO^+, which is thought to be the main precursor of ions found in flames.

Table 10. Rate data on CH_2 reactions.

Reference	$A/(cm^3/mol)^{n-1}\ s^{-1}$	b	E/kJ	T/K	Method
	$CH_2 + H \rightarrow CH + H_2$				
Peeters and Vinckier (1975)	2.4×10^{13}	...	...	2000	FL,TH,MS
Homann and Schweinfurth (1981)	5.0×10^{13}	...	...	298	FR,DI,MS
Roth and Löhr (1981)	3.0×10^{13}	0	0	1500–2570	ST,TH,RA
Recommended ($\log F = \pm 0.30$)	4.0×10^{13}	0	0	300–2500	...
	$CH_2 + O \rightarrow$ products				
Vinckier and Debruyn (1979b)	8.0×10^{13}	0	0	295–600	FR,DI,MS
Homann and Schweinfurth (1981)	5.0×10^{13}	...	...	298	FR,DI,MS
Recommended ($\log F = \pm 0.30$)	5.0×10^{13}	0	0	300–2000	...
	$CH_2 + O_2 \rightarrow$ products				
Peeters and Mahnen (1973b)	1.0×10^{14}	0	15.5	1200–1600	FL,TH,MS
Eberius *et al.* (1973)	1.5×10^{12}	0	0	298–480	FR,DI,MS
Laufer and Bass (1974)	9.0×10^{11}	...	...	298	SR,FP,GC
Peeters and Vinckier (1975)	1.2×10^{13}	0	0	1500–2200	FL,TH,MS
Pilling and Robertson (1977)	7.2×10^{11}	...	...	298	SR,FP,GC
Vinckier and Debruyn (1979b)	1.3×10^{13}	0	6.3	295–573	FR,DI,MS
Recommended ($\log F = \pm 0.40$)	1.3×10^{13}	0	6.3	300–2000	...
	$CH_2 + CH_3 \rightarrow C_2H_4 + H$				
Pilling and Robertson (1975)	3.0×10^{13}	...	...	298	SR,FP,GC
Laufer and Bass (1975)	6.0×10^{13}	...	...	298	SR,FP,GC
Olson and Gardiner (1978)	2.0×10^{13}	0	0	1800–2700	Review
Bhaskaran *et al.* (1979)	2.0×10^{13}	0	0	1700–2200	ST,TH,RA
Recommended ($\log F = \pm 0.40$)	4.0×10^{13}	0	0	300–2500	...
	$CH_2 + CH_2 \rightarrow (C_2H_2 + H_2?)$				
Braun *et al.* (1970)	3.2×10^{13}	...	...	298	SR,FP,UA
No recommendation					

$$CH + H_2 \rightarrow \text{products}, \; CH + O_2 \rightarrow \text{products}$$

Room-temperature measurements show these reactions to be very fast. Further measurements of the product distributions and the high-temperature rate coefficients are necessary before these reactions can be included with confidence in combustion models.

5.7. Reactions of CH_3OH and CH_3O/CH_2OH

Reactions of CH_3OH and CH_3O (and CH_2OH) are included in this review for completeness. These reactions are relatively unimportant for the combustion of hydrocarbons since CH_3OH is a negligible intermediate in this process, but the combustion of CH_3OH itself is of interest because of its role as an alternate fuel. CH_3O, however, is thought to be the product of $CH_3 + O_2$ and is thus an important intermediate in fuel-lean ignition processes.

Reactions of H, O, H, and CH_3 with CH_3OH

The reactions of H, O, and OH are responsible for the first attack on CH_3OH in flames (Figs. 45 to 47). In all cases CH_3O or CH_2OH is the primary product. Especially for the OH reaction there is a lack of data at high temperature.

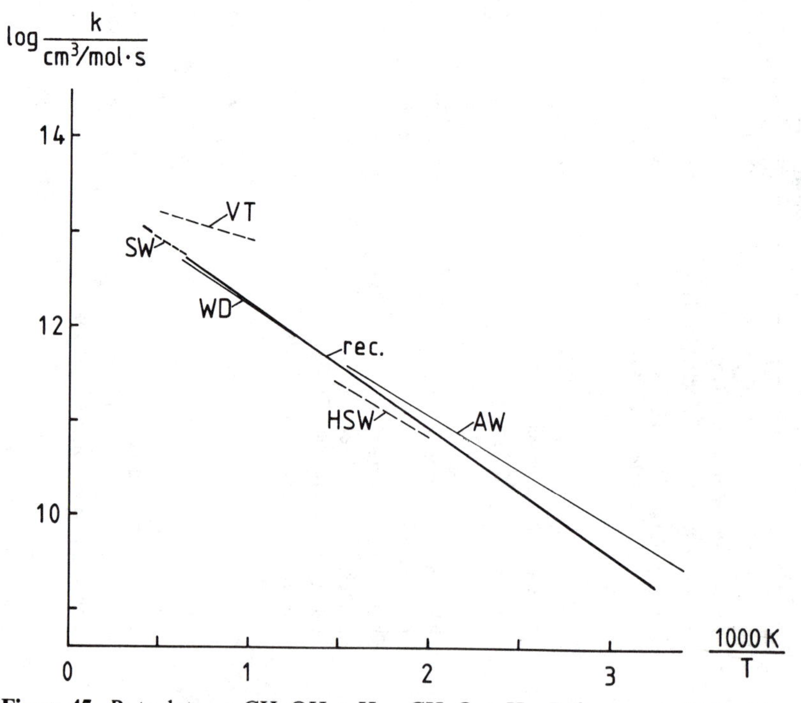

Figure 45. Rate data on $CH_3OH + H \rightarrow CH_3O + H_2$. References in Table 12.

Table 11. Rate data on CH reactions.

Reference	$A/(\mathrm{cm^3/mol})^{n-1}\,\mathrm{s^{-1}}$	b	E/kJ	T/K	Method
	$CH + O \rightarrow CO + H$				
Messing *et al.* (1980)	5.7×10^{13}	...	...	298	SR,FP,LF
Homann and Schweinfurth (1981)	2.0×10^{13}	...	...	298	FR,DI,MS
Recommended ($\log F = \pm 0.50$)	4.0×10^{13}	0	0	300–2000	...
	$CH + O \rightarrow CHO^+ + e$				
Peeters and Vinckier (1975)	1.7×10^{11}	0	0	2000–2400	FL,TH,SC
Vinckier (1979)	1.4×10^{10}	...	...	295	FR,DI,MS
Recommended ($\log F = \pm 0.50$)	2.5×10^{11}	0	7.1	300–2500	...
	$CH + H_2 \rightarrow$ products				
Braun *et al.* (1967)	6.2×10^{11}	...	...	298	SR,FP,UA
Bosnali and Perner (1971)	1.0×10^{13}	...	...	298	SR,PR,UA
Butler *et al.* (1980)	1.6×10^{13}	...	...	298	SR,FP,LF
No recommendation					
	$CH + O_2 \rightarrow$ products				
Bosnali and Perner (1971)	$<2.4 \times 10^{13}$	...	...	298	SR,PR,UA
Jachimowski (1977)	1.0×10^{13}	0	0	1815–2365	ST,TH,IE
Messing *et al.* (1979)	2.0×10^{13}	...	...	298	SR,FP,LF
Butler *et al.* (1980)	3.5×10^{13}	...	...	298	SR,FP,LF
Recommended ($\log F = \pm 0.50$)	2.0×10^{13}	0	0	300–2000	...

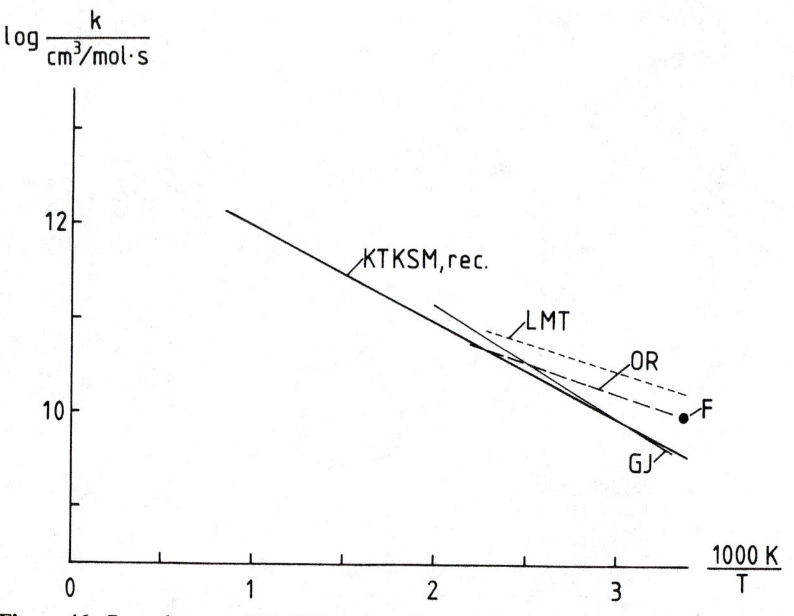

Figure 46. Rate data on $CH_3OH + O \rightarrow CH_3O + OH$. References in Table 12.

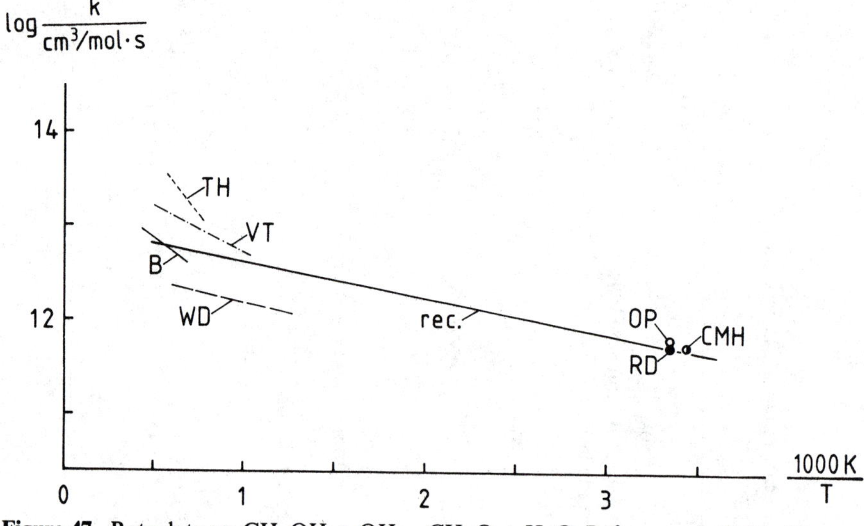

Figure 47. Rate data on $CH_3OH + OH \rightarrow CH_3O + H_2O$. References in Table 12.

Thermal decomposition of CH_3OH

This reaction mainly leads to $CH_3 + OH$ (Spindler and Wagner, 1981). The rate data are sufficient to allow the determination of the limiting high- and low-pressure rate coefficients k_1 and k_∞ (Fig. 48).

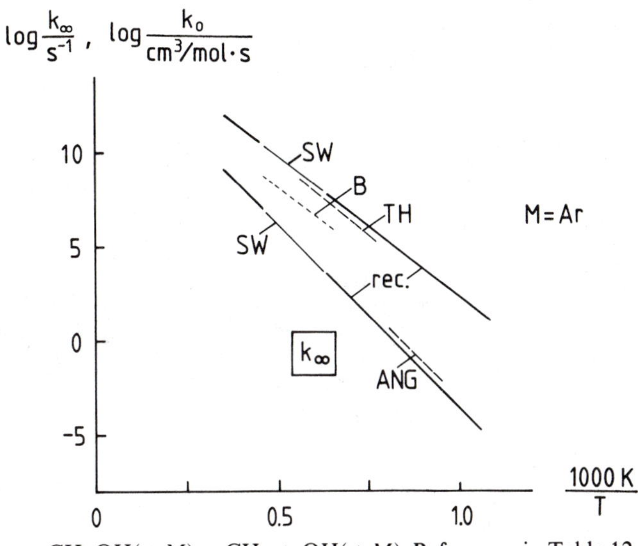

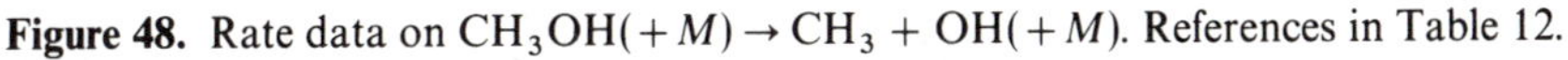
Figure 48. Rate data on $CH_3OH(+M) \rightarrow CH_3 + OH(+M)$. References in Table 12.

Reactions of CH_3O/CH_2OH

The reactions of this radical may be important in ignition processes (Gardiner *et al.*, 1981) since CH_3O is formed in the reaction of CH_3 with O_2 (see above). The reaction with H atoms is known to be fast, whereas that with O atoms is relatively slow. No information is available for the corresponding OH reaction. These radical–radical reactions compete with the reaction with O_2 (Fig. 49) and with the thermal decomposition (Fig. 50). In both cases, the high-temperature measurements show large discrepancies.

6. Reactions of C_2-hydrocarbons

As mentioned earlier (see introduction to Section 5), oxidation of C_2-species plays an important general role in hydrocarbon combustion. Central points of current modeling research are C_2H_5 reactions (Figs. 2 and 3) because of their rate-determining character and reactions of acetylene, which is of special interest in rich hydrocarbon combustion due to its roles in the processes of soot and (probably) non-Zeldovich NO formation (Warnatz *et al.*, 1982).

6.1. Reactions of C_2H_6

Reactions of H, O, and OH with C_2H_6

In all three cases, there are sufficient data to define curved Arrhenius plots (Figs. 51 to 53). For $H + C_2H_6 \rightarrow C_2H_5 + H_2$ the recommendation by Clark

Table 12. Rate data on CH_3OH reactions.

Reference	$A/(cm^3/mol)^{n-1}\ s^{-1}$	b	E/kJ	T/K	Method
	$CH_3OH + H \rightarrow CH_3O + H_2$				
Aders and Wagner (1971)	2.3×10^{13}	0	22.2	295–653	FR,DI,MS,ES
Westbrook and Dryer (1979)	3.2×10^{13}	0	29.3	800–1600	FR,TH,GC
Vandooren and van Tiggelen (1981)	3.4×10^{13}	0	10.9	1000–2000	FL,TH,MS
Hoyermann *et al.* (1981a)	1.3×10^{13}	0	22.0	500–680	FR,DI,MS
Spindler and Wagner (1982)	3.2×10^{13}	0	22.2	1600–2100	ST,TH,UA
Recommended ($\log F = \pm 0.30$)	4.0×10^{13}	0	25.5	300–2000	...
	$CH_3OH + O \rightarrow CH_3O + OH$				
LeFevre *et al.* (1972)	5.0×10^{12}	0	11.9	217–366	FR,DI,MS,ES
Owens and Roscoe (1976)	1.5×10^{12}	0	12.8	300–450	FR,DI,MS
Faubel (1977)	1.0×10^{10}	...	...	298	FR,DI,MS
Keil *et al.* (1981)	1.0×10^{13}	0	19.6	300–1000	SR,PH,RF
Grotheer and Just (1981)	3.4×10^{13}	0	22.5	300–1000	FR,DI,MS
Recommended ($\log F = \pm 0.40$)	1.0×10^{13}	0	19.6	300–1000	...
	$CH_3OH + OH \rightarrow CH_3O + H_2O$				
Bowman (1975b)	3.0×10^{13}	0	25.0	1545–2180	ST,TH,CL,IE
Campbell *et al.* (1976)	5.7×10^{11}	...	...	292	SR,PH,GC
Ravishankara and Davis (1978)	6.0×10^{11}	...	...	298	SR,FP,RF
Overend and Paraskevopoulos (1978)	6.4×10^{11}	...	...	298	SR,FP,RA
Westbrook and Dryer (1979)	4.0×10^{12}	0	8.4	800–1600	FR,TH,GC
Vandooren and van Tiggelen (1981)	4.8×10^{13}	0	18.8	1000–2000	FL,TH,MS
Recommended ($\log F = \pm 0.50$)	1.0×10^{13}	0	7.1	300–2000	...

	$CH_3OH + CH_3 \rightarrow CH_3O + CH_4$				
Kerr and Parsonage (1976)	2.3×10^{11}	0	41.0	350–550	Review
Spindler and Wagner (1982)	8.9×10^{12}	0	41.1	1600–2100	ST,TH,UA
No recommendation					
	$CH_3OH + M \rightarrow CH_3 + OH + M$ (k_1, M = Ar)				
Bowman (1975a)	4.0×10^{15}	0	285	1545–2180	ST,TH,CL,IE
Tsuboi and Hashimoto (1981)	6.0×10^{17}	0	310	1300–1750	ST,TH,IE
Spindler and Wagner (1982)	2.0×10^{17}	0	286	1600–2100	ST,TH,UA
Recommended ($\log F = \pm 0.50$)	2.0×10^{17}	0	286	1000–2500	...
	$CH_3OH \rightarrow CH_3 + OH$ (k_∞)				
Aronowitz *et al.* (1977)	6.0×10^{16}	0	381	1070–1225	FR,TH,GC
Spindler and Wagner (1982)	9.4×10^{15}	0	376	1600–2100	ST,TH,UA
Recommended ($\log F = \pm 0.50$)	9.4×10^{15}	0	376	1000–2500	...

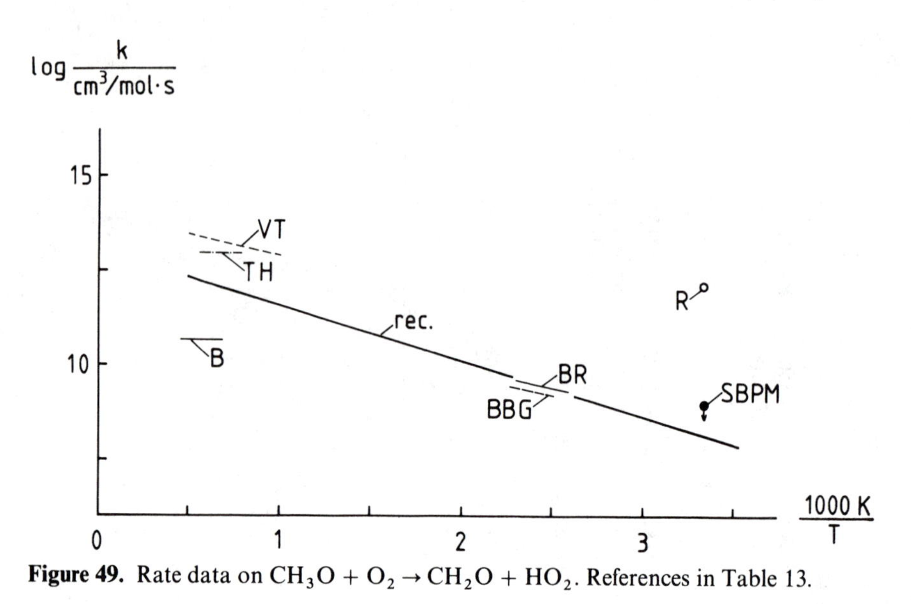

Figure 49. Rate data on $CH_3O + O_2 \rightarrow CH_2O + HO_2$. References in Table 13.

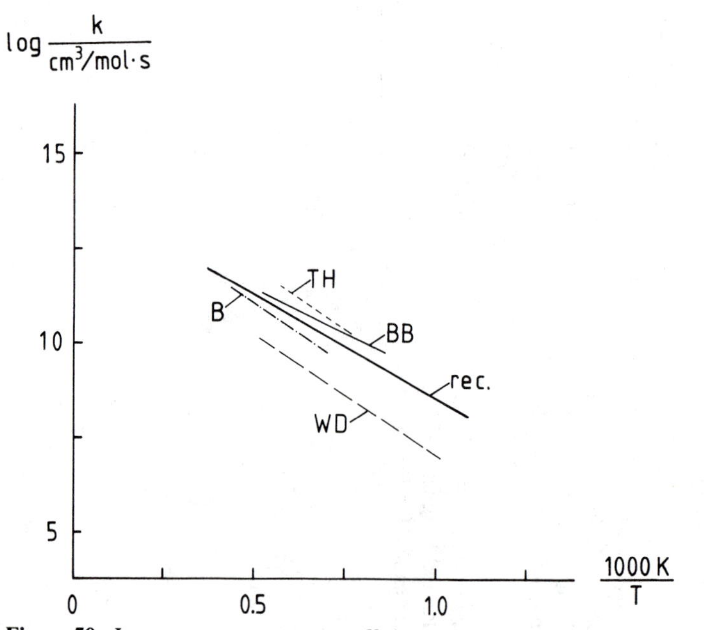

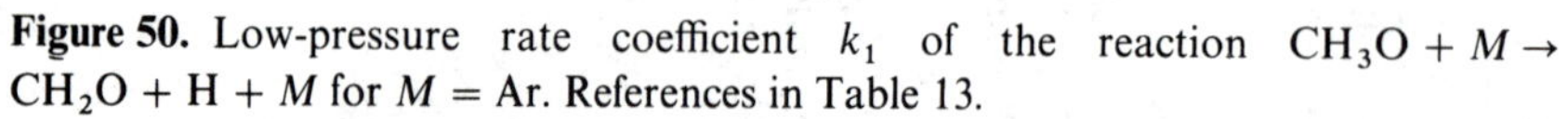
Figure 50. Low-pressure rate coefficient k_1 of the reaction $CH_3O + M \rightarrow CH_2O + H + M$ for $M = Ar$. References in Table 13.

Table 13. Rate data on CH_3O/CH_2OH reactions.

Reference	$A/(cm^3/mol)^{n-1}\,s^{-1}$	b	E/kJ	T/K	Method
	$CH_3O + H + CH_2O + H_2$				
Hoyermann *et al.* (1981b)	2.0×10^{13}	...	...	<298	FR,DI,MS
Recommended ($\log F = \pm 0.30$)	2.0×10^{13}	0	0	300–2000	...
	$CH_2OH + H \rightarrow CH_2O + H_2$				
Hoyermann *et al.* (1981b)	3.0×10^{13}	...	...	<298	FR,DI,MS
Recommended ($\log F = \pm 0.30$)	3.0×10^{13}	0	0	300–2000	...
	$CH_3O/CH_2OH + O \rightarrow$ products				
Hoyermann *et al.* (1981b)	$<5.0 \times 10^{12}$	...	...	<298	FR,DI,MS
No recommendation					
	$CH_3O/CH_2OH + O_2 \rightarrow CH_2O + HO_2$				
Bowman (1975a)	5.0×10^{10}	0	0	1545–2180	ST,TH,CL,IE
Barker *et al.* (1977)	3.2×10^{11}	0	16.7	396–442	SR,TH,P,GC
Batt and Robinson (1979)	1.0×10^{12}	0	20.0	383–433	SR,TH,GC
Sanders *et al.* (1980)	$<1.0 \times 10^{9}$	...	...	298	SR,PH,LF
Radford (1980)	1.2×10^{12}	...	...	300	FR,DI,LM
Vandooren and van Tiggelen (1981)	1.0×10^{14}	0	20.9	1000–2000	FL,TH,MS
Tsuboi and Hashimoto (1981)	1.0×10^{13}	0	0	1300–1750	ST,TH,IE
Recommended ($\log F = \pm 1.0$)	1.0×10^{13}	0	30.0	300–2000	...

Table 13 (*continued*).

Reference	$A/(\text{cm}^3/\text{mol})^{n-1}\,\text{s}^{-1}$	b	E/kJ	T/K	Method
	$CH_3O/CH_2OH + M \rightarrow CH_2O + H + M$ (k_1, M undefined)				
Bowman (1975a)	1.5×10^{14}	0	121	1545–2180	ST,TH,UA,IE
Brabbs and Brokaw (1975)	5.0×10^{13}	0	88	1200–1800	ST,TH,UE
Westbrook and Dryer (1979)	2.5×10^{13}	0	121	800–1600	FR,TH,GC
Tsuboi and Hashimoto (1981)	1.3×10^{15}	0	121	1300–1750	ST,TH,IE
Recommended ($\log F = \pm 1.0$)	1.0×10^{14}	0	105	1000–2200	...
	$CH_3O/CH_2OH \rightarrow CH_2O + H$ (k_∞)				
Batt and Robinson (1979)	1.6×10^{14}	0	105	393–473	SR,TH,GC
Recommended ($\log F = \pm 1.0$)	1.6×10^{14}	0	105	300–2000	...

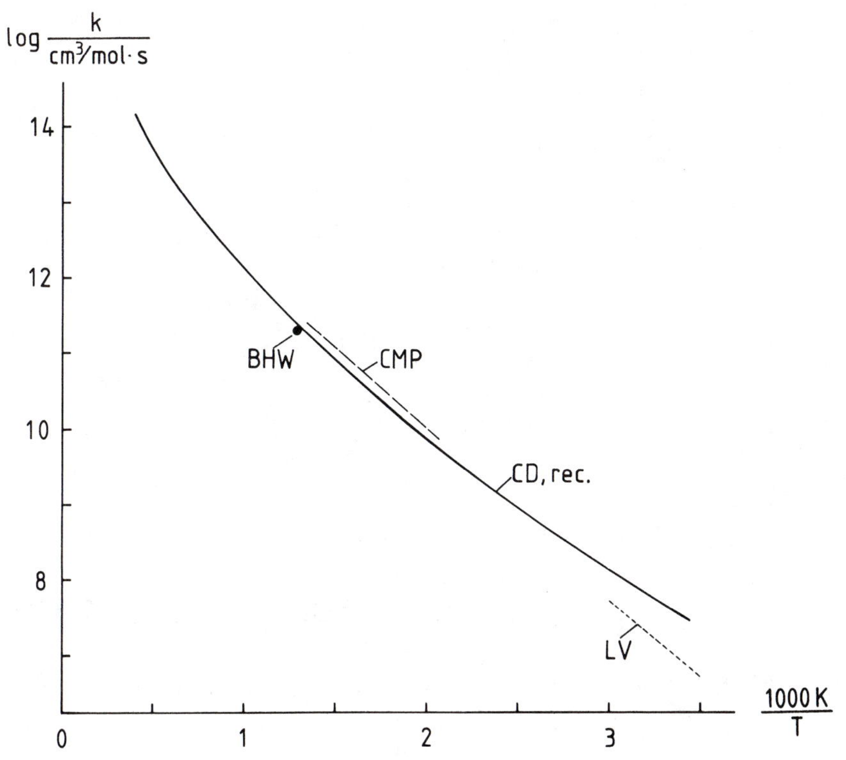

Figure 51. Rate data on $C_2H_6 + H \rightarrow C_2H_5 + H_2$. References in Table 14.

and Dove (1973) is adopted. Flame propagation is relatively insensitive to these rate coefficients even if C_2H_6 combustion itself is considered (Fig. 4).

Reactions of CH_3 and HO_2 with C_2H_6

These reactions (Fig. 54) are not important in determining rates of combustion. The rate coefficient of the HO_2 reaction is estimated from data on HO_2 attack on single C–H bonds (Baldwin *et al.*, 1977).

Thermal decomposition of C_2H_6

The reverse of this step has been discussed before (Section 5.2). There are sufficient rate data on the pressure dependence of this reaction to allow the determination of Arrhenius parameters for low- and high-pressure rate coefficients k_1 and k_∞ (Fig. 55). Examples of fall-off curves are given in Fig. 36.

6.2. Reactions of C_2H_5

Reactions of C_2H_5 radicals play a central part in hydrocarbon combustion because of the rate-determining character of the competition between thermal

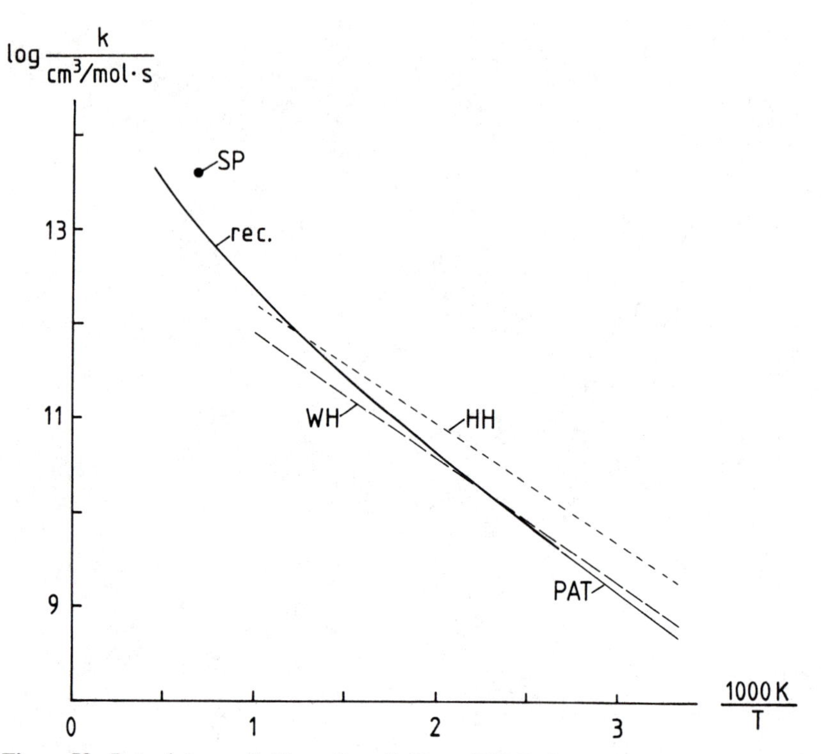

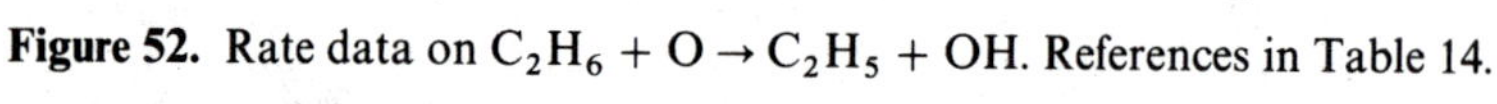
Figure 52. Rate data on $C_2H_6 + O \rightarrow C_2H_5 + OH$. References in Table 14.

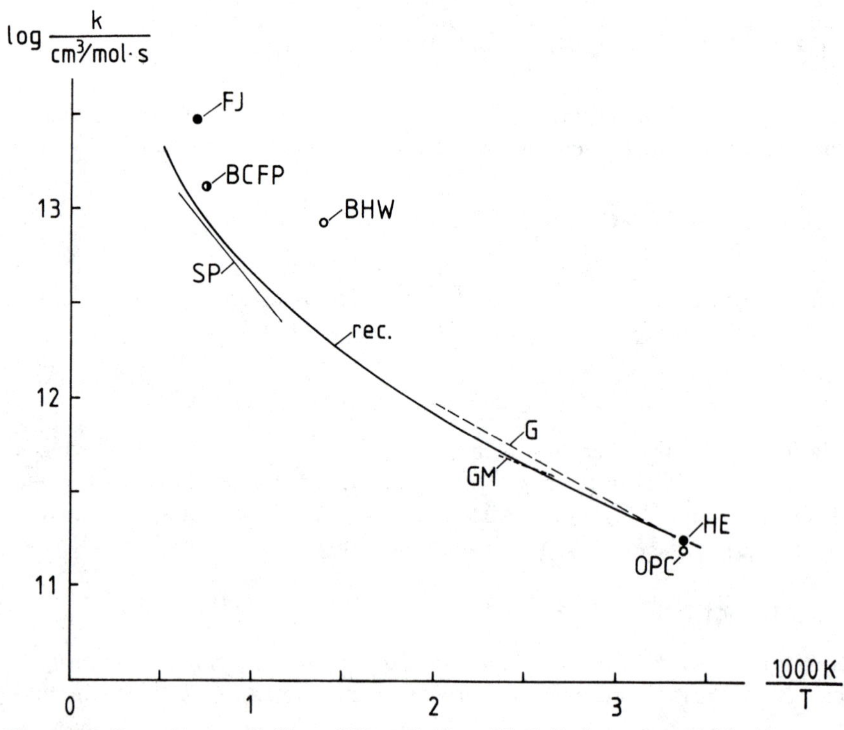

Figure 53. Rate data on $C_2H_6 + OH \rightarrow C_2H_5 + H_2O$. References in Table 14.

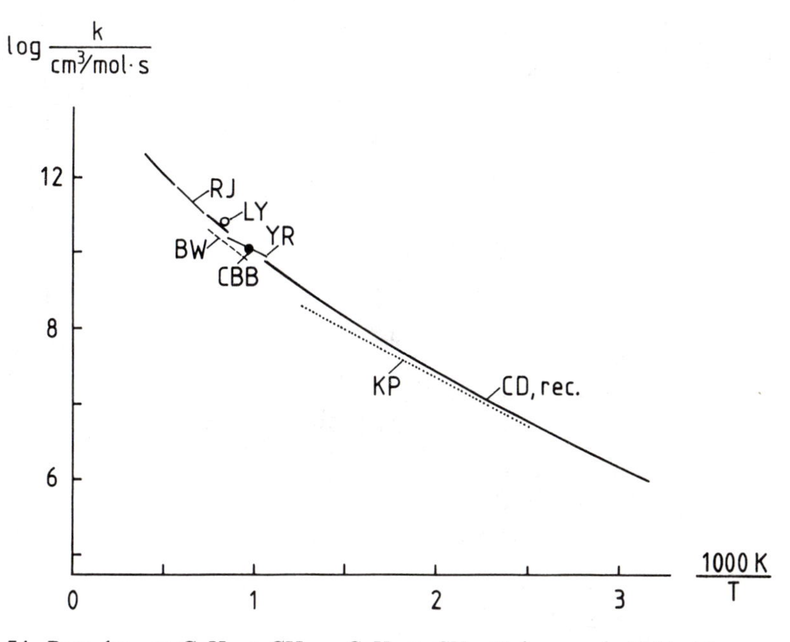

Figure 54. Rate data on $C_2H_6 + CH_3 \rightarrow C_2H_5 + CH_4$. References in Table 14.

decomposition, oxidation, and recombination or disproportionation of C_2H_5 (Figs. 1 and 4).

Disproportionation reactions ($C_2H_5 + H \rightarrow CH_3 + CH_3$, $C_2H_5 + C_2H_5 \rightarrow C_2H_4 + C_2H_6$)

These reactions have weak chain-terminating character because of the lower reactivity of CH_3 and the stable molecules compared with C_2H_5. For both reactions there are sufficient rate data (Fig. 56). The rate coefficient of C_2H_5 disproportionation is determined from measurements relative to C_2H_5 recombination (Fig. 57).

Oxidation reactions ($C_2H_5 + O \rightarrow$ products, $C_2H_5 + O_2 \rightarrow C_2H_4 + HO_2$)

The reaction with O atoms is relatively unimportant in flames due to their low concentration. A rate coefficient can be estimated from that of the corresponding reactions of O with CH_3 (Fig. 32) and with C_5H_{11} (Hoyermann and Sievert, 1979a). The products are mainly CH_3CHO and H (>80%, see Hoyermann and Sievert, 1979b). The reaction with O_2 (Fig. 58) is well investigated only at temperatures around 800 K. C_2H_4 and HO_2 are the main products, whereas OH formation (expected to be important in low-temperature oxidation) can be neglected (Baker *et al.*, 1971).

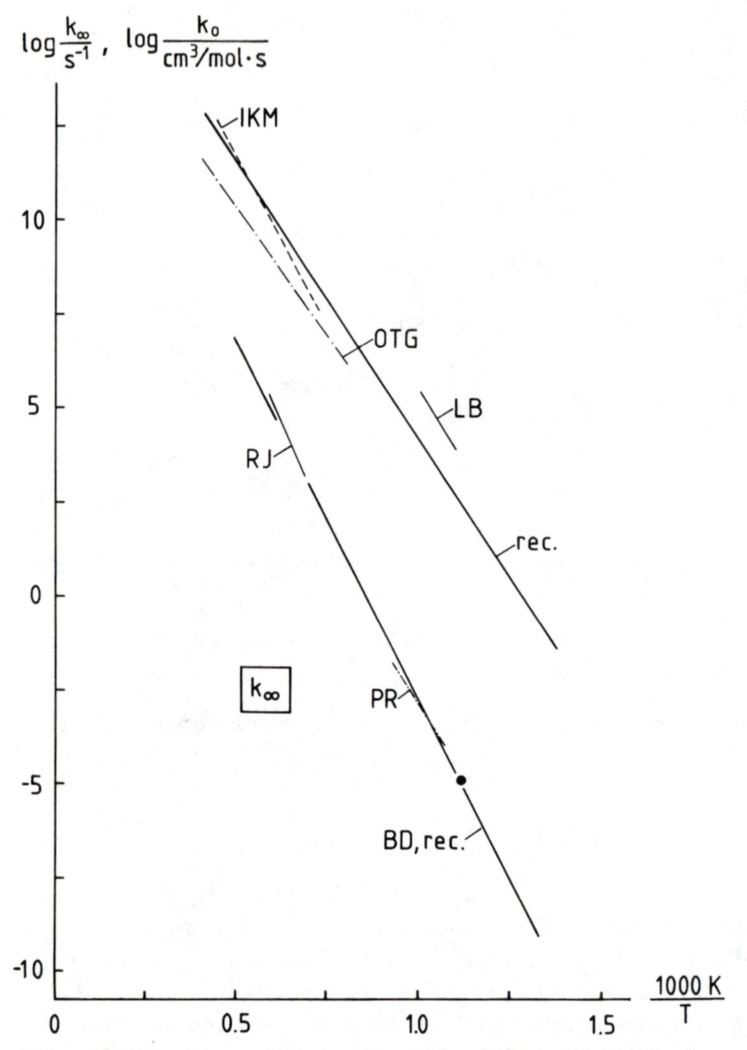

Figure 55. Rate data on $C_2H_6(+M) \rightarrow CH_3 + CH_3(+M)$. For the low-pressure rate coefficient k_1, $M = Ar$. References in Table 14. Fall-off curves are shown in Fig. 35.

Recombination reactions
($C_2H_5 + CH_3 \rightarrow C_3H_8$, $C_2H_5 + C_2H_5 \rightarrow C_4H_{10}$)

These reactions are slightly chain-terminating. Except for calculations based on the reverse reaction of C_3H_8 thermal decomposition, there is a lack of data for the recombination of C_2H_5 with CH_3, whereas the rate coefficient of C_2H_5 recombination is well known (Fig. 57). Some of the older values of this coefficient have been corrected due to defects of the thermochemistry used in data reduction (Walker, 1977).

Table 14. Rate data on C_2H_6 reactions.

Reference	$A/(cm^3/mol)^{n-1} s^{-1}$	b	E/kJ	T/K	Method
	$C_2H_6 + H \rightarrow C_2H_5 + H_2$				
Clark and Dove (1973a)	5.4×10^2	3.5	21.8	300–1800	Review
Camilleri *et al.* (1974)	1.9×10^{14}	0	40.9	503–753	FR,DI,GC
Baldwin *et al.* (1977)	1.3×10^{14}	0	40.6	···	Review
Lede and Villermaux (1978)	5.0×10^{13}	0	38.1	281–347	FR,DI,UA
Recommended ($\log F = \pm 0.30$)	5.4×10^2	3.5	21.8	300–2000	···
	$C_2H_6 + O \rightarrow C_2H_5 + OH$				
Westenberg and deHaas (1967)	1.8×10^{13}	0	25.5	300–1000	FR,DI,ES
Papadapoulos *et al.* (1971)	2.7×10^{13}	0	27.2	300–370	FR,DI,GC,CL
Smets and Peeters (1975)	4.0×10^{13}	···	···	1450	FL,TH,MS
Huie and Herron (1975)	3.0×10^{13}	0	24.1	···	Review
Recommended ($\log F = \pm 0.40$)	3.0×10^7	2.0	21.4	300–1500	···
	$C_2H_6 + OH \rightarrow C_2H_5 + H_2O$				
Fenimore and Jones (1963)	3.0×10^{13}	···	···	1400–1600	FL,TH,MS
Baldwin *et al.* (1970)	8.5×10^{12}	···	···	775	SR,TH,P,GC
Greiner (1970b)	1.1×10^{13}	0	10.2	300–500	SR,FP,UA
Smets and Peeters (1975)	6.5×10^{13}	0	23.5	900–1800	FL,TH,MS
Overend *et al.* (1975)	1.6×10^{11}	···	···	295	SR,FP,RA
Gordon and Mulac (1975)	3.5×10^{12}	0	6.9	381–416	SR,PR,UA
Howard and Evenson (1976b)	1.8×10^{11}	···	···	296	FR,DI,LM
Bradley *et al.* (1976)	1.3×10^{13}	···	···	1300	ST,TH,UA
Baldwin *et al.* (1977)	3.7×10^{12}	0	6.9	···	Review
Recommended ($\log F = \pm 0.30$)	6.3×10^6	2.0	2.7	300–2000	···

Table 14 (*continued*).

Reference	$A/(cm^3/mol)^{n-1} s^{-1}$	b	E/kJ	T/K	Method
	$C_2H_6 + HO_2 \rightarrow C_2H_5 + H_2O_2$				
Baldwin *et al.* (1977)	6.0×10^{12}	0	81.2	···	Review
Recommended ($\log F = \pm 0.50$)	6.0×10^{12}	0	81.2	300–1000	···
	$C_2H_6 + CH_3 \rightarrow C_2H_5 + CH_4$				
Clark and Dove (1973a)	0.55	4.0	34.7	300–1800	Review
Yampolskii and Rybin (1974)	3.0×10^{12}	0	47.0	980–1130	SR,TH,MS,GC
Kerr and Parsonage (1976)	5.6×10^{11}	0	48.6	400–800	Review
Roth and Just (1979)	0.55	4.0	34.7	1460–1700	ST,TH,RA
Bradley and West (1976b)	3.2×10^{13}	0	75.0	1055–1325	ST,TH,GC
Chen *et al.* (1976)	1.3×10^{10}	···	···	1038	SR,TH,GC
Lee and Yeh (1979)	6.5×10^{10}	···	···	1206	Review
Recommended ($\log F = \pm 0.30$)	0.55	4.0	34.7	300–2000	···
	$C_2H_6 + M \rightarrow CH_3 + CH_3 + M$ (k_1, M = Ar)				
Lin and Back (1966)	1.8×10^{22}	0	304	913–999	SR,TH,P
Izod *et al.* (1971)	2.4×10^{21}	0	368	1400–2200	ST,TH,IE
Olson *et al.* (1979)	1.0×10^{17}	0	269	1300–2500	ST,TH,LS,UA
Recommended ($\log F = \pm 1.0$)	1.0×10^{19}	0	285	800–2500	···
	$C_2H_6 \rightarrow CH_3 + CH_3$ (k_∞)				
Baulch and Duxbury (1980)	2.4×10^{16}	0	366	750–1500	Review
Roth and Just (1979)	2.7×10^{16}	0	362	1460–1700	ST,TH,RA
Pratt and Rogers (1979a)	1.0×10^{13}	0	305	941–1073	FR,TH,GC
Pacey and Wimalasena (1980)	1.1×10^{-5}	···	···	903	FR,TH,GC
Recommended ($\log F = \pm 0.50$)	2.4×10^{16}	0	366	750–2000	···

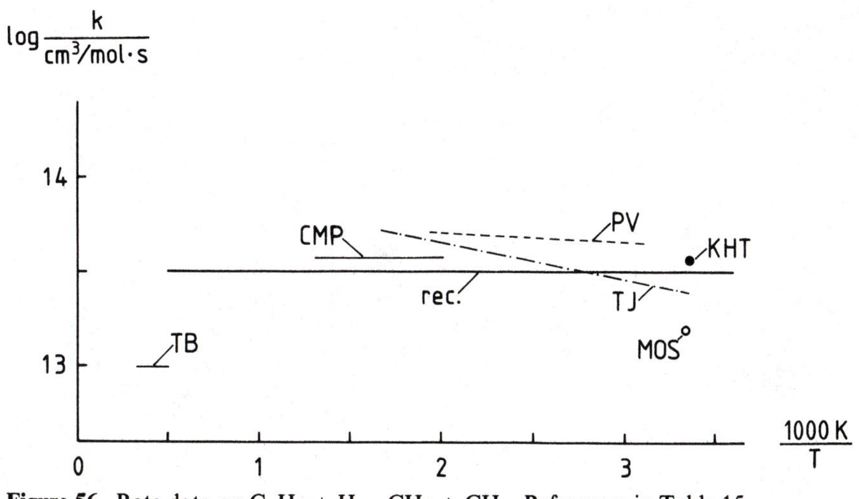

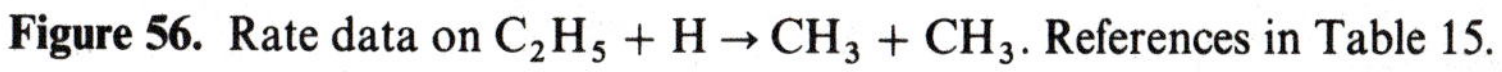

Figure 56. Rate data on $C_2H_5 + H \rightarrow CH_3 + CH_3$. References in Table 15.

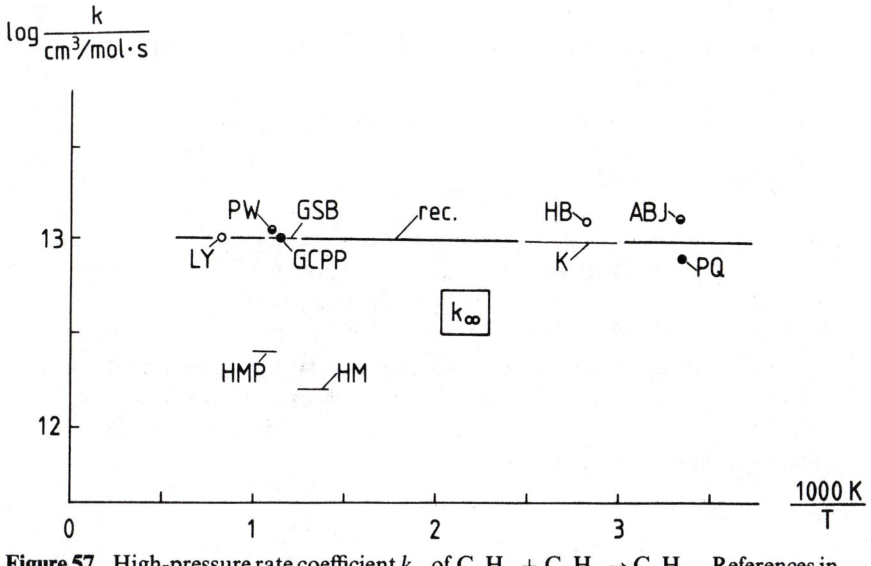

Figure 57. High-pressure rate coefficient k_∞ of $C_2H_5 + C_2H_5 \rightarrow C_4H_{10}$. References in Table 15.

Thermal decomposition of C_2H_5

This is an important and often rate-determining step because of the formation of H atoms (Fig. 4). The high-pressure rate coefficient k_∞ is well defined by a number of measurements and by determination from the reverse reaction (Fig. 59). Fall-off curves are shown in Fig. 60.

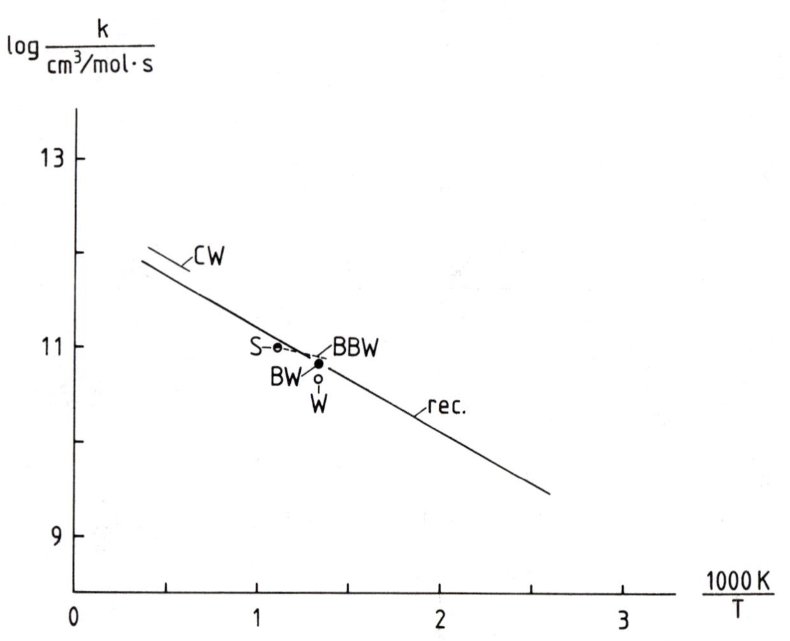

Figure 58. Rate data on $C_2H_5 + O_2 \rightarrow C_2H_4 + HO_2$. References in Table 15.

6.3. Reactions of C_2H_4

$$C_2H_4 + H\ (+M) \rightarrow C_2H_5\ (+M)$$

The reverse of this reaction has been discussed above, including the fall-off behavior (Fig. 60). There exist numerous measurements at room temperature (Fig. 61), but no determination of the activation energy. The review of Kerr and Parsonage (1972) recommends an Arrhenius expression derived from an estimated preexponential factor. This leads to a large rate coefficient for high temperature, in disagreement with values given by Baldwin *et al.* (1966) and with determinations from the reverse reaction, which are the basis of the Arrhenius expression recommended here.

$$C_2H_4 + H \rightarrow C_2H_3 + H_2$$

This is an important reaction path leading to acetylene from C_2H_4, C_2H_5, and C_2H_6 formed by CH_3 recombination (Section 5.2). Unfortunately, the available measurements show much scatter (Fig. 62), suggesting the necessity of further work on this important reaction.

$$C_2H_4 + O \rightarrow \text{products}$$

Probably because of the complexity of the subsequent reactions of the radicals formed in this reaction, it has attracted much attention. Thus, many

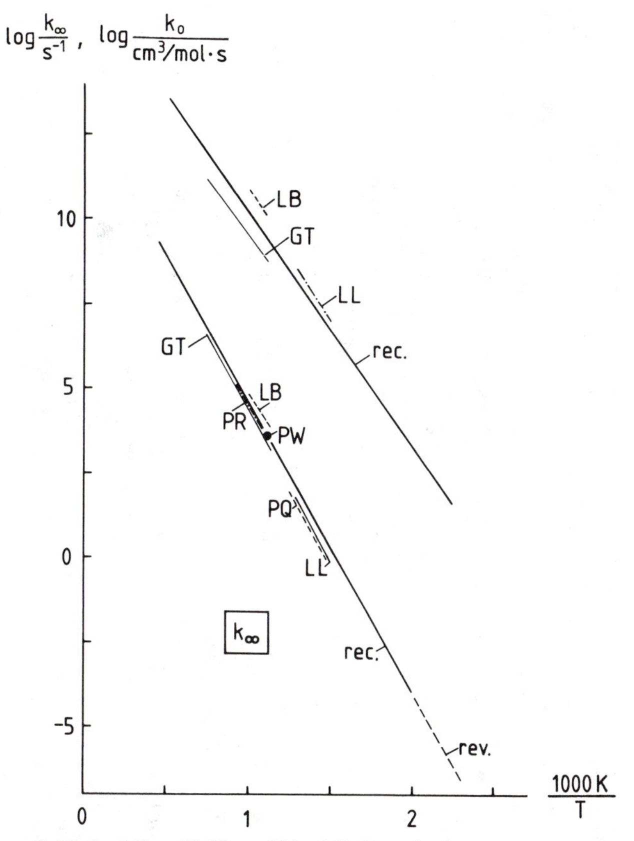

Figure 59. Rate data on $C_2H_5(+M) \rightarrow C_2H_4 + H(+M)$. For the low-pressure rate coefficient k_1, M is undefined. References in Table 15. Fall-off curves are given in Fig. 60.

measurements are available in the literature (Fig. 63). The reaction occurs via an addition complex, which can be stabilized at high pressure (Eusuf and Wagner, 1972; Gaedtke *et al.*, 1973). At single-collision conditions and high collision energies, this complex seems to decompose to form the vinoxy radical and H atoms, C_2H_3O (Buss *et al.*, 1981), whereas at normal or reduced pressures used in flame investigations the products are postulated to be CH_3 and CHO (Blumenberg *et al.*, 1977; Pruss *et al.*, 1974).

$C_2H_4 + OH \rightarrow$ products

There is a definite lack of rate information on this reaction which, together with $H + C_2H_4$, accounts for C_2H_4 consumption in flames. (The reaction of C_2H_4 with O is of less importance because of the lower concentrations of O atoms.) At low temperature, its pressure dependence shows that this is an addition reaction (Atkinson *et al.*, 1977; Gordon and Mulac, 1975). Accordingly, the rate coefficient has a small negative activation energy (Fig. 64).

Table 15. Rate data on C_2H_5 reactions.

Reference	$A/(cm^3/mol)^{n-1}\ s^{-1}$	b	E/kJ	T/K	Method
	$C_2H_5 + H \rightarrow CH_3 + CH_3$				
Kurylo *et al.* (1970a)	3.6×10^{13}	...	...	298	SR,FP,RA
Teng and Jones (1972)	1.1×10^{14}	0	3.6	202–603	FR,DI,GC
Michael *et al.* (1973)	1.5×10^{13}	...	...	298	SR,HG,RA
Camilleri *et al.* (1974)	3.7×10^{13}	0	0	503–753	FR,DI,GC
Pratt and Veltman (1976)	6.4×10^{13}	0	0.9	321–521	FR,DI,MS
Tabayashi and Bauer (1979)	1.0×10^{13}	0	0	1950–2770	ST,TH,LS
Recommended ($\log F = \pm 0.30$)	3.0×10^{13}	0	0	300–1500	...
	$C_2H_5 + O \rightarrow$ products				
Recommended ($\log F = \pm 0.30$)	5.0×10^{13}	0	0	300–2000	Estimate
	$C_2H_5 + O_2 \rightarrow C_2H_4 + HO_2$				
Sampson (1963)	1.0×10^{11}	...	...	903	FR,TH,GC
Baker *et al.* (1971)	2.2×10^{11}	0	5.8	713–896	SR,TH,P,GC
Cooke and Williams (1971)	3.2×10^{12}	0	21.0	1700–2400	ST,TH,UA,LS
Walker (1975)	5.5×10^{10}	...	...	753	SR,TH,P,GC
Baldwin and Walker (1981)	6.6×10^{10}	...	...	753	SR,TH,P,GC
Recommended ($\log F = \pm 0.50$)	2.0×10^{12}	0	20.9	700–2000	...
	$C_2H_5 + CH_3 \rightarrow C_3H_8$		(k_∞)		
Lifshitz and Frenklach (1975)	2.5×10^{12}	0	0	1050–1700	ST,TH,GC
Koike and Gardiner (1980)	7.2×10^{12}	0	0	1300–1700	ST,TH,IA
Recommended ($\log F = \pm 0.50$)	7.0×10^{12}	0	0	300–2000	...
	$C_2H_5 + C_2H_5 \rightarrow C_4H_{10}$		(k_∞)		
Hiatt and Benson (1972)	4.0×10^{11}	...	...	354	SR,TH,GC
Golden *et al.* (1973)	1.0×10^{13}	0	0	800–900	SR,TH,MS

Kerr (1973)	1.6×10^{13}	0	0	320–400	Review
Hughes *et al.* (1974)	2.5×10^{12}	0	0	895–981	SR,TH,GC
Hughes and Marshall (1975)	1.6×10^{12}	0	0	693–803	SR,TH,GC
Golden *et al.* (1976)	1.0×10^{13}	...	...	860	SR,TH,MS
Parks and Quinn (1976)	8.4×10^{12}	...	...	298	SR,PH,UE
Walker (1977)	1.0×10^{13}	0	0	300–1000	Review
Lee and Yeh (1979)	1.0×10^{13}	...	...	1200	SR,TH,GC
Pacey and Wimalasena (1980)	1.1×10^{13}	...	...	903	FR,TH,GC
Recommended ($\log F = \pm 0.20$)	1.0×10^{13}	0	0	300–1200	...
	$C_2H_5 + C_2H_5 \rightarrow C_2H_4 + C_2H_6$				
Kerr and Trotman-Dickenson (1960a)	$k_{disp}/k_{comb} = 0.15$			323–778	SR,PH,GC
Dixon *et al.* (1963)	$k_{disp}/k_{comb} = 0.14$			208–313	SR,PH,GC
Hooper *et al.* (1975)	$k_{disp}/k_{comb} = 0.145$			173–298	SR,PH,GC
Adachi *et al.* (1979)	$k_{disp}/k_{comb} = 0.139$			298	SR,FP,UA
Recommended ($\log F = \pm 0.25$)	1.4×10^{12}	0	0	300–1200	...
	$C_2H_5 + M \rightarrow C_2H_4 + H + M$ (k_1, M undefined)				
Lin and Back (1966)	1.8×10^{18}	0	136	913–999	SR,TH,P,GC
Loucks and Laidler (1967)	6.8×10^{17}	0	133	673–773	SR,HG,P,GC
Glänzer and Troe (1973)	1.0×10^{16}	0	126	900–1350	ST,TH,UA
Recommended ($\log F = \pm 1.0$)	1.0×10^{17}	0	130	700–1500	...
	$C_2H_5 \rightarrow C_2H_4 + H$ (k_∞)				
Purnell and Quinn (1962)	1.0×10^{13}	0	168	693–803	SR,TH,GC
Lin and Back (1966)	3.8×10^{13}	0	159	913–999	SR,TH,P,GC
Loucks and Laidler (1967)	2.7×10^{14}	0	171	673–773	SR,HG,P,GC
Glänzer and Troe (1973)	1.3×10^{13}	0	167	900–1350	ST,TH,UA
Pratt and Rogers (1979a)	3.2×10^{13}	0	175	941–1073	FR,TH,GC
Pacey and Wimalasena (1980)	4.0×10^{3}	...	...	903	FR,TH,GC
Recommended ($\log F = \pm 0.50$)	2.0×10^{13}	0	166	500–2000	...

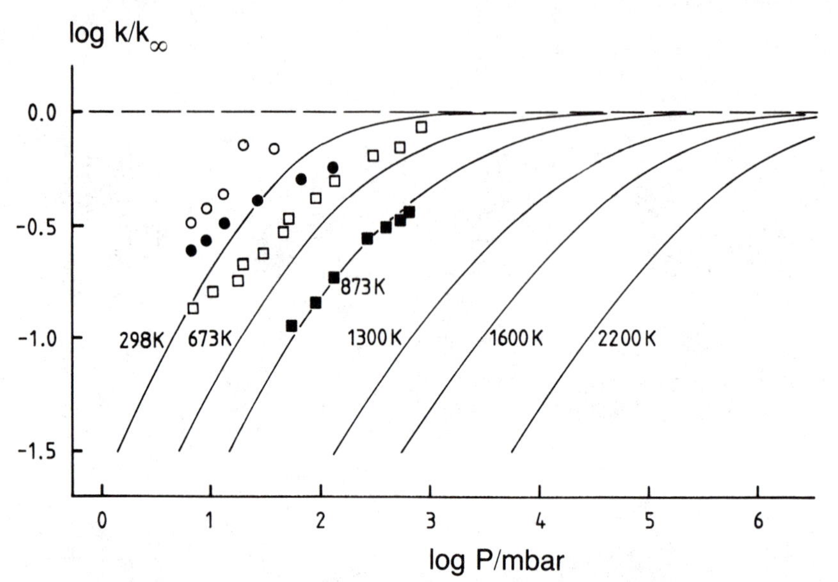

Figure 60. Fall-off curves for $H + C_2H_4 \rightarrow C_2H_5$ at different temperatures. Lines calculated as described in Section 1.2. Points: experimental values, ■: Lin and Back (1966) for $T = 873$ K, □: Loucks and Laidler (1967) for $T = 673$ K, ●: Braun and Lenzi (1968) for $T = 298$ K, ○: Kurylo *et al.* (1970) for $T = 298$ K.

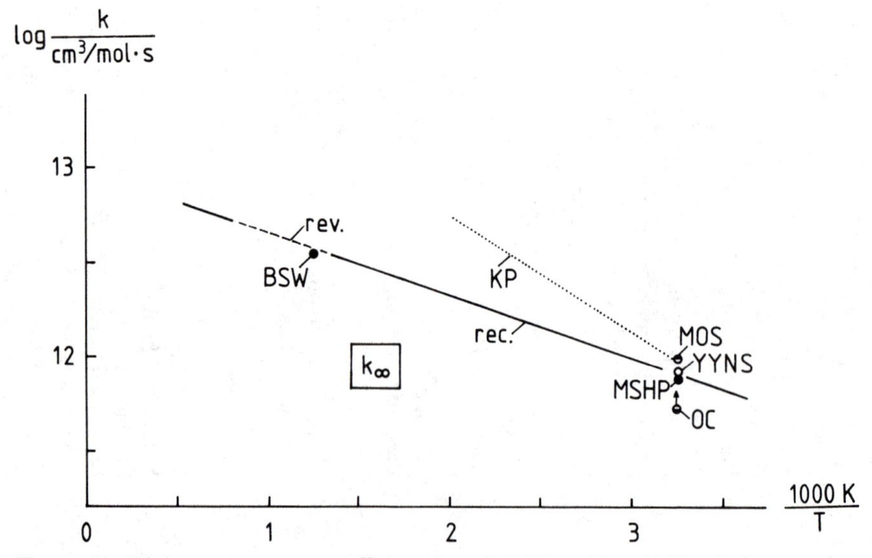

Figure 61. High-pressure rate coefficient k_∞ of $C_2H_4 + H \rightarrow C_2H_5$. References in Table 15. Fall-off curves are given in Fig. 60.

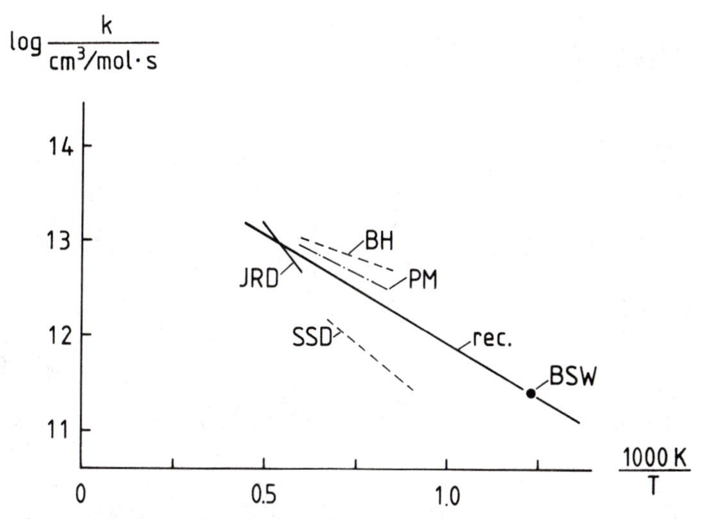

Figure 62. Rate data on $C_2H_4 + H \rightarrow C_2H_3 + H_2$. References in Table 16.

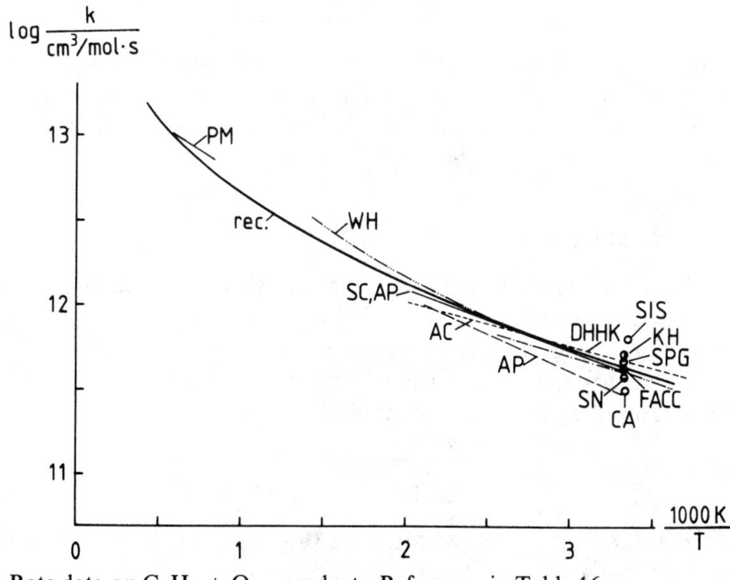

Figure 63. Rate data on $C_2H_4 + O \rightarrow$ products. References in Table 16.

However, this behavior is incompatible with the rate measurements at high temperature, which suggests an additional abstraction channel. The rate coefficient recommended at high temperature is chosen to be consistent with the low-temperature data and to give a reasonable preexponential factor for this exothermic reaction ($\Delta H^\circ_{298} = -45$ kJ).

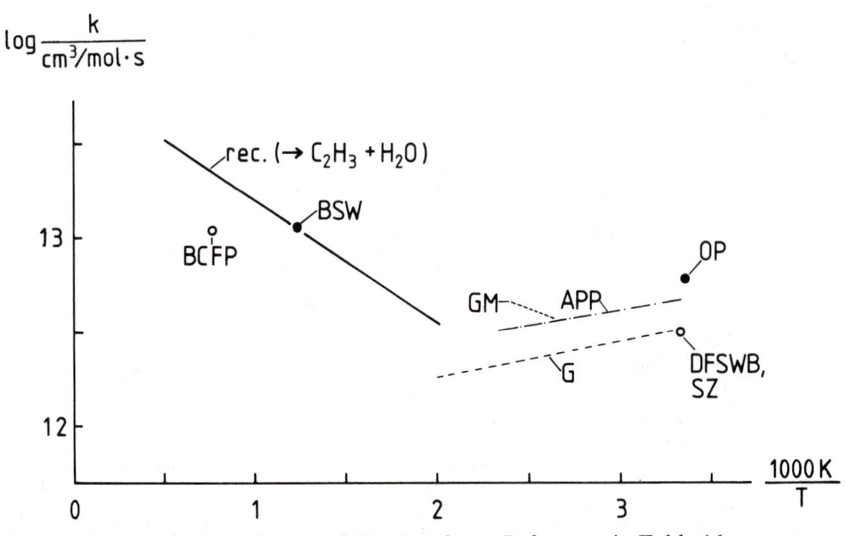

Figure 64. Rate data on C_2H_4 + OH → products. References in Table 16.

Thermal decomposition of C_2H_4 and the reaction with CH_3

These reactions are unimportant for flame propagation. Little information is available on the CH_3 reaction, whereas there exist sufficient data on both channels for the thermal decomposition of C_2H_4, which are important for the ignition of C_2H_4 mixtures.

6.4. Reactions of C_2H_3

In the decomposition of C_2H_3 there are three main competing reactions: thermal decomposition, reaction with H atoms, and reaction with O_2. Since all of these reactions are fairly fast at combustion temperatures and all lead to C_2H_2, determination of their individual rate coefficients is not especially important for hydrocarbon combustion.

$C_2H_3 + H \rightarrow C_2H_2 + H_2$

There is large scatter of the rate data (Fig. 65). For the determination of the recommended rate coefficients the older proposals by Benson and Haugen (1967) and Volpi and Zocchi (1966) have been disregarded.

$C_2H_3 + O \rightarrow$ products

This reaction is unimportant for flame propagation due to the low O-atom concentration in regions of considerable C_2H_3 production. Because it is a radical–radical reaction and has a large room-temperature rate coefficient, it can not have a substantial activation energy.

Table 16. Rate data on C_2H_4 reactions.

Reference	$A/(cm^3/mol)^{n-1}\ s^{-1}$	b	E/kJ	T/K	Method
	$C_2H_4 + H \rightarrow C_2H_5\ (k_\infty)$				
Baldwin *et al.* (1966)	3.5×10^{12}	...	...	813	SR,TH,P,GC
Kerr and Parsonage (1972)	9.3×10^{13}	0	11.7	...	Review
Michael *et al.* (1973)	9.7×10^{11}	...	...	298	SR,HG,RA
Milhelcic *et al.* (1975)	7.5×10^{11}	...	...	298	SR,PR,RA
Lee *et al.* (1978)	2.2×10^{13}	0	8.6	198–320	SR,FP,RF
Yoichi *et al.* (1978)	6.6×10^{11}	...	...	298	SR,PR,RA
Ishikawa *et al.* (1978)	6.6×10^{11}	...	...	298	SR,PR,RA
Oka and Cvetanovic (1979)	4.7×10^{11}	...	...	298	FR,HG,CL
Recommended ($\log F = \pm 0.30$)	1.0×10^{13}	0	6.3	300–2000	...
	$C_2H_4 + H \rightarrow C_2H_3 + H_2$				
Baldwin *et al.* (1966)	2.5×10^{11}	...	...	813	SR,TH,P,GC
Benson and Haugen (1967)	6.3×10^{13}	0	25.0	1200–1700	ST,TH,GC
Skinner *et al.* (1971)	1.6×10^{14}	0	58.6	1100–1500	ST,TH,GC
Peeters and Mahnen (1973b)	1.1×10^{14}	0	35.7	1200–1700	FL,TH,MS
Just *et al.* (1977)	7.6×10^{15}	0	104.0	1700–2000	ST,TH,RA
Recommended ($\log F = \pm 0.50$)	1.5×10^{14}	0	42.7	700–2000	...
	$C_2H_4 + O \rightarrow$ products				
Westenberg and deHaas (1969b)	8.2×10^{6}	2.0	1.3	195–715	FR,DI,ES
Atkinson and Cvetanovic (1971)	3.0×10^{11}	...	...	298	FR,HG,CL
Stuhl and Niki (1971)	3.8×10^{11}	...	...	300	SR,FP,CL
Davis *et al.* (1972)	3.3×10^{12}	0	4.7	232–500	SR,FP,RF
Atkinson and Cvetanovic (1972)	7.9×10^{12}	0	8.1	298–473	FR,HG,CL

Table 16 (*continued*).

Reference	$A/(\mathrm{cm}^3/\mathrm{mol})^{n-1}\,\mathrm{s}^{-1}$	b	E/kJ	T/K	Method
Stuhl and Niki (1972b)	3.8×10^{11}	...	...	298	SR,PH,CL
Kurylo and Huie (1973)	4.8×10^{11}	...	...	298	SR,FP,RF
Gaedtke *et al.* (1973)	7.0×10^{11}	...	...	300	SR,PH,UA,GC
Peeters and Mahnen (1973b)	2.3×10^{13}	...	...	1200–1700	FL,TH,MS
Furuyama *et al.* (1974)	4.3×10^{11}	...	...	299	FR,HG,CL
Slagle *et al.* (1974)	4.6×10^{11}	...	...	300	FR,DI,MS
Atkinson and Pitts (1974a)	3.4×10^{12}	0	5.3	300–392	SR,HG,CL
Atkinson and Pitts (1974b)	4.0×10^{11}	...	...	300	SR,HG,CL
Singleton and Cvetanovic (1976)	7.0×10^{12}	0	7.0	298–486	SR,HG,GC
Atkinson and Pitts (1977a)	5.5×10^{12}	0	6.2	298–439	SR,FP,CL
Sugawara *et al.* (1980)	6.0×10^{11}	...	...	296	SR,PR,RA
Recommended ($\log F = \pm 0.20$)	1.6×10^{9}	1.2	3.1	300–2000	...
	$C_2H_4 + OH \rightarrow$ products				
Baldwin *et al.* (1966)	1.1×10^{13}	...	...	813	SR,TH,P,GC
Greiner (1970a)	7.6×10^{11}	0	−3.6	300–500	SR,FP,UA
Smith and Zellner (1973)	3.1×10^{12}	...	...	298	SR,FP,RA
Davis *et al.* (1975)	3.2×10^{12}	...	...	300	SR,FP,RF
Gordon and Mulac (1975)	2.9×10^{13}	0	6.5	381–416	SR,PR,UA
Bradley *et al.* (1976)	1.1×10^{13}	...	...	1300	ST,TH,UA
Atkinson *et al.* (1977)	1.3×10^{12}	0	−3.2	300–425	ST,FP,RF
Overend and Paraskevopoulos (1977b)	6.0×10^{12}	...	...	296	SR,FP,RA
Recommended ($\log F = \pm 0.40$)	3.0×10^{13}	0	12.5	500–2000	

	$C_2H_4 + CH_3 \rightarrow C_2H_3 + CH_4$				
Kerr and Parsonage (1976)	4.2×10^{11}	0	46.5	350–650	Review
Chen *et al.* (1976)	1.0×10^{9}	···	···	1038	SR,TH,GC
Recommended ($\log F = \pm 0.30$)	4.2×10^{11}	0	46.5	300–1000	···
	$C_2H_4 + M \rightarrow C_2H_2 + H_2 + M$ (k_0, $M = Ar$)				
Just *et al.* (1977)	2.6×10^{17}	0	332.0	1700–2200	ST,TH,RA,IE
Tanzawa and Gardiner (1980b)	3.0×10^{17}	0	340.0	2000–2540	ST,TH,LS
Recommended ($\log F = \pm 0.50$)	2.6×10^{17}	0	332.0	1500–2500	···
	$C_2H_4 + M \rightarrow C_2H_3 + H + M$ (k_0, $M = Ar$)				
Just *et al.* (1977)	2.6×10^{17}	0	404.0	1700–2200	ST,TH,RA
Tanzawa and Gardiner (1980b)	3.1×10^{17}	0	400.0	2000–2540	ST,TH,LS
Recommended ($\log F = \pm 0.50$)	2.6×10^{17}	0	404.0	1500–2500	···

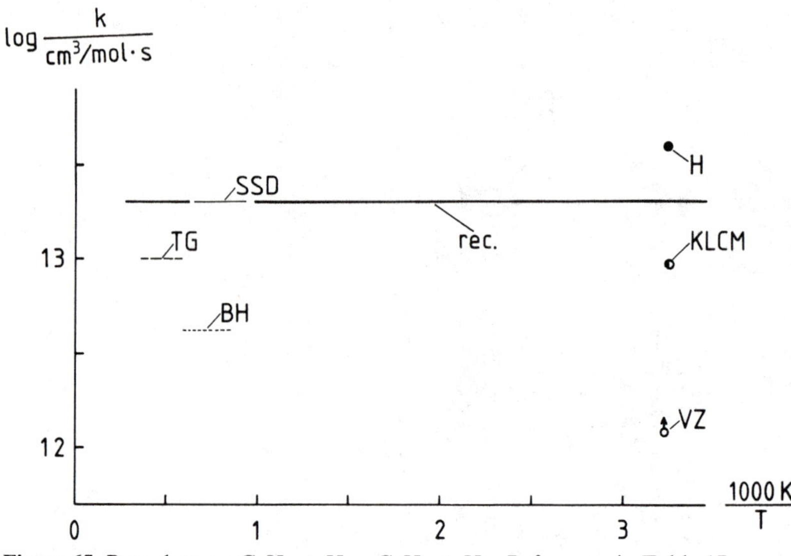

Figure 65. Rate data on $C_2H_3 + H \rightarrow C_2H_2 + H_2$. References in Table 17.

$$C_2H_3 + O_2 \rightarrow C_2H_2 + HO_2$$

The only mention of this reaction in the literature is an estimate of its rate coefficient used in shock tube work (Cooke and Williams, 1971). The basis of the recommendation given here is the fact that HO_2-forming reactions of a radical with O_2 always have a relatively small preexponential factor of about 10^{12} cm^3/mol s.

Thermal decomposition of C_2H_3

For this pressure-dependent reaction some rate data are available for the low-pressure rate coefficient k_0 at high temperature, whereas there is a lack of information about the high-pressure rate coefficient k_∞ (Fig. 66). Fall-off curves are presented in Fig. 67.

6.5. Reactions of C_2H_2

These reactions are generally important for fuel-rich hydrocarbon combustion (Warnatz *et al.*, 1982). Unfortunately, the situation is complicated by the fact that the reactions of H, O, and OH with acetylene have both abstraction and addition reaction channels.

$$C_2H_2 + H\ (+M) \rightarrow C_2H_3\ (+M)$$

The reverse of this addition reaction has been discussed above. There exist only low-temperature measurements in the literature extrapolation of which is

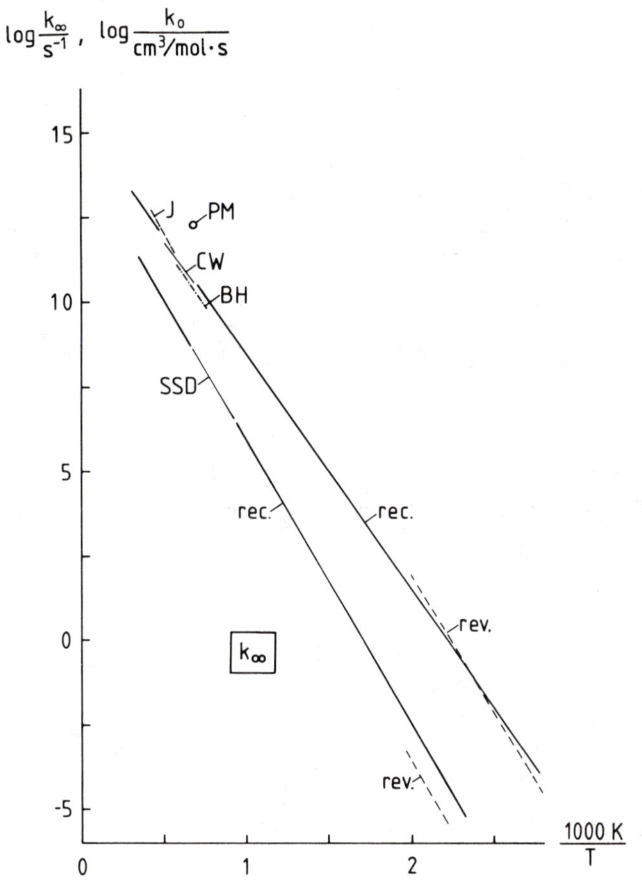

Figure 66. Rate data on $C_2H_3(+M) \rightarrow C_2H_2 + H(+M)$. References in Table 17. Fall-off curves are given in Fig. 67.

not consistent with the rate coefficient calculated by aid of equilibrium data from the reverse reaction. The source of this inconsistency may be an incorrect value of the heat of formation of the C_2H_3 radical. The data given by Hoyermann *et al.* (1968) have not been taken into consideration (Fig. 68) in evaluating the recommended high-pressure rate coefficient, since they do not extend far enough into the high-pressure region (see fall-off curves presented in Fig. 67).

$$C_2H_2 + H \rightarrow C_2H + H_2$$

Only indirect measurements in flames and shock tubes are available in the literature (Fig. 69). This reaction can partly account for the C_4H_2 formed in rich acetylene flames by $C_2H + C_2H_2 \rightarrow C_4H_2 + H$ (Warnatz *et al.*, 1982). The recommended rate coefficient has been chosen to be consistent with the

Table 17. Rate data on C_2H_3 reactions.

Reference	$A/(\mathrm{cm}^3/\mathrm{mol})^{n-1}\,\mathrm{s}^{-1}$	b	E/kJ	T/K	Method
	$C_2H_3 + H \rightarrow C_2H_2 + H_2$				
Volpi and Zocchi (1966)	1.0×10^{12}	⋯	⋯	313	FR,DI,MS
Benson and Haugen (1967b)	4.0×10^{12}	0	0	1200–1700	SR,TH
Skinner *et al.* (1971)	2.0×10^{13}	0	0	1100–1500	ST,TH,GC
Keil *et al.* (1976)	9.0×10^{12}	⋯	⋯	298	FR,DI,MS,RA
Olson and Gardiner (1978)	1.0×10^{13}	0	0	1800–2700	ST,TH,IA
Tanzawa and Gardiner (1980b)	1.0×10^{13}	0	0	2000–2540	ST,TH,LS
Hoyermann (1981)	4.0×10^{13}	⋯	⋯	298	FR,DI,MS
Recommended ($\log F = \pm 0.30$)	2.0×10^{13}	0	0	300–2500	⋯
	$C_2H_3 + O \rightarrow$ products				
Homann and Schweinfurth (1981)	2.0×10^{13}	⋯	⋯	298	FR,DI,MS
Hoyermann (1981)	3.3×10^{13}	⋯	⋯	298	FR,DI,MS
Recommended ($\log F = \pm 0.30$)	3.0×10^{13}	0	0	300–2000	⋯
	$C_2H_3 + O_2 \rightarrow C_2H_2 + HO_2$				
Cooke and Williams (1971)	1.6×10^{13}	0	42.0	1400–2400	ST,TH,UA,UE
Recommended ($\log F = \pm 0.50$)	1.0×10^{12}	0	0	1000–2000	⋯
	$C_2H_3 + M \rightarrow C_2H_2 + H + M$ (k_1, $M = Ar$)				
Benson and Haugen (1967b)	8.0×10^{14}	0	132.0	1200–1700	ST,TH,GC
Peeters and Mahnen (1973b)	2.0×10^{12}	⋯	⋯	1500	FL,TH,MS
Jachimowski (1977)	3.0×10^{16}	0	170.0	1815–2365	ST,TH,IE
Coats and Williams (1979)	1.6×10^{15}	0	132.0	1300–2000	ST,TH,CL,IE
Recommended ($\log F = \pm 1.0$)	3.0×10^{15}	0	134.0	500–2500	⋯
	$C_2H_3 \rightarrow C_2H_2 + H$ (k_∞)				
Skinner *et al.* (1971)	1.6×10^{14}	0	159.0	1100–1500	ST,TH,GC
Recommended ($\log F = \pm 1.0$)	1.6×10^{14}	0	159.0	500–2000	⋯

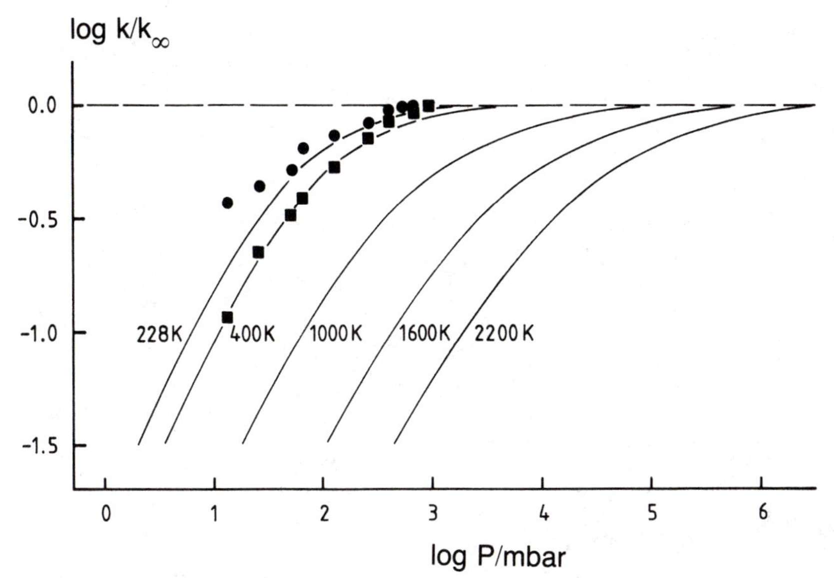

Figure 67. Fall-off curves for $C_2H_2 + H \rightarrow C_2H_3$ at different temperatures. Lines: calculated as described in Section 1.2. Points: experimental values, ■: Payne and Stief (1976) for $T = 400$ K, ●: Payne and Stief (1976) for $T = 228$ K.

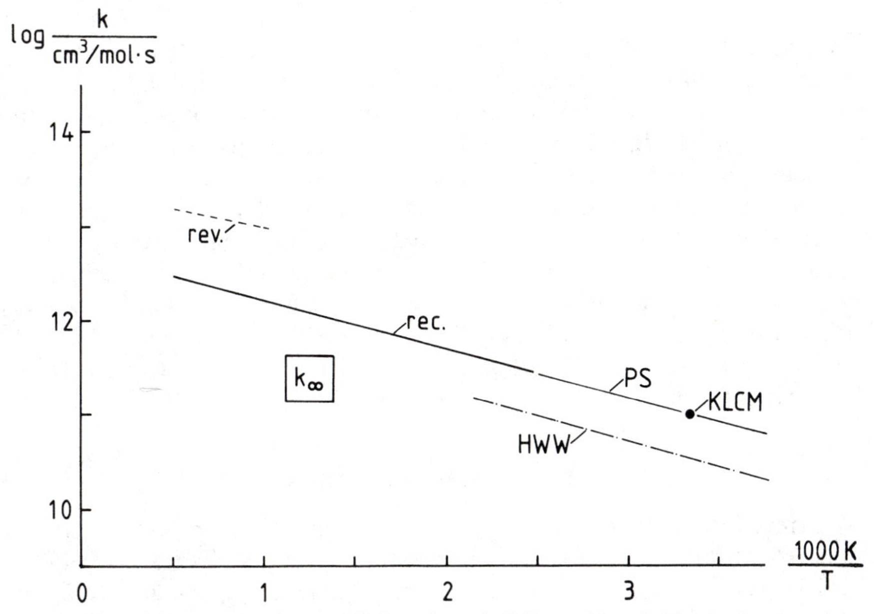

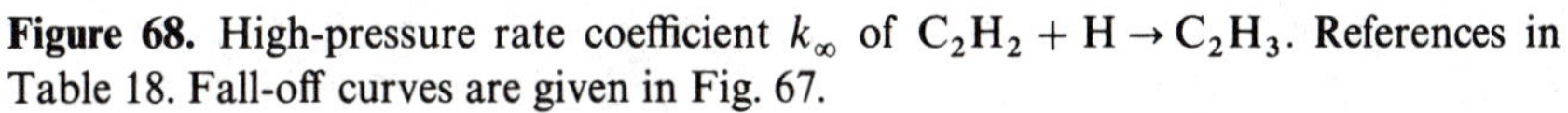
Figure 68. High-pressure rate coefficient k_∞ of $C_2H_2 + H \rightarrow C_2H_3$. References in Table 18. Fall-off curves are given in Fig. 67.

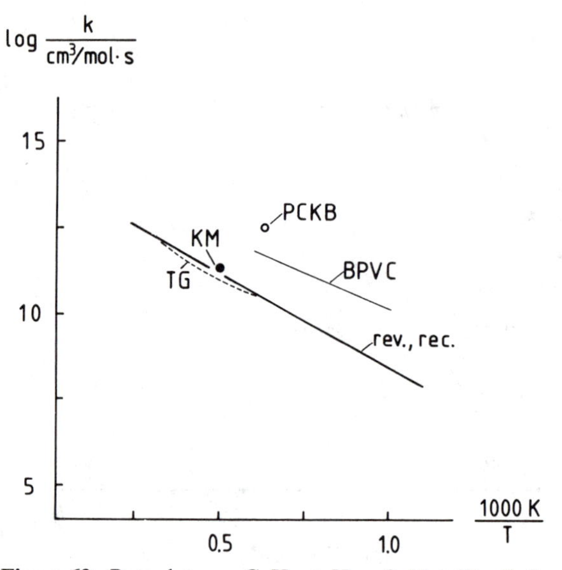

Figure 69. Rate data on $C_2H_2 + H \rightarrow C_2H + H_2$. References in Table 18.

heat of reaction $\Delta H^\circ_{298} = 87$ kJ. The thermochemical data (of C_2H) are exact enough to include rate coefficients determined from the reverse reaction.

$C_2H_2 + O \rightarrow$ products

At least at high temperature, the reaction of O atoms with acetylene has two reaction channels, one leading to $CH_2 + CO$ and one to $CHCO + H$, as can be shown by direct determination of the rate of formation of H atoms (Roth and Löhr, 1981). Low-temperature studies of the product spectrum of this reaction are conflicting, partly claiming CHCO and H atoms to be major primary products besides CH_2 and CO (Jones and Bayes, 1972; Kanofsky *et al.*, 1974) and partly claiming CH_2 and CO to be the only major primary products (Blumenberg *et al.*, 1977). Because of this unclarified situation, no recommendation of the low-temperature rate coefficient for the CHCO + H channel is given here. The rate coefficient for the $CH_2 + CO$ channel is well enough established (Fig. 70) to allow the identification of a non-Arrhenius behavior of its temperature dependence.

$C_2H_2 + OH \rightarrow$ products

This reaction has been extensively investigated only at low temperature, where the rate coefficient was found to be pressure dependent (Perry *et al.*, 1977; Michael *et al.*, 1981). At first glance, the high-temperature measurements and the high-pressure rate coefficient extrapolated from low temperature seem to match one another (Fig. 71). However, the high-temperature measurements

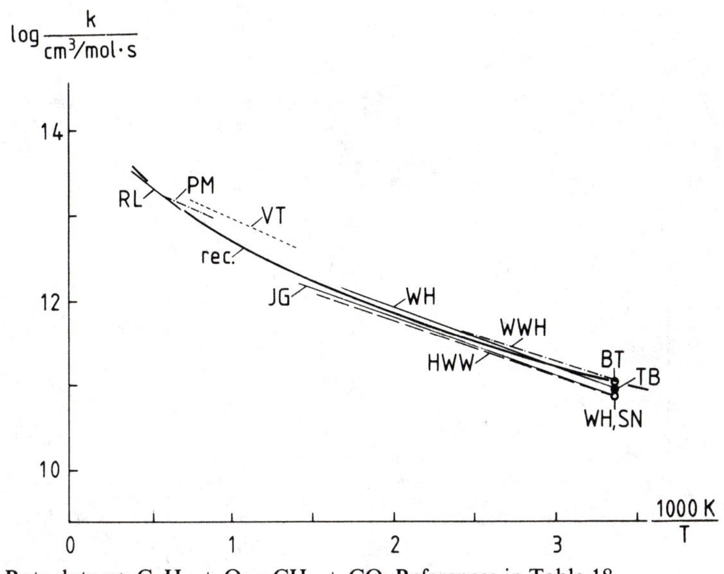

Figure 70. Rate data on $C_2H_2 + O \rightarrow CH_2 + CO$. References in Table 18.

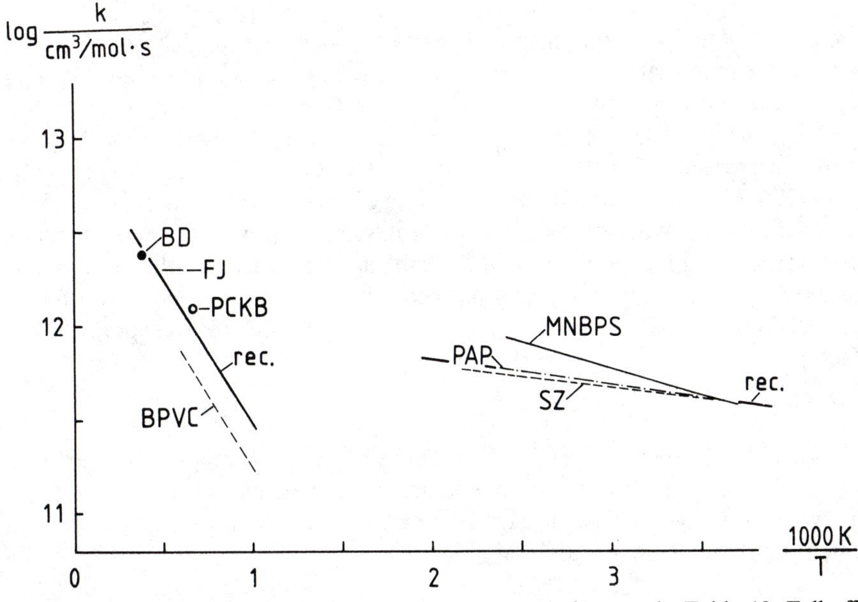

Figure 71. Rate data on $C_2H_2 + OH \rightarrow$ products. References in Table 18. Fall-off curves are given in Fig. 72.

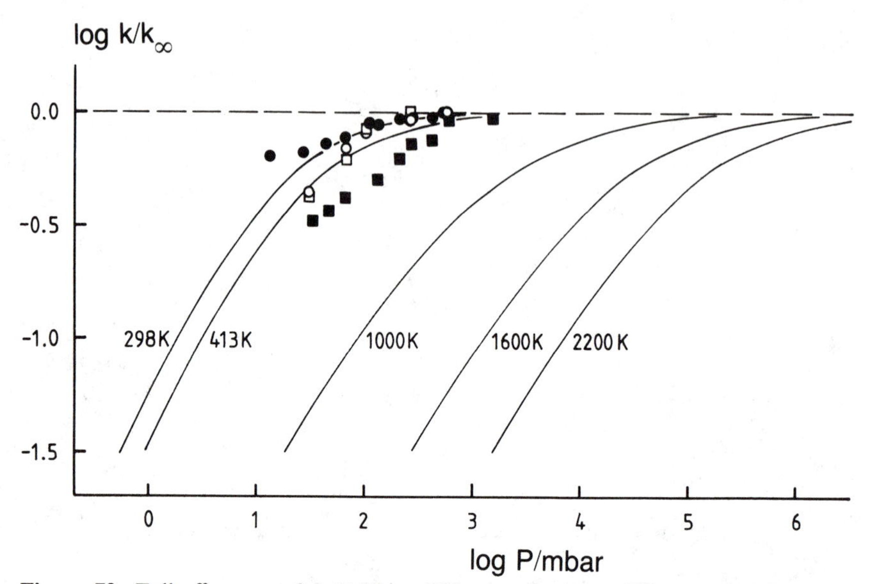

Figure 72. Fall-off curves for $C_2H_2 + OH \rightarrow$ products at different temperatures. Lines: calculated as described in Section 1.2. Points: experimental values, ■: Michael *et al.* (1981) for $T = 413$ K, ●: Michael *et al.* (1981) for $T = 298$ K, ○ Perry and Williamson (1982) for $T = 297$ K, □: Perry and Williamson (1982) for $T = 429$ K.

were carried out at experimental conditions where strong fall-off of the addition reaction rate coefficient (Fig. 72) would be expected. Therefore, other product channels must be open at high temperature.

The abstraction reaction $C_2H_2 + OH \rightarrow C_2H + H_2O$ has been postulated in the interpretation of flame studies by Porter *et al.* (1967) and by Browne *et al.* (1969). This step can also partly explain C_4H_2 formation via C_2H in rich acetylene flames (Warnatz *et al.*, 1982). The inferences drawn by Fenimore and Jones (1964) and Bar-Nun and Dove (1980) may indeed refer to this reaction. The recommended rate coefficient has been chosen to be consistent with the endothermicity ($\Delta H^\circ = 23.4$ kJ) and to give a reasonable preexponential factor.

$$C_2H_2 + C_2H \rightarrow C_4H_2 + H$$

This reaction provides C_4H_2 formation in rich flames (Warnatz *et al.*, 1982). No direct measurements at high temperature are available. Nevertheless, the activation energy of this step must be small, because the room-temperature rate coefficient is very large.

Thermal decomposition of C_2H_2

This reaction is unimportant in flame propagation because of its large activation energy. It does contribute to the initiation phase of shock-initiated combustion of acetylene.

Table 18. Rate data on C_2H_2 reactions.

Reference	$A/(cm^3/mol)^{n-1}\,s^{-1}$	b	E/kJ	T/K	Method
	$C_2H_2 + H + M \rightarrow C_2H_3 + M (k_1, M = Ar)$				
Hoyermann *et al.* (1968)	4.2×10^{17}	0	2.9	243–463	FR,DI,MS,ES
Recommended ($\log F = \pm 0.50$)	4.2×10^{17}	0	2.9	300–500	...
	$C_2H_2 + H \rightarrow C_2H_3 (k_\infty)$				
Hoyermann et al. (1968)	2.3×10^{12}	0	10.5	243–463	FR,DI,MS,ES
Keil *et al.* (1976)	9.5×10^{10}	...	...	298	FR,DI,MS,RA
Payne and Stief (1976)	5.5×10^{12}	0	10.1	193–400	SR,FP,RF
Recommended ($\log F = \pm 1.0$)	5.5×10^{12}	0	10.1	300–2000	...
	$C_2H_2 + H \rightarrow C_2H + H_2$				
Porter *et al.* (1967)	3.0×10^{12}	...	...	1600	FL,TH,GC,MS
Browne *et al.* (1969)	2.0×10^{14}	0	79.6	1000–1700	FL,TH,UA,GC
Tanzawa and Gardiner (1979)	7.8×10^{13}	3.2	2.1	1700–3400	ST,TH,LS,IE
Koike and Morinaga (1981)	2.8×10^{11}	...	...	2000	ST,TH,UA
Recommended ($\log F = \pm 0.50$)	6.0×10^{13}	0	99	300–3000	
	$C_2H_2 + O \rightarrow CH_2 + CO$				
Brown and Thrush (1967)	9.2×10^{10}	...	...	298	FR,DI,ES
Hoyermann *et al.* (1969)	1.2×10^{13}	0	12.6	243–673	FR,DI,MS,ES
Westenberg and deHaas (1969a)	2.0×10^{13}	0	13.4	195–616	FR,DI,ES
Bradley and Tse (1969)	1.1×10^{11}	...	...	298	FR,DI,ES
James and Glass (1969)	1.4×10^{13}	0	13.2	273–729	FR,DI,CL
Stuhl and Niki (1971)	7.8×10^{10}	...	...	300	SR,FP,CL
Peeters and Mahnen (1973b)	5.2×10^{13}	0	15.5	1200–1700	FL,TH,MS

Table 18 (*continued*).

Reference	$A/(cm^3/mol)^{n-1}\,s^{-1}$	b	E/kJ	T/K	Method
Homann *et al.* (1977)	1.2×10^{13}	0	11.3	220–500	FR,DI,MS
Vandooren and van Tiggelen (1977)	6.7×10^{13}	0	16.7	700–1430	FL,TH,MS
Westenberg and deHaas (1977)	7.7×10^{10}	...	...	297	SR,FP,RF
Roth and Löhr (1981)	1.2×10^{14}	0	27.5	1500–2750	ST,TH,RA
Wellmann (1981)	1.1×10^{13}	0	12.0	295–1333	FR,DI,MS
Recommended ($\log F = \pm 0.30$)	4.1×10^{8}	1.5	7.1	300–2500	...
	$C_2H_2 + O \rightarrow CHCO + H$				
Roth and Löhr (1981)	4.3×10^{14}	0	50.7	1500–2570	ST,TH,RA
Wellmann (1981)	2.0×10^{12}	...	...	1000	FR,DI,MS
Recommended ($\log F = \pm 0.50$)	4.3×10^{14}	0	50.7	1000–2500	...
	$C_2H_2 + OH \rightarrow$ products (k_∞, low temperature)				
Smith and Zellner (1973)	1.2×10^{12}	0	2.1	210–460	SR,FP,RA
Perry *et al.* (1977)	1.2×10^{12}	0	2.6	298–422	SR,FP,RF
Michael *et al.* (1981)	4.1×10^{12}	0	5.4	228–413	SR,FP,RF
Recommended ($\log F = \pm 0.40$)	3.0×10^{12}	0	4.6	300–500	...
	$C_2H_2 + OH \rightarrow$ products (high temperature)				
Fenimore and Jones (1964)	2.0×10^{12}	0	0	1700–2000	FL,TH,MS
Porter *et al.* (1967)	1.2×10^{12}	...	...	1600	FL,TH,MS,GC
Browne *et al.* (1969)	6.0×10^{12}	0	29.3	1000–1700	FL,TH,UA,GC
Bar-Nun and Dove (1980)	3.5×10^{12}	0	9.0	~2650	ST,TH,MS
Recommended ($\log F = \pm 0.50$)	2.0×10^{13}	0	29.3	1000–2000	...

	$C_2H_2 + CH_2 \rightarrow C_3H_3 + H$				
Vinckier and Debruyn (1979a)	7.8×10^{11}	···	···	295	FR,DI,MS
Homann and Schweinfurth (1981)	1.8×10^{12}	···	···	298	FR,DI,MS
Recommended ($\log F = \pm 0.60$)	1.8×10^{12}	···	···	>298	···
	$C_2H_2 + CH \rightarrow (C_3H_3\ ?)$				
Bosnali and Perner (1971)	4.5×10^{13}	···	···	298	SR,PR,UA
Homann and Schweinfurth (1981)	3.0×10^{13}	···	···	298	FR,DI,MS
Recommended ($\log F = \pm 0.30$)	3.0×10^{13}	0	0	>298	···
	$C_2H_2 + C_2H \rightarrow C_4H_2 + H$				
Lange and Wagner (1975)	3.0×10^{13}	···	···	320	FR,DI,MS
Tanzawa and Gardiner (1979)	4.0×10^{13}	0	0	1700–3400	ST,TH,LS,UE
Laufer and Bass (1979)	1.9×10^{13}	···	···	298	SR,PH,GC
Frank and Just (1980)	3.5×10^{13}	0	0	2300–2700	ST,TH,RA
Recommended ($\log F = \pm 0.30$)	3.5×10^{13}	0	0	300–2500	···
	$C_2H_2 + M \rightarrow C_2H + H + M\ (k_1,\ M = Ar)$				
Jachimowski (1977)	1.0×10^{14}	0	477	1815–2365	ST,TH,IE
Tanzawa and Gardiner (1979)	4.2×10^{16}	0	448	1700–3400	ST,TH,IE,LS
Bar-Nun and Dove (1980)	2.0×10^{10}	···	···	2650	ST,TH,MS
Frank and Just (1980)	3.6×10^{16}	0	446	1850–3000	ST,TH,RA
Recommended ($\log F = \pm 0.50$)	4.0×10^{16}	0	447	1500–3500	···

6.6. Reactions of C_2H

C_2H is probably the main precursor of C_4H_2 in fuel-rich flames, formed by the reaction $C_2H + C_4H_2 \rightarrow C_4H_2 + H$ mentioned above (Warnatz *et al.*, 1982). Unfortunately, there is lack of information on the processes competing with the reaction forming C_4H_2, some of which may play important parts in soot formation.

$$C_2H + O \rightarrow \text{products}$$

This is a very fast reaction, but unimportant in flames because of the low simultaneous C_2H and O-atom concentrations. The products of this reaction are unknown.

$$C_2H + H_2 \rightarrow C_2H_2 + H$$

The reverse of this reaction has been discussed before. The recommended rate coefficient takes into account the measurements available in the literature. Unfortunately, the reverse reaction is not known well enough to permit including rate coefficients determined from rate measurements of the reverse reaction.

$$C_2H + O_2 \rightarrow \text{products}$$

This reaction consumes C_2H in lean and moderately rich flames, preventing C_4H_2 formation under these conditions (Warnatz *et al.*, 1982). Unfortunately, the room-temperature rate coefficients are not in agreement and high-temperature rate coefficients not available. The recommended value is based on flame measurements of C_4H_2 formation rates (Warnatz *et al.*, 1982).

6.7. Reactions of CH_3CHO and CH_3CO

These reactions are relatively unimportant in flame propagation, since acetaldehyde is only a minor by-product formed by reactions of alkyl radicals with O atoms and of OH radicals with C_2H_4. Apart from the high-temperature range of the OH reaction, there are sufficient data available in the literature to specify rate coefficients for the reactions of CH_3CHO with H, O, and OH, and CH_3 and for its thermal decomposition (Figs. 73 to 75). The acetyl radical CH_3CO is the product of H, O, OH, and CH_3 attack on acetaldehyde. Because of its instability it mainly undergoes thermal decomposition (Figs. 76 and 77), though the reactions with H and O atoms also seem to be fast enough to contribute significantly.

6.8. Reactions of CH_2CO and CHCO

Ketene may be one of the products of the reactions of OH with C_2H_2 and of O with C_2H_3, but it is not believed to play an important part in combustion as far as is known. Rate data are available (Fig. 78) for $CH_2CO + H \rightarrow$ products

Table 19. Rate data on C_2H reactions.

Reference	$A/(cm^3/mol)^{n-1}\,s^{-1}$	b	E/kJ	T/K	Method
	$C_2H + O \rightarrow$ products				
Shaub and Bauer (1978)	1.4×10^{13}	0	13.2	1400–2600	ST,TH,GC
Homann and Schweinfurth (1981)	1.0×10^{3}	...	...	298	FR,DI,MS
Recommended ($\log F = \pm 0.30$)	1.0×10^{13}	0	0	300–2500	...
	$C_2H + H_2 \rightarrow C_2H_2 + H$				
Lange and Wagner (1975)	1.0×10^{11}	...	...	320	FR,DI,MS
Laufer and Bass (1979)	9.1×10^{10}	...	...	298	SR,PH,GC
Tanzawa and Gardiner (1980a)	2.0×10^{12}	...	...	2000	Review
Koike and Morinaga (1981)	7.5×10^{12}	0	0	1800–2500	ST,TH,UA
Recommended ($\log F = \pm 1.0$)	1.5×10^{13}	0	13.0	300–3000	...
	$C_2H + O_2 \rightarrow$ products				
Lange and Wagner (1975)	3.3×10^{12}	...	...	320	FR,DI,MS
Renlund *et al.* (1981)	1.3×10^{13}	...	...	300	SR,FP,CL
Recommended ($\log F = \pm 0.50$)	5.0×10^{13}	0	6.3	>300	...

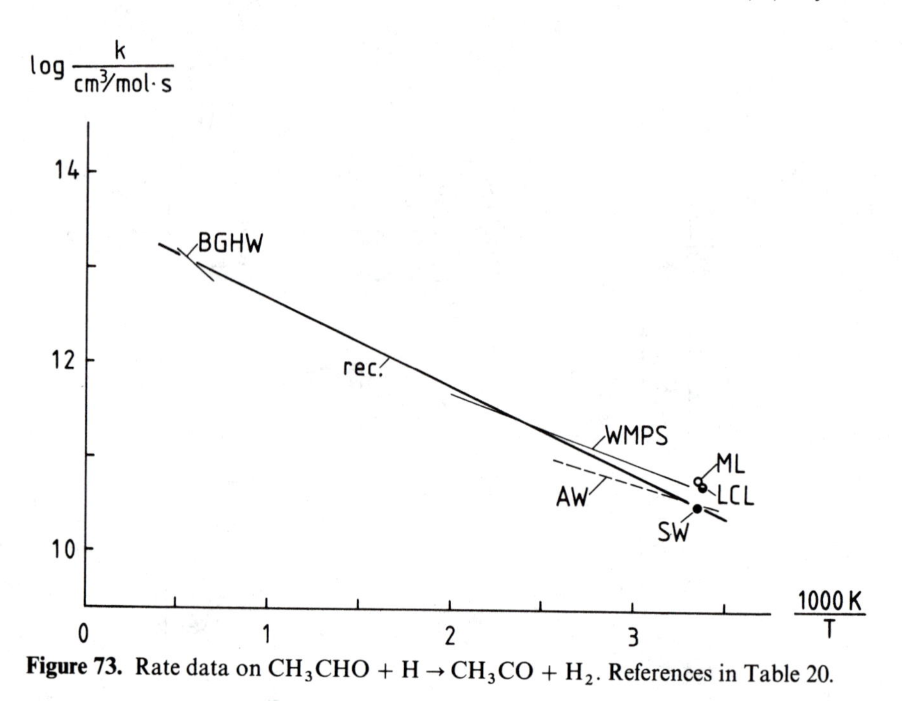

Figure 73. Rate data on $CH_3CHO + H \rightarrow CH_3CO + H_2$. References in Table 20.

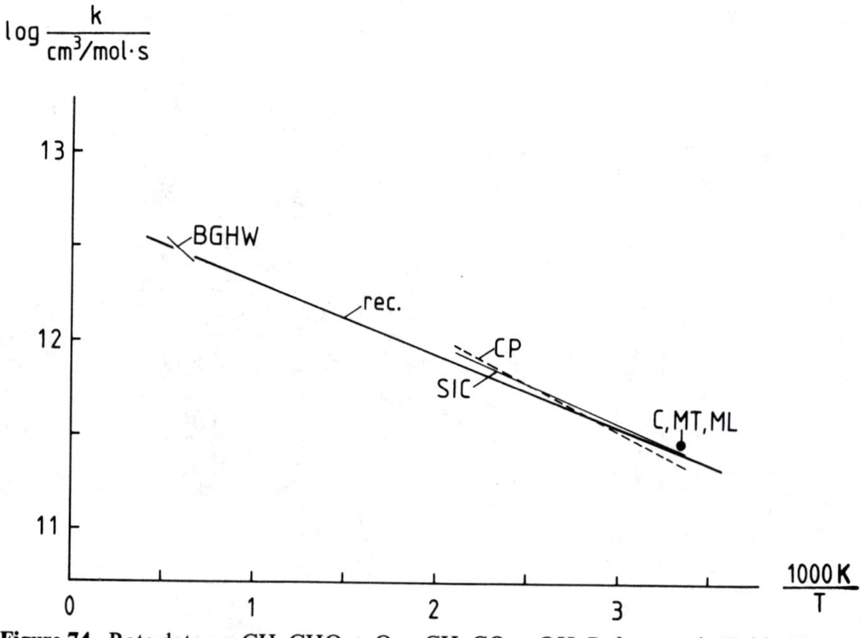

Figure 74. Rate data on $CH_3CHO + O \rightarrow CH_3CO + OH$. References in Table 20.

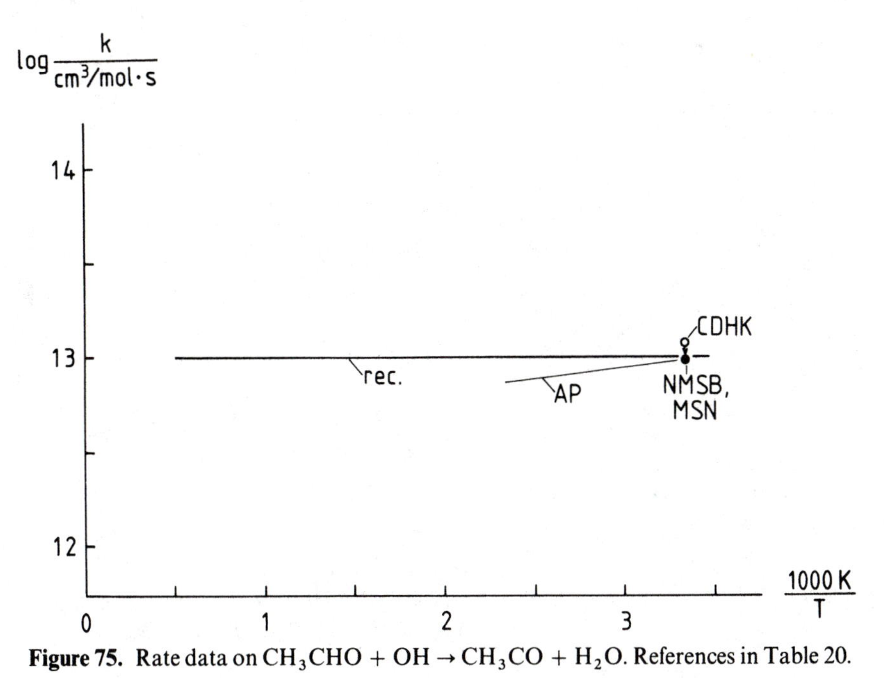

Figure 75. Rate data on $CH_3CHO + OH \rightarrow CH_3CO + H_2O$. References in Table 20.

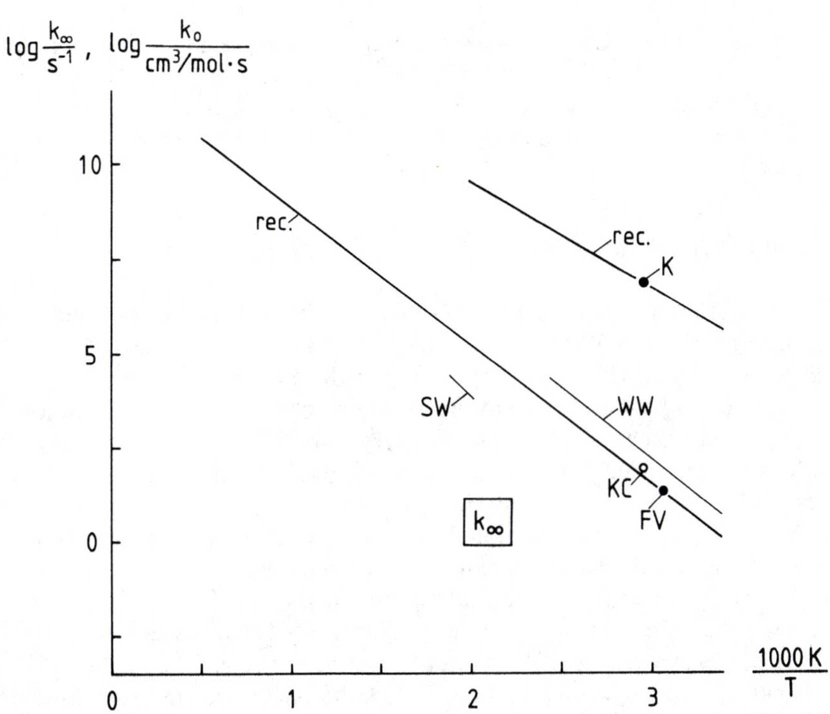

Figure 76. Rate data on $CH_3CO(+M) \rightarrow CH_3 + CO(+M)$. References in Table 21. Fall-off curves are given in Fig. 77.

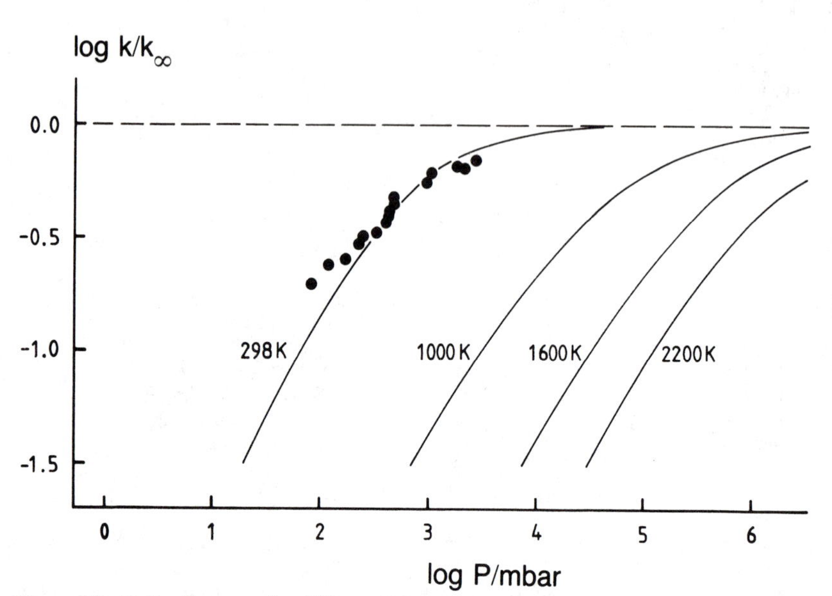

Figure 77. Fall-off curves for $CH_3 + CO \rightarrow CH_3CO$ at different temperatures. Lines: calculated as described in Section 1.2. Points: experimental values of Watkins and Ward (1974) for $T = 273$ K.

(probably $CH_3 + CO$) and for the thermal decomposition of ketene. On the other hand, there is little information on the reactions with O and OH. CHCO is formed in the reaction of O atoms with acetylene (see above). Rate data on its reactions are scarce and unreliable due to their indirect character.

7. Reactions of C_3- and C_4-hydrocarbons

From simulations of a number of laminar premixed flames of C_3- and C_4-hydrocarbons, it is known (Warnatz, 1981) that after the initial H-atom abstraction by H, O, or OH, the destiny of the alkyl radicals formed is the central point of interest. Due to the instability of C_3- and C_4-alkyl radicals with respect to thermal decomposition (see below), the competing reactions with O or O_2 and recombination or disproportionation reactions are unimportant in flame propagation.

Roughly, the current state of understanding of the combustion mechanism of these hydrocarbons can be reduced to the question of how fast CH_3 and C_2H_5 are formed in the decomposition of these radicals. Since C_2H_5 is more reactive than CH_3, flames of fuels producing C_2H_5 propagate faster than those of fuels producing CH_3 instead, as can be seen, e.g., by comparison of ethane– and methane–air flames around the stoichiometric composition (Warnatz, 1981). Similarly, in ignition processes the relative rates of formation of CH_3 and C_2H_5 play an important part, since C_2H_5 leads rapidly to chain

Table 20. Rate data on CH_3CHO reactions.

Reference	$A/(cm^3/mol)^{n-1}\ s^{-1}$	b	E/kJ	T/K	Method
	$CH_3CHO + H \rightarrow CH_3CO + H_2$				
Lambert *et al.* (1967)	5.2×10^{10}	...	...	297	FR,DI,MS
Aders and Wagner (1973)	2.6×10^{12}	0	10.9	295–389	FR,DI,MS,ES
Slemr and Warneck (1975)	3.2×10^{10}	...	...	298	FR,DI,MS
Whytock *et al.* (1976b)	1.3×10^{13}	0	13.8	298–500	SR,FP,RF
Beeley *et al.* (1977)	8.7×10^{13}	0	29.0	1550–1850	ST,TH,IE,UE
Michael and Lee (1977)	5.9×10^{10}	...	...	298	FR,DI,RF
Recommended ($\log F = \pm 0.30$)	4.0×10^{13}	0	17.6	300–2000	...
	$CH_3CHO + O \rightarrow CH_3CO + OH$				
Cvetanovic (1956)	3.0×10^{11}	...	...	300	SR,HG,GC,MS
Cadle and Powers (1967)	1.1×10^{13}	0	9.6	299–476	FR,DI,MS,GC
Morris *et al.* (1971)	9.0×10^{12}	...	...	300	FR,DI,MS
Mack and Thrush (1974a)	2.9×10^{11}	...	...	300	FR,DI,GC,ES
Beeley *et al.* (1977)	1.0×10^{13}	0	16.7	1550–1850	ST,TH,IE,UE
Michael and Lee (1977)	3.0×10^{11}	...	...	298	FR,DI,RF
Singleton *et al.* (1977)	7.2×10^{12}	0	8.2	298–472	SR,HG,CL
Recommended ($\log F = \pm 0.30$)	5.0×10^{12}	0	7.5	300–2000	...
	$CH_3CHO + OH \rightarrow CH_3CO + H_2O$				
Morris *et al.* (1971)	9.5×10^{12}	...	...	300	FR,DI,MS
Cox *et al.* (1976b)	1.2×10^{13}	...	...	298	FR,PH,GC,CL
Niki *et al.* (1978)	9.6×10^{12}	...	...	298	SR,PH,IE

Table 20 (*continued*).

Reference	$A/(cm^3/mol)^{n-1} s^{-1}$	b	E/kJ	T/K	Method
Atkinson and Pitts (1978)	4.1×10^{12}	0	−2.1	299–426	SR,FP,RF
Recommended ($\log F = \pm 0.40$)	1.0×10^{13}	0	0	300–2000	...
	$CH_3CHO + CH_3 \rightarrow CH_3CO + CH_4$				
Kerr and Parsonage (1976)	8.5×10^{10}	0	25.1	300–525	Review
Recommended ($\log F = \pm 0.30$)	8.5×10^{10}	0	25.1	300–525	...
	$CH_3CHO \rightarrow CH_3 + CHO$ (k_∞)				
Liu and Laidler (1968)	2.0×10^{15}	0	331.0	753–813	SR,TH,GC
Colket *et al.* (1975)	7.1×10^{15}	0	342.0	1000–1200	FR,TH,GC
Recommended ($\log F = \pm 0.50$)	2.0×10^{15}	0	331.0	500–2000	...

Table 21. Rate data on CH_3CO reactions.

Reference	$A/(cm^3/mol)^{n-1}\ s^{-1}$	b	E/kJ	T/K	Method
	$CH_3CO + H \rightarrow$ products				
Recommended ($\log F = \pm 1.0$)	2.0×10^{13}	0	0	>300	Estimate
	$CH_3CO + O \rightarrow CH_3 + CO_2$				
Mack and Thrush (1974a)	6.0×10^{12}	...	...	573	FR,DI,GC,ES
Recommended ($\log F = \pm 0.50$)	2.0×10^{13}	0	0	>300	...
	$CH_3CO + M \rightarrow CH_3 + CO + M$ (k_1, M undefined)				
O'Neal and Benson (1962)	4.7×10^{11}	0	52.4	399–568	SR,PH,GC
Kerr and Calvert (1965)	9.4×10^{6}	...	...	338	SR,PH,GC
Recommended ($\log F = \pm 0.50$)	1.2×10^{15}	0	52.4	300–500	...
	$CH_3CO \rightarrow CH_3 + CO$ (k_∞)				
O'Neal and Benson (1962)	2.0×10^{10}	0	62.8	399–568	SR,PH,GC
Kerr and Calvert (1965)	9.5×10^{1}	...	...	338	SR,PH,GC
Frey and Vinall (1973)	1.9×10^{1}	...	...	326	SR,PH,GC
Szirivicza and Walsh (1974)	2.0×10^{13}	0	91.0	498–525	SR,TH,GC
Watkins and Ward (1974)	3.2×10^{13}	0	72.0	260–413	SR,PH,GC
Recommended ($\log F = \pm 0.50$)	3.0×10^{12}	0	70.0	300–500	...

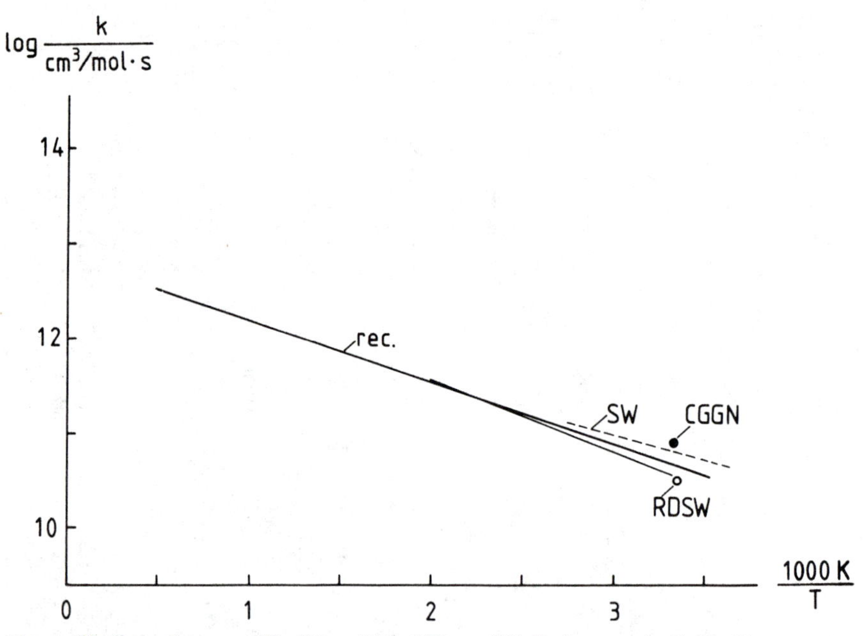

Figure 78. Rate data on $CH_2CO + H \rightarrow CH_3 + CO$. References in Table 22.

branching by thermal decomposition into C_2H_4 and H atoms whereas CH_3 is relatively unreactive (Gardiner *et al.*, 1981). The relative rates of formation of CH_3 and C_2H_5 are influenced by three processes.

(1) Attack of H, O, and OH (HO_2 attack is negligible) on alkanes and the resulting distribution of the different alkyl radical isomers formed;
(2) Thermal decomposition of the different alkyl radical isomers to form smaller alkyl radicals, or H atoms, by alkene elimination;
(3) Reactions of propene and butene, which are formed by alkyl radical thermal decomposition.

Finally, the decomposition reactions of C_3H_4 and C_4H_2 are discussed in this section because of the importance of these species in rich hydrocarbon flames (Warnatz *et al.*, 1982).

7.1. Thermal decomposition and attack of H,O, OH, and HO_2 on propane and butane

As in the combustion of CH_4 and C_2H_6, the abstraction reactions of H, OH, and (to a lesser extent) O with the hydrocarbon fuel provide the main fuel consumption in flame propagation, whereas the reaction with HO_2 is unimportant due to the low HO_2 concentration and the small rate coefficients. In contrast to CH_4 and C_2H_6 combustion, for higher hydrocarbons not only the global rate coefficients for the attack of H, O, and OH are of interest but

Table 22. Rate data on CH_2CO reactions.

Reference	$A/(cm^3/mol)^{n-1}\ s^{-1}$	b	E/kJ	T/K	Method
	$CH_2CO + H \rightarrow (CH_3 + CO\ ?)$				
Carr *et al.* (1968)	7.8×10^{10}	...	...	298	FR,DI,MS
Ridley *et al.* (1972)	3.2×10^{10}	...	...	297	SR,PH,RF
Slemr and Warneck (1975)	3.6×10^{12}	0	9.8	218–363	FR,DI,MS
Michael *et al.* (1979)	1.1×10^{13}	0	14.4	298–500	SR,FP,RF
Recommended ($\log F = \pm 0.30$)	7.0×10^{12}	0	12.6	300–500	...
	$CH_2CO + 0 \rightarrow (CHO + CHO\ ?)$				
Carr *et al.* (1968)	5.3×10^{11}	...	...	298	FR,DI,MS
Mack and Thrush (1974b)	3.4×10^{11}	...	...	293	FR,DI,GC,ES
Recommended ($\log F = \pm 0.50$)	2.0×10^{13}	0	9.6	300	...
	$CH_2CO + OH \rightarrow (CH_2O + CHO\ ?)$				
Vandooren and van Tiggelen (1977)	2.8×10^{13}	0	0	480–1000	FL,TH,MS
Faubel *et al.* (1977)	$>1.0 \times 10^{12}$	...	...	295	FR,DI,MS
Recommended ($\log F = \pm 0.50$)	1.0×10^{13}	0	0	300–1000	...
	$CH_2CO + M \rightarrow CH_2 + CO\ (k_1, M = Ar)$				
Wagner and Zabel (1971)	3.6×10^{15}	0	248	1300–2000	ST,TH,UA
Recommended ($\log F = \pm 0.50$)	3.6×10^{15}	0	248	1000–2000	...
	$CH_2CO \rightarrow CH_2 + CO\ (k_\infty)$				
Wagner and Zabel (1971)	3.0×10^{14}	0	297	1300–2000	ST,TH,UA
Recommended ($\log F = \pm 0.50$)	3.0×10^{14}	0	297	1000–2000	...

Table 23. Rate data on CHCO reactions.

Reference	$A/(\mathrm{cm^3/mol})^{n-1}\,\mathrm{s^{-1}}$	b	E/kJ	T/K	Method
	$CHCO + H \rightarrow CH_2 + CO$				
Roth and Löhr (1981)	3.0×10^{13}	0	0	1500–2750	ST,TH,RA
Wellmann (1981)	3.0×10^{12}	...	...	1000	FR,DI,MS
Recommended ($\log F = \pm 0.50$)	3.0×10^{13}	0	0	1000–2500	...
	$CHCO + O \rightarrow$ products				
Jones and Bayes (1973)	1.2×10^{12}	...	...	298	FR,DI,MS
Recommended ($\log F = \pm 1.0$)	1.2×10^{12}	0	0	298	...

also rate coefficients for the attack on specific C–H single bonds, because the formation rates of different alkyl isomers (e.g., n-C_3H_7 and i-C_3H_7 for n-C_4H_9 and s-C_4H_9 for n-butane) are different. Data for rate coefficients for attack on C–H single bonds can be derived from experiments with hydrocarbons with different distributions of primary, secondary, and tertiary C–H bonds; assuming additivity of these rates one can derive rate coefficients of reactions of higher hydrocarbons with H, O, OH, and HO_2. Results of this type of research are given in Table 24 for H, O, OH, and HO_2. The data for C_3-hydrocarbons are presented in Tables 25 to 28 and Figs. 79 to 92, for C_4-hydrocarbons in Tables 29 to 32 and Figs. 93 to 98.

Table 24. Arrhenius parameters for attack of H, O, OH, and HO_2 and C–H single bonds of alkanes (Baldwin *et al.*, 1977; Herron and Huie, 1969).

Reaction	$A/(\mathrm{cm^3/mol\ s})$	E/kJ
H + primary C–H	2.2×10^{13}	40.6
H + secondary C–H	5.0×10^{13}	34.9
H + tertiary C–H	8.7×10^{13}	29.3
O + primary C–H	5.0×10^{12}	5.76
O + secondary C–H	1.3×10^{13}	4.47
O + tertiary C–H	1.6×10^{13}	3.28
OH + primary C–H	6.1×10^{11}	6.90
OH + secondary C–H	1.4×10^{12}	3.60
OH + tertiary C–H	1.2×10^{12}	−0.79
HO_2 + primary C–H	1.0×10^{12}	81.2
HO_2 + secondary C–H	1.0×10^{12}	71.1
HO_2 + tertiary C–H	1.0×10^{12}	60.2

Table 25. Rate data on C_3H_8 reactions.

Reference	$A/(cm^3/mol)^{n-1}\ s^{-1}$	b	E/kJ	T/K	Method
	$C_3H_8 + H \rightarrow C_3H_7 + H_2$				
Yang (1963)	1.6×10^{12}	0.5	29.3	300–500	SR,PR,GC
Kazmi *et al.* (1963)	1.3×10^{14}	0	34.3	368–443	FR,TH,CP
Baldwin (1964)	4.5×10^{11}	...	...	793	SR,TH,P,GC
Baker *et al.* (1970b)	5.4×10^{11}	...	...	753	SR,TH,P,GC
Baldwin *et al.* (1977)	2.1×10^{14}	0	36.5	...	Review
Lede and Villermaux (1978)	6.3×10^{13}	0	32.7	281–347	FR,DI,UA
Recommended ($\log F = \pm 0.40$)	2.1×10^{14}	0	36.5	300–1000	...
	$C_3H_8 + O \rightarrow C_3H_7 + OH$				
Harker and Burton (1975)	3.9×10^{10}	...	...	329	FR,HG,UA
Huie and Herron (1975)	5.2×10^{13}	0	20.1	...	Review
Recommended ($\log F = \pm 0.50$)	5.2×10^{13}	0	20.1	300–1000	...
	$C_3H_8 + OH \rightarrow C_3H_7 + H_2O$				
Greiner (1970b)	7.2×10^{12}	0	5.6	300–500	SR,FP,UA
Bradley *et al.* (1973)	5.0×10^{11}	...	...	298	FR,DI,ES
Harker and Burton (1975)	1.2×10^{12}	...	...	329	FR,HG,UA
Gordon and Mulac (1975)	1.3×10^{12}	...	...	381	SR,PR,UA
Overend *et al.* (1975)	1.2×10^{12}	...	...	295	SR,FP,RA
Baldwin *et al.* (1977)	6.3×10^{12}	0	4.9	...	Review
Darnall *et al.* (1978)	9.6×10^{11}	...	...	300	SR,PH,GC
Recommended ($\log F = \pm 0.40$)	6.3×10^{12}	0	4.9	300–1000	...

Table 25 (*continued*).

Reference	$A/(cm^3/mol)^{n-1} s^{-1}$	b	E/kJ	T/K	Method
	$C_3H_8 + HO_2 \rightarrow C_3H_7 + H_2O_2$				
Baldwin *et al.* (1977)	6.2×10^{12}	0	74.0	...	Review
Recommended ($\log F = \pm 0.50$)	6.2×10^{12}	0	74.0	300–1000	...
	$C_3H_8 + CH_3 \rightarrow C_3H_7 + CH_4$				
Camilleri *et al.* (1975)	2.0×10^{12}	0	47.3	676–743	FR,TH,GC
Camilleri *et al.* (1975)	5.0×10^{15}	0	96.4	743–813	FR,TH,GC
Lifshitz and Frenklach (1975)	3.5×10^{12}	0	43.1	1050–1700	ST,TH,GC
Kerr and Parsonage (1976)	2.0×10^{11}	0	40.2	550–750	Review
Mintz and LeRoy (1978)	4.6×10^{13}	0	59.6	609–648	FR,TH,GC
Pratt and Rogers (1979b)	2.4×10^{11}	0	33.3	891–1079	FR,TH,GC
Durban and Marshall (1980)	2.6×10^{11}	0	40.7	323–453	SR,PH,GC
Recommended ($\log F = \pm 0.50$)	2.0×10^{12}	0	47.3	500–1000	...
	$C_3H_8 \rightarrow CH_3 + C_2H_5$ (k_∞)				
Lifshitz and Frenklach (1975)	3.2×10^{15}	0	347	1050–1700	ST,TH,GC
Bradley (1979)	2.0×10^{8}	0	157	1210–1680	ST,TH,GC
Pratt and Rogers (1979b)	5.3×10^{15}	0	350	891–1079	FR,TH,GC
Chiang and Skinner (1981)	2.5×10^{16}	0	366	1200–1450	ST,TH,RA
Recommended ($\log F = \pm 0.50$)	5.0×10^{15}	0	350	800–2000	...

Table 26. Rate data on C_3H_7 reactions.

Reference	$A/(\mathrm{cm^3/mol})^{n-1}\,\mathrm{s^{-1}}$	b	E/kJ	T/K	Method
	$i\text{-}C_3H_7 + H \rightarrow C_3H_8 (k_\infty)$				
Recommended ($\log F = \pm 0.50$)	2.0×10^{13}	0	0	300–2000	Estimate
	$C_3H_7 + O \rightarrow$ products				
Recommended ($\log F = \pm 0.30$)	4.5×10^{13}	0	0	300–2000	Estimate
	$i\text{-}C_3H_7 + O_2 \rightarrow C_3H_6 + HO_2$				
Baldwin *et al.* (1976)	1.2×10^{11}	...	...	713	SR,TH,P,GC
Recommended ($\log F = \pm 0.50$)	1.0×10^{12}	0	12.5	500–2000	...
	$n\text{-}C_3H_7 + O_2 \rightarrow C_3H_6 + HO_2$				
Baker *et al.* (1970a)	2.2×10^{10}	...	...	753	SR,TH,P,GC
Baker *et al.* (1971)	3.8×10^{10}	...	...	753	SR,TH,P,GC
Baldwin *et al.* (1971)	5.8×10^{10}	...	...	723	SR,TH,P
Baldwin *et al.* (1973)	1.8×10^{10}	...	...	723	SR,TH,GC
Recommended ($\log F = \pm 0.50$)	1.0×10^{12}	0	21.0	500–2000	...
	$i\text{-}C_3H_7 + i\text{-}C_3H_7 \rightarrow C_6H_{14} (k_\infty)$				
Hiatt and Benson (1972)	4.0×10^{11}	...	...	415	SR,TH,GC
Golden *et al.* (1973)	5.0×10^{12}	0	0	700–800	SR,TH,MS
Golden *et al.* (1974b)	3.0×10^{12}	0	0	683–808	SR,TH,MS
Parkes and Quinn (1975)	2.5×10^{12}	...	...	298	SR,PH,GC
Parkes and Quinn (1976)	4.8×10^{12}	...	...	298	SR,PH,UA
Recommended ($\log F = \pm 0.30$)	4.0×10^{12}	0	0	300–2000	...

Table 26 (*continued*).

Reference	$A/(\text{cm}^3/\text{mol})^{n-1}\,\text{s}^{-1}$	b	E/kJ	T/K	Method
	$n\text{-}C_3H_7 + n\text{-}C_3H_7 \rightarrow C_6H_{14}\ (k_\infty)$				
Whiteway and Masson (1956)	6.3×10^{12}	...	...	373	SR,PH,P
Walker (1977)	1.0×10^{13}	...	...	...	Estimate
Recommended ($\log F = \pm 0.30$)	1.0×10^{13}	0	0	300–2000	...
	$i\text{-}C_3H_7 + i\text{-}C_3H_7 \rightarrow C_3H_6 + C_3H_8$				
Kerr and Trotman-Dickenson (1959)	$k_{disp}/k_{comb} = 0.65$			293–774	SR,PH,GC
Papic and Laidler (1971)	$k_{disp}/k_{comb} = 0.63$			523–623	SR,HG,GC
Golden *et al.* (1974a)	$k_{disp}/k_{comb} = 1.16$			683–808	SR,TH,MS
Parkes and Quinn (1976)	$k_{disp}/k_{comb} = 0.65$			298	SR,PH,UE
McKay *et al.* (1977)	$k_{disp}/k_{comb} = 0.50$			518–573	SR,TH,GC
Mintz and LeRoy (1978)	$k_{disp}/k_{comb} = 0.57$			567–609	FR,TH,GC
Kirsch and Parkes (1979)	$k_{disp}/k_{comb} = 0.60$		0	302	SR,PH,GC
Recommended ($\log F = \pm 0.30$)	2.4×10^{12}	0	0	300–1000	...
	$n\text{-}C_3H_7 + i\text{-}C_3H_7 \rightarrow C_3H_6 + C_3H_8$				
Kerr and Calvert (1961)	$k_{disp}/k_{comb} = 0.16$			297–403	SR,PH,GC
Papic and Laidler (1971)	$k_{disp}/k_{comb} = 0.16$			523–623	SR,HG,GC
Mintz and LeRoy (1978)	$k_{disp}/k_{comb} = 0.14$			609–648	FR,TH,GC
Recommended ($\log F = \pm 0.30$)	1.6×10^{12}	0	0	300–1000	...
	$i\text{-}C_3H_7 \rightarrow C_3H_6 + H\ (k_\infty)$				
Kerr and Trotman-Dickenson (1959)	6.3×10^{13}	0	154	650–800	SR,PH,GC
Papic and Laidler (1971)	2.0×10^{14}	0	162	523–623	SR,HG,GC
Camilleri *et al.* (1975)	2.5×10^{13}	0	171	676–813	FR,TH,GC
Recommended ($\log F = \pm 0.50$)	2.0×10^{14}	0	162	500–1000	...

	$n\text{-}C_3H_7 \rightarrow CH_3 + C_2H_4\ (k_\infty)$				
Kerr and Trotman-Dickenson (1959)	4.0×10^{13}	0	123	650–800	SR,PH,GC
Calvert and Sleppy (1959)	2.8×10^{15}	0	146	471–549	SR,PH,GC,MS
Jackson and McNesby (1961)	8.0×10^{13}	0	130	569–693	SR,PH,MS
Kerr and Calvert (1961)	7.3×10^{14}	0	144	497–564	SR,PH,GC
Back and Takamuku (1964)	7.0×10^{8}	0	85.8	573–673	SR,HG,MS
Lin and Laidler (1966)	3.5×10^{13}	0	131	533–573	SR,TH,P,GC
Papic and Laidler (1971)	2.5×10^{14}	0	136	523–623	SR,HG,GC
Camilleri *et al.* (1975)	6.3×10^{12}	0	136	676–813	FR,TH,GC
Mintz and LeRoy (1978)	5.0×10^{12}	0	117	609–648	FR,TH,GC
Recommended ($\log F = \pm 0.50$)	3.0×10^{14}	0	139	500–1000	...
	$n\text{-}C_3H_7 \rightarrow C_3H_6 + H\ (k_\infty)$				
Kerr and Trotman-Dickenson (1959)	6.3×10^{13}	0	154	650–800	SR,PH,GC
Jackson and McNesby (1961)	1.2×10^{14}	0	155	569–693	SR,PH,MS
Back and Takamuku (1964)	3.1×10^{10}	0	113	573–673	SR,HG,MS
Mintz and LeRoy (1978)	4.1×10^{11}	0	120	567–609	FR,TH,GC
Recommended ($\log F = \pm 0.50$)	1.0×10^{14}	0	156	500–1000	...

Table 27. Rate data on C_3H_6 reactions.

Reference	$A/(cm^3/mol)^{n-1}\ s^{-1}$	b	E/kJ	T/K	Method
	$C_3H_6 + H \rightarrow i\text{-}C_3H_7\ (k_\infty)$				
Dodonov *et al.* (1969)	7.0×10^{11}	...	...	298	FR,DI,MS
Kerr and Parsonage (1972)	7.2×10^{12}	0	5.0	300–500	Review
Kurylo *et al.* (1971)	6.1×10^{12}	0	5.1	177–473	SR,FP,RF
Cowfer *et al.* (1971)	5.4×10^{11}	...	...	298	FR,DI,MS,RA
Wagner and Zellner (1972c)	5.4×10^{12}	0	5.2	195–390	FR,DI,MS,ES
Camilleri *et al.* (1975)	6.3×10^{12}	0	7.0	708–805	FR,TH,GC
Ishikawa *et al.* (1978)	9.0×10^{11}	...	...	298	SR,PR,RA
Yoichi *et al.* (1978)	9.0×10^{11}	...	...	298	SR,PR,RA
Oka and Cvetanovic (1979)	7.5×10^{11}	...	...	298	FR,HG,CL
Recommended ($\log F = \pm 0.30$)	4.0×10^{12}	0	4.0	300–1500	...
	$C_3H_6 + H \rightarrow n\text{-}C_3H_7\ (k_\infty)$				
Falconer *et al.* (1963)	6.0×10^{10}	...	...	298	SR,HG,GC
Lexton *et al.* (1971)	3.0×10^{10}	...	...	290	FR,DI,GC
Kerr and Parsonage (1972)	7.2×10^{12}	0	12.1	300–500	Review
Wagner and Zellner (1972c)	4.4×10^{12}	0	11.5	195–390	FR,DI,MS,ES
Yano (1977)	1.8×10^{12}	...	...	1200–1400	ST,TH,GC,MS
Recommended ($\log F = \pm 0.30$)	4.0×10^{12}	0	11.0	300–1500	...
	$C_3H_6 + O \rightarrow$ products (k_∞)				
Stuhl and Niki (1971)	2.2×10^{12}	...	...	300	SR,FP,CL
Kurylo (1972b)	2.5×10^{12}	0	0.3	201–424	SR,FP,FR
Gaedtke *et al.* (1973)	2.2×10^{12}	...	...	300	SR,PH,UA,GC
Atkinson and Pitts (1974b)	2.1×10^{12}	0	0	300–392	SR,HG,CL
Atkinson and Pitts (1974a)	2.0×10^{12}	...	...	300	SR,HG,CL

Furuyama *et al.* (1974)	2.0×10^{12}	...	...	299	FR,HG,CL
Singleton and Cvetanovic (1976)	7.6×10^{12}	0	3.0	298–480	SR,HG,GC
Atkinson and Pitts (1977a)	6.3×10^{12}	0	2.2	298–349	SR,FP,CL
Michael and Lee (1977)	2.4×10^{12}	...	...	298	FR,DI,RF
Sugawara *et al.* (1980)	2.8×10^{12}	...	...	296	SR,PR,RA
Recommended ($\log F = \pm 0.30$)	5.0×10^{12}	0	1.9	300–500	...
	$C_3H_6 + OH \rightarrow$ products				
Morris *et al.* (1971)	1.0×10^{13}	...	...	300	FR,DI,MS
Bradley *et al.* (1973)	3.0×10^{12}	...	...	298	FR,DI,ES
Stuhl (1973a)	8.7×10^{12}	...	...	298	SR,FP,RF
Simonaitis and Heicklen (1973)	8.2×10^{12}	0	0.4	373–473	SR,PH,GC
Gordon and Mulac (1975)	4.5×10^{14}	0	12.6	381–416	SR,PR,UA
Pastrana and Carr (1975)	3.0×10^{12}	...	...	300	FR,DI,UA
Atkinson and Pitts (1975)	2.5×10^{12}	0	−4.5	297–425	SR,FP,RF
Ravishankara *et al.* (1978)	1.6×10^{13}	...	...	298	SR,FP,RF
Nip and Paraskevopoulos (1979)	1.5×10^{13}	...	...	297	SR,FP,GC,RA
Recommended ($\log F = \pm 0.30$)	1.0×10^{13}	0	0	300–500	...
	$C_3H_6 + CH_3 \rightarrow C_3H_5 + CH_4$				
Kerr and Parsonage (1972)	1.4×10^{11}	0	36.8	350–600	Review
Recommended ($\log F = \pm 0.30$)	1.4×10^{11}	0	36.8	350–600	...

Table 28. Rate data on C_3H_4 reactions.

Reference	$A/(cm^3/mol)^{n-1}\ s^{-1}$	b	E/kJ	T/K	Method
	$CH_3CCH + H \rightarrow C_3H_5\ (k_\infty)$				
Brown and Thrush (1967)	2.4×10^{11}	...	...	298	FR,DI,ES
Wagner and Zellner (1972a)	1.2×10^{13}	0	9.6	195–503	FR,DI,ES,MS
Whytock *et al.* (1976a)	3.6×10^{13}	0	10.3	215–460	SR,FP,RF
Michael and Lee (1977)	3.8×10^{11}	...	...	298	FR,DI,RF
Recommended ($\log F = \pm 0.30$)	2.0×10^{13}	0	10.0	300–500	...
	$CH_2 = C = CH_2 + H \rightarrow C_3H_5\ (k_\infty)$				
Wagner and Zellner (1972b)	1.2×10^{13}	0	8.8	273–470	FR,DI,ES,MS
Recommended ($\log F = \pm -0.40$)	1.2×10^{13}	0	8.8	300–500	...
	$CH_3CCH + O \rightarrow$ products				
Brown and Thrush (1967)	4.0×10^{11}	...	...	298	FR,DI,ES
Herbrechtsmeier and Wagner (1974)	1.0×10^{13}	0	8.4	290–360	FR,DI,MS
Arrington and Cox (1975)	1.4×10^{13}	0	8.2	298–600	FR,DI,CL
Atkinson and Pitts (1977b)	1.2×10^{13}	0	7.3	297–439	SR,FP,CL
Wellmann (1981)	1.5×10^{13}	0	8.8	295–1333	FR,DI,MS
Recommended ($\log F = \pm 0.30$)	1.5×10^{13}	0	8.8	300–2000	...
	$CH_3CCH + OH \rightarrow$ products				
Bradley *et al.* (1973)	5.7×10^{11}	...	...	298	FR,DI,ES
Recommended ($\log F = \pm 0.30$)	5.0×10^{12}	0	5.4	>298	...
	$CH_2 = C = CH_2 + OH \rightarrow$ products				
Bradley *et al.* (1973)	2.7×10^{11}	...	...	298	FR,DI,ES
Recommended ($\log F = \pm 0.30$)	2.7×10^{11}	...	...	298	...

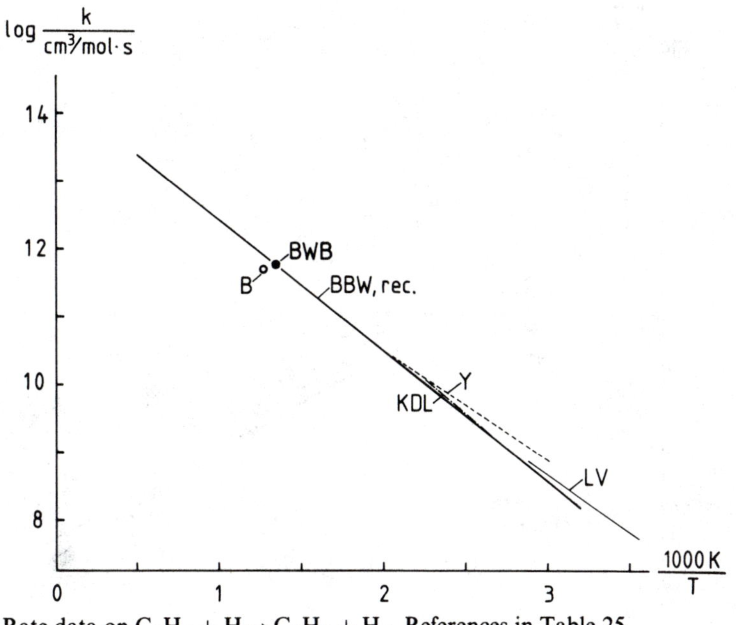

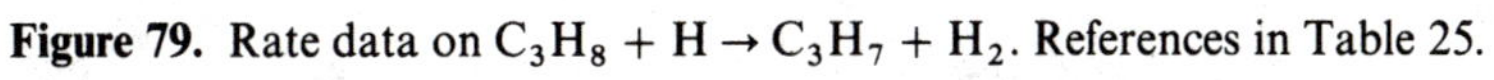
Figure 79. Rate data on $C_3H_8 + H \rightarrow C_3H_7 + H_2$. References in Table 25.

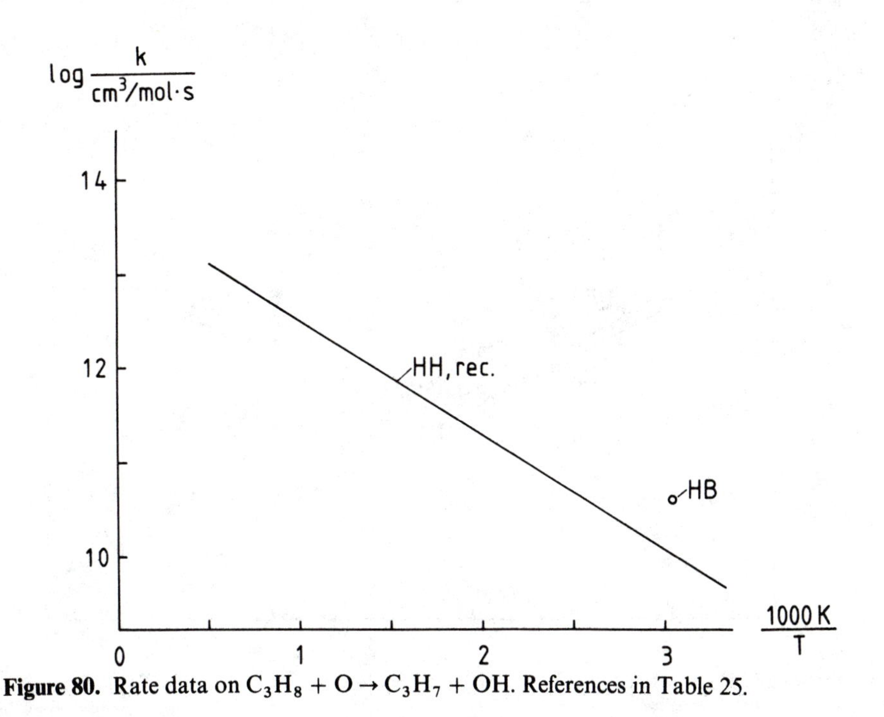

Figure 80. Rate data on $C_3H_8 + O \rightarrow C_3H_7 + OH$. References in Table 25.

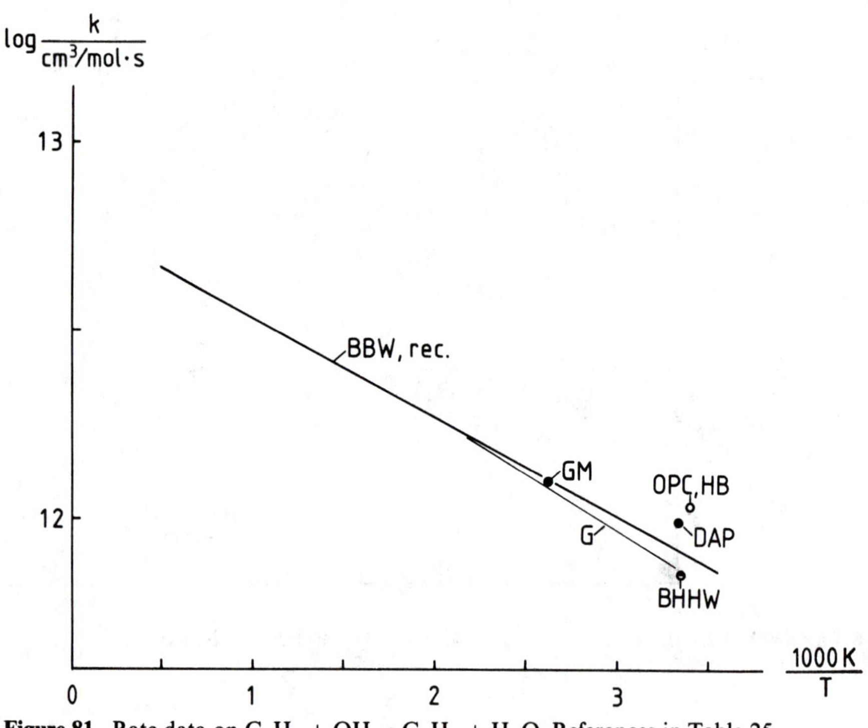

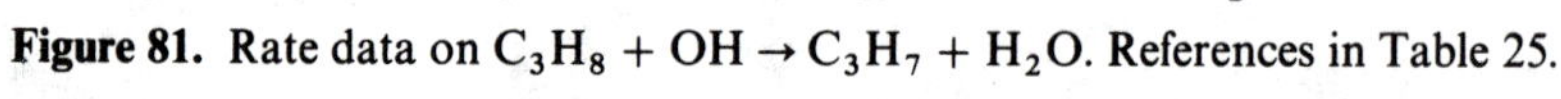
Figure 81. Rate data on $C_3H_8 + OH \rightarrow C_3H_7 + H_2O$. References in Table 25.

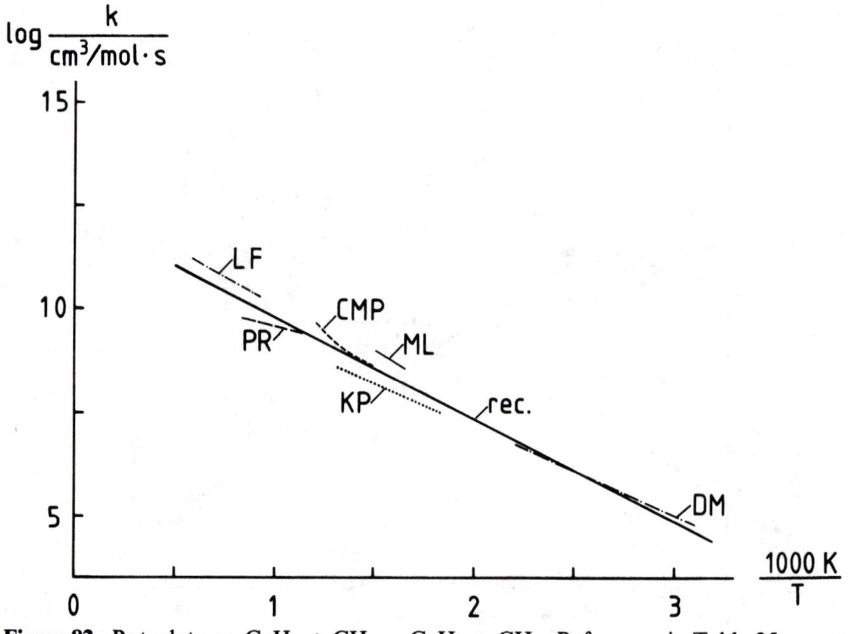

Figure 82. Rate data on $C_3H_8 + CH_3 \rightarrow C_3H_7 + CH_4$. References in Table 25.

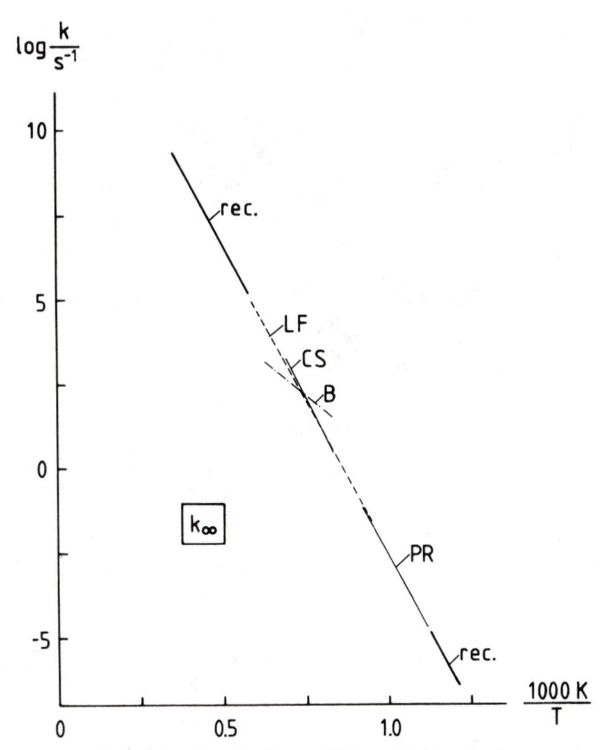

Figure 83. High-pressure rate coefficient k_∞ for $C_3H_8 \rightarrow CH_3 + C_2H_5$. References in Table 25.

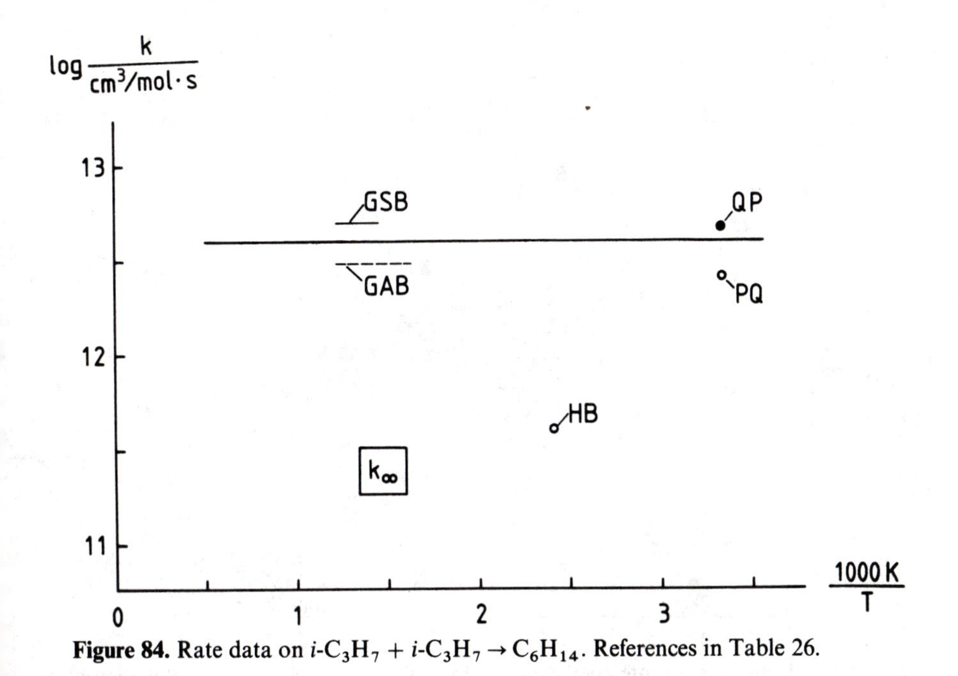

Figure 84. Rate data on $i\text{-}C_3H_7 + i\text{-}C_3H_7 \rightarrow C_6H_{14}$. References in Table 26.

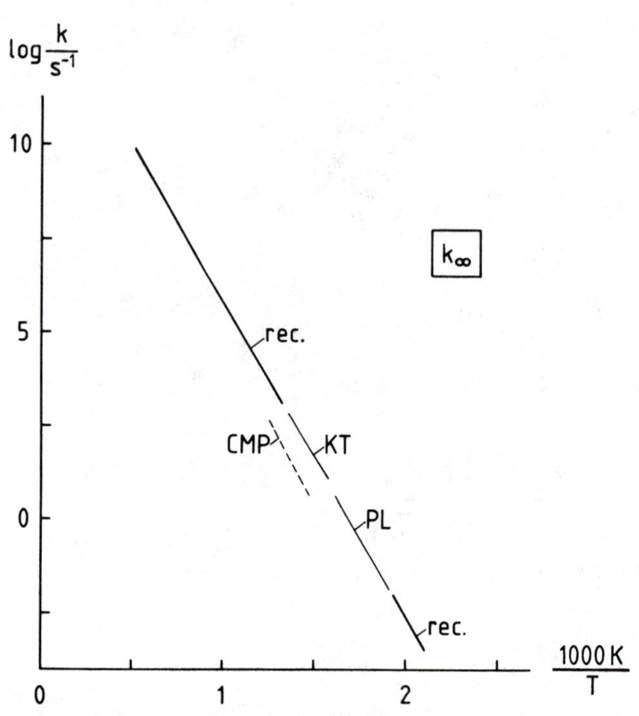

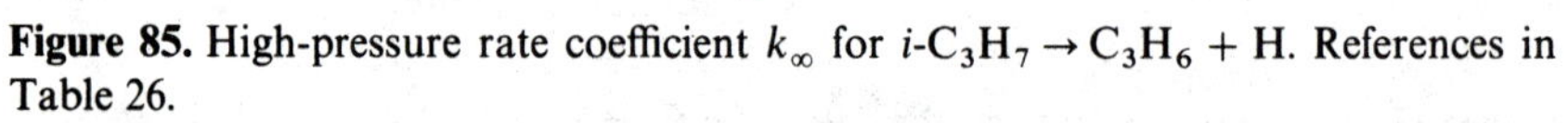

Figure 85. High-pressure rate coefficient k_∞ for $i\text{-}C_3H_7 \rightarrow C_3H_6 + H$. References in Table 26.

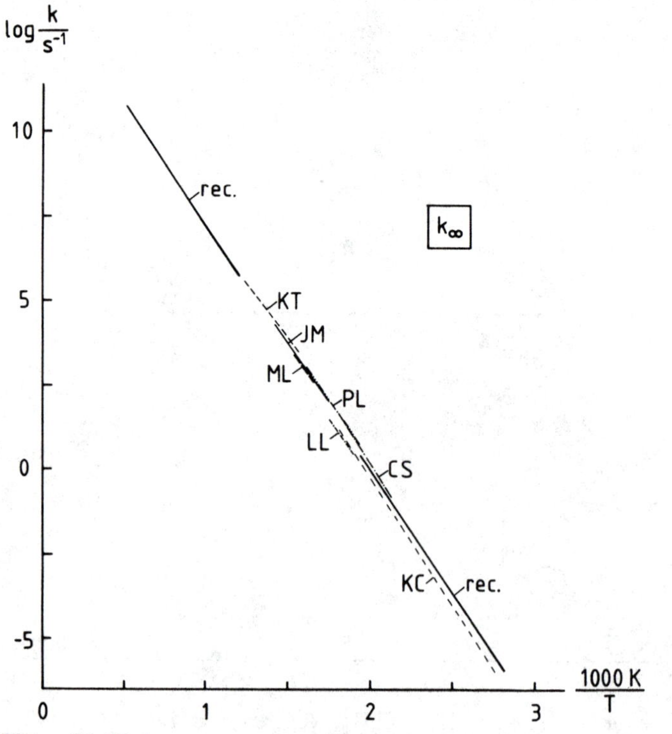

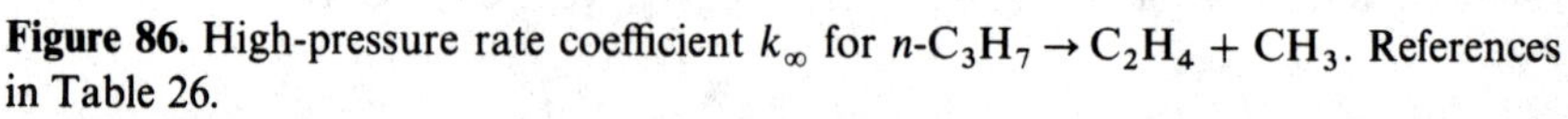

Figure 86. High-pressure rate coefficient k_∞ for $n\text{-}C_3H_7 \rightarrow C_2H_4 + CH_3$. References in Table 26.

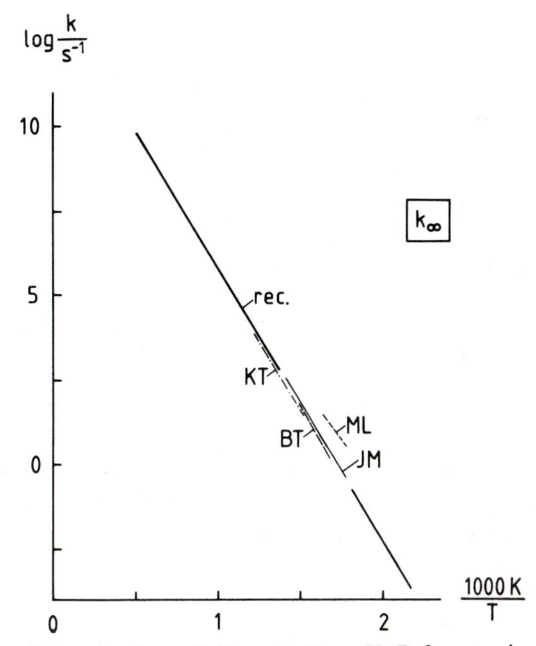

Figure 87. High-pressure rate coefficient k_∞ for n-$C_3H_7 \rightarrow C_3H_6 + H$. References in Table 26.

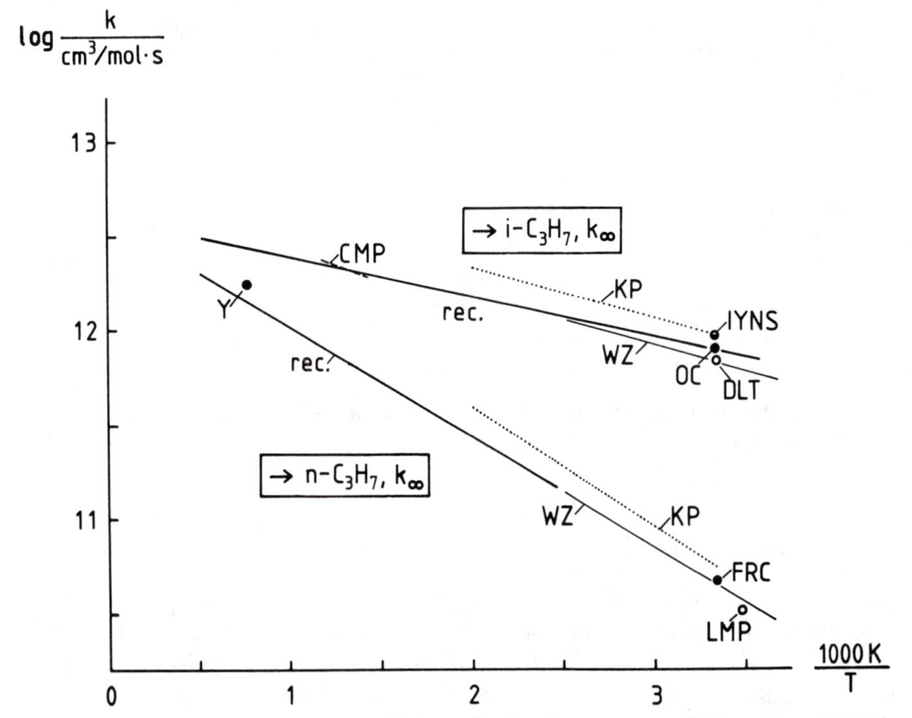

Figure 88. High-pressure rate coefficients k_∞ for the reactions $C_3H_6 + H \rightarrow i$-C_3H_7 and $C_3H_6 + H \rightarrow n$-C_3H_7. References in Table 27.

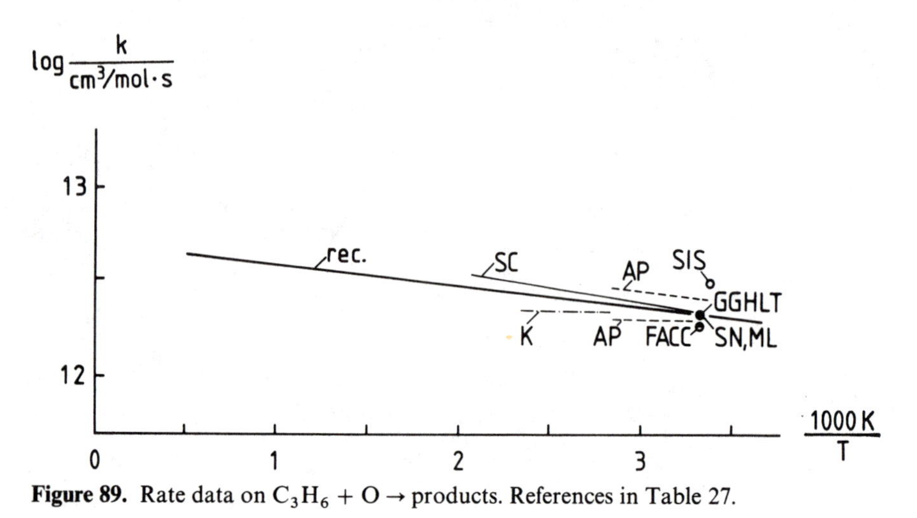

Figure 89. Rate data on $C_3H_6 + O \rightarrow$ products. References in Table 27.

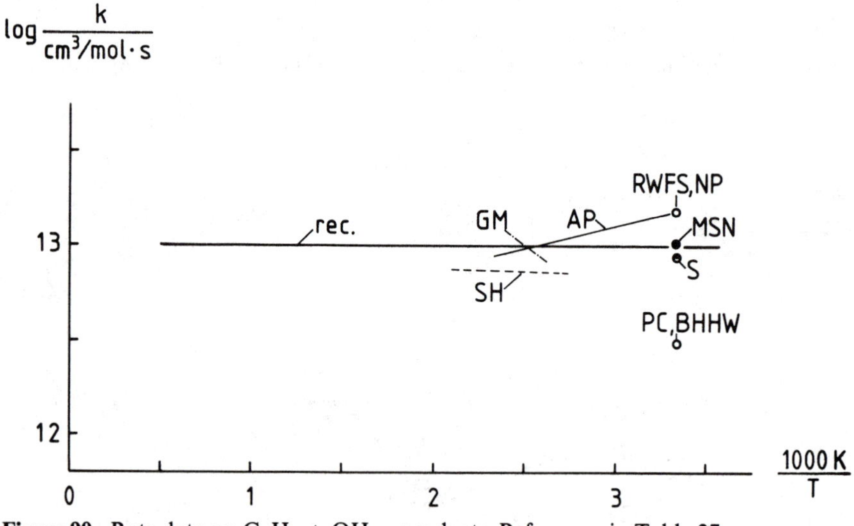

Figure 90. Rate data on $C_3H_6 + OH \rightarrow$ products. References in Table 27.

While thermal decomposition of alkanes is relatively unimportant in flame propagation, it does play a part in ignition processes. Sufficient rate data are available for C_3H_8, but not for n-C_4H_{10} and i-C_4H_{10}.

7.2. Thermal decomposition of C_3H_7 and C_4H_9

Data on these reactions (except for n-$C_3H_7 \rightarrow C_2H_4 + CH_3$) and for i-C_3H_7 recombination are scarce and show much scatter (see Tables 26 and 30 and Figs. 84 to 87). In all cases, the product distribution obeys the "one bond-removed rule" (Dryer and Glassman, 1979). At least the major reaction route is

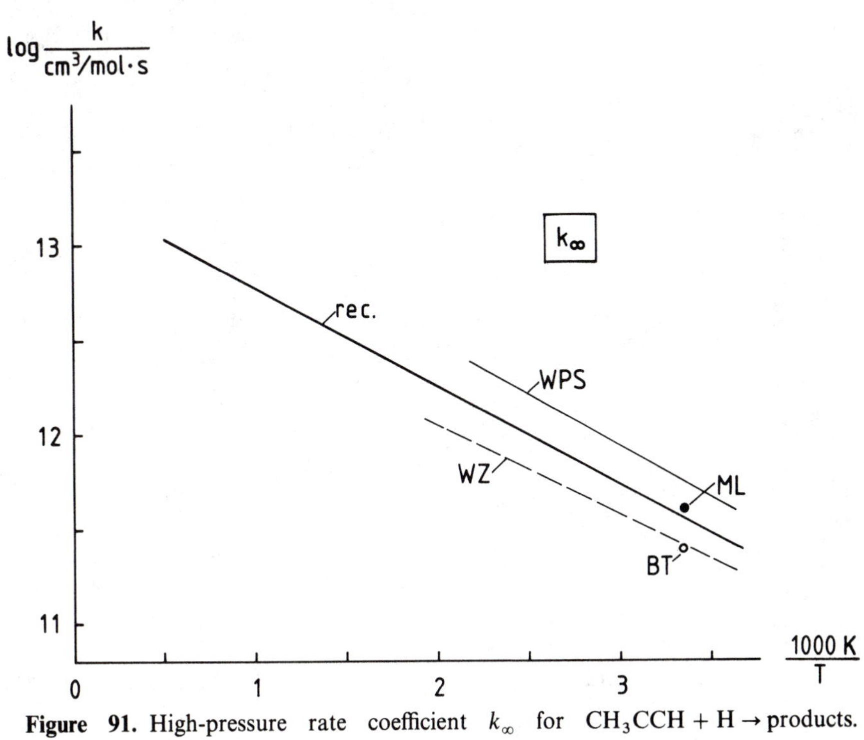

Figure 91. High-pressure rate coefficient k_∞ for $CH_3CCH + H \rightarrow$ products.
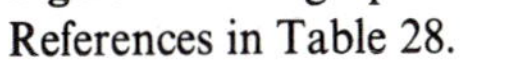
References in Table 28.

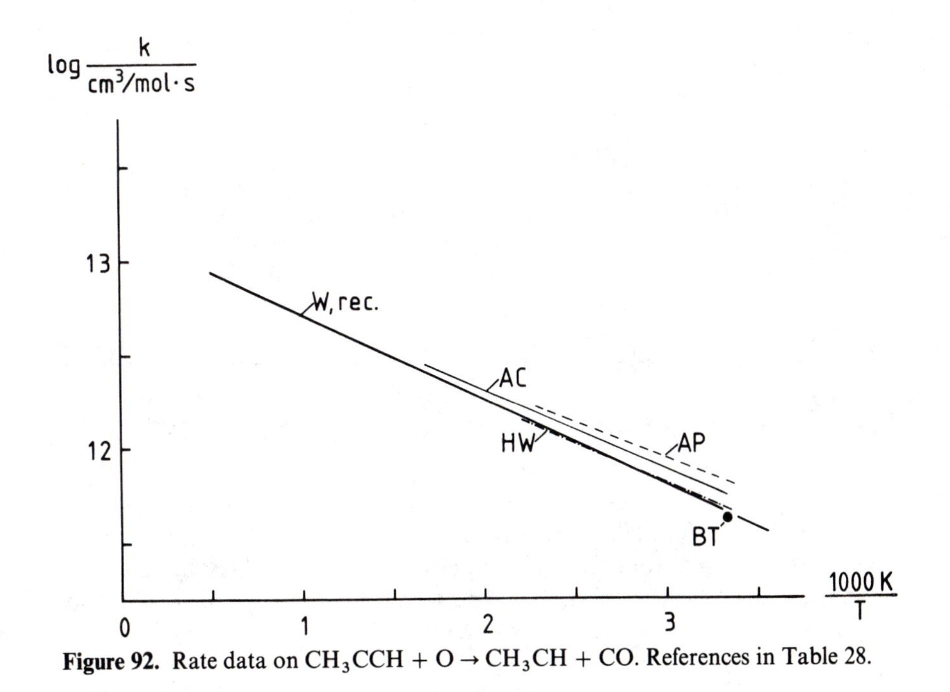

Figure 92. Rate data on $CH_3CCH + O \rightarrow CH_3CH + CO$. References in Table 28.

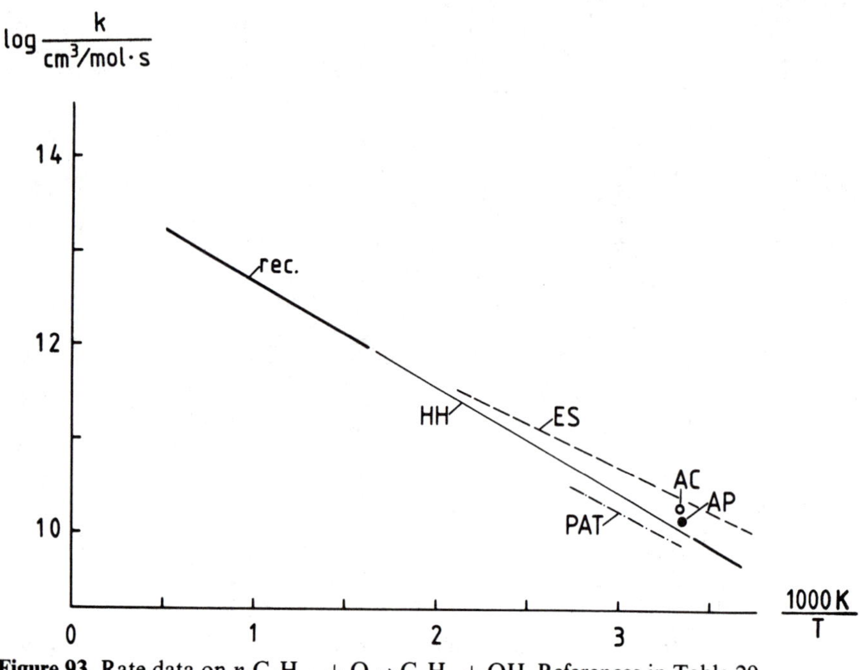

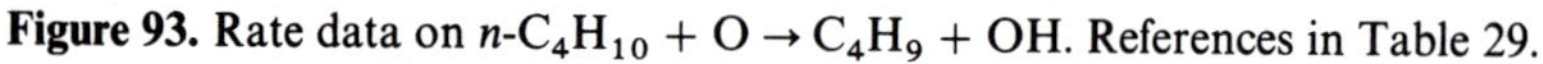
Figure 93. Rate data on $n\text{-}C_4H_{10} + O \rightarrow C_4H_9 + OH$. References in Table 29.

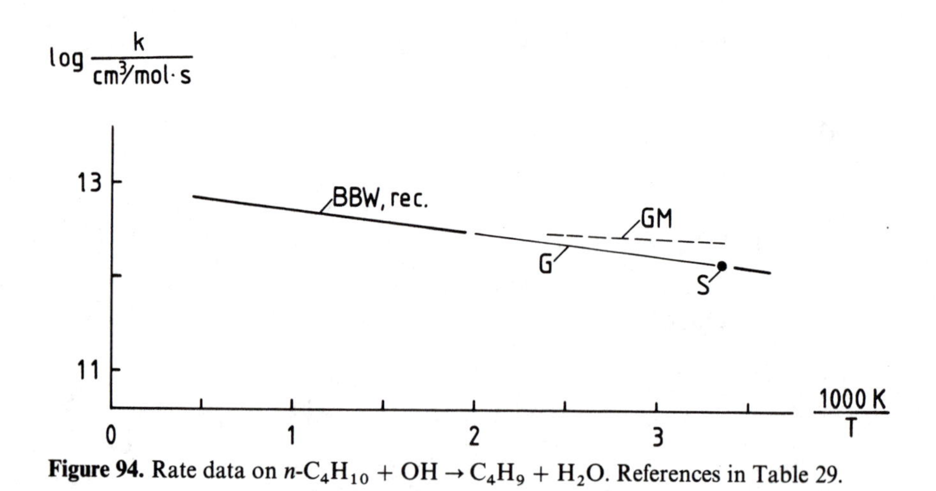

Figure 94. Rate data on $n\text{-}C_4H_{10} + OH \rightarrow C_4H_9 + H_2O$. References in Table 29.

correctly predicted in cases with more than one product channel (e.g., $n\text{-}C_3H_7$ thermal decomposition). This rule says that when a radical decomposes, a bond once removed from the unpaired-electron site is broken. In addition, when there is a choice between C–H and C–C bond rupture, the C–C bond usually is broken, due to the lower bond strength.

Table 29. Rate data on C_4H_{10} reactions.

Reference	$A/(cm^3/mol)^{n-1} s^{-1}$	b	E/kJ	T/K	Method
	$n\text{-}C_4H_{10} + H \rightarrow C_4H_9 + H_2$				
Yang (1963)	6.3×10^{11}	0.5	26.4	300–500	SR,PR,GC
Baker *et al.* (1970b)	8.8×10^{11}	...	...	753	SR,TH,P,GC
Baldwin *et al.* (1977)	3.1×10^{14}	0	35.9	...	Review
Recommended ($\log F = \pm 0.40$)	3.1×10^{14}	0	35.9	300–1000	...
	$i\text{-}C_4H_{10} + H \rightarrow C_4H_9 + H_2$				
Yang (1963)	4.0×10^{11}	0.5	19.7	300–500	SR,PR,GC
Baker *et al.* (1970b)	1.0×10^{12}	...	...	753	SR,TH,P,GC
Baldwin *et al.* (1977)	2.1×10^{14}	0	31.5	...	Review
Recommended ($\log F = \pm 0.40$)	2.1×10^{14}	0	31.5	300–1000	...
	$n\text{-}C_4H_{10} + O \rightarrow C_4H_9 + OH$				
Elias and Schiff (1960)	3.0×10^{13}	0	17.6	223–473	FR,DI,CL
Herron and Huie (1969)	6.2×10^{13}	0	21.4	250–600	FR,DI,MS
Papadopoulos *et al.* (1971)	1.7×10^{13}	0	19.0	300–365	FR,DI,CL
Atkinson and Cvetanovic (1971)	1.4×10^{10}	...	...	298	FR,HG,CL
Atkinson and Pitts (1974b)	1.9×10^{10}	...	...	301	SR,HG,CL
Huie and Herron (1975)	7.7×10^{13}	0	19.6	...	Review
Recommended ($\log F = \pm 0.40$)	6.2×10^{13}	0	21.4	300–1000	...
	$i\text{-}C_4H_{10} + O \rightarrow C_4H_9 + OH$				
Huie and Herron (1975)	4.7×10^{13}	0	16.3	...	Review
Recommended ($\log F = \pm 0.50$)	4.7×10^{13}	0	16.3	300–1000	...

Table 29 (*continued*).

Reference	$A/(\mathrm{cm^3/mol})^{n-1}\,\mathrm{s^{-1}}$	b	E/kJ	T/K	Method
	n-$C_4H_{10} + OH \rightarrow C_4H_9 + H_2O$				
Greiner (1970b)	8.5×10^{12}	0	4.4	300–500	SR,FP,UA
Stuhl (1973b)	1.4×10^{12}	...	...	298	SR,FP,RF
Gordon and Mulac (1975)	4.6×10^{12}	0	1.5	298–416	SR,PR,UA
Baldwin *et al.* (1977)	9.0×10^{12}	0	4.4	...	Review
Recommended ($\log F = \pm 0.40$)	9.0×10^{12}	0	4.4	300–1000	...
	i-$C_4H_{10} + OH \rightarrow C_4H_9 + H_2O$				
Baker *et al.* (1970b)	2.2×10^{13}	...	...	753	SR,TH,P,GC
Baldwin *et al.* (1977)	5.8×10^{12}	0	2.6	...	Review
Darnall *et al.* (1978)	1.5×10^{12}	...	...	300	SR,PH,GC
Butler *et al.* (1978)	9.6×10^{11}	...	...	304	SR,PH,GC
Recommended ($\log F = \pm 0.40$)	5.8×10^{12}	0	2.6	300–1000	...
	n-$C_4H_{10} + HO_2 \rightarrow C_4H_9 + H_2O_2$				
Baldwin *et al.* (1977)	8.0×10^{12}	0	72.9	...	Review
Recommended ($\log F = \pm 0.50$)	8.0×10^{12}	0	72.9	300–1000	...
	i-$C_4H_{10} + HO_2 \rightarrow C_4H_9 + H_2O_2$				
Baldwin *et al.* (1977)	4.4×10^{12}	0	64.0	...	Review
Recommended ($\log F = \pm 0.50$)	4.4×10^{12}	0	64.0	300–1000	...
	n-$C_4H_{10} + CH_3 \rightarrow C_4H_9 + CH_4$				
Kerr and Parsonage (1976)	4.0×10^{11}	0	40.2	350–750	Review
Pacey and Purnell (1972b)	2.5×10^{14}	0	76.2	886–951	FR,TH,GC
Recommended ($\log F = \pm 0.40$)	4.0×10^{11}	0	40.2	300–1000	...

	$i\text{-}C_4H_{10} + CH_3 \rightarrow C_4H_9 + CH_4$				
Kerr and Parsonage (1976)	9.5×10^{10}	0	33.1	550–750	Review
Konar *et al.* (1973)	1.0×10^{13}	0	58.9	770–855	SR,TH,GC
Recommended (log $F = \pm 0.40$)	9.5×10^{10}	0	33.1	300–1000	...
	$n\text{-}C_4H_{10} \rightarrow C_2H_5 + C_2H_5\ (k_\infty)$				
Purnell and Quinn (1962)	3.8×10^{18}	0	361	693–803	SR,TH,GC
Tsang (1969)	2.0×10^{16}	0	340	...	Review
Golden *et al.* (1974b)	2.5×10^{16}	0	339	~1100	SR,TH,MS
Recommended (log $F = \pm 0.50$)	2.0×10^{16}	0	340	500–2000	...
	$n\text{-}C_4H_{10} \rightarrow CH_3 + n\text{-}C_3H_7\ (k_\infty)$				
Purnell and Quinn (1962)	3.8×10^{18}	0	361	693–803	SR,TH,GC
Benson and O'Neal (1970)	4.0×10^{17}	0	360	...	Review
Golden *et al.* (1974b)	5.5×10^{15}	0	339	~1100	SR,TH,MS
Pratt and Rogers (1979c)	1.9×10^{15}	0	323	984–1065	FR,TH,GC
Recommended (log $F = \pm 0.50$)	2.0×10^{16}	0	340	500–2000	...
	$i\text{-}C_4H_{10} \rightarrow CH_3 + i\text{-}C_3H_7\ (k_\infty)$				
Brooks (1966)	1.2×10^{15}	0	270	823–853	SR,TH,P,GC
Tsang (1969)	4.0×10^{16}	0	347	...	Review
Konar *et al.* (1973)	6.3×10^{16}	0	342	770–855	SR,TH,GC
Golden *et al.* (1974b)	8.0×10^{16}	0	347	~1100	SR,TH,MS
Pratt and Rogers (1980)	8.4×10^{15}	0	358	970–1031	FR,TH,GC
Recommended (log $F = \pm 0.50$)	4.0×10^{16}	0	347	500–2000	...

Table 30. Rate data on C_4H_9 reactions.

Reference	$A/(\mathrm{cm^3/mol})^{n-1}\ \mathrm{s^{-1}}$ [1]	b	E/kJ	T/K	Method
	$C_4H_9 + O \rightarrow$ products				
Recommended ($\log F = \pm 0.30$)	4.0×10^{13}	0	0	300–2000	Estimate
	i-$C_4H_9 + O_2 \rightarrow C_4H_8 + HO_2$				
Slater and Calvert (1968)	9.9×10^{7}	...	...	313	SR,PH,GC
Baker *et al.* (1971)	2.3×10^{10}	...	...	753	SR,TH,P,GC
Recommended ($\log F = \pm 0.50$)	1.0×10^{12}	0	23.9	300–2000	...
	n-$C_4H_9 + O_2 \rightarrow C_4H_8 + HO_2$				
Baker *et al.* (1971)	1.6×10^{11}	...	...	753	SR,TH,P,GC
Baker *et al.* (1975)	2.7×10^{11}	...	...	753	SR,TH,P,GC
Lenhardt *et al.* (1980)	4.5×10^{12}	...	...	298	FR,FP,MS
Recommended ($\log F = \pm 0.50$)	1.6×10^{11}	...	...	753	...
	s-$C_4H_9 + O_2 \rightarrow C_4H_8 + HO_2$				
Baker *et al.* (1971)	1.7×10^{11}	...	...	753	SR,TH,P,GC
Baker *et al.* (1975)	1.7×10^{11}	...	...	753	SR,TH,P,GC
Lenhardt *et al.* (1980)	1.0×10^{13}	...	...	298	FR,FP,MS
Recommended ($\log F = \pm 0.50$)	1.7×10^{11}	...	...	753	...
	t-$C_4H_9 + O_2 \rightarrow C_4H_8 + HO_2$				
Evans and Walker (1979)	8.0×10^{11}	0	9.1	713–813	SR,TH,GC
Lenhardt *et al.* (1980)	1.4×10^{13}	...	...	298	FR,FP,MS
Recommended ($\log F = \pm 0.50$)	8.0×10^{11}	0	9.1	500–2000	...

	$t\text{-}C_4H_9 + t\text{-}C_4H_9 \rightarrow C_8H_{18}$ (k_∞)				
Golden *et al.* (1973)	6.0×10^{11}	...	...	650	SR,TH,MS
Golden *et al.* (1974b)	3.0×10^{11}	...	...	620	SR,TH,MS
Parkes and Quinn (1975)	1.2×10^{12}	...	...	298	SR,PH,UA
Choo *et al.* (1976)	6.3×10^{11}	...	...	700	SR,TH,MS
Marshall *et al.* (1976)	8.0×10^{11}	...	...	800	SR,TH,GC
Parkes and Quinn (1976)	2.3×10^{11}	0	5.8	298–423	SR,PH,UA
Bethune *et al.* (1981)	4.7×10^{12}	...	...	298	SR,PH,IR,GC
Recommended ($\log F = \pm 0.30$)	1.0×10^{12}	0	0	300–1000	...
	$n\text{-}C_4H_9 + n\text{-}C_4H_9 \rightarrow C_8H_{18}$ (k_∞)				
Walker (1971)	1.0×10^{13}	...	...	...	Estimate
No recommendation					
	$t\text{-}C_4H_9 + t\text{-}C_4H_9 \rightarrow C_4H_8 + C_4H_{10}$				
Choo *et al.* (1976)	$k_{disp}/k_{comb} = 0.33$			700	SR,TH,MS
McKay *et al.* (1977)	$k_{disp}/k_{comb} = 2.60$			483–515	SR,TH,GC
Mintz and Leroy (1978)	$k_{disp}/k_{comb} = 2.2\text{–}7.4$			573–673	FR,TH,GC
Bethune *et al.* (1981)	$k_{disp}/k_{comb} = 2.90$			298	SR,PH,IR,GC
Recommended ($\log F = \pm 0.50$)	2.5×10^{12}	0	0	300–1000	...
	$n\text{-}C_4H_9 + n\text{-}C_4H_9 \rightarrow C_4H_8 + C_4H_{10}$				
Kerr and Trotman-Dickenson (1960b)	$k_{disp}/k_{comb} = 0.7$			373	SR,PH,GC
Morganroth and Calvert (1966)	$k_{disp}/k_{comb} = 0.14$			430–520	SR,PH,GC
No recommendation					

Table 30 (*continued*).

Reference	$A/(\mathrm{cm^3/mol})^{n-1}\,\mathrm{s^{-1}}$	b	E/kJ	T/K	Method
	$i\text{-}C_4H_9 + i\text{-}C_4H_9 \rightarrow C_4H_8 + C_4H_{10}$				
Slater and Calvert (1968)	$k_{disp}/k_{comb} = 0.075$			298–441	SR,PH,GC
No recommendation					
	$n\text{-}C_4H_9 \rightarrow C_2H_4 + C_2H_5\ (k_\infty)$				
Kerr and Trotman-Dickenson (1960b)	1.6×10^{11}	0	92.0	477–689	SR,PH,GC
Morganroth and Calvert (1966)	3.7×10^{13}	0	120	430–520	SR,PH,GC
Recommended ($\log F = \pm 1.0$)	3.7×10^{13}	0	120	400–2000	...
	$s\text{-}C_4H_9 \rightarrow C_3H_6 + CH_3\ (k_\infty)$				
Gruver and Calvert (1956)	5.0×10^{11}	0	100	500–600	SR,PH,MS
Lin and Laidler (1967)	2.3×10^{14}	0	137	533–613	SR,TH,GC
Recommended ($\log F = \pm 1.0$)	2.3×10^{14}	0	137	500–2000	...
	$i\text{-}C_4H_9 \rightarrow C_3H_6 + CH_3\ (k_\infty)$				
Metcalfe and Trotman-Dickenson (1960)	6.3×10^{12}	0	109	552–790	SR,PH,GC
Slater and Calvert (1968)	9.0×10^{12}	0	130	543–598	SR,PH,GC
Recommended ($\log F = \pm 1.0$)	7.0×10^{12}	0	110	500–2000	...
	$t\text{-}C_4H_9 \rightarrow C_3H_6 + CH_3\ (k_\infty)$				
Birell and Trotman-Dickenson (1960)	1.0×10^{16}	0	193	742–797	SR,PH,GC
Recommended ($\log F = \pm 1.0$)	1.0×10^{16}	0	193	500–2000	...

Table 31. Rate data on C_4H_8 reactions.

Reference	$A/(cm^3/mol)^{n-1}\ s^{-1}$	b	E/kJ	T/K	Method
	$1\text{-}C_4H_8 + H \rightarrow C_4H_9\ (k_\infty)$				
Kerr and Parsonage (1972)	8.1×10^{11}	...	...	298	Review
Daby *et al.* (1971)	8.3×10^{11}	...	...	298	FR,DI,MS
Ishikawa *et al.* (1978)	1.2×10^{12}	...	...	298	SR,PR,RA
Yoichi *et al.* (1978)	1.2×10^{12}	...	...	298	SR,PR,RA
Oka and Cvetanovic (1979)	8.4×10^{11}	...	...	298	FR,HG,CL
Recommended ($\log F = \pm 0.30$)	1.0×10^{12}	...	...	298	...
	$trans\text{-}2\text{-}C_4H_8 + H \rightarrow C_4H_9\ (k_\infty)$				
Daby *et al.* (1971)	5.4×10^{11}	...	...	298	FR,DI,MS
Ishikawa *et al.* (1978)	6.6×10^{11}	...	...	298	SR,PR,RA
Yoichi *et al.* (1978)	6.6×10^{11}	...	...	298	SR,PR,RA
Oka and Cvetanovic (1979)	4.6×10^{11}	...	...	298	FR,HG,CL
Recommended ($\log F = \pm 0.30$)	5.5×10^{11}	...	...	298	...
	$cis\text{-}2\text{-}C_4H_8 + H \rightarrow C_4H_9\ (k_\infty)$				
Daby *et al.* (1971)	4.7×10^{11}	...	...	298	FR,DI,MS
Ishikawa *et al.* (1978)	6.0×10^{11}	...	...	298	SR,PR,RA
Yoichi *et al.* (1978)	6.0×10^{11}	...	...	298	SR,PR,RA
Oka and Cvetanovic (1979)	3.6×10^{11}	...	...	298	FR,HG,CL
Recommended ($\log F = \pm 0.30$)	4.5×10^{11}	...	...	298	...

Table 31 (*continued*).

Reference	$A/(\mathrm{cm^3/mol})^{n-1}\,\mathrm{s^{-1}}$	b	E/kJ	T/K	Method
	iso-$C_4H_8 + H \rightarrow C_4H_9$ (k_∞)				
Oka and Cvetanovic (1979)	2.4×10^{12}	...	...	298	FR,HG,CL
No recommendation					
	1-$C_4H_8 + O \rightarrow$ products				
Smith (1968)	7.3×10^{12}	...	3.3	305–410	SR,FP,UA
Atkinson and Cvetanovic (1972)	6.1×10^{12}	0	3.4	298–473	FR,HG,CL
Parkes and Quinn (1975)	1.2×10^{12}	...	...	298	SR,PH,UA
Choo *et al.* (1976)	6.3×10^{11}	...	...	700	SR,TH,MS
Marshall *et al.* (1976)	8.0×10^{11}	...	...	800	SR,TH,GC
Parkes and Quinn (1976)	2.3×10^{11}	0	5.8	298–423	SR,PH,UA
Bethune *et al.* (1981)	4.7×10^{12}	...	...	298	SR,PH,IR,GC
Recommended ($\log F = \pm 0.30$)	1.0×10^{12}	0	0	300–1000	...
	n-$C_4H_9 + n$-$C_4H_9 \rightarrow C_8H_{18}$ (k_∞)				
Walker (1971)	1.0×10^{13}	...	...	...	Estimate
No recommendation					
	t-$C_4H_9 + t$-$C_4H_9 \rightarrow C_4H_8 + C_4H_{10}$				
Choo *et al.* (1976)	$k_{disp}/k_{comb} = 0.33$			700	SR,TH,MS
McKay *et al.* (1977)	$k_{disp}/k_{comb} = 2.60$			483–515	SR,TH,GC
Mintz and LeRoy (1978)	$k_{disp}/k_{comb} = 2.2–7.4$ .4			573–673	FR,TH,GC
Bethune *et al.* (1981)	$k_{disp}/k_{comb} = 2.90$			298	SR,PH,IR,GC
Recommended ($\log F = \pm 0.50$)	2.5×10^{12}	0	0	300–1000	...

	n-C_4H_9 + n-C_4H_9 → C_4H_8 + C_4H_{10}				
Kerr and Trotman-Dickenson (1960b)	$k_{disp}/k_{comb} = 0.7$			373	SR,PH,GC
Morganroth and Calvert (1966)	$k_{disp}/k_{comb} = 0.14$			430–520	SR,PH,GC
No recommendation					
	1-C_4H_8 + O → products				
Smith (1968)	7.3×10^{12}	...	3.3	305–410	SR,FP,UA
Atkinson and Cvetanovic (1972)	6.1×10^{12}	0	3.4	298–473	FR,HG,CL
Furuyama *et al.* (1974)	9.4×10^{12}	0	3.4	298–473	FR,HG,CL
Singleton and Cvetanovic (1976)	7.2×10^{12}	0	2.8	298–480	SR,HG,GC
Atkinson and Pitts (1977a)	8.4×10^{12}	0	2.8	298–439	SR,FP,CL
Sugawara *et al.* (1980)	2.8×10^{12}	...	...	296	SR,PR,RA
Recommended ($\log F = \pm 0.30$)	8.0×10^{12}	0	3.4	300–500	...
	iso-C_4H_8 + O → products				
Smith (1968)	9.0×10^{12}	0	0.4	305–410	SR,FP,UA
Atkinson and Cvetanovic (1972)	6.3×10^{12}	0	0	298–473	FR, HG, CL
Furuyama *et al.* (1974)	9.8×10^{12}	0	0	298–473	FR,HG,CL
Singleton and Cvetanovic (1976)	8.7×10^{12}	0	−0.4	298–480	SR,HG,GC
Atkinson and Pitts (1977a)	1.1×10^{13}	0	0.4	298–439	SR,FP,CL
Recommended ($\log F = \pm 0.30$)	9.5×10^{12}	0	0	300–500	...
	cis-2-C_4H_8 + O → products				
Davis *et al.* (1973)	5.8×10^{12}	0	−1.3	268–443	SR,FP,RF
Furuyama *et al.* (1974)	9.0×10^{12}	...	...	298	FR,HG,CL
Singleton and Cvetanovic (1976)	6.7×10^{12}	0	−1.1	298–480	SR,HG,GC
Atkinson and Pitts (1977a)	7.3×10^{12}	0	−1.0	298–439	SR,FP,CL
Sugawara *et al.* (1980)	1.2×10^{13}	...	...	296	SR,PR,RA
Recommended ($\log F = \pm 0.30$)	7.0×10^{12}	0	−1.0	300–500	...

Table 31 (*continued*).

Reference	$A/(\mathrm{cm^3/mol})^{n-1}\,\mathrm{s^{-1}}$	b	E/kJ	T/K	Method
	trans-2-C_4H_8 + O → products				
Atkinson and Pitts (1977a)	1.4×10^{13}	0	−0.1	298–439	SR,FP,CL
Sugawara *et al.* (1980)	1.4×10^{13}	...	...	296	SR,PR,RA
Recommended ($\log F = \pm 0.30$)	1.4×10^{13}	0	0	300–500	...
	1-C_4H_8 + OH → products				
Morris and Niki (1971b)	2.5×10^{13}	...	...	298	FR,DI,MS
Atkinson and Pitts (1975)	4.6×10^{12}	0	−3.9	297–425	SR,FP,RF
Pastrana and Carr (1975)	9.0×10^{12}	...	...	300	FR,DI,UA
Ravishankara *et al.* (1978)	1.8×10^{13}	...	...	298	SR,FP,RF
Nip and Paraskevopoulos (1979)	2.0×10^{13}	...	...	297	SR,FP,RA
Recommended ($\log F = \pm 0.30$)	4.0×10^{12}	0	−3.9	300–500	...
	iso-C_4H_8 + OH → products				
Morris and Niki (1971b)	3.9×10^{3}	...	...	298	FR,DI,MS
Atkinson and Pitts (1975)	5.5×10^{12}	0	−4.2	297–425	SR,FP,RF
Recommended ($\log F = \pm 0.30$)	5.5×10^{12}	0	−4.2	300–400	...
	trans–2–C_4H_8 + OH → products				
Morris and Niki (1971b)	4.3×10^{13}	...	...	298	FR,DI,MS
Atkinson and Pitts (1975)	6.7×10^{12}	0	−4.6	297–425	SR,FP,RF
Pastrana and Carr (1975)	6.6×10^{12}	...	...	300	FR,DI,UA
Recommended ($\log F = \pm 0.30$)	6.7×10^{12}	0	−4.6	300–400	...
	cis-2-C_4H_8 + OH → products				
Morris and Niki (1971b)	3.7×10^{13}	...	...	298	FR,DI,MS
Atkinson and Pitts (1975)	6.3×10^{12}	0	−4.1	297–425	SR,FP,RF

Ravishankara *et al.* (1978)	2.6×10^{13}	...	...	298	SR,FP,RF
Recommended ($\log F = \pm 0.30$)	6.0×10^{12}	0	-4.1	300–400	...
	$1\text{-}C_4H_8 + CH_3 \rightarrow CH_4 + C_4H_7$				
Kerr and Parsonage (1976)	2.5×10^{11}	0	34.8	350–650	Review
Recommended ($\log F = \pm 0.30$)	2.5×10^{11}	0	34.8	300–700	...
	$\text{cis-2-}C_4H_8 + CH_3 \rightarrow CH_4 + C_4H_7$				
Kerr and Parsonage (1976)	1.8×10^{11}	0	33.9	350–650	Review
Recommended ($\log F = \pm 0.30$)	1.8×10^{11}	0	33.9	300–700	...
	$\text{trans-2-}C_4H_8 + CH_3 \rightarrow CH_4 + C_4H_7$				
Kerr and Parsonage (1976)	1.0×10^{12}	0	40.2	350–650	Review
Recommended ($\log F = \pm 0.30$)	1.0×10^{12}	0	40.2	300–700	...

Table 32. Rate data on C_4H_2 reactions.

Reference	$A/(cm^3/mol)^{n-1} s^{-1}$	b	E/kJ	T/K	Method
	$C_4H_2 + H \rightarrow C_4H_3(k_\infty)$				
Schwanebeck and Warnatz (1975)	1.3×10^{12}	...	...	298	FR,DI,MS
Laufer and Bass (1979)	6.0×10^{11}	...	...	298	SR,FP,GC,UA
Recommended ($\log F = \pm 0.50$)	6.5×10^{12}	0	4.2	>300	...
	$C_4H_2 + O \rightarrow C_3H_2 + CO$				
Niki (1967)	9.0×10^{11}	...	...	300	FR,DI,MS
Homann *et al.* (1975)	8.0×10^{13}	0	10.2	297–343	FR,DI,MS
Wellmann (1981)	2.7×10^{13}	0	7.2	295–1000	FR,DI,MS
Recommended ($\log F = \pm 0.30$)	2.7×10^{13}	0	7.2	300–1000	...
	$C_4H_2 + OH \rightarrow$ products				
Homann *et al.* (1982)	3.0×10^{13}	...	...	294	FR,DI,RF
Warnatz *et al.* (1982)	$\sim 3.0 \times 10^{13}$	...	...	~ 2000	FL,TH,MS
Recommended ($\log F = \pm 0.50$)	3.0×10^{13}	0	0	300–2000	...

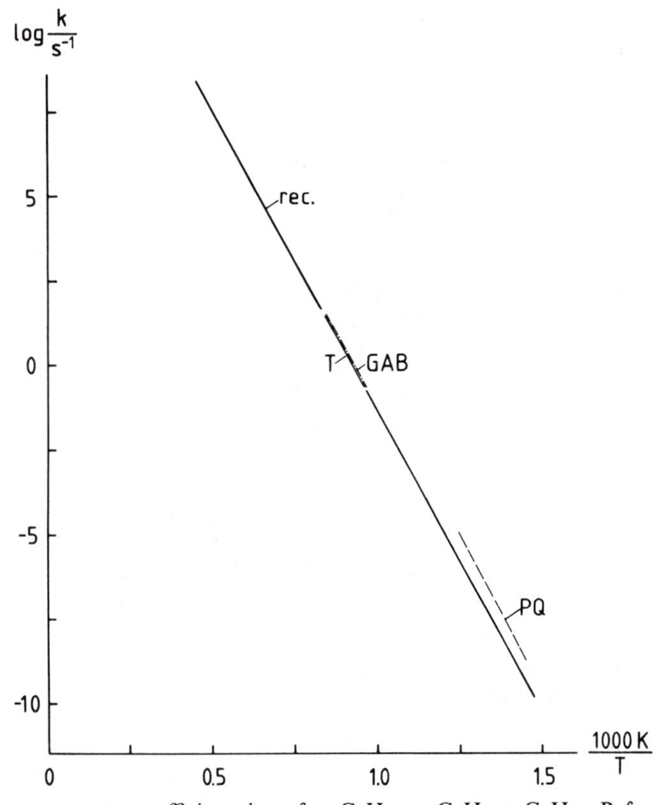

Figure 95. High-pressure rate coefficient k_∞ of n-$C_4H_{10} \rightarrow C_2H_5 + C_2H_5$. References in Table 29.

7.3. Reactions of propene and butene

There is no evidence that these reactions may play a rate-controlling part in high-temperature combustion or ignition processes. Since the available data are scarce (see Tables 27 and 31 and Figs. 88 to 90) and may be complicated by unresolved pressure dependences, more knowledge about the rate coefficients and products is desirable. Propene and butenes are important intermediates in the decomposition of propyl and butyl radicals, respectively, and must be consumed somehow in the reaction zones of flames of C_3- and higher hydrocarbons.

7.4. Reactions of C_3H_4 and C_4H_2

As mentioned in connection with the reactions forming C_3H_4 and C_4H_2, these species are important in fuel-rich high-temperature combustion. In particular, C_4H_2 plays a dominant part in the oxidation of C_2H_2 and seems to be important generally as a precursor of soot formation. Measurements in rich

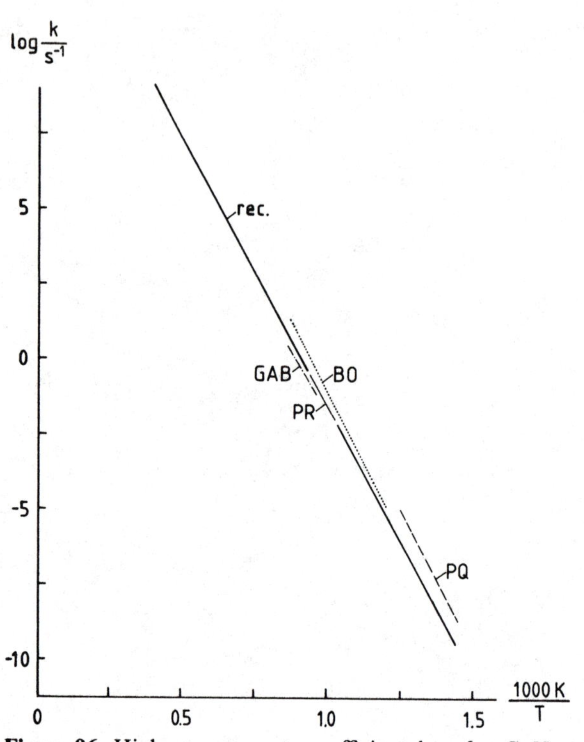

Figure 96. High-pressure rate coefficient k_∞ of n-$C_4H_{10} \rightarrow n$-$C_3H_7 + CH_3$. References in Table 29.

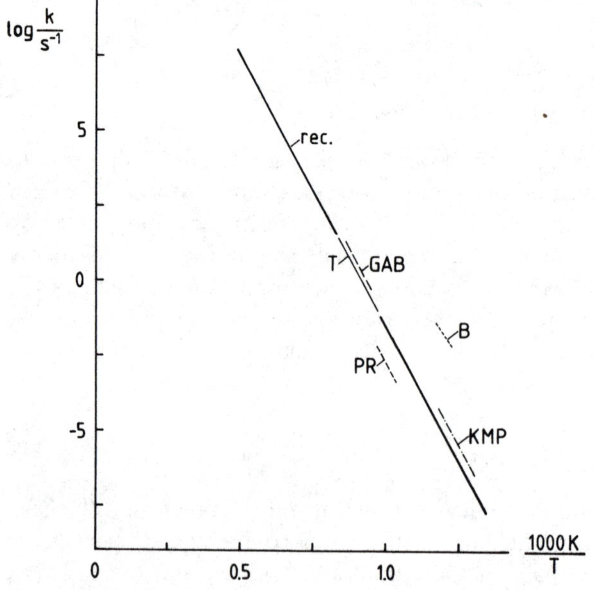

Figure 97. High-pressure rate coefficient k_∞ of i-$C_4H_{10} \rightarrow C_3H_7 + CH_3$. References in Table 29.

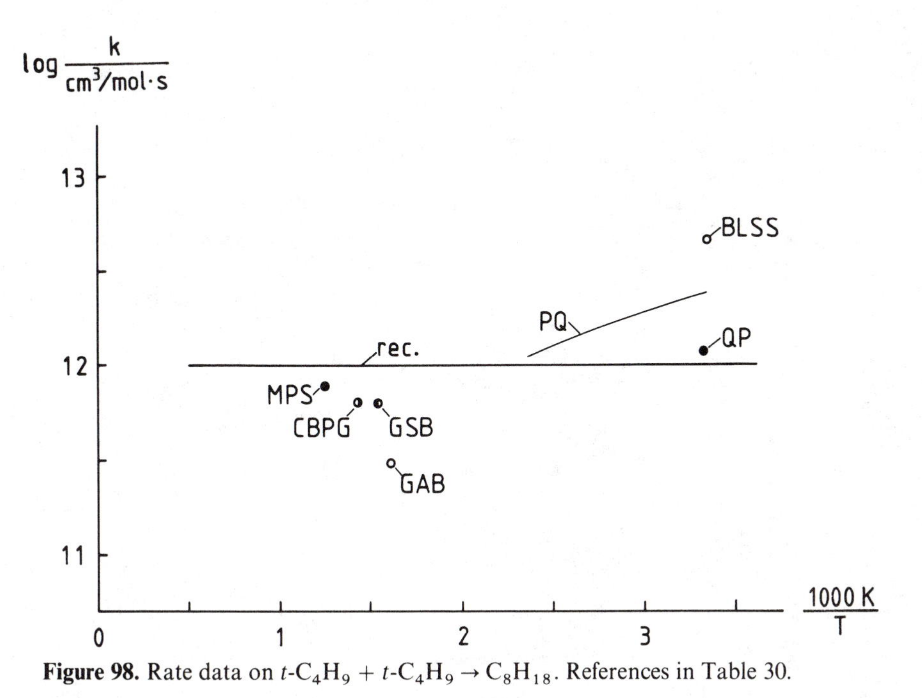

Figure 98. Rate data on $t\text{-}C_4H_9 + t\text{-}C_4H_9 \rightarrow C_8H_{18}$. References in Table 30.

C_2H_2/O_2 premixed flames near the soot limit show about 30% of the acetylene to be decomposed via C_4H_2 (Warnatz *et al.*, 1982, 1983). Also, this leads to consequences for the combustion of other hydrocarbons, since under rich conditions the greatest part of these fuels is oxidized via C_2H_2. Experimental data on the attack of H and OH (O-atom attack is unimportant in rich flames) on C_3H_4 and C_4H_2 are scarce (see Tables 28 and 32 and Figs. 91 and 92), with the least data being available on the most important reactions of OH with these species. Furthermore, there is lack of data on reactions leading from C_4H_2 to the formation of higher hydrocarbons, predominantly polyacetylenes.

8. Mechanism of small hydrocarbon combustion

The recommendations given in the preceding sections are summarized in Table 33. It should be taken into consideration that in many cases the listed rate coefficient parameters give expressions that are valid only for a limited temperature range (~1000–2500 K). It should also be noted that reverse reactions are listed separately for all cases where both directions may be important in combustion modeling. For computer programs that automatically take reverse reactions into account, only one of each pair should be included. Using both directions independently may lead to small numerical

Table 33. Recommended rate coefficient parameters for high temperature combustion in the C/H/O System.

Reaction		$A/(cm^3/mol)^{n-1}\,s^{-1}$	b	E/kJ
		H_2/O_2 reactions (HO_2/H_2O_2 reactions excluded)		
$H + O_2 \rightarrow OH + O$		1.2×10^{17}	−0.91	69.1
$OH + O \rightarrow O_2 + H$		1.8×10^{13}	0	0
$O + H_2 \rightarrow OH + H$		1.5×10^{7}	2.0	31.6
$OH + H_2 \rightarrow H_2O + H$		1.0×10^{8}	1.6	13.8
$H + H_2O \rightarrow OH + H_2$		4.6×10^{8}	1.6	77.7
$OH + OH \rightarrow H_2O + O$		1.5×10^{9}	1.14	0
$O + H_2O \rightarrow OH + OH$		1.5×10^{10}	1.14	72.2
$H + H + Ar \rightarrow H_2 + Ar$	(k_1)	6.4×10^{17}	−1.0	0
$H + H + H_2 \rightarrow H_2 + H_2$	(k_1)	9.7×10^{16}	−0.6	0
$H_2 + Ar \rightarrow H + H + Ar$	(k_1)	2.2×10^{14}	0	402
$H_2 + H_2 \rightarrow H + H + H_2$	(k_1)	8.8×10^{14}	0	402
$H + OH + H_2O \rightarrow H_2O + H_2O$	(k_1)	1.4×10^{23}	−2.0	0
$H_2O + H_2O \rightarrow H + OH + H_2O$	(k_1)	1.6×10^{17}	0	478
$O + O + Ar \rightarrow O_2 + Ar$	(k_1)	1.0×10^{17}	−1.0	0
$O_2 + Ar \rightarrow O + O + Ar$	(k_1)	1.2×10^{14}	0	451
		HO_2/H_2O_2 reactions		
$H + O_2 + H_2 \rightarrow HO_2 + H_2$	(k_1)	2.0×10^{18}	−0.8	0
$H + HO_2 \rightarrow OH + OH$		1.5×10^{14}	0	4.2
$H + HO_2 \rightarrow H_2 + O_2$		2.5×10^{13}	0	2.9
$O + HO_2 \rightarrow OH + O_2$		2.0×10^{13}	0	0
$OH + HO_2 \rightarrow H_2O + O_2$		2.0×10^{13}	0	0
$HO_2 + HO_2 \rightarrow H_2O_2 + O_2$		2.0×10^{12}	0	0

$OH + OH + N_2 \rightarrow H_2O_2 + N_2$	(k_1)	1.3×10^{22}	-2.0	0
$H_2O_2 + N_2 \rightarrow OH + OH + N_2$	(k_1)	1.2×10^{17}	0	190
$H + H_2O_2 \rightarrow H_2 + HO_2$		1.7×10^{12}	0	15.7
$H_2 + HO_2 \rightarrow H_2O_2 + H$		7.3×10^{11}	0	78.1
$H + H_2O_2 \rightarrow H_2O + OH$		1.0×10^{13}	0	15.0
$O + H_2O_2 \rightarrow OH + HO_2$		2.8×10^{13}	0	26.8
$OH + H_2O_2 \rightarrow H_2O + HO_2$		7.0×10^{12}	0	6.0
Reactions of CO/CO_2				
$CO + H + H_2 \rightarrow CHO + H_2$	(k_1)	6.9×10^{14}	0	7.0
$CO + O + CO \rightarrow CO_2 + CO$	(k_1)	5.3×10^{13}	0	-19.0
$CO + OH \rightarrow CO_2 + H$		4.4×10^{6}	1.5	-3.1
$CO + HO_2 \rightarrow CO_2 + OH$		1.5×10^{14}	0	98.7
$CO + O_2 \rightarrow CO_2 + O$		2.5×10^{12}	0	200
$CO_2 + H \rightarrow CO + OH$		1.6×10^{14}	0	110
$CO_2 + O \rightarrow CO + O_2$		1.7×10^{13}	0	220
Reactions of CH_4				
$CH_4 + H \rightarrow CH_3 + H_2$		2.2×10^{4}	3.0	36.6
$CH_4 + O \rightarrow CH_3 + OH$		1.2×10^{7}	2.1	31.9
$CH_4 + OH \rightarrow CH_3 + H_2O$		1.6×10^{6}	2.1	10.3
$CH_4 + Ar \rightarrow CH_3 + H + Ar$	(k_1)	2.0×10^{17}	0	370
$CH_4 \rightarrow CH_3 + H$	(k_∞)	1.0×10^{15}	0	420
Reactions of CH_3				
$CH_3 + H + Ar \rightarrow CH_4 + Ar$	(k_1)	8.0×10^{26}	-3.0	0
$CH_3 + H \rightarrow CH_4$	(k_∞)	6.0×10^{16}	-1.0	0

Table 33 (*continued*).

Reaction		$A/(cm^3/mol)^{n-1} s^{-1}$	b	E/kJ
$CH_3 + O \rightarrow CH_2O + H$		7.0×10^{13}	0	0
$CH_3 + H_2 \rightarrow CH_4 + H$		6.6×10^{2}	3.0	32.4
$CH_3 + O_2 \rightarrow CH_2O + H + O$		1.5×10^{13}	0	120
$CH_3 + CH_3 \rightarrow C_2H_6$	(k_∞)	2.4×10^{14}	−0.4	0
$CH_3 + CH_3 \rightarrow C_2H_5 + H$		8.0×10^{14}	0	111
$CH_3 + CH_3 \rightarrow C_2H_4 + H_2$		1.0×10^{16}	0	134
$CH_3 + M \rightarrow CH_2 + H + M$	(k_1)	1.0×10^{16}	0	379
		Reactions of CH_2O		
$CH_2O + H \rightarrow CHO + H_2$		2.5×10^{13}	0	16.7
$CH_2O + O \rightarrow CHO + OH$		3.5×10^{13}	0	14.7
$CH_2O + OH \rightarrow CHO + H_2O$		3.0×10^{13}	0	5.0
$CH_2O + CH_3 \rightarrow CHO + CH_4$		1.0×10^{11}	0	25.5
$CH_2O + Ar \rightarrow CHO + H + Ar$	(k_1)	5.0×10^{16}	0	320
		Reactions of CHO		
$CHO + H \rightarrow CO + H_2$		2.0×10^{14}	0	0
$CHO + O \rightarrow CO + OH$		3.0×10^{13}	0	0
$CHO + O \rightarrow CO_2 + H$		3.0×10^{13}	0	0
$CHO + OH \rightarrow CO + H_2O$		5.0×10^{13}	0	0
$CHO + O_2 \rightarrow CO + HO_2$		3.0×10^{12}	0	0
$CHO + Ar \rightarrow CO + H + Ar$	(k_1)	2.5×10^{14}	0	70.3
		Reactions of CH_2		
$CH_2 + H \rightarrow CH + H_2$		4.0×10^{13}	0	0
$CH_2 + O \rightarrow CO + H + H$		5.0×10^{13}	0	0
$CH_2 + O_2 \rightarrow CO_2 + H + H$		1.3×10^{13}	0	6.3
$CH_2 + CH_3 \rightarrow C_2H_4 + H$		4.0×10^{13}	0	0

Reaction				
Reactions of CH				
$CH + O \rightarrow CO + H$		4.0×10^{13}	0	0
$CH + O \rightarrow CHO^{+} + e$		2.5×10^{11}	0	7.1
$CH + O_2 \rightarrow$ products		2.0×10^{13}	0	0
Reactions of CH_3OH				
$CH_3OH + H \rightarrow CH_3O + H_2$		4.0×10^{13}	0	25.5
$CH_3OH + O \rightarrow CH_3O + OH$		1.0×10^{13}	0	19.6
$CH_3OH + OH \rightarrow CH_3O + H_2O$		1.0×10^{13}	0	7.1
$CH_3OH + Ar \rightarrow CH_3 + OH + Ar$	(k_1)	2.0×10^{17}	0	286
$CH_3OH \rightarrow CH_3 + OH$	(k_∞)	9.4×10^{15}	0	376
Reactions of CH_3O/CH_2OH				
$CH_3O + H \rightarrow CH_2O + H_2$		2.0×10^{13}	0	0
$CH_2OH + H \rightarrow CH_2O + H_2$		3.0×10^{13}	0	0
$CH_3O + O_2 \rightarrow CH_2O + HO_2$		1.0×10^{13}	0	30.0
$CH_3O + M \rightarrow CH_2O + H + M$	(k_1)	1.0×10^{14}	0	105
$CH_3O \rightarrow CH_2O$	(k_∞)	1.6×10^{14}	0	105
Reactions of C_2H_6				
$C_2H_6 + H \rightarrow C_2H_5 + H_2$		5.4×10^{2}	3.5	21.8
$C_2H_6 + O \rightarrow C_2H_5 + OH$		3.0×10^{7}	2.0	21.4
$C_2H_6 + OH \rightarrow C_2H_5 + H_2O$		6.3×10^{6}	2.0	2.7
$C_2H_6 + HO_2 \rightarrow C_2H_5 + H_2O_2$		6.0×10^{12}	0	81.2
$C_2H_6 + CH_3 \rightarrow C_2H_5 + CH_4$		0.55	4.0	34.7
$C_2H_6 + Ar \rightarrow CH_3 + CH_3 + Ar$	(k_1)	1.0×10^{19}	0	285
$C_2H_6 \rightarrow CH_3 + CH_3$	(k_∞)	2.4×10^{16}	0	366

Table 33 (*continued*).

Reaction		$A/(cm^3/mol)^{n-1} s^{-1}$	b	E/kJ
	Reactions of C_2H_5			
$C_2H_5 + H \rightarrow CH_3 + CH_3$		3.0×10^{13}	0	0
$C_2H_5 + O \rightarrow$ products		5.0×10^{13}	0	0
$C_2H_5 + O_2 \rightarrow C_2H_4 + HO_2$		2.0×10^{12}	0	20.9
$C_2H_5 + CH_3 \rightarrow C_3H_8$	(k_∞)	7.0×10^{12}	0	0
$C_2H_5 + C_2H_5 \rightarrow C_4H_{10}$	(k_∞)	1.0×10^{13}	0	0
$C_2H_5 + C_2H_5 \rightarrow C_2H_4 + C_2H_6$		1.4×10^{12}	0	0
$C_2H_5 + M \rightarrow C_2H_4 + H + M$	(k_1)	1.0×10^{17}	0	130
$C_2H_5 \rightarrow C_2H_4 + H$	(k_∞)	2.0×10^{13}	0	166
	Reactions of C_2H_4			
$C_2H_4 + H \rightarrow C_2H_5$	(k_∞)	1.0×10^{13}	0	6.3
$C_2H_4 + H \rightarrow C_2H_3 + H_2$		1.5×10^{14}	0	42.7
$C_2H_4 + O \rightarrow$ products		1.6×10^{9}	1.2	3.1
$C_2H_4 + OH \rightarrow C_2H_3 + H_2O$		3.0×10^{13}	0	12.5
$C_2H_4 + CH_3 \rightarrow C_2H_3 + CH_4$		4.2×10^{11}	0	46.5
$C_2H_4 + Ar \rightarrow C_2H_2 + H_2 + Ar$	(k_1)	2.6×10^{17}	0	332
$C_2H_4 + Ar \rightarrow C_2H_3 + H + Ar$	(k_1)	2.6×10^{17}	0	404
	Reactions of C_2H_3			
$C_2H_3 + H \rightarrow C_2H_2 + H_2$		2.0×10^{13}	0	0
$C_2H_3 + O \rightarrow$ products		3.0×10^{13}	0	0
$C_2H_3 + O_2 \rightarrow C_2H_2 + HO_2$		1.0×10^{12}	0	0
$C_2H_3 + Ar \rightarrow C_2H_2 + H + Ar$	(k_1)	3.0×10^{15}	0	134
$C_2H_3 \rightarrow C_2H_2 + H$	(k_∞)	1.6×10^{14}	0	159

	Reactions of C_2H_2			
$C_2H_2 + H + Ar \rightarrow C_2H_3 + Ar$	(k_1)	4.2×10^{17}	0	2.9
$C_2H_2 + H \rightarrow C_2H_3$	(k_∞)	5.5×10^{12}	0	10.1
$C_2H_2 + H \rightarrow C_2H + H_2$		6.0×10^{13}	0	99
$C_2H_2 + O \rightarrow CH_2 + CO$		4.1×10^{8}	1.5	7.1
$C_2H_2 + O \rightarrow CHCO + H$		4.3×10^{14}	0	50.7
$C_2H_2 + OH \rightarrow$ products	(k_∞)	3.0×10^{12}	0	4.6
$C_2H_2 + OH \rightarrow C_2H + H_2O$		1.0×10^{13}	0	29.3
$C_2H_2 + CH_2 \rightarrow C_3H_3 + H$		1.8×10^{12}	0	0
$C_2H_2 + CH \rightarrow C_3H_3$		3.0×10^{13}	0	0
$C_2H_2 + C_2H \rightarrow C_4H_2 + H$		3.5×10^{13}	0	0
$C_2H_2 + Ar \rightarrow C_2H + H + Ar$	(k_1)	4.0×10^{16}	0	0
	Reactions of C_2H			
$C_2H + O \rightarrow$ products		1.0×10^{13}	0	0
$C_2H + O_2 \rightarrow$ products		5.0×10^{13}	0	0
$C_2H + H_2 \rightarrow$ products		1.5×10^{13}	0	13
	Reactions of CH_3CHO			
$CH_3CHO + H \rightarrow CH_3CO + H_2$		4.0×10^{13}	0	17.6
$CH_3CHO + O \rightarrow CH_3CO + OH$		5.0×10^{12}	0	7.5
$CH_3CHO + OH \rightarrow CH_3CO + H_2O$		1.0×10^{13}	0	0
$CH_3CHO + CH_3 \rightarrow CH_3CO + CH_4$		8.5×10^{10}	0	25.1
$CH_3CHO \rightarrow CH_3 + CHO$	(k_∞)	2.0×10^{15}	0	331
	Reactions of CH_3CO			
$CH_3CO + H \rightarrow$ products		2.0×10^{13}	0	0
$CH_3CO + O \rightarrow$ products		2.0×10^{13}	0	0
$CH_3CO + M \rightarrow CH_3 + CO + M$	(k_1)	1.2×10^{15}	0	52.4
$CH_3CO \rightarrow CH_3 + CO$	(k_∞)	3.0×10^{12}	0	70.0

Table 33 (*continued*).

Reaction		$A/(cm^3/mol)^{n-1} s^{-1}$	b	E/kJ
		Reactions of CH_2CO		
$CH_2CO + H \rightarrow CH_3 + CO$		7.0×10^{12}	0	12.6
$CH_2CO + O \rightarrow CHO + CHO$		2.0×10^{13}	0	9.6
$CH_2CO + OH \rightarrow CH_2O + CHO$		1.0×10^{13}	0	0
$CH_2CO + Ar \rightarrow CH_2 + CO + Ar$	(k_1)	3.6×10^{15}	0	248
$CH_2CO \rightarrow CH_2 + CO$	(k_∞)	3.0×10^{14}	0	297
		Reactions of CHCO		
$CHCO + H \rightarrow CH_2 + CO$		3.0×10^{13}	0	0
$CHCO + O \rightarrow$ products		1.2×10^{12}	0	0
$CHCO + O_2 \rightarrow$ products		2.0×10^{13}	0	0
		Reactions of C_3H_8		
$C_3H_8 + H \rightarrow C_3H_7 + H_2$		2.1×10^{14}	0	36.5
$C_3H_8 + O \rightarrow C_3H_7 + OH$		5.2×10^{13}	0	20.1
$C_3H_8 + OH \rightarrow C_3H_7 + H_2O$		6.3×10^{12}	0	4.9
$C_3H_8 + HO_2 \rightarrow C_3H_7 + H_2O_2$		6.2×10^{12}	0	74.0
$C_3H_8 + CH_3 \rightarrow C_3H_7 + CH_4$		2.0×10^{12}	0	47.3
$C_3H_8 \rightarrow C_2H_5 + CH_3$	(k_∞)	5.0×10^{15}	0	350
		Reactions of C_3H_7		
$i\text{-}C_3H_7 + H \rightarrow C_3H_8$	(k_∞)	2.0×10^{13}	0	0
$C_3H_7 + O \rightarrow$ products		4.5×10^{13}	0	0
$i\text{-}C_3H_7 + O_2 \rightarrow C_3H_6 + HO_2$		1.0×10^{12}	0	12.5
$n\text{-}C_3H_7 + O_2 \rightarrow C_3H_6 + HO_2$		1.0×10^{12}	0	21.0
$i\text{-}C_3H_7 + i\text{-}C_3H_7 \rightarrow C_6H_{14}$	(k_∞)	4.0×10^{12}	0	0
$n\text{-}C_3H_7 + n\text{-}C_3H_7 \rightarrow C_6H_{14}$	(k_∞)	1.0×10^{13}	0	0

Reaction		A	b	E
$i\text{-}C_3H_7 + i\text{-}C_3H_7 \rightarrow C_3H_6 + C_3H_8$		2.4×10^{12}	0	0
$n\text{-}C_3H_7 + n\text{-}C_3H_7 \rightarrow C_3H_6 + C_3H_8$		1.6×10^{12}	0	0
$i\text{-}C_3H_7 \rightarrow C_3H_6 + H$	(k_∞)	2.0×10^{14}	0	162
$n\text{-}C_3H_7 \rightarrow CH_3 + C_2H_4$	(k_∞)	3.0×10^{14}	0	139
$n\text{-}C_3H_7 \rightarrow C_3H_6 + H$	(k_∞)	1.0×10^{14}	0	156
Reactions of C_3H_6				
$C_3H_6 + H \rightarrow i\text{–}C_3H_7$	(k_∞)	4.0×10^{12}	0	4.0
$C_3H_6 + H \rightarrow n\text{–}C_3H_7$	(k_∞)	4.0×10^{12}	0	11.0
$C_3H_6 + O \rightarrow$ products	(k_∞)	5.0×10^{12}	0	1.9
$C_3H_6 + OH \rightarrow$ products		1.0×10^{13}	0	0
$C_3H_6 + CH_3 \rightarrow C_3H_5 + CH_4$		1.4×10^{11}	0	36.8
Reactions of C_3H_4				
$CH_3CCH + H \rightarrow C_3H_5$	(k_∞)	2.0×10^{13}	0	10.0
$CH_2CCH_2 + H \rightarrow C_3H_5$	(k_∞)	1.2×10^{13}	0	8.8
$CH_3CCH + O \rightarrow$ products		1.5×10^{13}	0	8.8
$CH_3CCH + OH \rightarrow$ products		5.0×10^{12}	0	5.4
$CH_2CCH_2 + OH \rightarrow$ products		2.7×10^{11} (298 K)	…	…
Reactions of C_4H_{10}				
$i\text{-}C_4H_{10} + H \rightarrow C_4H_9 + H_2$		2.1×10^{14}	0	31.5
$n\text{-}C_4H_{10} + O \rightarrow C_4H_9 + OH$		6.2×10^{13}	0	21.4
$i\text{-}C_4H_{10} + O \rightarrow C_4H_9 + OH$		4.7×10^{13}	0	16.3
$n\text{-}C_4H_{10} + OH \rightarrow C_4H_9 + H_2O$		9.0×10^{12}	0	4.4
$i\text{-}C_4H_{10} + OH \rightarrow C_4H_9 + H_2O$		5.8×10^{12}	0	2.6
$n\text{-}C_4H_{10} + HO_2 \rightarrow C_4H_9 + H_2O_2$		8.0×10^{12}	0	72.9
$i\text{-}C_4H_{10} + HO_2 \rightarrow C_4H_9 + H_2O_2$		4.4×10^{12}	0	64.0
$n\text{-}C_4H_{10} + CH_3 \rightarrow C_4H_9 + CH_4$		4.0×10^{11}	0	40.2

Table 33 (*continued*).

Reaction		$A/(cm^3/mol)^{n-1}\,s^{-1}$	b	E/kJ
$i\text{-}C_4H_{10} + CH_3 \rightarrow C_4H_9 + CH_4$		9.5×10^{10}	0	33.1
$n\text{-}C_4H_{10} \rightarrow C_2H_5 + C_2H_5$	(k_∞)	2.0×10^{16}	0	340
$n\text{-}C_4H_{10} \rightarrow CH_3 + n\text{-}C_3H_7$	(k_∞)	2.0×10^{16}	0	340
$i\text{-}C_4H_{10} \rightarrow CH_3 + i\text{-}C_3H_7$	(k_∞)	4.0×10^{16}	0	347
Reactions of C_4H_9				
$C_4H_9 + O \rightarrow$ products		4.0×10^{13}	0	0
$i\text{-}C_4H_9 + O_2 \rightarrow C_4H_8 + HO_2$		1.0×10^{12}	0	23.9
$n\text{-}C_4H_9 + O_2 \rightarrow C_4H_8 + HO_2$		1.6×10^{11} (753 K)	...	...
$s\text{-}C_4H_9 + O_2 \rightarrow C_4H_8 + HO_2$		1.7×10^{11} (753 K)	...	...
$t\text{-}C_4H_9 + O_2 \rightarrow C_4H_8 + HO_2$		8.0×10^{11}	0	9.1
$t\text{-}C_4H_9 + t\text{-}C_4H_9 \rightarrow C_8H_{18}$	(k_∞)	1.0×10^{12}	0	0
$t\text{-}C_4H_9 + t\text{-}C_4H_9 \rightarrow C_4H_8 + C_4H_{10}$		2.5×10^{12}	0	0
$n\text{-}C_4H_9 \rightarrow C_2H_4 + C_2H_5$	(k_∞)	3.7×10^{13}	0	120
$s\text{-}C_4H_9 \rightarrow C_3H_6 + CH_3$	(k_∞)	2.3×10^{14}	0	137
$i\text{-}C_4H_9 \rightarrow C_3H_6 + CH_3$	(k_∞)	7.0×10^{12}	0	110
$t\text{-}C_4H_9 \rightarrow C_3H_6 + CH_3$	(k_∞)	1.0×10^{16}	0	193
Reactions of C_4H_8				
$1\text{-}C_4H_8 + H \rightarrow C_4H_9$	(k_∞)	1.0×10^{12} (298 K)	...	...
$trans\text{-}2\text{-}C_4H_8 + H \rightarrow C_4H_9$	(k_∞)	5.5×10^{11} (298 K)	...	...
$cis\text{-}2\text{-}C_4H_8 + H \rightarrow C_4H_9$	(k_∞)	4.5×10^{11} (298 K)	...	...
$1\text{-}C_4H_8 + O \rightarrow$ products		8.0×10^{12}	0	3.4
$iso\text{-}C_4H_8 + O \rightarrow$ products		9.0×10^{12}	0	0
$cis\text{-}2\text{-}C_4H_8 + O \rightarrow$ products		7.0×10^{12}	0	−1.0
$trans\text{-}2\text{-}C_4H_8 + O \rightarrow$ products		1.4×10^{13}	0	0

Reaction				
1-C_4H_8 + OH → products		4.0×10^{12}	0	−3.9
iso-C_4H_8 + OH → products		5.5×10^{12}	0	−4.2
cis-2-C_4H_8 + OH → products		6.0×10^{12}	0	−4.1
trans-2-C_4H_8 + OH → products		6.7×10^{12}	0	−4.6
1-C_4H_8 + CH_3 → C_4H_7 + CH_4		2.5×10^{11}	0	34.8
cis-2-C_4H_8 + CH_3 → C_4H_7 + CH_4		1.8×10^{11}	0	33.9
trans-2-C_4H_8 + CH_3 → C_4H_7 + CH_4		1.0×10^{12}	0	40.2
Reactions of C_4H_2				
C_4H_2 + H → C_4H_3	(k_x)	6.5×10^{12}	0	4.2
C_4H_2 + O → C_3H_2 + CO		2.7×10^{13}	0	7.2
C_4H_2 + OH → products		3.0×10^{13}	0	0

errors in computing the temperature in parts of the flame where both directions proceed rapidly, since there are inevitable small inconsistencies between the thermochemistry implied by k_f/k_r and the thermochemistry used otherwise in the program to compute energy equations.

9. Acknowledgments

Preparation of this chapter was supported by grants from the Karl-Winnacker Stiftung, the Deutsche Forschungsgemeinschaft, and the North Atlantic Treaty Organization. The author wishes to thank all who have provided assistance in collecting and discussing the data presented here, in particular, Prof. Dr. H. Gg. Wagner, Prof. Dr. K. H. Homann and Prof. Dr. G. Dixon-Lewis. The manuscript for this chapter was completed while the author was on sabbatical leave at the Combustion Research Facility, Sandia National Laboratories, Livermore, California.

10. References

Adachi, H., Basco, N., & James, D. G. L. (1979). Int. J. Chem. Kin. **11,** 995.

Adachi, H., Basco, N., & James, D. G. L. (1980). Int. J. Chem. Kin. **12,** 949.

Aders, W.-K. & Wagner, H. Gg. (1971). Z. Physik. Chem. NF **74,** 224.

Aders, W.-K. & Wagner, H. Gg. (1973). Ber. Bunsenges. Phys. Chem. **77,** 332.

Ahumada, J. J., Michael, J. V., & Osborne, D. T. (1972). J. Chem. Phys. **57,** 3736.

Albers, E. A. *et al.* (1971). 13th Symposium (International) on Combustion, p. 81.

Aronowitz, D., Naegeli, D. W., & Glassman, I. (1977) J. Phys. Chem. **81,** 2555.

Arrington, C. A. & Cox, D. J. (1975). J. Phys. Chem. **798,** 2584.

Atkinson, R. & Cvetanovic, R. J. (1971). J. Chem. Phys. **55,** 659.

Atkinson, R. & Cvetanovic, R. J. (1972). J. Chem. Phys. **56,** 432.

Atkinson, R. & Pitts, J. N. (1974a). Chem. Phys. Letters **27,** 467.

Atkinson, R. & Pitts, J. N. (1974b). J. Phys. Chem. **78,** 1780.

Atkinson, R. & Pitts, J. N. (1975). J. Chem. Phys. **63,** 3591.

Atkinson, R. & Pitts, J. N. (1977a). J. Chem. Phys. **67,** 38.

Atkinson, R. & Pitts, J. N. (1977b). J. Chem. Phys. **68,** 2492.

Atkinson, R. & Pitts, J. N. (1978). J. Chem. Phys. **68,** 3581.

Atkinson, R., Hansen, D. A., & Pitts, J. N. (1975). J. Chem. Phys. **62,** 3284.

Atkinson, R., Perry, R. A., & Pitts, J. N. (1977). J. Chem. Phys. **66,** 1197.

Atri, G. M. *et al.* (1977). Comb. Flame **30,** 1.

Avery, H. E. & Cvetanovic, R. J. (1965). J. Chem. Phys. **43,** 3727.

Azatyan, V. V., Aleksandrov, E. N., & Konstantinov, P. V. (1971). Vses. Konf. Kinet. Mekh. Gazofazn. Reakts. 2nd, 33.

Back, R. A. & Takamuku, S. (1964). J. Amer. Chem. Soc. **86,** 2558.

Baker, R. R., Baldwin, R. R., & Walter, R. W. (1970a). Trans. Faraday Soc. **66,** 3016.

Baker, R. R., Baldwin, R. R., & Walker, R. W. (1970b). Trans. Faraday Soc. **66,** 3812.

Baker, R. R., Baldwin, R. R., & Walker, R. W. (1971). 13th Symposium (International) on Combustion, p. 291.

Baker, R. R. *et al.* (1975). J. Chem. Soc. Faraday I **71,** 736.

Baldwin, R. R. (1964). Trans. Faraday Soc. **60,** 527.

Baldwin, R. R. & Cowe, D. W. (1962). Trans. Faraday Soc. **58,** 1768.

Baldwin, A. C. & Golden, D. M. (1978). Chem. Phys. Letters **55,** 350.

Baldwin, R. R. & Balker, R. W. (1979). 17th Symposium (International) on Combustion, p. 525.

Baldwin, R. R. & Walker, R. W. (1981). 18th Symposium (International) on Combustion, p. 819.

Baldwin, R. R., Simmons, R. F., & Walker, R. W. (1966). Trans. Faraday Soc. **65,** 806.

Baldwin, R. R., Hopkins, D. E., & Walker, R. W. (1970). Trans. Faraday Soc. **66,** 189.

Baldwin, R. R. *et al.* (1971). 13th Symposium (International) on Combustion, p. 251.

Baldwin, R. R. *et al.* (1972). Int. J. Chem. Kin. **4,** 277.

Baldwin, R. R., Walker, R. W., Yorke, D. A. (1973). J. Chem. Soc. Faraday I **69,** 826.

Baldwin, R. R., Cleugh, C. J., & Walker, R. J. (1976). J. Chem. Soc. Faraday I **72,** 1715.

Baldwin, R. R., Bennett, J. P., & Walker, R. W. (1977). 16th Symposium (International) on Combustion, p. 1041.

Barker, J. R. *et al.* (1970). J. Chem. Phys. **52,** 2079.

Barker, J. R., Benson, S. W., & Golden, D. M. (1977). Int. J. Chem. Kin. **9,** 31.

Bar-Nun, A. & Dove, J. E. (1980). Proc. 12th International Symp. on Shock Tubes and Waves, Magnes Press, Jerusalem, p. 457.

Basevich, V. Y., Kogarko, S. M., & Furman, G. A. (1971). Izv. Akad. Nauk SSSR, Ser. Khim **1971,** 1406.

Batt, I. & Robinson, G. N. (1979). Int. J. Chem. Kin. **11,** 1045.

Baulch, D. L. & Duxbury, J. (1980). Comb. Flame **37,** 313.

Baulch, D. L., *et al.* (1972). *Evaluated Kinetic Data for High Temperature Reactions, Vol. 3: Homogeneous Gas Phase Reactions in the* O_2-O_3 *System, the* CO-CO_2-H_2 *System, and of the Sulfur-containing Species*, Butterworths, London.

Baulch, D. L., Drysdale, D. D., & Horne, D. G. (1973). *Evaluated Kinetic Data for High Temperature Reactions, Vol. 2: Homogeneous Gas Phase Reactions in the* H_2-N_2-O_2 *System*, Butterworths, London.

Baulch, D. L. *et al.* (1976). *Evaluated Kinetic Data for High Temperature Reactions, Vol. 1: Homogeneous Gas Phase Reactions of the* H_2-O_2 *System*, Butterworths, London.

Beeley, P. *et al.* (1977). 16th Symposium (International) on Combustion, p. 1013.

Bennett, J. E., & Mile, B. (1973). J. Chem. Soc. Faraday I **69,** 1389.

Benson, S. W. (1976). *Thermochemical Kinetics*, John Wiley and Sons, New York.

Benson, S. W. & Haugen, G. R. (1967a). J. Phys. Chem. **71,** 1735.

Benson, S. W. & Haugen, G. R. (1967b). J. Phys. Chem. **71,** 4404.

Benson, S. W. & O'Neal, H. E. (1970). *Kinetic Data on Gas Phase Unimolecular Reactions*, U.S. Government Printing Office, Washington, D.C.

Berlie, M. R. & LeRoy, D. J. (1953). Disc. Faraday Soc. **14,** 50.

Bethune, D. S. *et al.* (1981). J. Chem. Phys. **75,** 2231.

Bhaskaran, K. A., Frank, P., & Just, Th. (1979). 12th Shock Tube Symposium, Jerusalem.

Biermann, H. W., Zetzsch, C., & Stuhl, F. (1978). Ber. Bunsenges. Phys. Chem. **82,** 633.

Biordi, J. C., Papp, J. F., & Lazzara, C. P. (1974). J. Chem. Phys. **61,** 741.

Biordi, J. C., Lazzara, C. P., & Papp, J. F. (1975). 15th Symposium (International) on Combustion, p. 917.

Biordi, J. C., Lazzara, C. P., & Papp, J. F. (1976). Comb. Flame **26,** 57.

Birell, R. N. & Trotman-Dickenson, A. F. (1960). J. Chem. Soc. **1960,** 4218.

Blumenberg, B., Hoyermann, K., & Sievert, R. (1977). 16th Symposium (International) on Combustion, p. 841.

Bosnali, M. W. & Perner, D. (1971). Z. Naturforsch. **26A,** 1768.

Bowman, C. T. (1970). Comb. Sci. Technol. **2,** 161.

Bowman, C. T. (1975a). 15th Symposium (International) on Combustion, p. 869.

Bowman, C. T. (1975b). Comb. Flame **25,** 343.

Brabbs, T. A. & Brokaw, R. S. (1975). 15th Symposium (International) on Combustion, p. 893.

Bradley, J. N. (1974). Proc. Roy. Soc. **A337,** 199.

Bradley, J. N. (1979). J. Chem. Soc. Faraday I **75,** 2819.

Bradley, J. N. & Tse, R. S. (1969). Trans. Faraday Soc. **65,** 2685.

Bradley, J. N. & West, K. O. (1976a). J. Chem. Soc. Faraday I **72,** 8.

Bradley, J. N. & West, K. O. (1976b). J. Chem. Soc. Faraday I **72,** 558.

Bradley, J. N. *et al.* (1973). J. Chem. Soc. Faraday I **69,** 1889.

Bradley, J. N. *et al.* (1976). Int. J. Chem. Kin. **8,** 549.

Braun, W., McNesby, J. R., & Bass, A. M. (1967). J. Chem. Phys. **46,** 2071.

Braun, W., Bass, A. M., & Pilling, M. (1970). J. Chem. Phys. **52,** 5131.

Breen, J. E. & Glass, G. P. (1971). Int. J. Chem. Kin. **3,** 145.

Brennen, W. R. *et al.* (1965). J. Chem. Phys. **43,** 2569.

Breshears, W. D. & Bird, P. F. (1973). 14th Symposium (International) on Combustion, p. 211.

Brooks, C. T. (1966), Trans. Faraday Soc. **62,** 935.

Brown, J. M. & Thrush, B. A. (1967). Trans. Faraday Soc. **63,** 630.

Browne, W. G. *et al.* (1969). 12th Symposium (International) on Combustion, p. 1035.

Burcat, A. *et al.* (1973). Int. J. Chem. Kin. **5,** 345.

Burrows, J. P., Harris, G. W., & Thrush, B. A. (1977). Nature **267,** 233.

Buss, R. J. *et al.* (1981). Xth International Conference on Photochemistry (abstracts).

Butler, R., Solomon, I. J., & Snelson, A. (1978). Chem. Phys. Letters **54,** 19.

Butler, J. W. *et al.* (1980). *ACS Symposium Series 134,* The American Chemical Society, Washington, D.C.

Cadle, R. D. & Powers, J. W. (1967). J. Phys. Chem. **71,** 1702.

Calvert, J. G. & Sleepy, W. C. (1959). J. Amer. Chem. Soc. **81,** 1544.

Camilleri, P., Marshall, R. M., & Purnell, J. H. (1974). J. Chem. Soc. Faraday I **70,** 1434.
Camilleri, P., Marshall, R. M., & Purnell, J. H. (1975). J. Chem. Soc. Faraday I **71,** 1491.
Campbell, I. M. & Handy, B. J. (1975). J. Chem. Soc. Farad. Trans. I **71,** 2097.
Campbell, I. M. & Handy, B. J. (1977). Chem. Phys. Letters **47,** 475.
Campbell, I. M. & Handy, B. J. (1978). J. Chem. Soc. Faraday I **74,** 316.
Campbell, I. M., McLaughlin, D. F., & Handy, B. J. (1976). Chem. Phys. Lett. **38,** 362.
Campbell, I. M., Rogerson, J. S., & Handy, B. J. (1978). J. Chem. Soc. Faraday I **74,** 2672.
Carr, R. W. *et al.* (1968). J. Chem. Phys. **49,** 846.
Chan, W. H. *et al.* (1977). Chem. Phys. Letters **45,** 240.
Chang, J. S. & Barker, J. R. (1979). J. Phys. Chem. **83,** 3059.
Chang, J. S. & Kaufman, F. (1978). J. Phys. Chem. **82,** 1683.
Chen, C.-J. & Back, M. H. (1975). Can. J. Chem. **53,** 3580.
Chen, C.-J. & McKenney, D. J. (1972). Can. J. Chem. **50,** 992.
Chen, C.-J., Back, M. H., & Back, R. A. (1976). Can. J. Chem. **54,** 3175.
Cheng, J.-T. & Yeh, C.-T. (1977). J. Phys. Chem. **81,** 1982.
Cherian, M. A. *et al.* (1981). 18th Symposium (International) on Combustion, p. 385.
Chiang, C.-C. & Skinner, G. B. (1979). 12th Shock Tube Symposium, Jerusalem.
Chiang, C.-C. & Skinner, G. B. (1981). 18th Symposium (International) on Combustion, p. 915.
Choo, K. Y. *et al.* (1976). Int. J. Chem. Kin. **8,** 45.
Clark, T. C. & Dove, J. E. (1973a). Can. J. Chem. **51,** 2147.
Clark, T. C. & Dove, J. E. (1973b). Can. J. Chem. **51,** 2155.
Clark, T. C., Izod, T. P. J., & Kistiakowsky, G. B. (1971a) J. Chem. Phys. **54,** 1295.
Clark, T. C., Izod, T. P. J., & Matsuda, S. (1971b) J. Chem. Phys. **55,** 4644.
Clark, J. A. & Quinn, C. P. (1976). Trans. Faraday Soc. **72,** 706.
Clyne, M. A. A. & Down, S. (1974). J. Chem. Soc. Faraday II **70,** 253.
Coats, C. M. & Williams, A. (1979). 17th Symposium (International) on Combustion, p. 611.
Colket, M. B., Naegeli, D. W., & Glassman, I. (1975). Int. J. Chem. Kin. **7,** 223.
Colket, M. B., Naegeli, D. W., & Glassman, I. (1977). 16th Symposium (International) on Combustion, p. 1023.
Cooke, D. F. & Williams, A. (1971). 13th Symposium (International) on Combustion, p. 757.
Cowfer, J. A. *et al.* (1971). J. Phys. Chem. **75,** 1584.
Cox, R. A. & Burrows, J. P. (1979). J. Phys. Chem. **83,** 2560.
Cox, R. A., Derwent, R. G., & Holt, P. M. (1976a) J. Chem. Soc. Faraday I **72,** 2031.
Cox, R. A. et al. (1976b). J. Chem. Soc. Faraday I **72,** 2061.
Cullis, C. F., Hucknall, D. J., & Shepherd, J. V. (1973). Proc. Roy. Soc. A**335,** 525.
Cvetanovic, R. J. (1956). Can. J. Chem. **34,** 775.
Daby, E. E., Niki, H., & Weinstock, B. (1971). J. Phys. Chem. **75,** 1601.
Darnall, K. R., Atkinson, R., & Pitts, J. N. (1978). J. Phys. Chem. **82,** 1581.

Davis, D. D. *et al.* (1972). J. Chem. Phys. **56,** 4848.

Davis, D. D., Huie, R. E., & Herron, J. T., (1973). J. Chem. Phys. **59,** 628.

Davis, D. D., Wong, W., & Schiff, R. (1974a). J. Phys. Chem. **78,** 463.

Davis, D. D., Fischer, S., & Schiff, R. (1974b). J. Chem. Phys. **61,** 2213.

Davis, D. D. *et al.* (1975). J. Chem. Phys. **63,** 1707.

Day, M. J., Thompson, K., & Dixon-Lewis, G. (1973). 14th Symposium (International) on Combustion, p. 47.

Dean, A. M. & Kistiakowsky, G. B. (1971). J. Chem. Phys. **54,** 1718.

Dean, A. M. & Steiner, D. C. (1977). J. Chem. Phys. **66,** 598.

Dean A. M. *et al.* (1979). 17th Symposium (International) on Combustion, p. 577.

Dean, A. M., Johnson, R. L., & Steiner, D. C. (1980). Comb. Flame **37,** 41.

Demore, W. B. & Tschuikow-Roux. E. (1974). J. Phys. Chem. **78,** 1447.

Dingledey, D. P. & Calvert, J. G. (1963). J. Amer. Chem. Soc. **85,** 856.

Dixon, P. S., Stefani, A. P., & Swarc, M. (1963). J, Amer. Chem. Soc. **85,** 2551.

Dixon-Lewis, G. (1983). Comb. Sci. Tech. (in press).

Dixon-Lewis, G. & Rhodes, P. (1975). Deuxieme Symposium European sur la Combustion, p. 473.

Dixon-Lewis, G., Greenberg, J. B., & Goldsworthy, F. A. (1975). 15th Symposium (International) on Combustion, p. 717.

Dodonov, A. F., Lavrovskaya, G. K., & Tal'roze, V. L. (1969). Kinet. Katal. **10,** 477.

Dryer, F. L. & Glassman, I. (1979). In: *Alternative Hydrocarbon Fuels, Combustion and Kinetics* C. T. Bowman & J. Birkeland Ed., AIAA, New York.

Dubinsky, R. N. & McKenney, D. J. (1975). Can. J. Chem. **53,** 3531.

Durban, P. C. & Marshall, R. M. (1980). Int. J. Chem. Kin. **12,** 1031.

Eberius, K. H., Hoyermann, K., & Wagner, H. Gg. (1973). 14th Symposium (International) on Combustion, p. 147.

Elias, L. (1963). J. Chem. Phys. **38,** 989.

Elias, L. & Schiff, H. I. (1960). Can. J. Chem. **38,** 1657.

Ernst, J., Wagner, H. Gg., & Zellner, R. (1977). Ber. Bunsenges. Phys. Chem. **81,** 1270.

Ernst, J., Wagner, H. Gg., & Zellner, R. (1978). Ber. Bunsenges. Phys. Chem. **82,** 409.

Eubank. C. S., Gardiner, W. C., & Simmie, J. M. (1981). First International Specialists Meeting of the Combustion Institute, Bordeaux (in press).

Eusuf, M. & Wagner, H. Gg. (1972). Ber. Bunsenges. Phys. Chem. **76,** 437.

Evans. G. A. & Walker, R. W. (1979). J. Chem. Soc. Faraday I **75,** 1458.

Eyre, J. A., Hikita, T., & Dorfman, L. M. (1970). J. Chem. Phys. **53,** 1281.

Falconer, W. E. & Sunder, W. A. (1972). Int. J. Chem. Kin. **4,** 315.

Falconer, W. E., Rabinovitch, B. S., & Cvetanovic, R. J. (1963). J. Chem. Phys. **39,** 40.

Faubel, C. (1977). Dissertation, Universität Göttingen.

Faubel, C., Wagner, H. Gg., & Hack, W. (1977). Ber. Bunsenges. Phys. Chem. **81,** 689.

Felder, W. & Fontijn, A. (1979). Chem. Phys. Lett. **67,** 53.

Felder, W. & Fontijn, A. (1981). 18th Symposium (International) on Combustion, p. 797.

Fenimore, C. P. (1969). 12th Symposium (International) on Combustion, p. 463.
Fenimore, C. P. & Jones, G. W. (1963). 9th Symposium (International) on Combustion, p. 597.
Fenimore, C. P. & Jones, G. W. (1964). J. Chem. Phys. **41,** 1887.
Frank, P. & Just, Th. (1980). Comb. Flame **38,** 231.
Frey, H. M. & Vinall, I. C. (1973). Int. J. Chem. Kin. **5,** 523.
Friswell, N. J. & Sutton, M. M. (1972). Chem. Phys. Letters **15,** 108.
Furuyama, S. *et al.* (1974). Int. J. Chem. Kin. **6,** 741.
Gaedtke, H. *et al.* (1973). 14th Symposium (International) on Combustion, p. 295.
Gardiner, W. C. & Walker, B. F. (1968). J. Chem. Phys. **48,** 5279.
Gardiner, W. C. *et al.* (1973). 14th Symposium (International) on Combustion, p. 61.
Gardiner, W. C., Mallard, W. G., & Owen, J. H. (1974). J. Chem. Phys. **60,** 2290.
Gardiner, W. C. (1975). 15th Symposium (International) on Combustion, p. 857.
Gardiner, W. C. *et al.* (1981). In: Flames, Lasers, and Reactive Systems J. R. Bowen, N. Manson, A. K. Oppenheim & R. I. Soloukhin Ed., AIAA, New York.
Gay, I. D. *et al.* (1965). J. Chem. Phys. **43,** 4017.
Glass, G. P. *et al.* (1965). J. Chem. Phys. **42,** 608.
Golden, D. M., Spokes, G. N., & Benson, S. W. (1973). Angew. Chem. **85,** 602.
Golden, D. M. (1974a). J. Amer. Chem. Soc. **96,** 1645.
Golden, D. M., Alfassi, Z. B., & Beadle, P. C. (1974b). Int. J. Chem. Kin. **6,** 359.
Golden, D. M. (1976). Int. J. Chem. Kin. **8,** 381.
Goldfinger, P. *et al.* (1965). Trans. Faraday Soc. **61,** 1933.
Gordon, S. & Mulac, W. A. (1975). Int. J. Chem. Kin. Symp. **1,** 289.
Gorse, R. A. & Volman, D. H. (1972). J. Photochem. **1,** 1.
Gorse, R. A. & Volman, D. H. (1974). J. Photochem. **3,** 115.
Graham, R. A. *et al.* (1979). J. Phys. Chem. **83,** 1563.
Gray, P. & Herod, A. A. (1968). Trans. Faraday Soc. **64,** 2723.
Greiner, N. R. (1970a). J. Chem. Phys. **53,** 1284.
Greiner, N. R. (1970b). J. Chem. Phys. **53,** 1070.
Grotheer, H. H. & Just, Th. (1981). (to be published).
Gruver, J. T. & Calvert, J. G. (1956). J. Amer. Chem. Soc. **78,** 5208.
Hack, W. (1977). Ber. Bunsenges, Phys. Chem. **81,** 1118.
Hack, W., Hoyermann, K., & Wagner, H. Gg. (1974). Z. Naturforsch. **A29,** 1236.
Hack, W., Hoyermann, K., & Wagner, H. Gg. (1975). Int. J. Chem. Kin., Symposium No. 1 , p. 329.
Hack, W., Wagner, H. Gg., & Hoyermann, K. (1978a). Ber. Bunsenges. Phys. Chem. **82,** 713.
Hack, W., Preuss, A. W., & Wagner, H. Gg. (1978b). Ber. Bunsenges. Phys. Chem. **82,** 1167.
Hack, W. *et al.* (1979). Ber. Bunsenges. Phys. Chem. **83,** 1275.
Halstead, M. P. *et al.* (1969). Proc. Roy. Soc. A **310,** 525.
Halstead, M. P. *et al.* (1970). Proc. Roy. Soc. A **316,** 575.

Hamilton, E. J. & Lii, R.-R. (1977). Int. J. Chem. Kin. **9,** 875.

Hardy, W. A. *et al.* (1974). Ber. Bunsenges. Phys. Chem. **78,** 76.

Hardy, J. E., Gardiner, W. C., & Burcat, A. (1978). Int. J. Chem. Kin. **10,** 503.

Harker, A. B. & Burton, C. S. (1975). Int. J. Chem. Kin. **7,** 907.

Harris, G. W. & Pitts, J. N. (1979). J. Chem. Phys. **70,** 2581.

Hartig, R., Troe, J., & Wagner, H. Gg. (1971). 13th Symposium (International) on Combustion, p. 147.

Havel, J. J. (1974). J. Amer. Chem. Soc. **96,** 530.

Heffington, W. M. *et al.* (1977). 16th Symposium (International) on Combustion, p. 997.

Heicklen, J. & Meagher, J. F. (1974). J. Photochem. **3,** 455.

Held, A. M. *et al.* (1977). Can. J. Chem. **55,** 4128.

Heller, C. A. & Gordon, A. S. (1960). J. Phys. Chem. **64,** 390.

Herbrechtsmeier, P. & Wagner, H. Gg. (1974). Z. Physik. Chem. NF **93,** 143.

Herron, J. T. & Huie, R. E. (1969). J. Phys. Chem. **73,** 3327.

Herron, J. T. & Penzhorn, R. D. (1969). J. Phys. Chem. **73,** 191.

Hiatt, R. & Benson, S. W. (1972). J. Amer. Chem. Soc. **94,** 6886.

Hidaka, Y., Eubank, C. E., & Gardiner, W. C., Jr. (1982). J. Mol. Sci. **3,** 135.

Himme, B. (1977). Dissertation, Universität Göttingen.

Hochanadel, C. J., Ghormley, J. A., & Ogren, P. J. (1972). J. Chem. Phys. **56,** 4426.

Hochanadel, C. J., Sworski, T. J., & Ogren, P. J. (1980a). J. Phys. Chem. **84,** 3274.

Hochanadel, C. J., Sworsky, T. J., & Ogren, P. J. (1980b). J. Phys. Chem. **84,** 231.

Holt, P. M. & Kerr, J. A. (1977). Int. J. Chem. Kin. **9,** 185.

Homann, K. H. & Schweinfurth, H. (1981). Ber. Bunsenges. Phys. Chem. **85,** 569.

Homann, K. H., Warnatz, J., & Wellmann, C. (1977). 17th Symposium (International) on Combustion, p. 853.

Homann, K. H., Schwanebeck, W., & Warnatz, J. (1975). Ber. Bunsenges. Phys. Chem. **79,** 536.

Homer, J. B. & Hurle, I. R. (1969). Proc. Roy. Soc. Ser. A **314,** 585.

Hooper, D. G., Simon, M., & Back, M. H. (1975). Can. J. Chem. **53,** 1237.

Howard, C. J. (1976). J. Chem. Phys. **65,** 4771.

Howard, C. J. & Evenson, K. M. (1976a). J. Chem. Phys. **64,** 4303.

Howard, C. J. & Evenson, K. M. (1976b). J. Chem. Phys. **64,** 197.

Howard, M. J. & Smith, I. W. M. (1980). Chem. Phys. Letters **69,** 40.

Hoyermann, K. (1981). (private communication).

Hoyermann, K. & Sievert, R. (1979a). 17th Symposium (International on Combustion, p. 517.

Hoyermann, K. & Sievert, R. (1979b). Ber. Bunsenges. Phys. Chem. **83,** 732.

Hoyermann, K. & Sievert, R. (1979c). Ber. Bunsenges. Phys. Chem. **83,** 933.

Hoyermann, K., Wagner, H. Gg., & Wolfrum, J. (1968). Ber. Bunsenges. Phys. Chem. **72,** 1004.

Hoyermann, K., Wagner, H. Gg., & Wolfrum, J. (1969). Z. Physik. Chem. NF **63,** 193.

Hoyermann, K., Sievert, R., & Wagner, H. Gg. (1981a) (to be published).

Hoyermann, K. *et al.* (1981b). 18th Symposium (International) on Combustion, p. 831.

Hughes, D. G. & Marshall, R. M. (1975). J. Chem. Soc. Faraday I **71,** 413.

Hughes, D. G., Marshall, R. M., & Purnell, J. H. (1974). J. Chem. Soc. Faraday I **70,** 594.

Huie, R. E. & Herron, J. T. (1975). *Progress in Reaction Kinetics,* Vol. 8, Part 1, Pergamon Press, Oxford p. 1.

Huie, R. E., Herron, J. T., & Davis, D. D. (1971). J. Phys. Chem. **75,** 3902.

Huie, R. E., Herron, J. T., & Davis, D. D. (1972). J. Phys. Chem. **76,** 3311.

Ishikawa, Y. *et al.* (1978). Bull. Chem. Soc. Jpn. **51,** 2488.

Izod, T. P., Kistiakowsky, G. B., & Matsuda, S. (1971). J. Chem. Phys. **55,** 4425.

Jachimowski, C. J. (1974). Comb. Flame **23,** 233.

Jachimowski, C. J. (1977). Comb. Flame **29**, 55.

Jackson, W. M. & McNesby, J. R. (1961). J. Amer. Chem. Soc. **83,** 4891.

Jackson, W. M. & McNesby, J. R. (1962). J. Chem. Phys. **36,** 2272.

James, G. S. & Glass, G. P. (1969). J. Chem. Phys. **50,** 2268.

JANAF Thermochemical Tables (1971). D. R. Stull & H. Prophet, Project Directors, National Bureau of Standards, Washington, D.C.

Jones, I. T. N. & Bayes, K. D. (1972). J. Amer. Soc. **94,** 6869.

Jones, I. T. N. & Bayes, K. D. (1973). 14th Symposium (International) on Combustion, p. 277.

Just, Th., Roth, P., & Damm, R. (1977). 16th Symposium (International) on Combustion, p. 961.

Kanofsky, J. R., Lucas, D., & Gutman, D. (1973). 14th Symposium (International) on Combustion, p. 285.

Kanofsky, J. R. *et al.* (1974). J. Phys. Chem. **78,** 311.

Kazmi, H. A., Diefendorf, R. J., & LeRoy, D. J. (1963). Can. J. Chem. **41,** 690.

Keil, D. G. *et al.* (1976). Int. J. Chem. Kin. **8,** 825.

Keil, D. G. *et al.* (1981). (to be published).

Kerr, J. A. (1973). In: *Free Radicals,* Vol. 1, J. K. Kochi, Ed., John Wiley, New York.

Kerr, J. A. & Calvert, J. G. (1961). J. Amer. Chem. Soc. **83,** 3391.

Kerr, J. A. & Calvert, J. G. (1965). J. Phys. Chem. **69,** 1022.

Kerr, J. A. & Parsonage, M. J. (1972). *Evaluated Kinetic Data on Gas Phase Addition Reactions: Reactions of Atoms and Radicals with Alkenes, Alkynes, and Aromatic Compounds,* Butterworths, London.

Kerr, J. A. & Parsonage, M. J. (1976). *Evaluated Kinetic Data on Gas Phase Hydrogen Transfer: Reactions of Methyl Radicals,* Butterworths, London.

Kerr, J. A. & Trotman-Dickenson, A. F. (1959). Trans. Faraday Soc. **55,** 921.

Kerr, J. A. & Trotman-Dickenson, A. F. (1960a). J. Chem. Soc. **1960,** 1611.

Kerr, J. A. & Trotman-Dickenson, A. F. (1960b). J. Chem. Soc. **1960,** 1602.

Kerr, J. A. & Trotman-Dickenson, A. F. (1961). Progr. React. Kin. **1,** 107.

Kirsch, L. J. & Parkes, D. A. (1979). J. Chem. Soc. Faraday I **75,** 2694.

Klemm, R. B. (1979). J. Chem. Phys. **71,** 1987.

Klemm, R. B., Payne, W. A., & Stief, L. J. (1975). Int. J. Chem. Kin., Symposium No. **1,** 61.

Klemm, R. B., Skolnik, E. G., & Michael, J. V. (1980). J. Chem. Phys. **72,** 1256.

Klemm, R. B. *et al.* (1981). 18th Symposium (International) on Combustion, p. 785.

Knox, J. H. & Dalgleish, D. G. (1969). Int. J. Chem. Kin. **1,** 69.

Kochubei, V. F. & Moin, F. B. (1973). Ukr. Khim. Zh. **39,** 888.

Koike, T. & Gardiner, W. C. (1980). J. Phys. Chem. **84,** 2005.

Koike, T. & Morinaga, K. (1981). Bull. Chem. Soc. Japn. **54,** 529.

Konar, R. S., Marshall, R. M., & Purnell, J. H. (1973). Int. J. Chem. Kin. **5,** 1007.

Kondratiev, V. N. (1972). *Rate Constants of Gas Phase Reactions, Reference Book,* R. M. Fristrom, Ed., Office of Standard Reference Data, National Bureau of Standards, Washington, D.C.

Kurylo, M. J. (1972a). J. Phys. Chem. **76,** 3518.

Kurylo, M. J. (1972b). Chem. Phys. Letters **14,** 117.

Kurylo, M. J. & Huie, R. E. (1973). J. Chem. Phys. **58,** 1258.

Kurylo, M. J. & Timmons, R. B. (1969). J. Chem. Phys. **50,** 5076.

Kurylo, M. J., Hollinden, G. A., & Timmons, R. B. (1970a). J. Chem. Phys. **52,** 1773.

Kurylo, M. J., Peterson, N. C., & Braun, W. (1970b). J. Chem. Phys. **53,** 2776.

Kurylo, M. J., Peterson, N. C., & Braun, W. (1971). J. Chem. Phys. **54,** 4662.

Laidler, K. J. & Liu, M. T. H. (1967). Proc. Roy. Soc. A **297,** 365.

Lalo, C. & Vermeil, C. (1980). J. Chim. Phys.-Chim. Biol. **77,** 131.

Lambert, R. M., Christie, M. I., & Linnett, J. W. (1967). Chem. Commun. **1967,** 388.

Lange, W. & Wagner, H. Gg. (1975). Ber. Bunsenges. Phys. Chem. **79,** 165.

Laufer, A. H. & Bass, A. M. (1974). J. Phys. Chem. **78,** 1344.

Laufer, A. H. & Bass, A. M. (1975). J. Phys. Chem. **79,** 1635.

Laufer, A. H. & Bass, A. M. (1979). J. Phys. Chem. **83,** 310.

Lede, J. & Villermaux, J. (1978). Can. J. Chem. **56,** 392.

Lee, J. H. *et al.* (1978). J. Chem. Phys. **68,** 1817.

Lee, W.-M. & Yeh, C.-T. (1979). J. Phys. Chem. **83,** 771.

LeFevre, H. F., Meagher, J. F., & Timmons, R. B. (1972). Int. J. Chem. Kin. **4,** 103.

Lenhardt, T. M. McDade, C. E., & Bayes, K. D. (1980). J. Chem. Phys. **72,** 304.

Lewis, R. S. & Watson, R. T. (1980). J. Phys. Chem. **84,** 3495.

Lexton, M. J., Marshall, R. M., & Purnell, J. H. (1971). Proc. Roy. Soc. A **324,** 433.

Lifshitz, A. & Frenklach, M. (1975). J. Phys. Chem. **79,** 686.

Light, G. C. & Matsumoto, J. H. (1980). Int. J. Chem. Kin. **12,** 451.

Lii, R.-R. *et al.* (1979). J. Phys. Chem. **83,** 1803.

Lii, R.-R. *et al.* (1980a). J. Phys. Chem. **84,** 819.

Lii, R.-R, Sauer, M. C., & Gordon, S. (1980b). J. Phys. Chem. **84,** 817.

Lin, M. C. & Back, M. H. (1966). Can. J. Chem. **44,** 2357.

Lin, M. C. & Laidler, K. J. (1966). Can. J. Chem. **44,** 2927.

Lin, M. C. & Laidler, K. J. (1967). Can. J. Chem. **45,** 1315.

Lin, M. C., Shortridge, R. G., & Umstead, M. E. (1976). Chem. Phys. Lett. **37,** 279.

Liu, M. T. H. & Laidler, K. J. (1968). Can. J. Chem. **46,** 479.

Lloyd, A. C. *et al.* (1976). J. Phys. Chem. **80,** 789.

Loucks, L. F. & Laidler, K. J. (1967). Can. J. Chem. **45,** 2795.

Mack, G. P. R. & Thrush, B. A. (1973). J. Chem. Soc. Faraday I **69,** 208.

Mack, G. P. R. & Thrush, B. A. (1974a). J. Chem. Soc. Faraday I **70,** 178.

Mack, G. P. R. & Thrush, B. A. (1974b). J. Chem. Soc. Faraday I **70,** 187.

Mallard, W. G. & Owen, J. H. (1974). Int. J. Chem. Kin. **6,** 753.

Manthorne, K. C. & Pacey, P. D. (1978). Can. J. Chem. **56,** 1307.

Marshall, R. M. & Purnell, J. H. (1972). J. Chem. Soc. Chem. Comm. **13,** 764.

Marshall, R. M., Purnell, J. H., & Storey, P. D. (1976). J. Chem. Soc. Faraday I **72,** 85.

Matsui, Y. & Nomaguchi, T. (1979). Japanese J. Appl. Phys. **18,** 181.

McKay, G., Turner, J. M. C., & Zare, F. (1977). J. Chem. Soc. Faraday I **73,** 803.

McKnight, C., Niki, H., & Weinstock, B. (1967). J. Chem. Phys. **65,** 5219.

Messing, I., Sadowsky, C. M., & Filseth, S. V. (1979). Chem. Phys. Letters **66,** 95.

Messing, I. *et al.* (1980). Chem. Phys. Lett. **74,** 56.

Metcalfe, E. L. & Trotman-Dickenson, A. F. (1960). J. Chem. Soc. **1960,** 5072.

Michael, J. V. & Lee, J. H. (1977). Chem. Phys. Letters **51,** 303.

Michael, J. V., Osborne, D. T., & Suess, G. N. (1973). J. Chem. Phys. **58,** 2800.

Michael, J. V. *et al.* (1979). J. Chem. Phys. **70,** 5222.

Michael, J. V. *et al.* (1981) (to be published).

Milhelcic, D. *et al.* (1975). Ber. Bunsenges. Phys. Chem. **79,** 1230.

Mintz, K. J. & LeRoy, D. J. (1978). Can. J. Chem. **56,** 941.

Morganroth, W. E. & Calvert, J. G. (1966). J. Amer. Chem. Soc. **88,** 5387.

Morris, E. D. & Niki, H. (1971a). J. Chem. Phys. **55,** 1991.

Morris, E. D. & Niki, H. (1971b). J. Phys. Chem. **75,** 3640.

Morris, E. D. & Niki, H. (1972). Int. J. Chem. Kin. **5,** 47.

Morris, E. D., Stedman, D. H., & Niki, H. (1971). J. Amer. Chem. Soc. **93,** 3570.

Nadtochenko, V. A., Sarkisov, O. M., & Vedeneev, V. I. (1979a). Dokl. Akad. Nauk. SSSR **224,** 152.

Nadtochenko, V. A., Sarkisov, O. M., & Vedeneev, V. I. (1979b). Izv. Adad. Nauk. SSSR, Ser. Khim. **1979,** 651.

Niki, H. (1967). J. Chem. Phys. **47,** 3102.

Niki, H., Daby, E. E., & Weinstock, B. (1969a). J. Chem. Phys. **48,** 5729.

Niki, H., Daby, E. E., & Weinstock, B. (1969b). 12th Symposium (International) on Combustion, p. 277.

Niki, H. *et al.* (1978). J. Phys. Chem. **82,** 132.

Niki, H. *et al.* (1980). Chem. Phys. Lett. **72,** 43.

Nip, W. S. & Paraskevopoulos, G. (1979). J. Phys. Chem. **71,** 2170.

Oka, K. & Cvetanovic, R. J. (1979). Can. J. Chem. **57,** 777.

Okabe, H. (1981). J. Chem. Phys. **75,** 2772.

Olson, D. B. & Gardiner, W. C. (1978). Comb. Flame **32,** 151.

Olson, D. B., Tanzawa, T., & Gardiner, W. C. (1979). Int. J. Chem. Kin. **11,** 23.

O'Neal, H. E. & Benson, S. W. (1962). J. Chem. Phys. **36,** 2196.

Osif, T. L., Simonaitis, R., & Heicklen, J. (1975). J. Photochem. **4,** 233.

Overend, R. & Paraskevopoulos, G. (1977a). Chem. Phys. Letters **49,** 109.

Overend, R. & Paraskevopoulos, G. (1977b). J. Chem. Phys. **67,** 674.

Overend, R. & Paraskevopoulos, G. (1978). J. Phys. Chem. **82,** 1329.

Overend, R. P., Paraskevopoulos, G., & Cvetanovic, R. J. (1975). Can. J. Chem. **53,** 3374.

Owens, C. M. & Roscoe, J. M. (1976). Can. J. Chem. **54,** 984.

Pacey, P. D. & Purnell, J. H. (1972a). J. Chem. Soc. Faraday I **68,** 1462.

Pacey, P. D. & Purnell, J. H. (1972b). Int. J. Chem. Kin. **4,** 657.

Pacey, P. D. & Wimalasena, J. H. (1980). Chem. Phys. Letters **76,** 433.

Papadopoulos, C., Ashmore, P. G., & Tyler, B. J. (1971). 13th Symposium (International) on Combustion, p. 281.

Papic, M. M. & Laidler, K. J. (1971). Can. J. Chem. **49,** 549.

Parkes, D. A. (1980). Chem. Phys. Letters **76,** 527.

Parkes, D. A. & Quinn, C. P. (1975). Chem. Phys. Lett. **33,** 483.

Parkes, D. A. & Quinn, C. P. (1976). J. Chem. Soc. Faraday I **72,** 1952.

Parkes, D. A., Paul, D. M., & Quinn, C. P. (1976). J. Chem. Soc. Faraday I **72,** 1935.

Pastrana, A. & Carr, R. W. (1974). Int. J. Chem. Kin. **6,** 587.

Pastrana, A. V. & Carr, R. W. (1975). J. Phys. Chem. **79,** 765.

Paukert, T. T. & Johnston, H. S. (1972). J. Chem. Phys. **56,** 2824.

Payne, W. A. & Stief, L. J. (1976). J. Chem. Phys. **64,** 1150.

Peeters, J. & Mahnen, G. (1973a). 14th Symposium (International) on Combustion, p. 133.

Peeters, J. & Mahnen, G. (1973b). Combustion Institute European Symposium 1973, p. 53.

Peeters, J. & Vinckier, C. (1975). 15th Symposium (International) on Combustion, p. 969.

Perry, R. A. & Williamson, D. (1982). Chem. Phys. Lett. (in press).

Perry, R. A., Atkinson, R., & Pitts, J. N. (1977). J. Chem. Phys. **67,** 5577.

Peters, N. & Warnatz, J., Ed. (1982) *Numerical Methods in Laminar Flame Propagation.* A GAMM Workshop, Vieweg, Wiesbaden.

Pilling, M. J. & Robertson, J. A. (1975). Chem. Phys. Letters **33,** 336.

Pilling, M. J. & Robertson, J. A. (1977). Trans. Faraday Soc. I **73,** 968.

Place, x. & Weinbrg, x. (1966).

Porter, R. P. (1967). 11th Symposium (International) on Combustion, p. 907.

Pratt, G. L. & Rogers, D. (1979a). J. Chem. Soc. Faraday I **75,** 1089.

Pratt, G. L. & Rogers, D. (1979b). J. Chem. Soc. Faraday I **75,** 1101.

Pratt, G. L. & Rogers, D. (1979c). J. Chem. Soc. Faraday I **75,** 2688.

Pratt, G. L. & Rogers, D. (1980). J. Chem. Soc. Faraday I **76,** 1694.

Pratt, G. L. & Veltman, I. (1974). J. Chem. Soc. Faraday I **70,** 1840.

Pratt, G. L. & Veltman, I. (1976). J. Chem. Soc. Faraday I **72,** 1733.

Pruss, F. J., Slagle, I. R., & Gutman, D. (1974). J. Phys. Chem. **78,** 663.

Purnell, J. H. & Quinn, C. P. (1962). Proc. Roy. Soc. A **270,** 267.
Radford, H. E. (1980). Chem. Phys. Lett. **71,** 195.
Ravinshankara, A. R. & Davis, D. D. (1978). J. Phys. Chem. **82,** 2852.
Ravinshankara, A. R. *et al.* (1978). Int. J. Chem. Kin. **10,** 783.
Reilly, J. P. *et al.* (1978). J. Chem. Phys. **69,** 4381.
Renlund, A. M. *et al.* (1981). Chem. Phys. Lett. **84,** 293.
Ridley, B. A. *et al.* (1972). J. Chem. Phys. **57,** 520.
Robinson, P. J. & Holbrook, K. A. (1972). *Unimolecular Reactions*, Wiley, London.
Roth, P. & Just, Th. (1975). Ber. Bunsenges. Phys. Chem. **79,** 682.
Roth, P. & Just, Th. (1977). Ber. Bunsenges. Phys. Chem. **81,** 572.
Roth, P. & Just, Th. (1979). Ber. Bunsenges. Phys. Chem. **83,** 577.
Roth, P. Löhr, R. (1981). Ber. Bunsenges. Phys. Chem. **85,** 153.
Roth, P., Barner, U., & Löhr, R. (1979). Ber. Bunsenges. Phys. Chem. **83,** 929.
Rowland, F. S., Lee, P. S. T., & Montague, D. C. (1972). Disc. Faraday Soc. **53,** 111.
Sampson, R. J. (1963). J. Chem. Soc. **1963,** 5095.
Sanders, N. *et al.* (1980). Chem. Phys. **49,** 17.
Schecker, H. G. & Jost, W. (1969). Ber. Bunsenges. Phys. Chem. **73,** 521.
Schott, G. L. (1973). Comb. Flame **21,** 357.
Schott, G. L., Getzinger, R. W., & Seitz, W. A. (1974). Int. J. Chem. Kin. **6,** 921.
Schwanebeck, W. & Warnatz, J. (1975). Ber. Bunsenges. Phys. Chem. **79,** 530.
Seery, D. J. & Bowman, C. T. (1970). Comb. Flame **14,** 37.
Sepehrad, A., Marshall, R. M., & Purnell, H. (1979). J. Chem. Soc. Faraday I **75,** 835.
Shannon, T. W. & Harrison, A. G. (1963). Can. J. Chem. **41,** 2455.
Shaub, W. M. & Bauer, S. H. (1978). Comb. Flame **32,** 35.
Shibuya, K. *et al.* (1977). J. Phys. Chem. **81,** 2292.
Sie, B. K. T., Simonaitis, R., & Heicklen, J. (1976). Int. J. Chem. Kin. **8,** 85.
Simonaitis, R. & Heicklen, J. (1973). Int. J. Chem. Kin. **5,** 231.
Singleton, D. L. & Cvetanovic, R. J. (1976). J. Amer. Chem. Soc. **98,** 6812.
Singleton, D. L., Irwin, R. S., & Cvetanovic, R. J. (1977). Can. J. Chem. **55,** 3321.
Skinner, G. B. & Ball, W. E. (1960). J. Phys. Chem. **64,** 1025.
Skinner, G. B., Sweet, R. C., & Davis, S. K. (1971). J. Phys. Chem. **75,** 1.
Slack, M. W. (1977). Comb. Flame **28,** 241.
Slagle, I. R., Pruss, F. J., & Gutman, D. (1974). Int. J. Chem. Kin. **6,** 111.
Slater, D. H. & Calvert, J. G. (1968). Adv. Chem. Series **76,** 58.
Slemr, F. & Warneck, P. (1975). Ber. Bunsenges. Phys. Chem. **79,** 152.
Smets, B. & Peeters, J. (1975). Deuxieme Symposium Europeen sur la Combustion, p. 38. The Combustion Institute, Pittsburgh.
Smith, I. W. M. (1968). Trans. Faraday Soc. **64,** 378.
Smith, I. W. M. & Zellner, R. (1973). J. Chem. Soc. Faraday II **69,** 1617.
Smith, I. W. M. & Zellner, R. (1974). J. Chem. Soc. Faraday II **70,** 1045.
Smith, R. H. (1978). Int. J. Chem. Kin. **10,** 519.

Spindler, K. G. & Wagner, H. Gg. (1982). Ber. Bunsenges. Phys. Chem. **86,** 119.

Stuhl, F. (1973a). Ber. Bunsenges. Phys. Chem. **77,** 674.

Stuhl, F. (1973b). Z. Naturforsch. **A28,** 1383.

Stuhl, F. & Niki, H. (1971). J. Chem. Phys. **55,** 3954.

Stuhl, F. & Niki, H. (1972a). J. Chem. Phys. **57,** 3671.

Stuhl, F. & Niki, H. (1972b). J. Chem. Phys. **57,** 5403.

Sugawara, K., Ishikawa, Y., & Sato, S. (1980). Bull. Chem. Soc. Jpn. **53,** 1344.

Sworsky, T. J., Hochanadel, C. J., & Ogren, P. J. (1980). J. Phys. Chem. **84,** 129.

Szirovicza, L. & Walsh, R. (1974). J. Chem. Soc. Faraday I **70,** 33.

Tabayashi, K. & Bauer, S. H. (1979). Comb. Flame **34,** 63.

Tanzawa, T. & Gardiner, W. C. (1979). 17th Symposium (International) on Combustion, p. 563.

Tanzawa, T. & Gardiner, W. C. (1980a). J. Phys. Chem. **84,** 236.

Tanzawa, T. & Gardiner, W. C. (1980b). Comb. Flame **38,** 241.

Temps, F. & Wagner, H. Gg. (1982). Ber. Bunsenges. Phys. Chem. **86,** 2.

Teng, L. & Jones, W. E. (1972). J. Chem. Soc. Faraday I **68,** 1267.

Thrush, B. A. & Wilkinson, J. P. T. (1979). Chem. Phys. Lett. **66,** 441.

Trainor, D. W. & von Rosenberg, C. W. (1974). J. Chem. Phys. **61,** 1010.

Trainor, D. W. & von Rosenberg, D. W. (1975). 15th Symposium (International) on Combustion, p. 755.

Trenwith, A. B. (1966). Trans. Faraday Soc. **62,** 1538.

Trenwith, A. B. (1979). J. Chem. Soc. Faraday I **75,** 614.

Troe, J. (1969). Ber. Bunsenges. Phys. Chem. **73,** 946.

Tsang, W. (1969). Int. J. Chem. Kin. **1,** 245.

Tsuboi, T. (1976). Jpn. J. Appl. Phys. **15,** 159.

Tsuboi, T. (1978). Jpn. J. Appl. Phys. **17,** 709.

Tsuboi, T. & Hashimoto, K. (1981). Comb. Flame **42,** 61.

Tsuboi, T. & Wagner, H. Gg. (1974). 15th Symposium (International) on Combustion, p. 883.

Tully, F. P. & Ravishankara, A. R. (1980). J. Phys. Chem. **84,** 3126.

Umstead, M. E., Shortridge, R. G., & Lin, M C. (1977). Chem. Phys. **20,** 271.

Vandooren, J. & van Tiggelen, P. J. (1977). 16th Symposium (International) on Combustion, p. 1133.

Vandooren, J. & van Tiggelen, P. J. (1981). 18th Symposium (International) on Combustion, p. 473.

Vandooren, J., Peeters, J., & van Tiggelen, P. J. (1975). 15th Symposium (International) on Combustion, p. 745.

Vardanyan, I. A. *et al.* (1974). Comb. Flame **22,** 153.

Veyret, B. & Lesclaux, R. (1981). J. Phys. Chem. **85,** 1918.

Vinckier, C., Debruyn, W. (1979a). 17th Symposium (International) on Combustion, p. 623.

Vinckier, C. & Debruyn, W. (1979b). J. Phys. Chem. **83,** 2057.

Vinckier, C. (1979). J. Phys. Chem. **83,** 1234.

Volpi, G. G. & Zocchi, F. (1966). J. Chem. Phys. **44,** 4010.

Wagner, H. Gg. & Zabel, F. (1971). Ber. Bunsenges. Phys. Chem. **75,** 114.

Wagner, H. Gg. & Zellner, R. (1972a). Ber. Bunsenges. Phys. Chem. **76,** 518.

Wagner, H. Gg. & Zellner, R. (1972b). Ber. Bunsenges. Phys. Chem. **76,** 667.

Wagner, H. Gg. & Zellner, R. (1972c). Ber. Bunsenges. Phys. Chem. **76,** 440.

Walkauskas, L. P. & Kaufman, F. (1975). 15th Symposium (International) on Combustion, p. 691.

Walker, R. W., Ed. (1975). *Reaction Kinetics*, Vol. 1, The Chemical Society, London p. 161.

Walker, R. W., Ed. (1977). *Gas Kinetics and Energy Transfer*, Vol. 2, The Chemical Society, London p. 296.

Warnatz, J. (1978). Ber. Bunsenges. Phys. Chem. **83,** 950.

Warnatz, J. (1981). 18th Symposium (International) on Combustion, p. 369.

Warnatz, J. (1982). Comb. Sci. Technol. **26,** 203.

Warnatz, J. (1983). In: *Soot in Combustion Systems*, J. Lahaye & G. Prado, Ed., Plenum, London.

Warnatz, J. *et al.* (1982). 19th Symposium (International) on Combustion (in press).

Washida, N. (1980). J. Chem. Phys. **73,** 1665.

Washida, N. & Bayes, K. D. (1976). Int. J. Chem. Kin. **8,** 777.

Washida, N., Martinez, R. I., & Bayes, K. D. (1974). Z. Naturforsch. A **29,** 251.

Watkins, K. W. & Ward, W. W. (1974). Int. J. Chem. Kin. **6,** 855.

Wellmann, C. (1981). Dissertation, TH Darmstadt.

Westbrook, C. K. & Dryer, F. L. (1979). Comb. Sci. Technol. **20,** 125.

Westbrook, C. K. *et al.* (1977). J. Phys. Chem. **81,** 2542.

Westenberg, A. A. & de Haas, N. (1967). J. Chem. Phys. **46,** 490.

Westenberg, A. A. & de Haas, N. (1969a). J. Phys. Chem. **73,** 1181.

Westenberg, A. A., de Haas, N. (1969b). 12th Symposium (International) on Combustion, p. 289.

Westenberg, A. A. & de Haas, N. (1969c). J. Chem. Phys. **50,** 707.

Westenberg, A. A. & de Haas, N. (1972a). J. Phys. Chem. **76,** 1586.

Westenberg, A. A. & de Haas, N. (1972b). J. Phys. Chem. **76,** 2213.

Westenberg, A. A. & de Haas, N. (1972c). J. Phys. Chem. **76,** 2215.

Westenberg, A. A. & de Haas, N. (1973a). J. Chem. Phys. **58,** 4061.

Westenberg, A. A. & de Haas, N. (1973b). J. Chem. Phys. **58,** 4066.

Westenberg, A. A. & de Haas, N. (1977). J. Chem. Phys. **66,** 4900.

Westley, F. (1976). *Chemical Kinetics of the Gas Phase Combustion of Fuels* (*A Bibliography on the Rates and Mechanisms of Oxidation of Aliphatic* C_1 *to* C_{10} *Hydrocarbons and their Oxygenated Derivatives*), National Bureau of Standards, Washington, D.C.

Westley, F. (1979). Table of Recommended Rate Constants for Chemical Reactions Occurring in Combustion. National Bureau of Standards, Washington, D.C.

Whiteway, S. G. & Masson, C. R. (1956). J. Chem. Phys. **25,** 233.

Whytock, D. A., Payne, W. A., & Stief, L. J. (1976a). J. Chem. Phys. **65,** 191.

Whytock, D. A. *et al.* (1976b). J. Chem. Phys. **65,** 4871.

Wong, W. & Davis, D. D. (1974). Int. J. Chem. Kin. **6,** 401.

Yampolskii, Y. P. & Rybin, V. M. (1974). React. Kin. Catal. Letters **1,** 321.

Yang, K. (1963). J. Phys. Chem. **67,** 562.

Yano, T. (1977). Int. J. Chem. Kin. **9,** 725.

Yoichi, Y. *et al.* (1978). Bull. Chem. Soc. Jpn. **51,** 2488.

Zaslonko, I. S. & Smirnov, V. N. (1979). Fiz.-Khim. Protsessy Gazov. Kondens, Fazakh **1979,** 19.

Zellner, R. (1979). J. Phys. Chem. **83,** 18.

Zellner, R. & Steinert, W. (1976). Int. J. Chem. Kin. **8,** 397.

Zellner, R. & Wagner, H. Gg. (1980). The Sixth International Symposium on Gas Kinetics, University of Southampton, p. 7.

Zellner, R., Erler, K., & Field, D. (1977). 16th Symposium (International) on Combustion, p. 939.

Chapter 6

Survey of Rate Constants in the N/H/O System

Ronald K. Hanson
Siamak Salimian

1. Introduction

Current interest in high-temperature N/H/O kinetics stems primarily from the practical importance of combustion-generated emissions of nitrogen oxides (NO_x). The recognition of air pollution as a problem of societal concern has prompted a concentrated research effort on the kinetics of NO_x formation and decomposition and has led to a considerable expansion in the fundamental data base for rate constants of elementary reactions in the N/H/O system. The objectives of this survey are twofold: (1) to provide critical rate constant evaluations of reactions for which high-temperature data have been recently acquired, with due consideration given to previous evaluations, particularly the widely used survey at Leeds University (Baulch, *et al.*, 1973); and (2) to provide an extensive compilation of N/H/O reactions and rate constants including both nonevaluated and evaluated reactions, that is, reactions for which limited or no rate data are available currently as well as reactions which have undergone critical reviews previously or in this study.

The list of reactions critically evaluated has been limited to reactions that are of current interest in combustion modeling and have received sufficient attention in recent years to justify an initial or new assessment of the rate constant. The authors admit that this accounting is biased by their own prejudices regarding reactions of interest and the accuracy and reliability of various measurements. The evaluations are further limited to homogeneous reactions and, generally, to temperatures above about 1000 K.

The philosophy employed in these evaluations was to favor those studies which were more direct, that is, more effective in isolating the elementary reaction-rate parameters of interest. (Ideally, of course, the directness of the measurements should be demonstrated in each study by an appropriate sensitivity analysis.) Also, more recent experimental studies were generally preferred if they employed more sensitive or otherwise improved experimental techniques, including data recording and processing methods. The use of detailed kinetic modeling by computer, more common in recent studies, also was considered desirable in that such an approach tends to reduce errors which result from the use of overly simplified kinetic approximations.

The rate constant data selected for final analysis were largely obtained from shock tube experiments, although data from premixed flames and static and flow reactors were also included. The experimental techniques employed were numerous, including various forms of emission and absorption spectroscopy, laser schlieren deflection, molecular beam techniques, and chemical analysis. Those rate constant data that were rejected from consideration in selecting a recommended expression were excluded for several possible reasons, including the use of insensitive or questionable experimental techniques, excessive scatter, or large differences from other results considered to be more reliable.

The rate constant parameters listed are based entirely on experimental measurements and should in most cases be considered only as pragmatic representations of experimental data. By this we mean that no physical interpretation was consistently imposed on the selected preexponential factor or activation energy. In only a few cases, where explicitly noted, was theory used as an aid in specifying the form of the recommended rate constant expression. Error limits for the recommended expressions, when cited, typically reflect a combination of statistical precision of measurements and our estimate of possible systematic errors. These limits are generally expressed in terms of minimum and maximum rate constant factors, i.e., f and F where $k_{min} = fk$ and $k_{max} = Fk$.

The authors would like to acknowledge that preparation of this chapter has been a source of both pleasure and frustration. It was a pleasure to become more keenly aware of the state of knowledge of N/H/O system kinetics, but it was frustrating trying to deal in a fair and complete way with the ever-expanding literature. We apologize to those workers who feel slighted by our review.

We would also like to record our observation of the favorable trend toward improved accuracy in rate constant determinations that was evident in recent work and deserves comment. These advances may be attributed primarily to improved experimental techniques, particularly the development of new, sensitive detection schemes and data recording methods, and to the expanding role of the computer in both the design of experiments and in data reduction. The growing use of computers to handle extensive reaction mechanisms, and thereby account for the effects of secondary reactions more accurately than by approximate analytical methods, allows the experimentalist to optimize the conditions of his experiment to be most sensitive to the rate constants of interest as well as to reduce data with minimal error.

The authors hope that this survey will prove useful to specialists and nonspecialists interested in N/H/O kinetics and that it may aid in identifying areas where further research is desirable.

2. Organization

The results of this survey are presented in four sections. The first three provide critical reviews of rate constant data for 18 reactions in the N/O, N/H, and N/H/O systems. Arrhenius plots and tabulations of data from selected studies are included together, in most cases, with a recommended rate constant expression. In the fourth section, a more complete listing of N/H/O reactions and rate constants is given, primarily for the convenience of modelers and others interested in a more complete mechanism governing NO_x kinetics in N/H/O systems. These rate constants have been drawn from a variety of sources and include present evaluations, previous evaluations by the Leeds group, and a large number of estimates. The reader is warned to use such rate constant tabulations with caution; in some cases the uncertainties may be quite large.

The rate constant expressions are presented in the forms

$$k = A \exp(-\theta/T)$$

or

$$k = AT^B \exp(-\theta/T)$$

The reactions considered are all bimolecular, so that the units of k are $cm^3\ mol^{-1}\ s^{-1}$; the units of θ are K. Results for activation energies are sometimes cited in kcal or kJ units. For purposes of converting them to K the following conversion factors were employed:

$$R = 8.314\ J\ K^{-1}\ mol^{-1} = 1.987\ cal\ K^{-1}\ mol^{-1}$$

$$1\ cal = 4.184\ J$$

In those few cases in which high-temperature data were available for both the forward and reverse reactions, all the data were considered together in establishing the recommended expression.

Thermochemical data used in this survey and for computing ΔH_0° values for the evaluated reactions were taken from the JANAF (1971) tables except for the ΔH_{f298}° values for NH, NH_2, and HO_2, which were taken from Binkley and Melius (1982), Chase *et al.* (1982), and Howard (1980), respectively, and combined with the JANAF $H_0^\circ - H_{298}^\circ$ values to give ΔH_{f0}° values of 365.7, 193.2, and 13.4 kJ, respectively, for these species.

3. N/O reaction survey

This section provides a summary of six reactions in the N/O system for which new rate constant data have been reported since the 1973 Leeds evaluations by Baulch *et al.* The reactions considered are: $O + N_2 \rightarrow N + NO$, $O + NO \rightarrow N + O_2$, $NO + M \rightarrow N + O + M$, $N_2O + M \rightarrow N_2 + O + M$, $O + N_2O \rightarrow 2NO$ and $O + N_2O \rightarrow N_2 + O_2$.

3.1. $O + N_2 \rightarrow N + NO$

The formation of NO in high-temperature N_2/O_2 systems, including the postflame regions of air-breathing combustors, occurs largely through the "Zeldovich reactions"

$$O + N_2 \rightarrow N + NO \qquad \Delta H_0^\circ = 314 \text{ kJ} \tag{1}$$

$$N + O_2 \rightarrow O + NO \qquad \Delta H_0^\circ = -134 \text{ kJ} \tag{2}$$

and

$$N + OH \rightarrow H + NO \qquad \Delta H_0^\circ = -204 \text{ kJ} \tag{3}$$

In spite of its rate-limiting significance, k_1, the rate constant for reaction (1), only recently has been determined in a direct manner. There were several early studies involving this reaction, as reported by Baulch *et al.* (1969, 1973), but these studies were handicapped either by uncertainties in the data, sensitivity to the rate constants of other participating reactions, or the use of overly simplified kinetic models. In view of the particular importance of this rate constant to NO_x modeling, a brief review of these early studies is worthwhile.

The original formulation and verification of the Zeldovich mechanism is attributed to Zeldovich (1946), who studied explosions of gases within a combustion bomb. By assuming equilibrium between O and O_2, that O and N were in steady state, and that only reactions (1) and (2) were important, Zeldovich was able to obtain an approximate Arrhenius expression for k_1. Another early study was performed by Glick *et al.* (1957), who used a single-pulse shock tube technique. Their study was intended as an investigation of the overall reaction $N_2 + O_2 \rightarrow 2NO$, and an explicit expression for k_1 was not reported. They did, however, report a steric factor and activation energy for k_1 based on a model that assumed N-atom steady state and O/O_2 partial equilibrium.

In a subsequent theoretical study, Duff and Davidson (1959) used Glick's data to obtain an explicit expression for k_1. Their more complete kinetic analysis indicated that the expression for k_1 derived from Glick's parameters should be increased by approximately 35% at 3000 K. Later, Wray and Teare (1962) performed a series of shock tube experiments to study NO kinetics. Although they obtained reasonable agreement between their [NO] data and a kinetic model that employed Duff and Davidson's estimate of k_1, they state that k_1 could not be accurately determined from their experiments.

In 1969, Baulch *et al.* published their first recommendation for k_1, based on low-temperature data for the reverse reaction that had been obtained by Clyne and Thrush (1961a, 1961b). The suggested value for k_1 was approximately twice that obtained by Duff and Davidson. Baulch *et al.* evidently preferred this expression because they believed the determinations of k_{-1} to be more accurate than the determinations of k_1. Baulch *et al.* reported their expression for k_1 to be accurate within a factor of 2 in the temperature range 300–1600 K.

In more recent work, Newhall and Shahed (1970) measured NO formation rates in a combustion wave study and obtained reasonable agreement between data and theory using Duff and Davidson's expression. Their major conclusion was not that they had determined k_1, but that NO is formed in the post-flame region via the Zeldovich mechanism. They imply that the expressions they used for k_1 and k_2 may be off by a factor of 2. Livesey *et al.* (1971) performed a study of NO kinetics using premixed propane–oxygen flames to which N_2 had been added. They determined a value of k_1 at only one temperature, but the extracted value was sensitive to k_2. A shock tube study was performed by Bowman (1971), who obtained reasonable agreement between data and theory using the value of k_1 that was then recommended by Baulch *et al.* (1969). Although uncertainties in Bowman's data for NO precluded a direct, accurate determination of k_1, the data can be taken as a corroboration of the Zeldovich mechanism and as evidence that the value of k_1 employed is correct within a factor of about 2. Thompson *et al.* (1972) studied NO formation using premixed methane–air flames and a tubular combustor. They concluded that use of the value for k_1 suggested by Baulch *et al.* (1969) gave satisfactory results, although they claimed that increasing k_1 by 70% would yield even better agreement between data and theory.

Subsequently, the Leeds group (Baulch *et al.*, 1973) revised their recommendation for k_1 downward by about a factor of 2 to agree better with the results of Duff and Davidson (1959), Wray and Teare (1962), and Newhall and Shahed (1970), in spite of the apparent disagreement with the reverse rate data of Clyne and Thrush (1961a, 1961b). Baulch *et al.* (1973) suggested a factor of 2 uncertainty in k_1 in the temperature range 2000–5000 K.

Several subsequent investigations raised doubts about the 1973 Leeds suggestion for k_1. Waldman *et al.* (1974) studied methane–air combustion in a perfectly stirred reactor. In order to obtain reasonable agreement between experiment and theory they had to increase the value of k_1 suggested by Baulch *et al.* (1973) by a factor of 2.4. Leonard *et al.* (1976) used a flat flame burner to study both thermal and fuel–nitrogen NO formation in methane–air and propane–air flames. Although scatter in the data, uncertainty in the measurement of $[H_2]$, and uncertainty in the NO binary diffusion coefficients precluded a precise determination of k_1, the investigators obtained reasonable agreement between data and theory using a value of k_1 approximately 1.7 times that suggested by Baulch *et al.* (1973).

Table 1. Rate constant data for $O + N_2 \rightarrow N + NO$.

Reference	$\text{Log}(A/\text{cm}^3\,\text{mol}^{-1}\,\text{s}^{-1})$	θ/K	T range
Harris *et al.* (1976)	14.26[a]	38 000	2120–2480
Blauwens *et al.* (1977)	13.78	38 000	1880–2350
Monat *et al.* (1979)	14.26	38 370	2380–3850
Seery and Zabielski (1980)	14.25	38 370	2120–2230
Baulch *et al.* (1973) (evaluation)	13.88	38 000	2000–4000
This evaluation	14.26	38 370	2000–4000

[a] See text for discussion.

The most recent determinations of k_1, in three flame studies by Harris *et al.* (1976), Blauwens *et al.* (1977), and Seery and Zabielski (1980), and in one shock tube study by Monat *et al.* (1979), are regarded as most reliable. Results from these four studies are tabulated in Table 1 and plotted in Fig. 1 together with the 1973 Leeds recommendation.

The study by Harris *et al.* (1976) utilized the same approach and was conducted in the same laboratory as the earlier work of Livesy *et al.* (1971). Premixed laminar $CH_4/O_2/N_2$ flames were burned at atmospheric pressure, and the rate constant k_1 was inferred from comparisons of computed and measured NO profiles in the postflame region of fuel lean flames. NO was measured by chemiluminescent analysis of probe-sampled gases, while OH was determined by line absorption using a water vapor lamp. Temperature was found by an OH emission technique. Concentrations of the radical species H and O, needed for the computer calculations of NO formation, were inferred by partial equilibrium arguments which show that the O-atom concentration is proportional to the square of the OH concentration. Unfortunately, Harris *et al.* employed a low value of the OH oscillator strength ($f(0, 0) = 7.7 \times 10^{-4}$) relative to the now accepted value of $f(0, 0) = 1.10 \pm 0.3 \times 10^{-3}$ (Smith and Crosley, 1981), thereby introducing a proportional error in their determination of OH. This suggests that the value of [O] inferred from partial equilibrium was high by about a factor of 2. Since, to a good approximation, the rate of NO formation is given at early times by

$$d[NO]/dt = 2k_1[O][N_2] \tag{4}$$

the rate coefficient inferred by Harris *et al.* may have been low by a factor of 2. The expression originally reported by Harris *et al.* has therefore been multiplied by 2 in Table 1 and in Fig. 1. The corrected expression agrees within 15% with the results of Monat *et al.* (1979).

Blauwens *et al.* (1977) investigated low-pressure hydrocarbon/O_2/N_2 flames using molecular beam sampling and mass spectrometric detection of NO, O, OH, H, and other species. Temperature was inferred from the measured concentrations and an equilibrium constant assuming partial

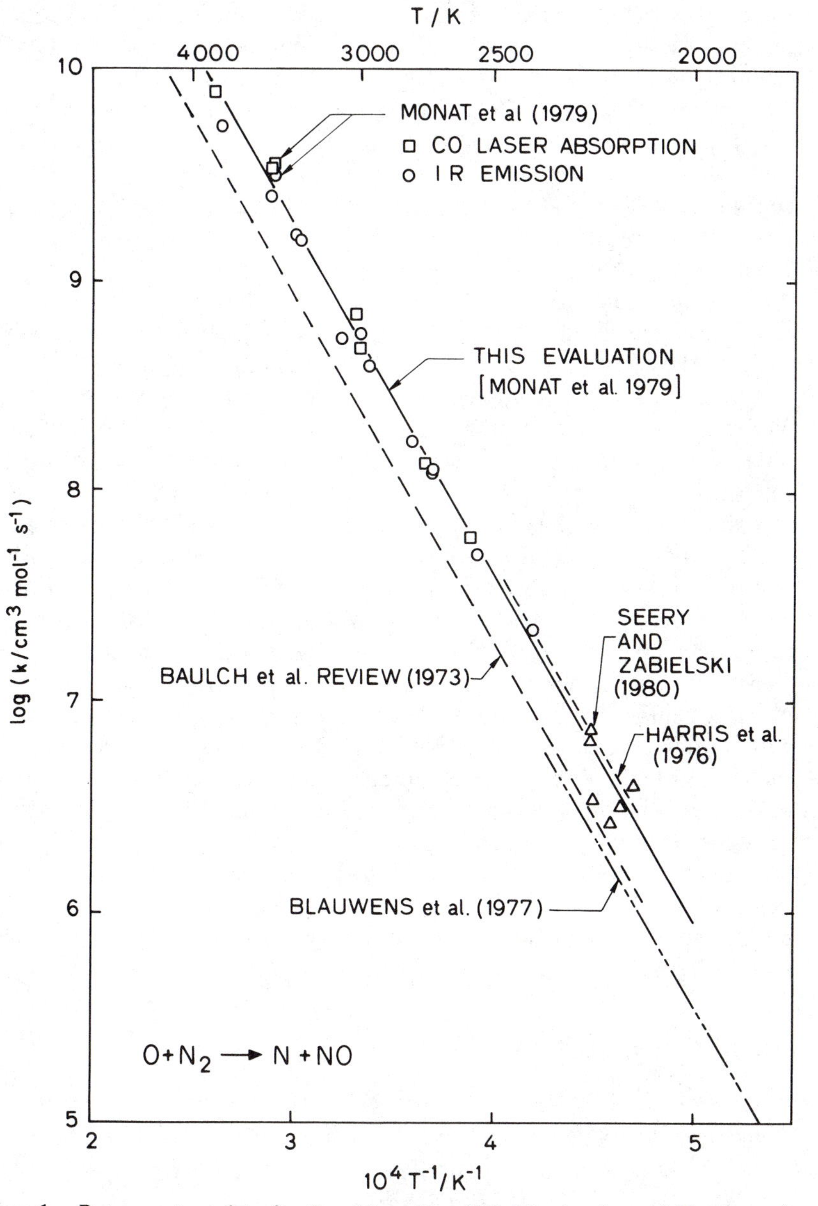

Figure 1. Rate constant data for $O + N_2 \rightarrow N + NO$. The k values of Harris *et al.* have been multipled by 2 (see text).

equilibrium. Although this was primarily a study of "prompt" NO formation, k_1 was also determined using the measured profiles of NO and O and the simplified kinetic expression (4) above. These workers report an uncertainty of only $\pm 30\%$ in the rate constant, but this seems unrealistically low in view of the possible measurement errors in absolute concentration and temperature and the uncertain influence of diffusion on the flow field. Blauwen's results falls below our recommendation by a factor of about 2.5 at 2000 K.

Seery and Zabielski (1980) utilized low-pressure CH_4–air flames to infer k_1. Their data included NO and N_2 composition profiles by molecular beam sampling mass spectrometry and microprobe sampling with chemiluminescent analysis, temperature profiles by thermocouples, and OH profiles by resonance absorption. Partial equilibrium was assumed in order to infer O-atom concentrations in the postflame region. The simplified expression (4) was used to infer k_1 at several flame locations. Although the scatter in the k_1 data show an average deviation of $\pm 28\%$ about the mean, the mean agrees within 4% with the best-fit expression reported by Monat *et al.* (1979).

The last study to be reviewed here is the work by Monat *et al.* (1979). This was a shock tube study covering the temperature range 2380–3850 K. Test gas mixtures of N_2, O_2, N_2O, and Kr were heated by incident shock waves, and the concentration of NO was monitored by two independent spectroscopic techniques: thermal emission at 5.3 μm and absorption of CO laser radiation at 5.17 μm. N_2O was used as a source of O atoms, and the test gas composition was optimized by computer simulations to emphasize the sensitivity of the measurements to k_1. Values for k_1 were inferred from comparisons of measured and calculated NO profiles using a detailed kinetic model.

Monat's determination of k_1 is regarded as the most direct and reliable measurement presently available. As his expression gives excellent agreement with the flame measurements of Seery and Zabielski and of Harris *et al.* (after the factor of 2 correction), it is adopted here without modification as the recommended expression in the temperature range 2000–4000 K. The estimated uncertainty factors are $f = 0.65$ and $F = 1.35$ based on Monat's uncertainty analysis. It is worth noting that extrapolation of Monat's expression to lower temperatures provides good agreement with the magnitude and temperature dependence of a recent determination of k_{-1} (Clyne and McDermid, 1975).

3.2. O + NO → N + O_2

The rate-limiting reaction for NO decomposition in stoichiometric or fuel-lean combustion gases at high temperatures is generally

$$O + NO \rightarrow N + O_2 \qquad \Delta H_0^\circ = 134 \text{ kJ} \tag{1}$$

In 1973 Baulch *et al.* recommended a rate constant for this reaction based primarily on data for the reverse rate. Two recent studies of the forward rate taken together with work done prior to the Leeds evaluation now provide a somewhat different, consistent expression for the forward rate constant.

Of the work done prior to 1973 (see Baulch *et al.*, 1973), the preferred measurements are those of Kaufman and Decker (1959), Wray and Teare (1962), and Clark *et al.* (1969). The data of Kaufman and Decker were obtained using a static reactor filled with NO/O_2 mixtures at reduced pressure. Reaction progress was monitored by uv spectrophotometry in the temperature range 1575–1660 K. Equilibrium of O atoms with O_2 was assumed. These results are preferred over the earlier work reported by Kaufman and Kelso (1955), which was based on measurements of Vetter (1949).

Wray and Teare (1962) studied NO decomposition over the temperature range 3000–8000 K in a variety of $NO/O_2/Ar$ mixtures using a shock tube. The NO concentration history was monitored by uv absorption at 127 nm, and a detailed kinetic mechanism was employed to fit the data. Only the data acquired in $NO/O_2/Ar = 0.5\%/0.25\%/99.25\%$ mixtures at temperatures near 5000 K were sufficiently insensitive to other rate constants to enable a determination of k_1.

Clark *et al.* (1969) obtained a single data point for k_1 at 3000 K in a reflected shock wave study of N_2O decomposition using time-of-flight mass spectrometric detection of NO, O_2, and O. It was observed that small but measurable decreases in NO concentration and increases in O_2 concentration occurred following pyrolysis of N_2O. These changes were attributed to reactions (1) and (2)

$$N + NO \rightarrow N_2 + O \tag{2}$$

Since reaction (2) is exothermic and fast, Clark *et al.* assumed the simple relationship

$$d[NO]/dt = -2k_1[O][NO]$$

in inferring their value for k_1.

Rate constant determinations after 1973 were reported by Hanson, Flower, and Kruger (1974) and by McCullough, Kruger, and Hanson (1977). Both of these studies were completed in the same laboratory, but using different facilities and experimental techniques. Hanson *et al.* heated mixtures of NO and N_2O dilute in Ar or Kr to temperatures of 2500–4100 K using incident shock waves and monitored NO decay by ir emission at 5.3 μm. The rate constant was inferred by fitting the observed decay of NO using a detailed kinetic model. The use of N_2O as a controlled source of O atoms effectively minimized the effects of other rate coefficients, so that this determination of k_1 should be considered as direct. The stated uncertainty in k_1 was $\pm 30\%$.

McCullough *et al.* heated NO/Ar mixtures to temperatures in the range 1750–2100 K in an alumina flow tube reactor and monitored the fractional decomposition of NO as a function of flow rate (residence time) in the reactor using a commercial chemiluminescent analyzer. Experiments at low temperatures were also performed but the data were excluded because of the influence of surface reactions. The high-temperature central section of the reactor was packed with small pieces of alumina to promote uniform flow, and pulsed tracer experiments were conducted to determine deviations from plug flow. A detailed kinetic and flow model was used, with some simplifications to reduce computing time, to calculate the fractional removal of NO versus flow rate. A

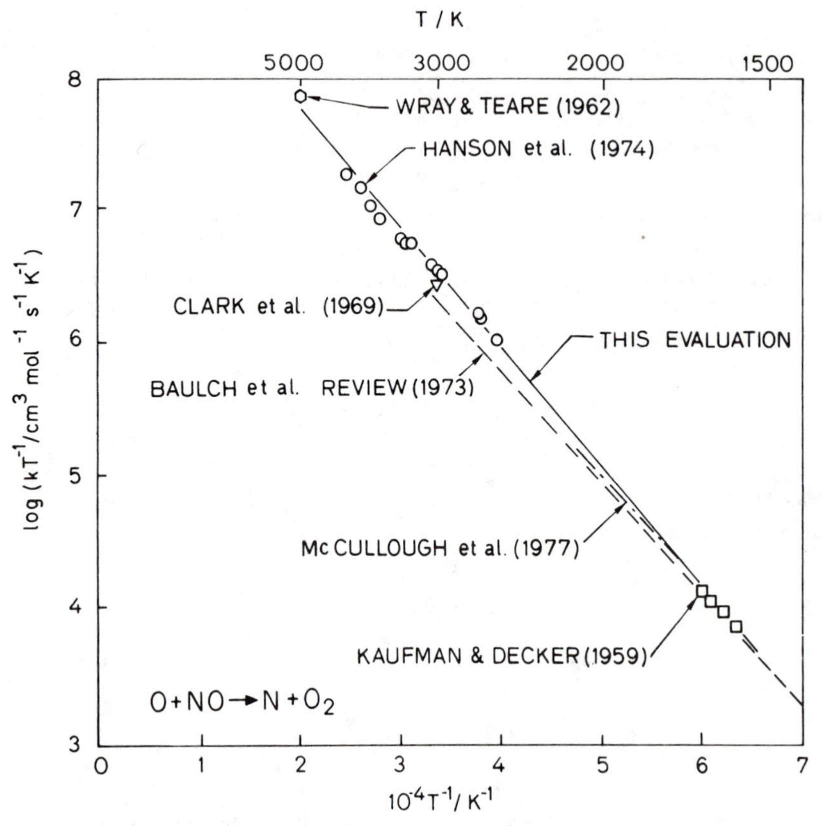

Figure 2. Rate constant data for $O + NO \rightarrow N + O_2$.

kinetic sensitivity analysis showed sufficient sensitivity to reaction (3)

$$NO + NO \rightarrow N_2O + O \tag{3}$$

that the data were fit by treating both k_1 and k_3 as variables.

A recommended expression for the rate constant (see Fig. 2 and Table 2) has been established by fitting the data of these five studies. The least-squares analysis utilized two data points (at the extreme temperatures) for each of the data sets reported by Kaufman and Decker (1959), Hanson *et al.* (1974), and McCullough *et al.* (1977). Single points were used to represent the data of Wray and Teare (1962) and of Clark *et al.* (1969). The resulting expression, recommended for use in the temperature range 1500–5000 K, is

$$k_1 = 3.81 \times 10^9\, T \exp(-20\,820/T)\ \text{cm}^3\ \text{mol}^{-1}\ \text{s}^{-1}$$

with uncertainty factors of $f = 0.7$ and $F = 1.3$. This expression provides good agreement with the reverse rate constant recommended by Baulch *et al.* (1973) at 1500 K and above, but it is inconsistent with the Baulch *et al.* expression at lower temperatures due to a differing temperature dependence.

Table 2. Rate constant data for $O + NO \rightarrow N + O_2$.

Reference	$\text{Log}(A/\text{cm}^3\ \text{mol}^{-1}\ \text{s}^{-1}\ \text{K}^{-B})$	B	θ/K	T range
Kaufman and Decker (1959)	12.56	0	19 900	1575–1665
Wray and Teare (1962)	Log k = 11.56	See Fig. 2		5000
Clark *et al.* (1969)	Log k = 9.95	See Fig. 2		3000
Baulch *et al.* (1973) (evaluation, based on reverse rate)	9.18	1.0	19 500	1000–3000
Hanson *et al.* (1974)	9.37	1.0	19 450	2500–4100
McCullough *et al.* (1977)	9.24	1.0	19 450	1750–2100
This evaluation	9.58	1.0	20 820	1500–5000

3.3. $NO + M \rightarrow N + O + M$

Thermal dissociation of nitric oxide is significant only at very high temperatures and generally may be excluded from detailed kinetic modeling of NO_x in combustion flows. The reaction is highly endothermic

$$NO + M \rightarrow N + O + M \qquad \Delta H_0^\circ = 628 \text{ kJ} \tag{1}$$

and has been investigated in only a few studies, all employing shock tube techniques. Although three studies were completed prior to 1973, by Freedman and Daiber (1961), Wray and Teare (1962), and Camac and Feinberg (1967), only the study by Wray and Teare employed gas mixtures and temperatures appropriate for inferring k_1 without excessive ambiguity (see Baulch *et al.*, 1973, for their review of these early studies). Two more recent shock tube studies have been reported, by Myerson (1973) and by Koshi *et al.* (1979), but the disagreement between these studies is quite substantial. As of this writing, no recommendation for this rate constant can be made.

Wray and Teare inferred k_1 by fitting profiles of NO behind incident shock waves in 0.5, 10, and 50% NO in Ar mixtures with temperatures of 4000–6700 K. NO was measured by uv absorption at 127 nm, and a seven-reaction mechanism was used in the calculations. Values of k_1 for M = Ar, N_2, and O_2 were assumed equal in the analysis, as were values of k_1 for M = NO, O, and N. Unfortunately, other reactions also significantly influenced the computed profiles, and no actual comparisons of measured and computed NO profiles were shown.

Myerson's investigation employed the ARAS (atomic resonance absorption spectroscopy) technique to monitor the formation of O atoms behind incident shock waves. Mixtures studied were 0.1, 1.0, and 10% NO in Ar; temperatures varied from 2600 to 6300 K, although only a portion of this range was suitable for determining the dissociation rate. A five-reaction model was employed to match computed and measured rates of O-atom formation, but only two reactions were found to be significant:

$$NO + Ar \rightarrow N + O + Ar \tag{1}$$

and

$$NO + NO \rightarrow N_2O + O \tag{2}$$

Analysis of all the data led to values for both rate constants, k_1 and k_2, although only the data for temperatures greater than 4150 K were plotted as determinations of k_1. No claim was made as to the value of k_1 with $M = NO$, although an effect should have been observed in the 10% NO mixtures if the efficiency of NO relative to Ar is as high as suggested by Wray and Teare (about 20) and by Koshi *et al.* (about 22; see below).

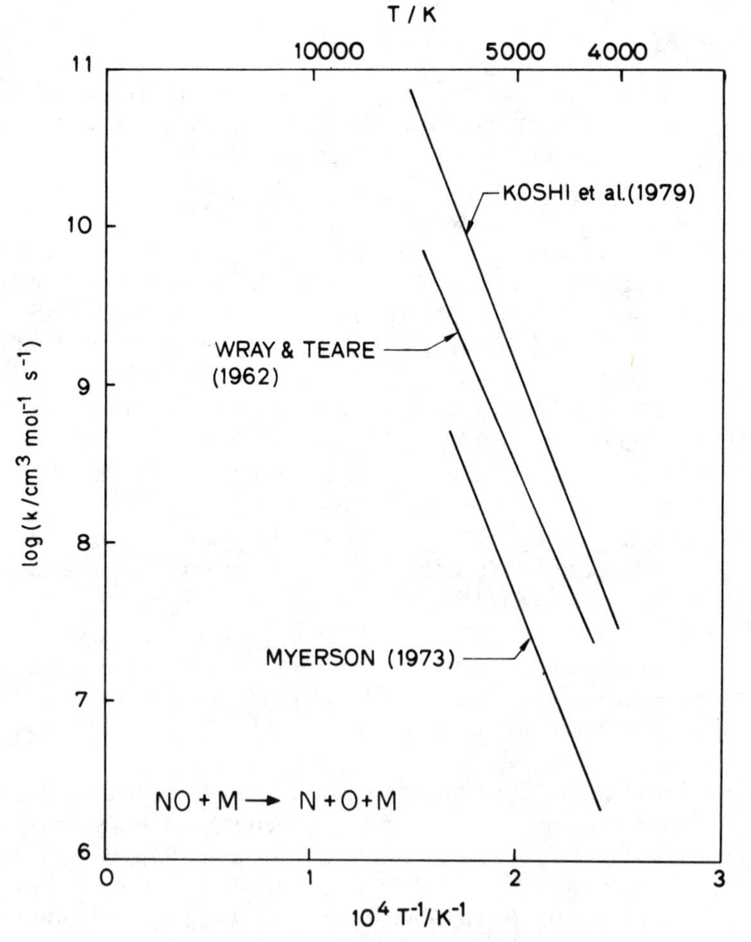

Figure 3. Rate constant data for $NO + M \rightarrow N + O + M$.

Myerson's value for k_1 falls below that of Wray and Teare by a factor of about 7 and below that of Koshi *et al.* by about 30 at 5000 K. This discrepancy, together with the extremely low activation energy reported for k_2 (29 kcal, to be compared with 64 kcal recommended by Baulch *et al.*, 1973), suggests the possibility of a significant error in Myerson's study.

The most recent determination of k_1 was reported by Koshi *et al.* (1979) based on incident shock wave experiments in NO/Ar mixtures (1–5% NO) at temperatures of 4000–7500 K. Detection of NO was by ir emission at two wavelengths (4.87 and 5.22 μm), with the objective of determining the vibrational temperature during NO decomposition. The analysis of the data allowed for coupling between vibrational relaxation and dissociation. Results were reported for both vibrational nonequilibrium and equilibrium rate constants; the latter have been used in Fig. 3 and Table 3. The vibrational temperatures inferred show disagreement with the coupled relaxation and dissociation model, possibly due to nonequilibrium within the vibrational levels monitored.

The discrepancy between the rate constants inferred in these three studies is unusually large, with Koshi's equilibrium rate constant at 5000 K being 30 times that of Myerson. (The ratio is 16 if the nonequilibrium rate constant is used.) Accordingly, no recommendation is made in this survey for the dissociation rate constant.

3.4. $N_2O + M \rightarrow N_2 + O + M$

The dissociation of nitrous oxide

$$N_2O(^1\Sigma) + M \rightarrow N_2(^1\Sigma) + O(^3P) + M \qquad \Delta H_0^\circ = 161 \text{ kJ} \tag{1}$$

is a reaction of broad interest to kineticists: it plays a role in various models of NO_x formation and decomposition; it forms the basis of increasingly used techniques for the controlled generation of O atoms in studies of elementary oxidation reactions; and finally, because the rate constant can be measured over the entire pressure range from the low- to the high-pressure limits, it is a convenient reaction for evaluation of unimolecular reaction theories. It is now

Table 3. Rate constant data for NO + $M \rightarrow$ N + O + M.

Reference	Log(A/cm^3 mol^{-1} s^{-1} K^{-B})	B	θ/K	T range	M
Myerson (1973)	14.14	0	74 680	4150–6000	Ar
Wray and Teare	20.60	−1.5	75 490	4200–6700	Ar, N_2, O_2
(1962)	$k \cong 20k_{Ar}$			4000–6600	NO, O, N
Koshi *et al.* (1979)	15.72	0	76 000	4350–7000	Ar
	$k \cong 22k_{Ar}$				NO
This evaluation	No recommendation				

widely accepted that this spin-hindered channel is indeed the principal reaction path which occurs. Evaluations of the limiting first- and second-order rate constants for this reaction have been presented previously by Baulch *et al.* (1973). Attention here is focused on evaluation of recent determinations of the second-order (low-pressure) rate constant, k_1.

The analysis of high-temperature N_2O decomposition measurements, which have been conducted in shock tubes, generally requires consideration of the additional reactions

$$O + N_2O \rightarrow NO + NO \qquad \Delta H_0^\circ = -153 \text{ kJ} \tag{2}$$

and

$$O + N_2O \rightarrow N_2 + O_2 \qquad \Delta H_0^\circ = -332 \text{ kJ} \tag{3}$$

At sufficiently high temperatures and low N_2O concentrations, these secondary reactions may have only minor effects, and the rate constant k_1 can be inferred directly from the rate of decay of N_2O. At another extreme, when O atoms are in a steady state, the observed rate of N_2O decomposition is given by

$$d[N_2O]/dt = -2k_1[N_2O][M] \tag{4}$$

where the factor of 2 arises from the fast reaction of O atoms [produced by reaction (1)] to remove N_2O by reactions (2) and (3). Most recent studies have utilized computer analyses with detailed kinetics models to account for the varying influence of these secondary reactions in determining k_1. This approach avoids some of the uncertainties inherent in earlier analyses which employed simplified, limiting-case models.

There have been several studies of k_1 since the Leeds survey was published in 1973. The results from seven studies have been selected as suitable for providing an improved rate constant expression. The results from the studies of Dove *et al.* (1975), Dean and Steiner (1977), Monat *et al.* (1977), Roth and Just (1977), Endo *et al.* (1979), Sulzmann *et al.* (1980), and Louge and Hanson (1981) are plotted in Fig. 4 together with the previous recommendation by Baulch *et al.* (1973) based on the expression reported by Olschewski *et al.* (1966). Table 4 provides a summary of the results and conditions of these various studies together with the parameters specifying the new recommended expression.

The seven studies utilized different experimental approaches. Dove, Nip, and Teitelbaum (1975) utilized laser schlieren measurements behind incident shock waves in N_2O/Ar mixtures (8% N_2O) to determine k_1 in the temperature range 2160–3590 K. Postshock total concentrations were sufficiently low, in the range 1.6×10^{-6} to 2.5×10^{-6} mol cm^{-3}, to neglect differences between the observed or apparent second-order rate constant and k_1. Detailed analysis of the complex laser schlieren signal using a model including reactions (1–3) led to values of $k_2 + k_3$ as well as k_1. Although the scatter in their data is relatively large, the results are in good agreement with the rate constant recommendation of this survey.

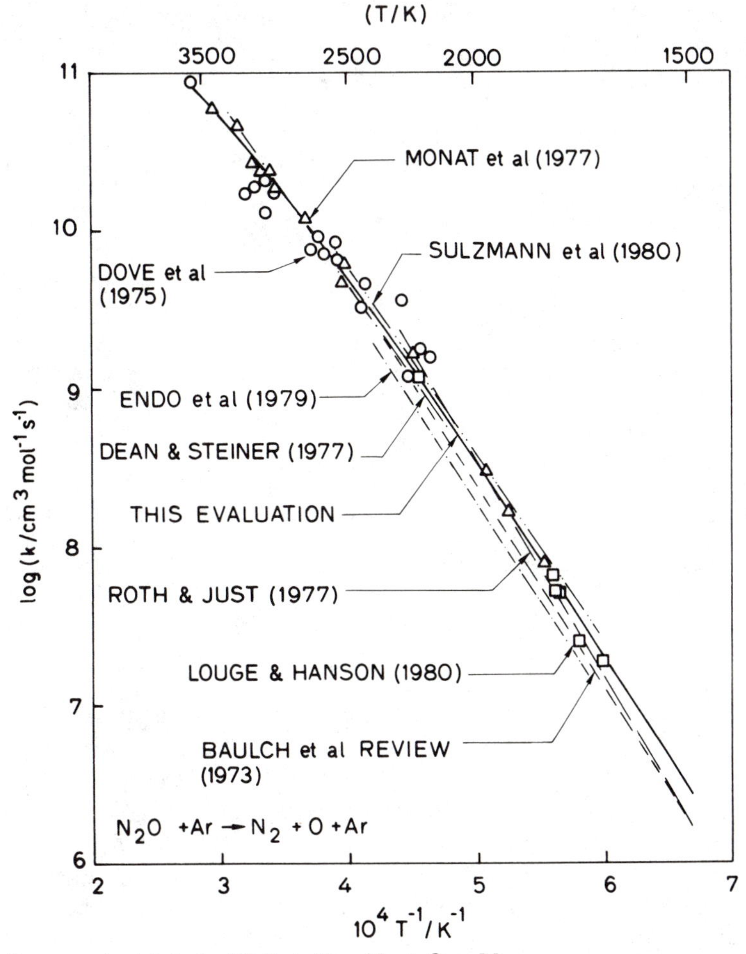

Figure 4. Rate constant data for $N_2O + M \rightarrow N_2 + O + M$.

Subsequent work by Dean (1976) and Dean and Steiner (1977) utilized reflected shock waves with detection of N_2O by ir emission (4.65 μm) and O atoms by CO + O chemiluminescence (450 nm) to infer both k_1 and $k_2 + k_3$. The analysis was based on detailed modeling considering reactions (1)–(3) together with the reaction

$$CO + O + M \rightarrow CO_2 + M$$

The mixtures were nominally N_2O in Ar (0.5–1% N_2O) and N_2O/CO/Ar (1% N_2O, 2–4% CO); the temperature was 2100–3200 K and the postshock total concentrations were about 8×10^{-6} mol cm^{-3}. It should be noted that the N_2O and O-atom measurements were made in separate studies (Dean, 1976, and Dean and Steiner, 1977, respectively). The latter study contains analysis of the complete data set and is considered to represent the final

Table 4. Rate constant data for $N_2O + M \rightarrow N_2 + O + M$.

Reference	Log(A/cm^3 mol^{-1} s^{-1} K^{-B})	B	θ/K	T range	M
Dove *et al.* (1975)	See Fig. 4			2160–3590	Ar
Dean and Steiner (1977)	14.43	0	27 180	2100–3200	Ar
Monat *et al.* (1977)	14.15	0	25 810	1810–3360	Ar, Kr, N_2
Roth and Just (1977)	15.16	0		1500–2250	Ar
Endo *et al.* (1979)	14.59	0	28 870	1700–2400	Ar[a]
Sulzmann *et al.* (1980)	14.57	0	27 660	1680–2560	Ar
Louge and Hanson (1981)	See Fig. 4			1660–2200	Ar
Baulch *et al.* (1973) (evaluation based on Olschewski *et al.*, (1966)	14.70	0	29 190	1500–2500	Ar
This evaluation	23.84	−2.5	32 710	1500–3600	Ar, Kr

[a] Endo *et al.* (1979) have also determined rate constants for M = He, Ne, Kr, Xe, N_2 and CF_4.

conclusions of the work. In subsequent work, Dean, Steiner, and Wang (1978) utilized a slightly modified value of k_1 (10% change) to fit data of experiments in $H_2/N_2O/CO/Ar$ mixtures. The results reported by Dean and Steiner (1977) are utilized in this survey. As may be seen in Fig. 4, these results are in excellent agreement with the recommended rate constant.

The study by Monat, Hanson, and Kruger (1977) utilized incident shock waves in N_2O/Ar (0.7–2.0% N_2O) and various $N_2O/N_2/O_2$ inert gas mixtures (nominally 2% N_2O, 21% N_2, 3% O_2, and the balance either Ar or Kr) to investigate N_2O decomposition and NO formation in the temperature range 1815–3365 K. Postshock total concentrations were typically below 3×10^{-6} mol cm^{-3}. Measurements of N_2O (ir emission at 4.5 μm) and NO (ir emission at 5.3 μm) concentration profiles were made, and a detailed kinetic analysis based on a nine-reaction mechanism was employed to infer k_1. No systematic difference was observed between M = Ar, Kr, or N_2, although the mole fraction of N_2 was too low to allow an accurate assessment of k_1 for $M = N_2$. The results for k_1 were found to be relatively insensitive to all other rate constants at high temperatures and only moderately sensitive to the specification of k_2 and k_3 at the low-temperature end of the experiments. Combined use of the N_2O decay and NO formation data, together with the assumption that $k_2 \cong k_3$, led to a reliable determination of k_1 over the full temperature range. A conservative estimate of the absolute uncertainty in k_1 was given as ±50%, although the scatter in the data was well below this level. Monat's best-fit expression also is in excellent agreement with the current recommendation for k_1.

Roth and Just (1977) utilized atomic resonance absorption spectroscopy to monitor O atom formation behind reflected shock waves in $N_2O/CH_4/Ar$ mixtures and were able to infer k_1 at somewhat lower temperatures ($1500 < T < 2250$ K) than other workers. Postshock concentrations were below about 1.6×10^{-5} mol cm^{-3} and, as in most other studies, no corrections were applied to the inferred second-order rate constant.

Endo *et al.* (1979) studied N_2O decomposition behind reflected shock waves in several heat bath gases (M = He, Ne, Ar, Kr, Xe, N_2, and CF_4). The temperature range covered was 1700–2400 K, with corresponding total concentrations of 4×10^{-5} to 6×10^{-6} mol cm^{-3}. N_2O was monitored in absorption at 230 nm, and k_1 was inferred assuming O atom steady state and applying Eq. (4). Their results for M = Ar are about 10–30% below the results of other studies and our current recommendation, the difference possibly being due to the use of the steady-state approximation or to the neglect of fall-off in the rate constant at their higher pressures. The reader is directed to the complete paper by Endo *et al.* for results on the variation of the rate constant with other heat bath gases.

Sulzmann, Kline, and Penner (1980) studied the thermal decomposition of N_2O behind reflected shock waves in N_2O/Ar mixtures (2% N_2O) at temperatures of 1685–2560 K and pressures of 1.7–4.6 atm. Postshock total concentrations were typically below 1.5×10^{-5} mol cm^{-3}. N_2O and NO time histories were measured by ir emission (4.5 μm) and γ-band absorption (226 nm), respectively. The data analysis utilized various algebraic expressions based on the reactions (1)–(3) to infer k_1, $k_2 + k_3$, and k_3/k_2. The ratio k_3/k_2, which has received considerable debate in the literature, was found to be 1.09 ± 0.10 in the temperature range studied. This value is consistent with the recommendation by Baulch *et al.* (1973) of 1.0 ± 0.2 ($1000 < T < 2800$ K). Although the scatter in their data for k_1 is relatively large, the best-fit expression reported by Sulzmann *et al.* is in excellent agreement with the rate constant recommended in this survey.

Finally, in recent unpublished work, Louge and Hanson (1981) carried out a limited series of incident shock wave experiments in N_2O/Ar mixtures and inferred k_1 by comparing N_2O time histories measured by ir emission (4.5 μm) with those computed using a detailed kinetic model. The effects of shock attenuation, boundary layers, finite electronic response and shock transit times, and fall-off were all analyzed, and small corrections were applied where necessary. Their results also agree well with the current recommendation for k_1.

Results for k_1 from these seven recent studies are tabulated in Table 4 and plotted in Fig. 4 together with the Leeds recommendation [which was based on the data of Olschewski *et al.* (1966)] and the recommendation of this survey. The data in Fig. 4 show sufficient curvature that a fit of the form

$$k_1 = AT^B \exp(-E_\circ/RT)$$

was employed for the recommended rate constant. The threshold energy $E_\circ$

was fixed at 65 kcal mol^{-1} according to the recommendation of Troe (1975). The parameter B was selected on the basis of unimolecular theory (Troe, 1975) by the approximation

$$B = 3/2 - S \cong -2.5$$

where S was evaluated at a mean temperature of 2500 K according to the simple relation (Troe, 1975) in terms of the tabulated enthalpy for N_2O

$$S = \frac{H°(T) - H°(0)}{RT} - 2.5 \cong 4.0$$

Thus, there was only one free parameter in the fit, namely, A, which was selected by a visual best fit to the data of these recent studies. The final expression

$$k_1 = 6.9 \times 10^{23}\ T^{-2.5} \exp(32\,710/T)\ \mathrm{cm^3\ mol^{-1}\ s^{-1}}$$

is recommended for use in the temperature range 1500–3600 K with estimated uncertainty factors of $f = 0.8$ and $F = 1.2$ in the middle temperature range, increasing to about $f = 0.6$ and $F = 1.4$ at the extremes in temperature.

With the exception of the work by Louge and Hanson, none of the studies surveyed corrected their data for pressure effects to obtain the low-pressure limiting rate constant (Chapter 4). We can estimate part of the error incurred by this neglect from the Lindemann–Hinshelwood relation

$$\frac{k}{k_\infty} = \frac{1}{1 + k_\infty/k_\circ[M]}$$

where k_∞ and $k_\circ[M]$ are the high- and low-pressure limiting values of a unimolecular rate constant k, which in the present case is $k_1[M]$. This relation can be rearranged to

$$k_\circ[M]/k = 1 + k_\circ[M]/k_\infty = \frac{1}{1 - k/k_\infty}$$

Consider first the experiments by Dove *et al.* (1975) and Monat *et al.* (1977), which were done at total concentrations less than about 3×10^{-6} mol cm^3. Substitution of this upper concentration, together with the present recommended expression for k_1 and the value for k_∞ based on the results of Olschewski *et al.* (1966)

$$k_\infty = 1.3 \times 10^{11} \exp(-30\,000/T)\ \mathrm{s^{-1}}$$

leads to $k_1[M]/k_{1\infty} < 0.026$ at 2000 K, and thus $k_\circ[M]/k = k_\circ[M]/k_1[M] = 1/(1 - .026) \cong 1.026$. The Lindemann–Hinshelwood correction for the difference between the rate constant inferred and the true low-pressure value is thus small relative to other sources of error. Although the concentrations were somewhat higher in the other studies, the Lindemann–Hinshelwood corrections are still less than 15–25%. Further fall-off corrections to k/k_∞ (See Section 3 of Chapter 4.) require additional effort for their calculation.

3.5. $O + N_2O \rightarrow NO + NO$

The reaction of O atoms with N_2O has two significant channels

$$O + N_2O \rightarrow NO + NO \qquad \Delta H_0^\circ = -153 \text{ kJ} \tag{1}$$

and

$$O + N_2O \rightarrow N_2 + O_2 \qquad \Delta H_0^\circ = -332 \text{ kJ} \tag{2}$$

There has been considerable debate regarding their rate constants (see, for example, Baulch *et al.*, 1973) and particularly the ratio k_1/k_2, but the weight of recent work shows rather convincingly that the ratio is approximately 1.0 at 2000 K. The rate coefficient k_1, in particular, now seems well established on the basis of forward and reverse rate measurements.

Work on reaction (1) prior to 1973 has been thoroughly discussed by Baulch *et al.* (1973). Recent work of a reasonably direct nature has been reported by Nip (1974), Monat *et al.* (1977), Monat (1977), McCullough *et al.* (1977), and Sulzmann *et al.* (1980). Other recent work, yielding data for $k_1 + k_2$, has been reported by Dean and Steiner (1977) and by Dove *et al.* (1975).

The studies by Monat *et al.*, Monat, and Sulzmann *et al.* each involved combined measurements of NO and N_2O in shock wave experiments. Sulzmann *et al.* utilized reflected shock waves in $N_2O/Ar(2\%\ N_2O)$ mixtures, detection of N_2O by infrared emission (4.52 μm) and NO by uv absorption (226 nm), and an analytical kinetic model which led to values for $k_1 + k_2$ and k_1/k_2. The resulting value for k_1 ($1680 < T < 2000$ K) has a rather large uncertainty but is in good agreement with the work at higher temperatures by Monat (1977) and the expression recommended by Baulch *et al.* for the temperature range 1200–2000 K.

Monat conducted incident shock wave experiments in mixtures of N_2O and various diluent gases (Ar, Kr, N_2, and O_2). Infrared emission was used to monitor both N_2O (4.5 μm) and NO (5.3 μm), and a detailed kinetic model was employed to infer k_1 as well as the N_2O dissociation rate constant. In these experiments, N_2O decomposed to form O atoms by the reaction

$$N_2O + M \rightarrow N_2 + O + M \tag{3}$$

and as the O atoms were produced they reacted with residual N_2O to form either NO by reaction (1) or $N_2 + O_2$ by reaction (2). The result was formation of NO to a nearly constant level, and it was this level of NO which primarily determined k_1. Monat's (1977) results for k_1 (which differ slightly from the results reported initially by Monat *et al.*, 1977) were dependent over the full temperature range studied on the value of k_3 assumed, but his value of k_3 is in excellent agreement with our recommended expression. The results for k_1 were also somewhat sensitive to the value of k_2 assumed at low temperatures. We estimate overall uncertainty factors of $f = 0.5$ and $F = 1.6$ for Monat's k_1.

Another measurement of the forward rate constant k_1 was reported by Nip (1974), who utilized mass spectrometric sampling behind incident shock waves in $N_2O/Ar/Kr$ mixtures. Analysis of measured values of NO, O_2, N_2O, and N_2 led to values for $k_1 + k_2$ and k_1/k_2 in the temperature range $2000 < T < 3200$ K. Nip's result for k_1 is approximately 0.5 that of Monat and has a somewhat lower activation energy than found by either Monat or Sulzmann *et al.*

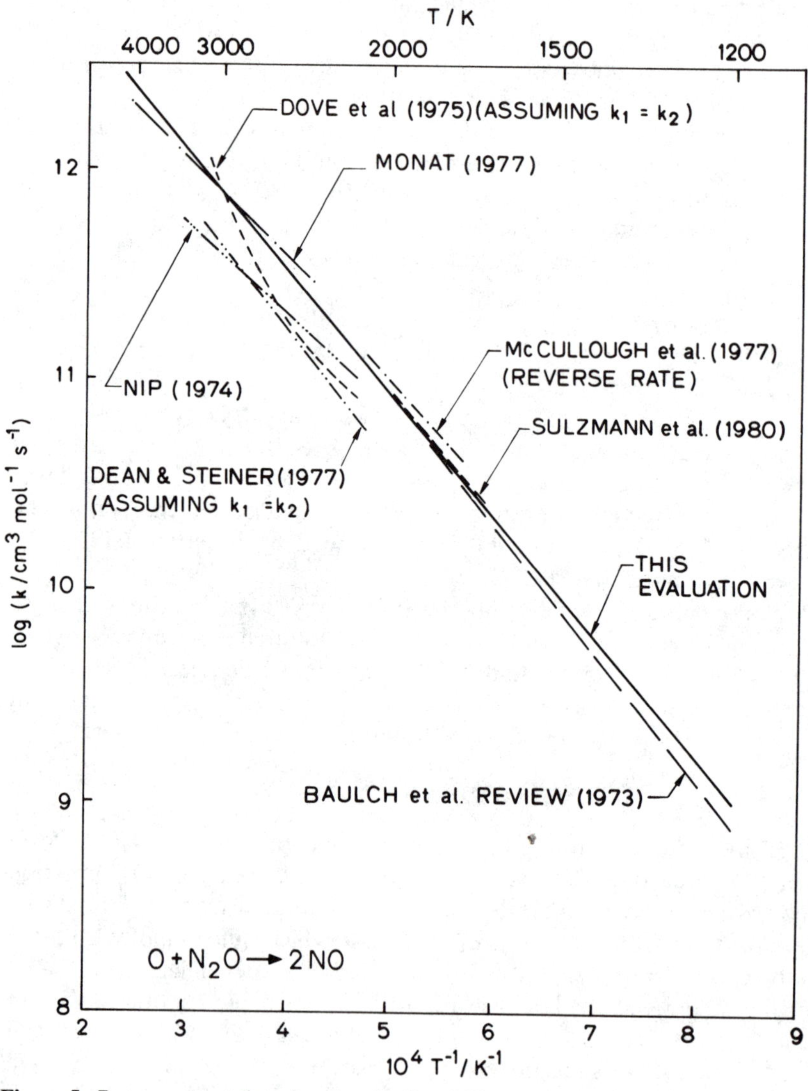

Figure 5. Rate constant data for $O + N_2O \rightarrow NO + NO$.

Results for the reverse rate constant have been obtained in the temperature range 1750–2100 K by McCullough *et al.* (1977) using a flow tube reactor. The fractional decomposition of NO in two NO/Ar mixtures (1 and 5% NO) was monitored as a function of flow rate (residence time) using a chemiluminescent analyzer. A detailed kinetic and flow model was used, and a sensitivity analysis was performed. McCullough's result

$$k_{-1} = 1.80 \times 10^{12 \pm 0.3} \exp(32\,110/T) \text{ cm}^3 \text{ mol}^{-1} \text{ s}^{-1}$$

was converted to a value for k_1, as shown in Fig. 5 and Table 5, and is in good agreement with the results of Sulzmann *et al.* and Monat.

A recommended expression for k_1 was established based on these recent studies and previous work. The least-squares analysis utilized two data points (at the extreme temperatures) for each of the data sets reported by McCullough *et al.* (1977), Sulzmann *et al.* (1980), and Monat (1977). Nip's temperature dependence seems inconsistent with other work and so his data have been represented by a single point at the low-temperature end of his study. Previous work was represented by including two points at the extreme ends of the expression for k_1 by Baulch *et al.* (1973) [which was based on the expression reported by Fenimore and Jones (1962)]. The resulting expression (see Table 5) is considered useful for the temperature range 1200–4100 K, and the estimated uncertainty factors are $f = 0.5$ and $F = 1.5$.

For convenience, Fig. 5 also indicates values for k_1 extracted from the work by Dove *et al.* (1975) and Dean and Steiner (1977) assuming that $k_1 = k_2$. Dove's results show excellent agreement with the recommended expression, thereby providing further evidence for $k_2 \cong k_1$. Dean and Steiner's value, however, is unaccountably low, suggesting that k_2 may be less than k_1 despite the results quoted above indicating otherwise.

Table 5. Rate constant data for $O + N_2O \rightarrow NO + NO$.

Reference	$\text{Log}(A/\text{cm}^3 \text{ mol}^{-1} \text{ s}^{-1})$	θ/K	T range
Nip (1974)	13.07	10 180	2000–3200
Dove *et al.* (1975) (assuming $k_1 = k_2$)	See Fig. 5		2160–3100
Monat (1977)	13.49	10 970	2380–4080
McCullough *et al.* (1977) (from reverse rate)	13.75	12 650	1750–2100
Dean and Steiner (1977) (assuming $k_1 = k_2$)	13.67	14 070	2100–3200
Sulzmann *et al.* (1980)	13.61	12 350	1680–2000
Baulch *et al.* (1973) (evaluation)	14.00	14 100	1200–2000
This evaluation	13.84	13 400	1200–4100

3.6. $O + N_2O \rightarrow N_2 + O_2$

Nearly all the available kinetic data for this rate constant have been obtained in the form of the sum $k_1 + k_2$ or the ratio k_1/k_2, where

$$O + N_2O \rightarrow N_2 + O_2 \qquad \Delta H_0^\circ = -332 \text{ kJ} \tag{1}$$

$$O + N_2O \rightarrow NO + NO \qquad \Delta H_0^\circ = -153 \text{ kJ} \tag{2}$$

Studies of these reactions prior to 1973 have been thoroughly reviewed in the Leeds survey (Baulch *et al.*, 1973). High-temperature results reported since then are listed in Table 6 and shown in Fig. 6. The most direct measurements were those of Sulzmann *et al.* (1980) and Nip (1974). In both of these studies, the rate constant sum $(k_1 + k_2)$ and the ratio (k_1/k_2) were determined, leading to values for k_1 and k_2.

Sulzmann *et al.* measured both N_2O (ir emission at 4.52 μm) and NO (uv absorption at 226 nm) behind reflected shocks in N_2O/Ar mixtures and reduced their data using a complex analytical model. They found $k_1/k_2 = 1.09 \pm 0.10$ over the temperature range 1680–2560 K, and their inferred values for both k_1 and k_2 are in good agreement with the Baulch *et al.* recommendations, although their stated uncertainties are rather large. Nip's experiments were conducted behind reflected shocks in $N_2O/Ar/Kr$ mixtures using mass spectrometric measurements of N_2O, N_2, O_2, and NO. An analytical model was used to reduce the data. His results for k_1 fall well below the Leeds recommendation (by about a factor of 3 at 2500 K), but are within the combined stated uncertainties of the two studies.

Additional results for k_1 can be inferred if the assumption that $k_1 = k_2$ is applied to the data of Dove *et al.* (1975) and of Dean and Steiner (1977), who both measured $k_1 + k_2$. The high-temperature portion of Dove's curve for k_1, determined by this approximation, is preferred since it yields good agreement with our evaluation for k_2 as shown in Fig. 5.

Table 6. Rate constant data for $0 + N_2O \rightarrow O_2 + N_2$.

Reference	$\text{Log}(A/\text{cm}^3\ \text{mol}^{-1}\ \text{s}^{-1})$	θ/K	T range
Nip (1974)	13.98	16 340	2000–3200
Dove *et al.* (1975) (assuming $k_1 = k_2$)	See. Fig. 6		2160–3100
Dean and Steiner (1977) (assuming $k_1 = k_2$)	13.67	14 070	2100–3200
Sulzmann *et al.* (1980)	13.65	12 350	1680–2000
Baulch *et al.* (1973) (evaluation)	14.0	14 100	1200–2000
This evaluation	14.0	14 100	1200–3200

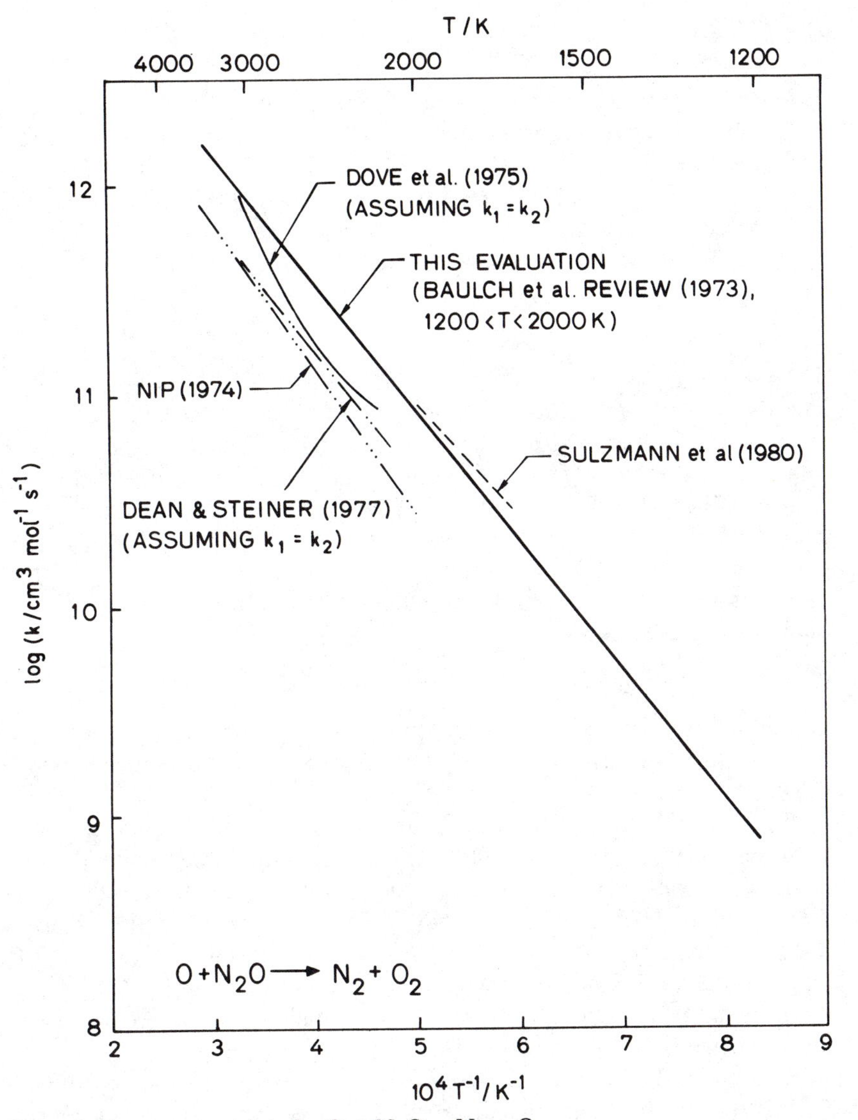

Figure 6. Rate constant data for $O + N_2O \rightarrow N_2 + O_2$.

The new data for k_1 from these four studies thus fall into two groups, with values which fall either nearly on or a factor of 2 to 3 below the Leeds recommendation (Baulch *et al.*, 1973), which was based on a k_1/k_2 ratio of 1.0. The strength of the evidence for $k_1/k_2 \cong 1.0$, together with the uncertainty in the data for k_1 that fall below the Leeds recommendation, argue in favor of retaining the Leeds expression without modification. We recommend uncertainty factors of $f = 0.4$ and $F = 1.5$ over the temperature range 1200–3200 K.

4. N/H reaction survey

This section provides a summary of four reactions in the N/H system for which new (since 1973) rate constant data are available. The reactions considered are: $NH_3 + M \rightarrow NH_2 + H + M$, $NH_3 + M \rightarrow NH + H_2 + M$, $H + NH_3 \rightarrow NH_2 + H_2$, and $H + NH_2 \rightarrow NH + H_2$. Reactions involving these species are relevant to the kinetics of fuel–nitrogen NO_x and to the removal of NO from flue gases by selective reaction with NH_3 or other NH_i compounds.

4.1. $NH_3 + M \rightarrow NH_2 + H + M$

The thermal decomposition of ammonia has been the subject of several experimental studies, from which it is clear that the overall kinetic mechanism under the conditions used is complex and that the observed rates of decomposition of NH_3 are strongly influenced by secondary reactions. Most early studies (prior to 1973; see the survey by Baulch *et al.*, 1973) were conducted with relatively high levels of NH_3, which led to overestimates of the rate constant and underestimates of the activation energy for the unimolecular process

$$NH_3 + M \rightarrow NH_2 + H + M \qquad \Delta H_0^\circ = 448 \text{ kJ} \qquad (1)$$

Of the early studies, the most direct appears to be that of Henrici (1966) who used uv absorption to monitor NH_3 decay behind incident shock waves in mixtures of 0.03–0.6% NH_3 in Ar.

More recently, there have been four shock tube studies conducted with sufficiently high dilution and sufficiently low pressure to determine the low-pressure, second-order rate constant k_1 directly. Dove and Nip (1979) studied NH_3 pyrolysis behind reflected shock waves (0.14–6.0% NH_3 in Kr) using mass spectrometric analysis of sampled gases to measure NH_3, NH_2, NH, and N_2. A detailed kinetic mechanism was developed and utilized to infer k_1. Their results are in reasonable agreement with other recent studies, which were all conducted in NH_3/Ar mixtures.

Holzrichter and Wagner (1981) studied NH_3 decomposition in argon (0.01–0.3% NH_3) behind incident and reflected shocks over a sufficient concentration range (9×10^{-7} to 2×10^{-4} mol cm^3) to infer both the low pressure (see Table 7 and Fig. 7) and high-pressure rate constants, with

$$k_\infty = 10^{15.74} \exp(-54\,250/T) \text{ s}^{-1}$$

The rate constants were inferred from the initial slope of the NH_3 decay as monitored by uv absorption. The uncertainty factors reported ($f \cong 0.4$ and $F \cong 1.7$) are relatively large, reflecting in part the difficulties inherent in initial slope measurements.

Yumura and Asaba (1981) monitored H-atom formation behind incident shocks using atomic resonance absorption ($\lambda = 121.6$ nm) to infer k_1 over

Table 7. Rate constant data for $NH_3 + M \rightarrow NH_2 + H + M$.

Reference	Log(A/cm^3 mol^{-1} s^{-1})	θ/K	T range	M
Henrici (1966)	16.40	45 800	2085–3100	Ar
Dove and Nip (1979)	16.08	45 800	2500–3000	Kr
Roose *et al.* (1980)	16.40	47 200	2200–3450	Ar
Holzrichter and Wagner (1981)	16.60	47 270	2200–3300	Ar
Yamura and Asaba (1981)	16.25	46 350	1740–3050	Ar
Baulch *et al.* (1973), (evaluation)	15.96	42 400	2000–3000	Ar
This evaluation (Roose *et al.* expression)	16.40	47 200	1740–3450	Ar

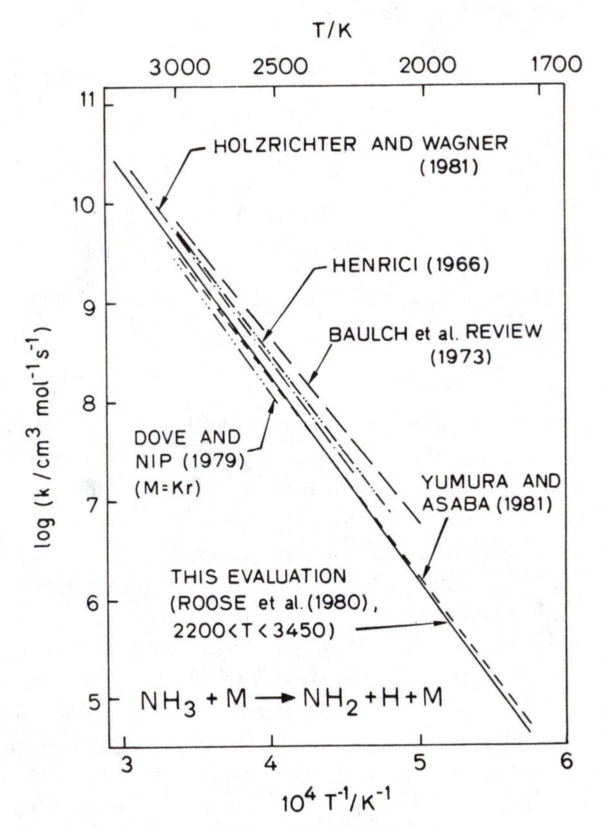

Figure 7. Rate constant data for $NH_3 + M \rightarrow NH_2 + H + M$.

the wide temperature range 1740–3050 K. This detection scheme was very sensitive, allowing use of highly diluted NH_3/Ar mixtures (10–10^3 ppm NH_3). A detailed kinetic mechanism was employed, and corrections were made for boundary layer and wall adsorption effects. The primary limitation of this study was the factor of 2 uncertainty in the H-atom calibration, resulting in a factor of 2 uncertainty in k_1.

In another recent study, Roose *et al.* (1980) measured k_1 behind incident shocks in NH_3/Ar mixtures (200–1200 ppm NH_3). Three species were monitored in emission: NH_3 (3 and 10.5 μm), NH_2 (540 nm), and NH (336 nm), with NH_2 the principal diagnostic for determining k_1. The primary limitation in this study was the NH_2 calibration, which had a stated uncertainty of less than $\pm 40\%$, leading to estimated uncertainty factors for k_1 of $f = 0.6$ and $F = 1.5$.

We recommend Roose's expression for the second-order rate constant, with M = Ar, because of the small scatter in the k_1 data and the relatively smaller uncertainty factors. This rate constant is in excellent agreement with that of Yumura and Asaba and is well within the uncertainty limits of Holzrichter and Wagner's k_1. Uncertainty factors of $f = 0.6$ and $F = 1.7$ are suggested for the full temperature range 1740–3450 K.

4.2. $NH_3 + M \rightarrow NH + H_2 + M$

A possible second channel in the unimolecular decomposition of ammonia is the spin-hindered reaction

$$NH_3(^1A) + M \rightarrow NH(^3\Sigma) + H_2(^1\Sigma) + M \qquad \Delta H_0^\circ = 405 \text{ kJ} \qquad (1)$$

Although reaction (1) is thermochemically more favorable than the spin-allowed path discussed in the previous section

$$NH_3(^1A) + M \rightarrow NH_2(^2B) + H(^2S) + M \qquad \Delta H_0^\circ = 448 \text{ kJ} \qquad (2)$$

we would expect a lower A factor and at least as large an activation energy for k_1 as observed for k_2. Recently, Holzrichter and Wagner (1981) concluded $k_1 \ll k_2$ on the basis of an observed induction time in emission from NH in shock-heated, dilute NH_3/Ar mixtures. Roose *et al.* (1980) and Roose (1981) attempted to measure k_1 behind incident shock waves using NH emission from highly dilute NH_3/Ar mixtures. Upper bounds on k_1 were established (see Fig. 8) for $2220 < T < 3430$ K based on an assessment of the maximum initial slope of the NH record. In one case, at $T = 2800$ K, an estimate of k_1 was made from a comparison of an NH emission record with a detailed computer simulation; the result was $k_2/k_1 \cong 40$. The uncertainty in this determination is large, due primarily to significant uncertainties in the NH calibration and in selecting the best-fit value of k_1. A change in the NH calibration was responsible for the difference between the most recent (Roose, 1981) and the initial (Roose *et al.*, 1980) reported values for k_1.

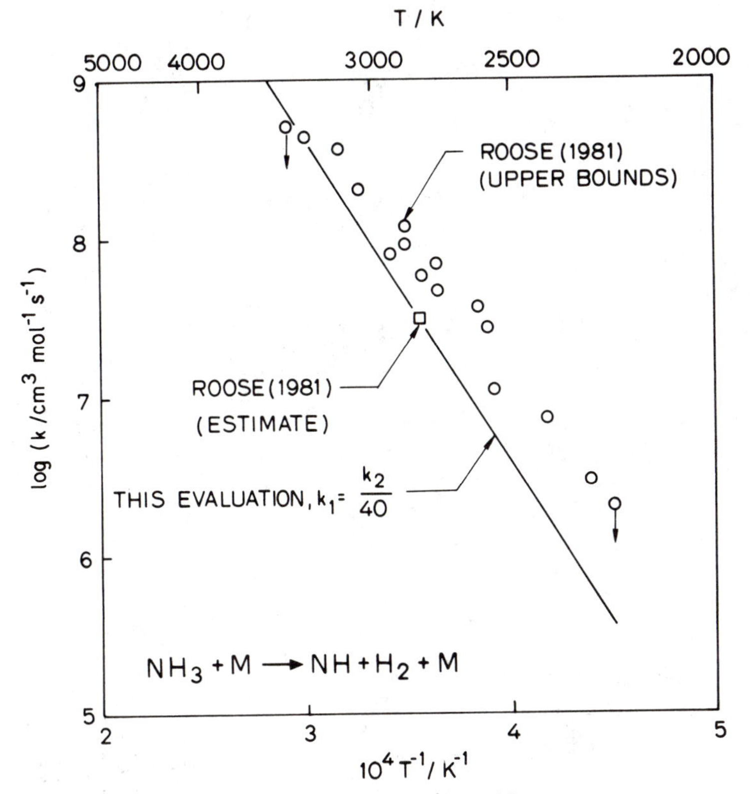

Figure 8. Rate constant data for $NH_3 + M \rightarrow NH + H_2 + M$.

It is safe to conclude that reaction (1) may generally be neglected in modeling NH_3 kinetics. If an estimate of k_1 is required, we recommend using, with caution,

$$k_1 \cong k_2/40 = 6.3 \times 10^{14} \exp(-47\,000/T) \text{ cm}^3 \text{ mol}^{-1} \text{ s}^{-1}$$

in the temperature range 2200–3500 K (Table 8). Estimated uncertainty factors are $f = 0.1$ and $F = 3$.

4.3. $H + NH_3 \rightarrow NH_2 + H_2$

The elementary reaction

$$H + NH_3 \rightarrow NH_2 + H_2 \qquad \Delta H_0^\circ = 16 \text{ kJ} \tag{1}$$

plays a significant role in the high-temperature chemistry of ammonia. Although early work on this reaction at low temperatures suggested a low reaction rate, three recent shock tube studies of a fairly direct nature indicate a substantial rate constant at high temperatures (Table 9 and Fig. 9).

Table 8. Rate constant data for $NH_3 + M \rightarrow NH + H_2 + M$.

Reference	$\text{Log}(A/\text{cm}^3\ \text{mol}^{-1}\ \text{s}^{-1})$	θ/K	T range	M
Roose (1981)	See Fig. 8 upper bound		2220–3430	Ar
	7.5		2800	Ar
This evaluation	14.8	47 000	2200–3500	Ar

Table 9. Rate constant data for $H + NH_3 \rightarrow NH_2 + H_2$.

Reference	$\text{Log}(A/\text{cm}^3\ \text{mol}^{-1}\ \text{s}^{-1})$	θ/K	T range
Dove and Nip (1974)	13.44	8 760	1500–2150
Holzrichter (1980)	12.15	0	2000
Yumura and Asaba (1981)	14.10	10 820	1860–2480
This evaluation (Yumura and Asaba expression)	14.10	10 820	1500–2500

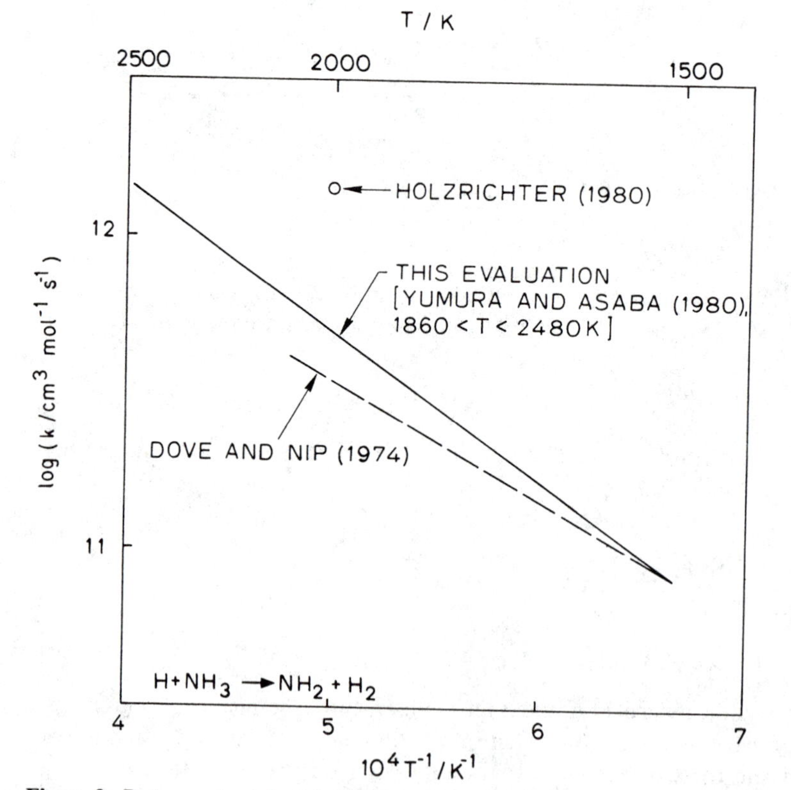

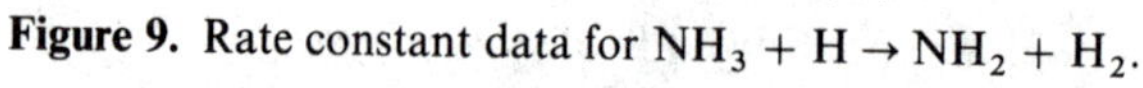

Figure 9. Rate constant data for $NH_3 + H \rightarrow NH_2 + H_2$.

Dove and Nip (1974) inferred the rate constant in the temperature range 1500–2150 K by monitoring the rate of decay of NH_3 in an excess of H atoms. The NH_3 time history was measured behind reflected shock waves using time-of-flight mass spectrometry of sampled gases. The H atoms were generated by shock heating rich H_2/O_2/inert gas mixtures with a small amount of NH_3 added. The H-atom concentration was calculated assuming partial equilibrium, and this assumption was checked indirectly by induction time measurements. Their final expression for k_1 (see Table 9), obtained by averaging results in different mixtures, exhibits relatively little scatter ($10^{0.16}$ was the standard deviation) but must be considered uncertain by about a factor of 2 due to experimental error and the possible influence of secondary reactions. .

Yumura and Asaba (1981) determined the rate constant by monitoring the steady-state level of H atoms behind incident shocks in NH_3/Ar mixtures using atomic resonance absorption at 121.6 nm. For the dilute mixtures used (<2000 ppm NH_3) the principal H-atom formation reaction was

$$NH_3 + Ar \rightarrow H + NH_2 + Ar \tag{2}$$

while the removal was due primarily to reaction (1). Thus, to a reasonable approximation,

$$[H]_{ss} \cong k_2[Ar]/k_1 \tag{3}$$

The argon concentration was known, and k_2 was determined in the same study so that H-atom measurements could be converted directly to values for k_1. A sensitivity analysis was performed to determine the range of conditions for which Eq. (3) should be valid. The final result and the standard deviation of the data were given as

$$k_1 = 10^{14.1 \pm 1.04} \exp(-10\,820 \pm 5180/T) \text{ cm}^3 \text{ mol}^{-1} \text{ s}^{-1}$$

although the scatter in the data and the error bars shown indicate a higher level of precision than quoted. The major uncertainties in this determination are associated with the H-atom calibration, the possible influences of secondary reactions, and a relatively large correction for boundary layer effects due to the use of long test times in a small tube.

Holzrichter (1980) studied reaction (1) in $N_2H_4/H_2/Ar$ mixtures by monitoring NH_3 profiles in absorption (at 210 nm) at temperatures near 2000 K. N_2H_4 provided a quick source of NH_2 radicals which reacted with H_2 to produce NH_3 through reaction (−1)

$$NH_2 + H_2 \rightarrow H + NH_3 \tag{-1}$$

leading eventually to a state of partial equilibrium for this reaction. A 13-reaction kinetic model was used to fit the NH_3 record leading to values for both k_1 (1.4×10^{12} cm^3 mol^{-1} s^{-1}) and k_{-1} (8.9×10^{11} cm^3 mol^{-1} s^{-1}). The implied value for the equilibrium constant of reaction (1) is about a factor of 2 less than the calculated equilibrium constant [using a value of

193.3 kJ for the heat of formation of NH_2 (Chase *et al.*, 1982)], suggesting an uncertainty of at least a factor of 2 should be assigned to Holzrichter's value for k_1.

In view of the uncertainties involved, the results of these three studies are in reasonable agreement. We recommend the intermediate expression by Yumura and Asaba (1981) with estimated uncertainty factors for the temperature range 1500–2500 K of $f = 0.4$ and $F = 2.5$.

4.4. $H + NH_2 \rightarrow NH + H_2$

The formation of NH at high temperatures is expected to proceed by means of radical reactions with NH_2, including

$$H + NH_2 \rightarrow NH + H_2 \qquad \Delta H_0^\circ = -44 \text{ kJ} \tag{1}$$

which plays a role in various models of NH_3 decomposition. This is a difficult reaction to study directly, and only recently have rate constant determinations been reported. There have been three separate shock tube investigations.

Dove and Nip (1979) studied this reaction as part of their investigation of ammonia pyrolysis. Using mass spectrometric analysis, they monitored the rates of formation of NH behind reflected shock waves in NH_3/Kr mixtures with and without added H_2 (6% NH_3 in Kr; 6% NH_3, 1.2% H_2 in Kr). The difference in the early time rates of formation was attributed to the NH-removal reaction

$$NH + H_2 \rightarrow NH_2 + H \tag{-1}$$

The rate constant k_{-1} was found to be approximately 2×10^{12} cm^3 mol^{-1} s^{-1} in the temperature range 2600–2800 K, with a stated uncertainty of a factor of 3. Using recent data for the heats of formation of NH (Binkley and Melius, 1982) and NH_2 (Chase *et al.*, 1982) to compute the equilibrium constant, we obtain the result $k_1 = 1.9 \times 10^{13}$ cm^3 mol^{-1} s^{-1} in this temperature range. The value of k_1 reported by Dove and Nip (1979) is larger by a factor of about 1.2 due to their use of a different equilibrium constant.

In another study of ammonia pyrolysis, Roose (1981) monitored NH behind incident shocks using uv emission (336 nm) and was able to infer k_1 from the NH concentration record at early times. The initial formation of NH was attributed to the slow reaction (Section 4.1).

$$NH_3 + M \rightarrow NH + H_2 + M \tag{2}$$

Very quickly, however, the production of NH is controlled by the combined effects of the dominant initiation reaction

$$NH_3 + M \rightarrow NH_2 + H + M \tag{3}$$

and the subsequent reactions (1) and (4)

$$NH_2 + M \rightarrow NH + H + M \tag{4}$$

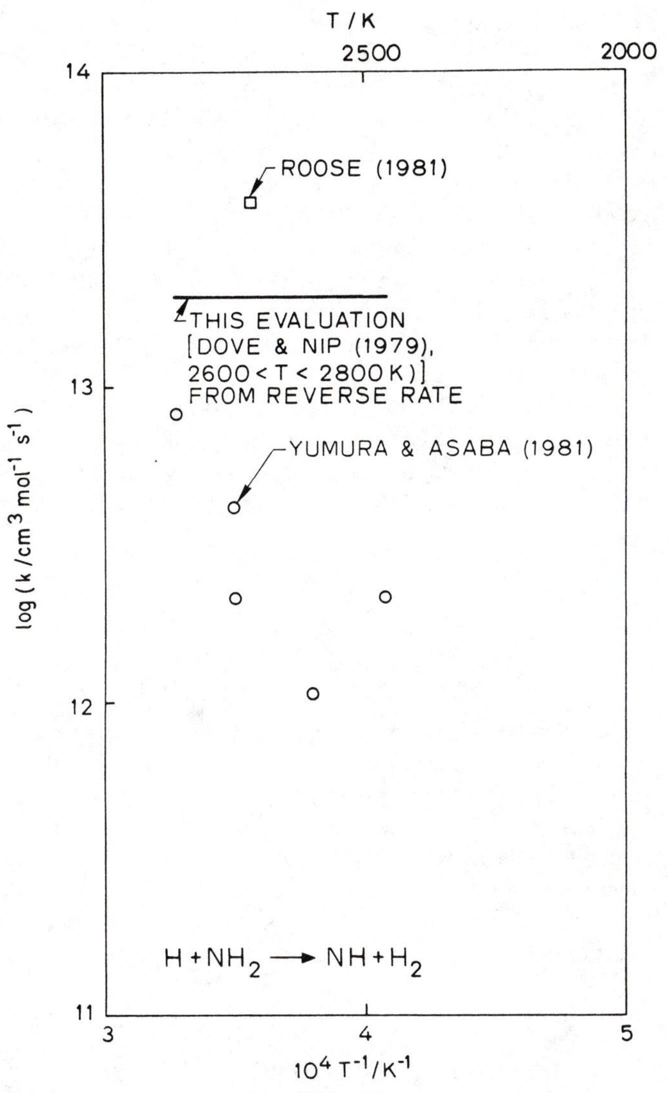

Figure 10. Rate constant data for $NH_2 + H \rightarrow NH + H_2$.

Table 10. Rate constant data for $H + NH_2 \rightarrow NH + H_2$.

Reference	$\text{Log}(A/\text{cm}^3\ \text{mol}^{-1}\ \text{s}^{-1})$	θ/K	T range
Dove and Nip (1979)	13.37	0	2600–2800
Roose (1981)	13.59	0	2800
Yumura and Asaba (1981)	See Fig. 10		2450–3020
This evaluation	13.28	0	2400–3000

The rate constant k_3 is known to good accuracy (determined in the same study), and k_4 has recently been determined (see Table II in Yumura and Asaba, 1981). Hence, the observed rate of NH formation could be readily converted through computer simulations to a value for k_1. The principal objectives of this study were determinations of k_2 and k_3, so that only one experiment (at 2800 K) was analyzed to infer k_1. The value inferred in his most recent analysis (Roose, 1981), $k_1 = 3.9 \times 10^{13}$ cm^3 mol^{-1} s^{-1}, is in good agreement with the findings of Dove and Nip (1979) and is to be preferred over the value reported previously (Roose *et al.*, 1980) using a different calibration for the NH emission. The uncertainty in k_1, which is due primarily to the NH calibration and sensitivity to other rate coefficients, is estimated as at least a factor of 3.

In a third shock tube study of ammonia pyrolysis, Yumura and Asaba (1981) monitored H atoms in dilue NH_3/Ar mixtures using atomic resonance absorption and were able to infer k_1 through comparisons with computer simulations. The critical reactions in the simulation were reactions (1), (3), and (4) together with

$$H + NH_3 \rightarrow NH_2 + H_2 \tag{5}$$

$$NH + NH \rightarrow N_2 + 2H \tag{6}$$

$$NH + M \rightarrow N + H + M \tag{7}$$

Results were obtained for the temperature range 2450–3020 K as shown in Fig. 10. The data exhibit significant scatter and generally fall more than a factor of 10 below the results of Roose (1981) and Dove and Nip (1979).

The uncertainties in these three studies of k_1 are substantial and prohibit a reliable recommendation. As an estimate, to be used with caution, we suggest a temperature-independent value of $k_1 = 1.9 \times 10^{13}$ cm^3 mol^{-1} s^{-1}, from Dove and Nip (1979), for the temperature range 2400–3000 K. Estimated uncertainty factors are $f = 0.1$ and $F = 4$ (Table 10).

5. N/H/O reaction survey

This section provides evaluations for eight reactions in the N/H/O system for which new (since 1973) data are available. The reactions considered are: $NH_3 + OH \rightarrow NH_2 + H_2O$, $NH_3 + O \rightarrow NH_2 + OH$, $HO_2 + NO \rightarrow NO_2 + OH$, $H + NO \rightarrow N + OH$, $NH + NO \rightarrow N_2O + H$, $H + N_2O \rightarrow N_2 + OH$, $NH_2 + O_2 \rightarrow$ products, and $NH_2 + NO \rightarrow$ products (including $N_2O + H_2$, $N_2 + H_2O$, $N_2 + H + OH$). These reactions are relevant to various aspects of NO_x kinetics in combustion systems, including the kinetics of fuel–nitrogen NO_x and the kinetics of NO removal from combustion gases by selective reaction with NH_3 or other NH_i compounds.

5.1. $NH_3 + OH \rightarrow NH_2 + H_2O$

The reaction of hydroxyl radicals with ammonia

$$NH_3 + OH \rightarrow NH_2 + H_2O \qquad \Delta H_0^\circ = -46\ \text{kJ} \tag{1}$$

is an important step in the oxidation of ammonia and plays a critical role in the removal of NO from combustion gases by injection of NH_3 (the Thermal DeNOx process). Although data for this rate constant are now available from both high- and low-temperature experiments, in this evaluation we have concentrated on the results obtained above 900 K which are of interest in combustion chemistry.

Silver and Kolb (1980) used a high-temperature flow reactor together with discharge flow techniques for generating OH radicals. OH was measured by resonance lamp fluorescence and the reaction was carried out with excess NH_3. The rate constant k_1 was obtained from the OH time history at room temperature and in the temperature range 575–1075 K.

Fujii *et al.* (1981) measured the induction times of $NH_3/O_2/H_2/Ar$ mixtures behind reflected shock waves using uv absorption to monitor NH_3. A detailed kinetic model was proposed and values for k_1 were obtained (in the temperature range 1360–1830 K) from a study of the sensitivity of the induction times to k_1. The authors reported an Arrhenius fit to their data and the data of Hack *et al.* (1974) at lower temperatures (298–670 K). The mechanism suggested by Fujii *et al.* (1981) is speculative, and their assumed products for the reaction between NH_2 and O_2 (the rate-limiting step in their system; See section 5.7) has been questioned by other investigators on theoretical grounds (Benson, 1981; Binkley and Melius, 1982; Pouchan and Chaillet, 1982). Assumption of the suggested alternative products for this reaction would lead to several subsequent reactions which were not included in the kinetic model of Fujii *et al.* (1981).

Niemitz *et al.* (1981) used reflected shock waves to study $NH_3/H_2O/Ar$ mixtures in the temperature range 1080–1695 K. A flash photolysis technique was used to produce OH radicals from H_2O. Measurements of NH_3 by ultraviolet absorption and OH by resonance lamp absorption, together with detailed kinetic modeling, yielded a value for k_1 at 1350 K. The authors reported a non-Arrhenius fit to their data and the previous data of Zellner and Smith (1974), Smith and Zellner (1975), and Perry *et al.* (1976).

Recently, Salimian *et al.* (1983) shock heated $NH_3/N_2O/Ar$ mixtures and monitored NH_3 and N_2O by infrared emission and OH by laser-based absorption spectroscopy (at 306.5 nm). A detailed kinetic model was employed and the sensitivities of the early-time NH_3 and OH profiles to the reactions of O and OH with NH_3 were investigated, with the following expression for k_1 inferred in the temperature range 1750–2060 K

$$k_1 = 10^{13.9 \pm 0.2} \exp(-4225/T)\ \text{cm}^3\ \text{mol}^{-1}\ \text{s}^{-1}$$

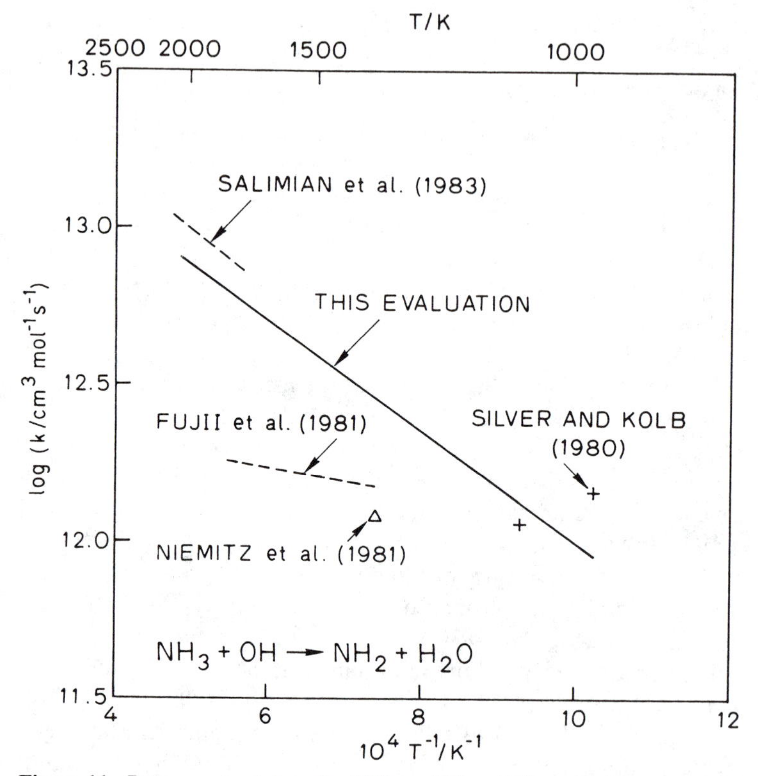

Figure 11. Rate constant data for $NH_3 + OH \rightarrow NH_2 + H_2O$.

Figure 11 shows the high-temperature data (above 900 K) for reaction 1. Our recommended expression for k_1 (see Table 11) is a least-squares Arrhenius fit to the data point of Niemitz *et al.* (1981), the two data points of Silver *et al.* (1980) at temperatures above 900 K, and two points from the expression of Salimian *et al.* (1983) at their extreme temperatures. The results of Fujii *et al.*

Table 11. Rate constant data for $NH_3 + OH \rightarrow NH_2 + H_2O$.

Reference	$\text{Log}(A/\text{cm}^3\ \text{mol}^{-1}\ \text{s}^{-1}\ \text{K}^{-B})$	B	θ/K	T range
Silver and Kolb (1980)	See Fig. 11			2600–2800
	12.51	0	1070	294–1075
Fujii *et al.* (1981)	12.50[a]	0	1010	1360–1830
Niemitz *et al.* (1981)	Log k = 12.08			1350
	8.90[b]	1.05	350	300–1400
Salimian *et al.* (1983)	13.9	0	4225	1750–2060
This evaluation	13.76	0	4055	900–2100

[a] An Arrhenius fit to the data of Fujii *et al.* (1981) and Hack *et al.* (1974).
[b] A fit to the data of Niemitz *et al.* (1981), Zellner and Smith (1974), Smith and Zellner (1975) and Perry *et al.* (1976).

(1981) have been excluded due to the uncertainty in their analysis. Suggested uncertainty factors for the recommended expression are $F = 2$ and $f = 0.5$ in the temperature range 900–2100 K. We caution the reader against extrapolation of this expression outside the quoted temperature range in view of the apparent non-Arrhenius behavior of this reaction (as discussed by Salimian *et al.*, 1983). Additional reliable data, particularly at high temperatures, is needed to refine the temperature dependence of this important reaction.

5.2. $NH_3 + O \rightarrow NH_2 + OH$

The reaction of O atoms with ammonia

$$NH_3 + O \rightarrow NH_2 + OH \qquad \Delta H_0^\circ = 24 \text{ kJ} \tag{1}$$

is of potential importance in the high-temperature oxidation of ammonia. Although there have been several low-temperature determinations of k_1, only one study (Salimian *et al.*, 1983) has been conducted at temperatures above 1000 K.

Salimian *et al.* (1983) shock heated $NH_3/N_2O/Ar$ mixtures in the temperature range 1750–2060 K and monitored NH_3 and N_2O by infrared emission and OH by laser absorption spectroscopy at 306.5 nm. At early times the chemistry of the $NH_3/N_2O/Ar$ system is dominated by the simple mechanism:

$$N_2O + M \rightarrow N_2 + O + M \tag{2}$$

$$NH_3 + O \rightarrow NH_2 + OH \tag{1}$$

$$NH_3 + OH \rightarrow NH_2 + H_2O \tag{3}$$

Using an accepted value for k_2, the authors were able to fit NH_3 and OH signals to extract values for k_1 and k_3:

$$k_1 = 10^{13.3 \pm 0.3} \exp(-4450/T) \text{ cm}^3 \text{ mol}^{-1} \text{ s}^{-1}$$

$$k_3 = 10^{13.9 \pm 0.2} \exp(-4225/T) \text{ cm}^3 \text{ mol}^{-1} \text{ s}^{-1}$$

Extrapolation of this Arrhenius expression for k_1 provides reasonable agreement with available data above 500 K [see Baulch *et al.* (1973)] but

Table 12. Rate constant data for $NH_3 + O \rightarrow NH_2 + OH$.

Reference	$\text{Log}(A/\text{cm}^3 \text{ mol}^{-1} \text{ s}^{-1})$	θ/K	T range
Salimian *et al.* (1983)	13.34	4470	1750–2060
This evaluation (Salimian *et al.* expression)	13.34	4470	1000–2100

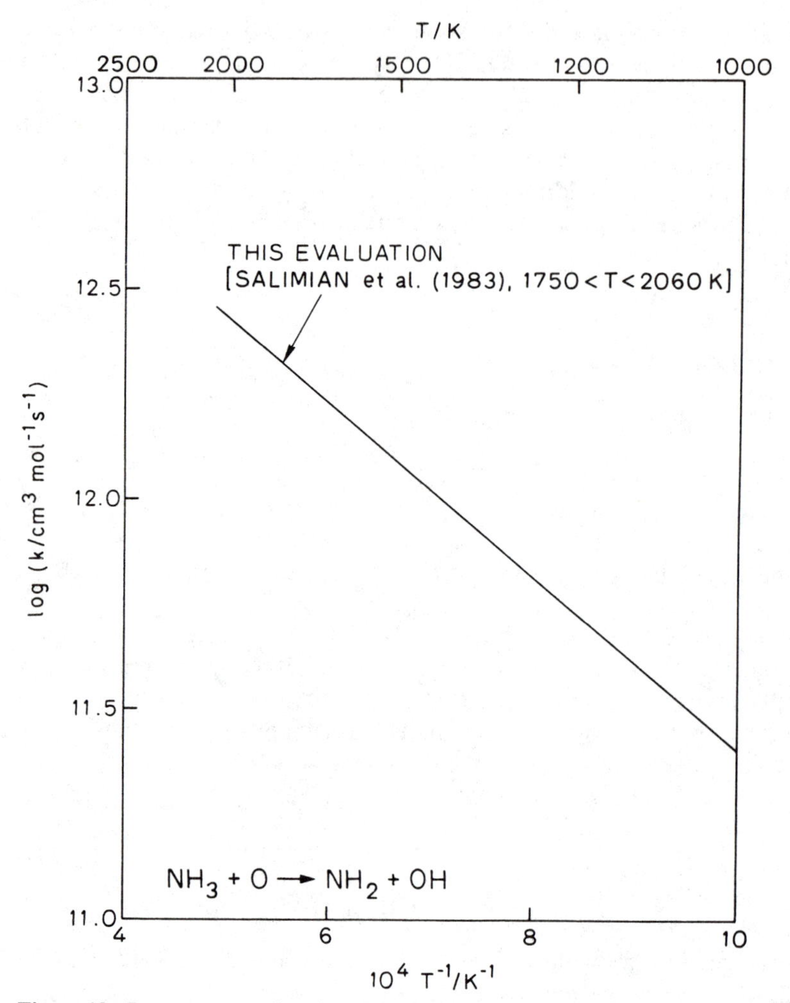

Figure 12. Rate constant data for $NH_3 + O \rightarrow NH_2 + OH$.

disagrees with data at lower temperatures, suggesting the possibility of non-Arrhenius character for reaction (1).

Our recommendation for this rate constant (see Table 12 and Fig. 12) is the expression reported by Salimian *et al.* (1983). Recommended uncertainty factors for the temperature range 1000–2100 K are $f = 0.5$ and $F = 2$.

5.3. $HO_2 + NO \rightarrow NO_2 + OH$

The reaction

$$HO_2 + NO \rightarrow NO_2 + OH \qquad \Delta H_0^\circ = -28 \text{ kJ} \tag{1}$$

is known to play an important role in atmospheric chemistry and in the removal of NO from combustion gases by injection of NH_3/H_2 (Thermal

Table 13. Rate constant data for $HO_2 + NO \rightarrow NO_2 + OH$.

Reference	$\mathrm{Log}(A/\mathrm{cm}^3\ \mathrm{mol}^{-1}\ \mathrm{s}^{-1})$	θ/K	T range
Glänzer and Troe (1975)			
(from reverse rate)[a]	See Fig. 13		1340–1760
Howard (1980)	See Fig. 13		
	12.32	−240	230–1270
This evaluation	12.32	−240	1000–2000

[a] See text

DeNOx process) or NH_3/H_2O_2 mixtures. Data for this reaction are available both at high and low temperatures; this review concentrates on data above 1000 K (Table 13).

Glänzer and Troe (1975) measured HO_2 and NO_2 by uv absorption in $HNO_3/NO_2/Ar$ mixtures behind reflected shock waves in the temperature range 1340–1760 K. The rate constant of the reverse reaction, k_{-1}, was determined from the half-time-to-peak of the HO_2 signal together with a detailed kinetic model. The forward rate constant k_1 was determined from k_{-1}

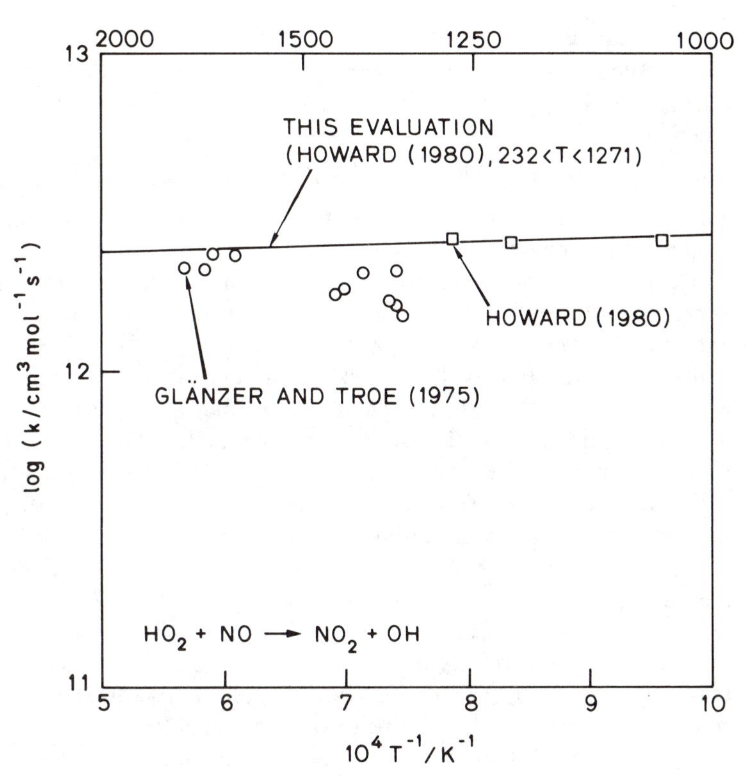

Figure 13. Rate constant data for $HO_2 + NO \rightarrow NO_2 + OH$.

and the equilibrium constant of the reaction based on thermochemical data available at the time (JANAF, 1971).

Howard (1980) has recently determined k_1 and k_{-1} in the temperature range 230–1270 K using a flow reactor and laser magnetic resonance measurements of HO_2, OH, NO, and NO_2. The data for k_1 and k_{-1} were also used to obtain the equilibrium constant ($K = k_1/k_{-1}$). Calculated values of K together with the thermochemical data of OH, NO, and NO_2 were used to obtain an improved heat of formation for $HO_2(\Delta H^\circ_{f,298}(HO_2) = 2.5 \pm 0.6$ kcal/mol) which differs significantly from the value previously suggested by JANAF ($\Delta H^\circ_{f,298}(HO_2) = 5 \pm 2$ kcal/mol).

We have used the data on k_{-1} obtained by Glänzer and Troe (1975) and recalculated their k_1 (assuming $\Delta H^\circ_{f,298}(HO_2) = 2.5$ kcal/mol). These data together with Howard's data above 1000 K are shown in Fig. 13. The uncertainty in the data of Glänzer and Troe is about 30% based on our estimate of the combined uncertainty in k_{-1} and the heat of formation of HO_2. In contrast the data obtained by Howard are direct measurement of k_1 with less scatter and uncertainty. In this evaluation, we recommend the expression obtained by Howard (1980) with uncertainty factors of $f = 0.7$ and $F = 1.2$ for the temperature range 1000 to 2000 K.

5.4. H + NO → N + OH

The formation and decomposition of nitric oxide in combustion systems is often controlled by the extended Zeldovich mechanism, which includes the reaction

$$H + NO \rightarrow N + OH \qquad \Delta H^\circ_0 = 204 \text{ kJ} \tag{1}$$

This reaction is of particular importance in modeling NO decomposition under fuel-rich combustion conditions.

The first direct measurements of the rate constant for reaction (1) did not appear until 1975 when four different groups (Bradley and Craggs, Duxbury and Pratt, Flower *et al.*, and Koshi *et al.*) presented results obtained from high-temperature (2200–4500 K) shock tube experiments. More recently, McCullough *et al.* (1977) have reported measurments of k_1 at somewhat lower temperatures (1750–2040 K) using a flow tube reactor, the data of Koshi *et al.* have been further analyzed by Ando and Asaba (1976), and Flower *et al.* (1977) reported additional measurements leading to a slight revision in their k_1. All of these studies have shown that NO decomposition in high-temperature NO/H_2/inert gas mixtures is described well by the extended Zeldovich mechanism and that reaction (1) is the rate-controlling step for sufficiently large H_2: NO ratios.

The most reliable results are considered to be those of Flower *et al.* (1977), Duxbury and Pratt (1975), and McCullough *et al.* (1977). Flower *et al.* conducted experiments behind incident shocks in NO/H_2/inert gas mixtures ($T = 2400$–4200 K; NO : $H_2 = 1:2$ to $2:1$) and monitored ir emission from

Table 14. Rate constant data for H + NO → N + OH.

Reference	$\text{Log}(A/\text{cm}^3\ \text{mol}^{-1}\ \text{s}^{-1})$	θ/K	T range
Bradley and Craggs (1975)	14.54	23 940	2530–3020
Duxbury and Pratt (1975)	14.41	24 560	2200–3250
Ando and Asaba (1976)	13.70	24 510	2400–3500
Flower *et al.* (1977)	14.35	25 410	2400–4200
MuCullough *et al.* (1977)	14.24	24 760	1750–2040
This evaluation	14.23	24 560	1750–4200

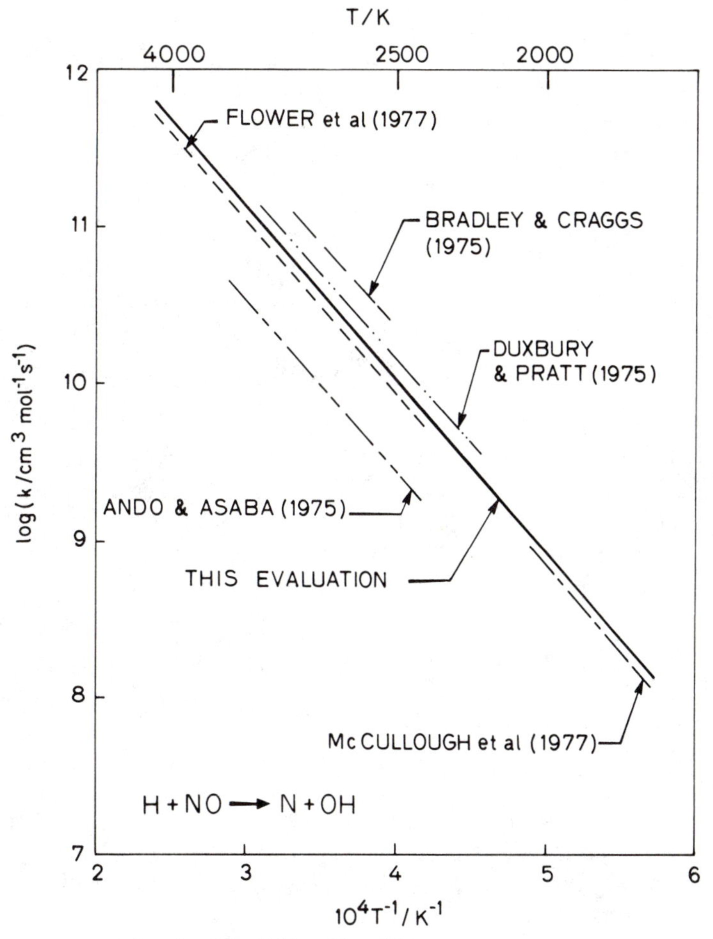

Figure 14. Rate constant data for H + NO → N + OH.

NO and H_2O at 5.3 and 6.3 μm, respectively. The rate constants inferred from comparisons of the measured concentration histories with those computed using a detailed mechanism were shown to be reasonably insensitive to the rate constants specified for other reactions. Duxbury and Pratt monitored uv absorption of OH and NO behind incident shock waves in dilute $NO/H_2/Ar$ mixtures ($T = 2200$–3250 K). Detailed modeling of both OH and NO profiles yielded results for k_1 with a standard deviation of 18%. The rate constant reported was based on a combined fit to their high temperature data and the low-temperature data of Campbell and Thrush (1968) for the reverse rate.

The experiments by McCullough *et al.* (1977) were performed in an alumina packed-bed flow tube reactor. Dilute $NO/H_2/Ar$ mixtures were heated to temperatures in the range 1750–2040 K, and the fractional decomposition of NO was monitored as a function of flow rate using a chemiluminescent analyzer. A detailed flow and kinetic model (including surface reactions) was used to infer k_1. A careful error analysis, including sensitivity to other rate constants, yielded error limits of $\pm 46\%$.

The recommended expression for k_1 (see Table 14 and Fig. 14) was obtained by a least-squares analysis of the expressions reported by Duxbury and Pratt, Flower *et al.*, and McCullough *et al.* Estimated uncertainty factors are $f = 0.6$ and $F = 1.4$ for the temperature range 1750–4200 K.

5.5. $NH + NO \rightarrow N_2O + H$

There have been relatively few determinations of the rate constant for the reaction

$$NH + NO \rightarrow N_2O + H \qquad \Delta H_0^\circ = -154\ \text{kJ} \tag{1}$$

This reaction plays a role in the conversion of NO to N_2 under some conditions, and the reverse reaction may be a significant step in the formation of NH_3 in N_2O–hydrogen systems. Recent work to determine k_1 includes two shock tube studies and one flame investigation.

Nip (1974) utilized time-of-flight mass spectrometry behind reflected shocks in $N_2O/H_2/O_2$/inert gas mixtures to infer k_{-1} in the temperature range 1600–2100 K. The sum of the rate constants $k_{-1} + k_2$ for the reactions

$$H + N_2O \rightarrow NO + NH \tag{-1}$$

and

$$H + N_2O \rightarrow N_2 + OH \tag{2}$$

was determined directly from the measured rate of decay of N_2O assuming a partial equilibrium value for the H-atom concentration. The influence of other reactions was neglected, based in part on estimates made for recognized interfering reactions. Measurements of the relative formation of NO and N_2 were used to determine k_{-1}/k_2 and (since $k_{-1} + k_2$ was also known) k_{-1}, with the result

$$k_{-1} = 10^{14.28} \exp(-17\,360/T)\ \text{cm}^3\ \text{mol}^{-1}\ \text{s}^{-1}$$

We have converted Nip's expression for k_{-1} to k_1 using Binkley and Melius' (1982) value for the heat of formation of NH ($\Delta H^\circ_{f,0} = 87.4$ kcal/mol).

Roose *et al.* (1981) determined k_1 from incident shock wave experiments in $NO/NH_3/Ar$ and $NO/NH_3/H_2/Ar$ mixtures ($T = 1760$–2850 K). The species monitored (spectroscopically) were NO, N_2O, NH_3, NH_2, and NH. The formation of N_2O as an intermediate species was attributed to two principal reactions, reaction (1) and the parallel reaction

$$NH_2 + NO \rightarrow N_2O + H_2 \tag{3}$$

so that

$$k_1[NH] + k_3[NH_2] = [\dot{N_2O}]_{formation}/[NO].$$

The value of $[\dot{N_2O}]_{formation}$ was inferred from the observed slope of the N_2O formation, $[\dot{N_2O}]_{net}$, by applying a relatively small correction for N_2O removal reactions, assumed to be primarily reaction (2). The H-atom concentration required was inferred from the rate of decay of NH_3 using an accepted rate constant for $H + NH_3 \rightarrow H_2 + NH_2$. Mixtures with and without H_2 were used to vary NH/NH_2 [and hence the relative influence of reactions (1) and (3)], so that both k_1 and k_3 could be determined.

With regard to reaction (3), it may be noted that the reaction path leading to N_2O does not necessarily produce H_2 as a co-product but may also produce 2H. Roose's results do not distinguish between these possibilities, and hence k_3 should probably be viewed as the rate coefficient for $NH_2 + NO \rightarrow N_2O + \cdots$. More recently, Roose (1981) has improved his NH_2 and NH calibration procedure, leading to a slightly modified expression (which supersedes his earlier results)

$$k_1 = 10^{13.35} \exp(-10\,600/T)\ \text{cm}^3\ \text{mol}^{-1}\ \text{s}^{-1}$$

Flame measurements of k_1 of a somewhat less direct nature have been reported recently by Peterson (1981). He utilized 0.1-atm, premixed $H_2/O_2/$ Ar/pyridine flames, and an extensive array of species measurements and data reduction procedures to infer k_1 over the temperature range 670–1480 K. The scatter in k_1 was substantial, as might be expected in such a complex system, and the resulting best-fit expression (see Table 15) was based on the combined data of this study and the low-temperature (300 K) data of Gordon *et al.* (1971) and Hansen *et al.* (1976).

Table 15. Rate constant data for the reaction $NH + NO \rightarrow N_2O + H$.

Reference	$\text{Log}(A/\text{cm}^3\ \text{mol}^{-1}\ \text{s}^{-1})$	θ/K	T range
Nip (1974)			
(from reverse rate)	12.03	230	1600–2100
Roose (1981)	13.35	10 600	1760–2850
Peterson (1981)	12.09	−960	670–1480
This evaluation	No recommendation		

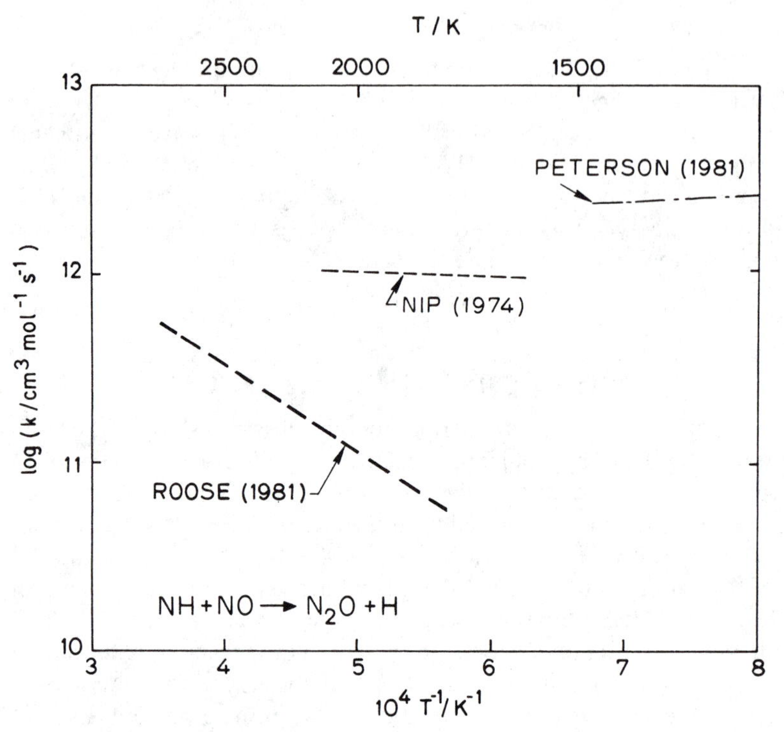

Figure 15. Rate constant data for $NH + NO \rightarrow N_2O + H$.

The large differences between the rate constants reported in these three studies (see Fig. 15) prohibit recommendation of a specific expression for k_1. We prefer the shock tube determinations as more direct, but the activation energies reported appear to be inconsistent with data at low temperatures. Clearly, further work is needed on this reaction. For the present, we choose to make no recommendation for k_1 but note that the expression based on Nip's reverse rate determination

$$k_1 = 10^{12.03} \exp(-230/T)\ \mathrm{cm^3\ mol^{-1}\ s^{-1}}$$

which falls nearly midway between the results of Peterson and Roose, may be useful as an approximation if confined to a temperature range of about 1600–3000 K.

5.6. $H + N_2O \rightarrow N_2 + OH$

The reaction of hydrogen atoms with nitrous oxide

$$H + N_2O \rightarrow N_2 + OH \qquad \Delta H_0^\circ = -263\ \mathrm{kJ} \tag{1}$$

plays an important role in N_2O/H_2 flames. The alternative path

$$H + N_2O \rightarrow NO + NH \qquad \Delta H_0^\circ = 154\ \mathrm{kJ} \tag{2}$$

is endothermic and is expected to be substantially slower than reaction (1). Baulch *et al.* (1973) reviewed studies of k_1 prior to 1973; here we have reviewed more recent measurements at temperatures above 1000 K.

Nip (1974) studied $N_2O/H_2/O_2$/inert gas mixtures behind reflected shock waves in the temperature range 1600–2100 K. The partial equilibrium state of rich H_2/O_2 mixtures provided a source of known amounts of H atoms for reaction with N_2O. Time-of-flight mass spectrometry was used to monitor N_2O, N_2, and NO. The rate constants k_1 and k_2 were determined by measuring the sum of the rate constants $(k_1 + k_2)$ from the N_2O decay rate and the ratio of the rate constants (k_1/k_2) from the $[N_2]/[NO]$ ratio.

Albers *et al.* (1975) added excess N_2O to H/He mixtures which were passed through an isothermal flow reactor in the temperature range 720–1110 K. ESR and mass spectrometry methods were used to measure H, OH, and N_2O. The rate constant k_1 was inferred from the decay of H radicals. Since measurable NO concentrations were not detected in the mass spectrometer, it was concluded that reaction (1) is dominant over reaction (2).

Balakhnine *et al.* (1977) utilized supersonic molecular beam sampling and a mass spectrometer to measure radical and stable species concentrations in the reaction zone of a lean, low-pressure N_2O/H_2 flame. The formation of N_2 in the flame front was attributed to reaction (1) together with reactions (3) and (4)

$$O + N_2O \rightarrow N_2 + O_2 \tag{3}$$

$$N_2O + M \rightarrow N_2 + O + M \tag{4}$$

Using N_2 concentration measurements together with their own determinations of k_3 and k_4, the rate constant k_1 was determined in the temperature range 1000–1700 K.

Dean *et al.* (1978) measured infrared emission of CO_2 and visible emission due to the reaction $O + CO \rightarrow CO_2 + h\nu$ in $H_2/O_2/CO/Ar$ and $H_2/N_2O/CO/Ar$ mixtures behind reflected shock waves. A detailed kinetic model was suggested that reproduced the CO_2 and visible emission profiles in the $H_2/O_2/CO/Ar$ mixtures. This model, together with suggested important N_2O reactions, provided a model for the $H_2/N_2O/CO/Ar$ mixtures. The rate constant k_1 was determined by matching measured and computed visible and CO_2 emission profiles in these mixtures.

Table 16. Rate constant data for $H + N_2O \rightarrow N_2 + OH$.

Reference	$\text{Log}(A/\text{cm}^3\ \text{mol}^{-1}\ \text{s}^{-1})$	θ/K	T range
Nip (1974)	14.32	13 140	1600–2100
Albers *et al.* (1975)	See Fig. 16		1014–1111
	14.34	8 710	718–1111
Balakhnine *et al.* (1977)	13.78	6 590	1000–1700
Dean *et al.* (1978)	15.26	13 590	2000–2850
Baulch *et al.* (1973)	13.88	7 600	700–2500
This evaluation	13.88	7 600	700–2500

Table 16 and Fig. 16 show recent results for k_1 and the expression recommended by Baulch *et al.* (1973). The experiments by Albers *et al.* (1975) provide the only direct measurements of k_1. The results obtained by Balakhnine *et al.* (1977) are influenced by their values for k_3 and k_4 which differ from those recommended in this evaluation (see Tables 6 and 4, respectively). The results of Dean *et al.* (1978) depend heavily on their kinetic model, and no sensitivity analysis was provided. The data obtained by Nip fall well below and appear to be inconsistent with other available data. We believe there is insufficient evidence to justify a modification of the expression found by Baulch *et al.* (1973), and accordingly we recommend

$$k = 10^{13.88} \exp(-7\,600/T) \text{ cm}^3 \text{ mol}^{-1} \text{ s}^{-1}$$

in the temperature range 700–2800 K with the estimated uncertainty factors $f = 0.6$ and $F = 1.6$.

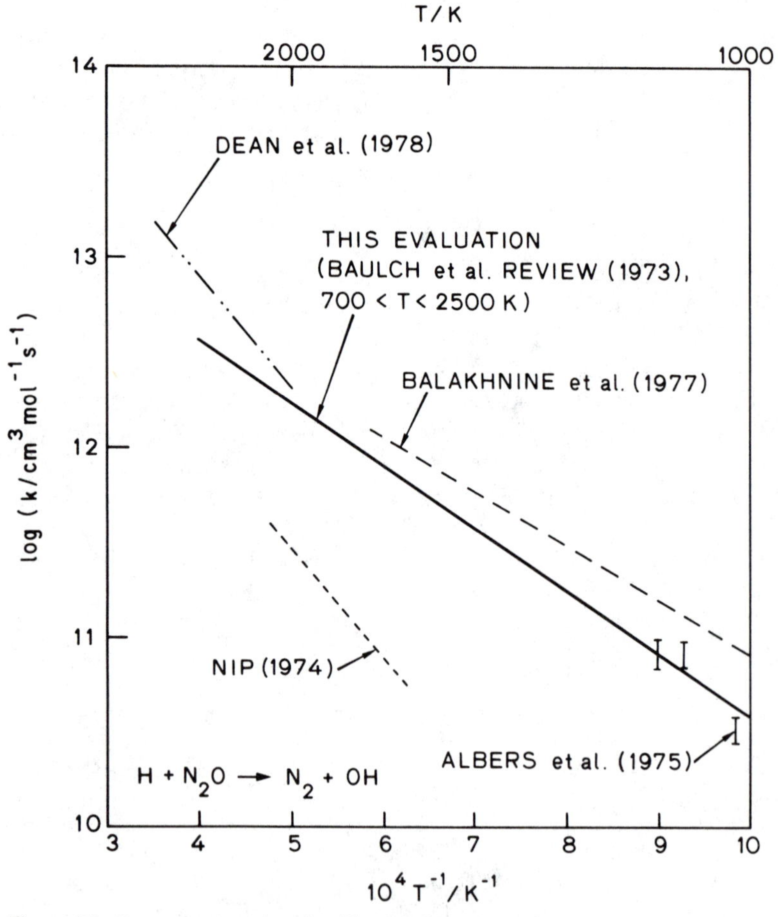

Figure 16. Rate constant data for $H + N_2O \rightarrow N_2 + OH$.

5.7. $NH_2 + O_2 \rightarrow$ products

The reaction between NH_2 and O_2 is believed to be important in the oxidation of ammonia. Recently, Binkley and Melius (1982) performed theoretical calculations on this reaction and discussed the following candidate paths [ΔH°_{f0} values for H_2NO and NH_2O_2 are taken from Binkley and Melius (1982)]:

$$NH_2 + O_2 \rightarrow H_2NO + O \qquad \Delta H = 133\,\text{kJ} \tag{1}$$

$$NH_2 + O_2 \rightarrow NO + H_2O \qquad \Delta H = -342\,\text{kJ} \tag{2}$$

$$NH_2 + O_2 \rightarrow HNO + OH \qquad \Delta H = -\ 52\,\text{kJ} \tag{3}$$

$$NH_2 + O_2 \rightarrow NH_2O_2 \qquad \Delta H = -\ 32\,\text{kJ} \tag{4}$$

$$NH_2 + O_2 \rightarrow NH + HO_2 \qquad \Delta H = 186\,\text{kJ} \tag{5}$$

Their calculations indicate that the most probable path at high temperatures is reaction (1).

Recently, Fujii *et al.* (1981) reviewed previous work on NH_3 oxidation and conducted new experiments using uv absorption to determine induction times for NH_3 removal in various NH_3, O_2, H_2, H_2O, and Ar mixtures. The experiments were conducted using reflected shock waves over a wide range of temperature and pressure ($760 < T < 2300$ K; $1 < P < 9.7$ atm). A 13-reaction mechanism was employed to model the ammonia induction times and to argue that reaction path (3) was consistent with their observations. Values of k_3 were inferred for the temperature range 1550–1800 K, but some dependence on pressure, attributed to the formation of NH_2O_2 as an intermediate complex, was observed. At low pressures, of the order of 1 atm, the reaction appeared to proceed directly with the apparent elementary reaction rate constant

$$k_3\,(\text{low pressure}) = 10^{12.25} \exp(-7\,550/T)\ \text{cm}^3\,\text{mol}^{-1}\,\text{s}^{-1}$$

The mechanism suggested by Fujii *et al.* (1981) is speculative, and additional measurements are needed to verify their model.

Dean *et al.* (1981) have recently studied NH_3 oxidation using a flow tube reactor ($T \cong 1300$ K, and $P \cong 1.2$ atm) with measurements of NH_3 and NO. Reaction (3) was employed in modeling the NO profiles, yielding the resulting rate constant

$$k_3\,(1300\ \text{K}) = 5 \times 10^8\ \text{cm}^3\,\text{mol}^{-1}\,\text{s}^{-1}$$

The results for k_3 obtained by Fujii *et al.* (1981) and Dean *et al.* (1981) are shown in Fig. 17 and Table 17. The results are sufficiently different, and the methods sufficiently indirect, that we are unable to make a specific recommendation for k_3. In addition, it should be noted that the reaction paths assumed by these workers differ from the more recent recommendations of Binkley and Melius (1982). Further work is needed both to verify the reaction products and the rate constant.

Table 17. Rate constant data for $NH_2 + O_2 \rightarrow HNO + OH$.

Reference	$\text{Log}(A/\text{cm}^3\ \text{mol}^{-1}\ \text{s}^{-1})$	θ/K	T range
Fujii *et al.* (1981)	12.25	7 550	1550–1800
Dean *et al.* (1981)	8.70	0	1300
This evaluation	No recommendation		

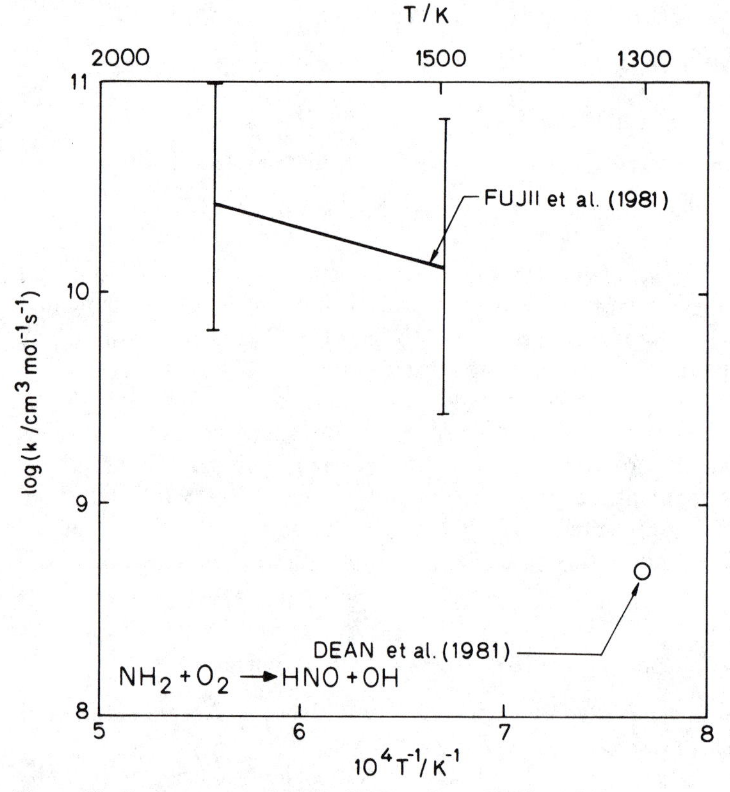

Figure 17. Rate constant data for $NH_2 + O_2 \rightarrow HNO + OH$.

5.8. $NH_2 + NO \rightarrow$ products

The efficiency of the "Thermal DeNOx" process (Lyon, 1975) for conversion of NO to N_2 in combustion systems has stimulated considerable interest in the reaction between NH_2 and NO. It is now generally accepted that this reaction proceeds directly, but controversy continues regarding the probable elementary reaction channels, their individual rate constants, and the overall rate constant at high temperatures. Miller *et al.* (1981) have recently reviewed this reaction from a theoretical and modeling viewpoint; here we emphasize recently reported experimental data.

Several investigators have taken the reaction channels to be

$$NH_2 + NO \rightarrow N_2 + H + OH \qquad \Delta H_0^\circ = -28 \text{ kJ} \tag{1}$$

and

$$NH_2 + NO \rightarrow N_2 + H_2O \qquad \Delta H_0^\circ = -522 \text{ kJ} \tag{2}$$

Experiments at low temperatures (Lesclaux *et al.*, 1975; Hack *et al.*, 1979; Hancock *et al.*, 1975; Gehring *et al.*, 1973) are in good agreement with regard to the overall rate constant for reaction between NH_2 and NO, k_{ov}, and clearly establish an inverse temperature dependence for the rate constant as indicated in Fig. 18a. Both Lesclaux and Hack write reaction (2) as the dominant channel, although the products of the reaction were not measured. On the basis of their observations of H_2O (vibrationally excited) and their failure to detect H atoms, Gehring *et al.* concluded that channel (2) is dominant. However, it is possible that other reaction channels could lead to similar observations with regard to H_2O and H (see Miller *et al.*, 1981).

High-temperature experiments have been carried out by Lyon (1976), Silver *et al.* (1980), Silver (1981), Roose *et al.* (1981), and Roose (1981). Lyon conducted flow tube reactor experiments at 1255 K in $NO/NH_3/O_2/He$ mixtures, and measured NO, NO_2, and NH_3 at fixed reaction time with variable NH_3/NO ratios. His observations were consistent with a simple six-reaction model including both reaction channels (1) and (2) with

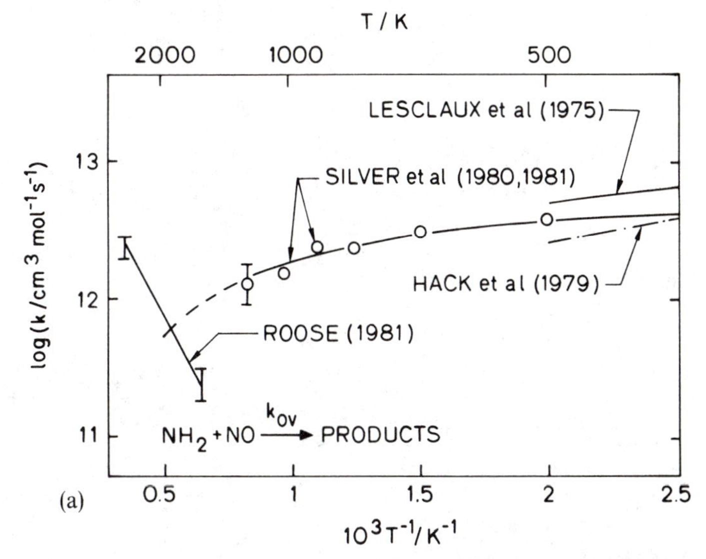

Figure 18 (a)–(c) (opposite). Rate constant data for $NH_2 + NO \rightarrow$ products. (a) The overall reaction $NH_2 + NO \rightarrow$ products; (b) $NH_2 + NO \rightarrow N_2O + H_2$; (c) The branching ratio $\alpha = k_1/(k_1 + k_2) \cong k_1/k_{ov}$.

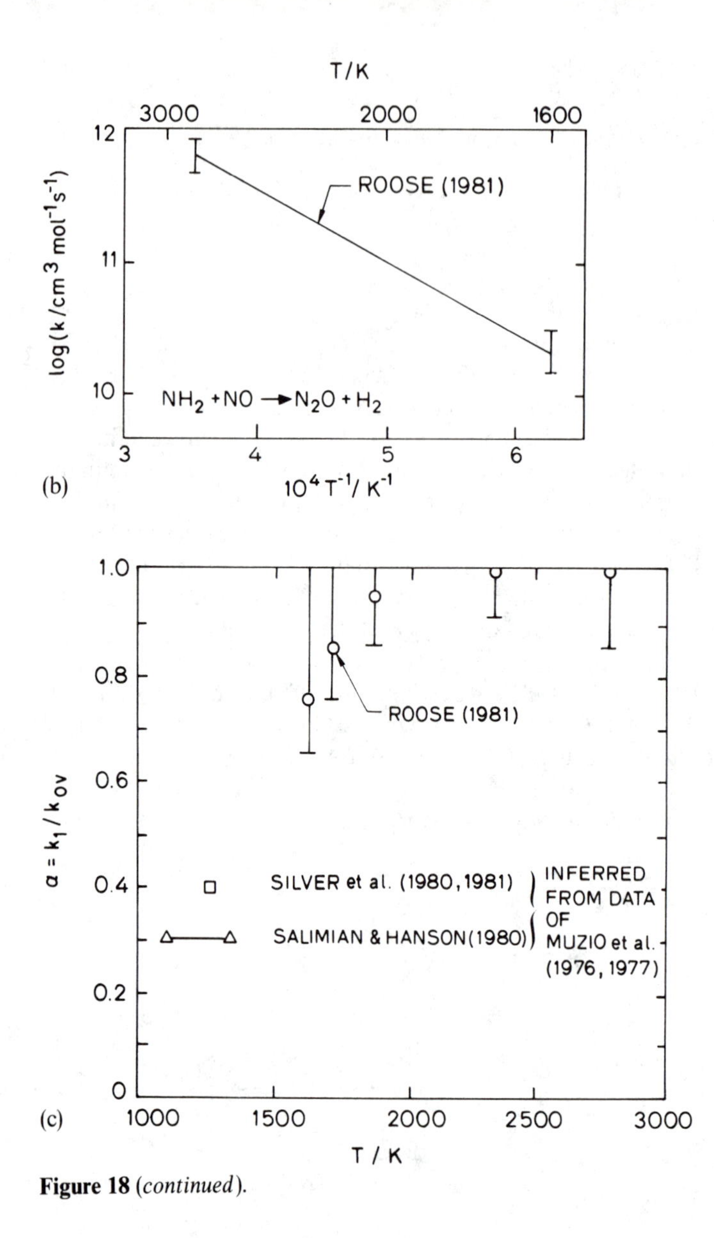

Figure 18 (*continued*).

Table 18a. Rate constant data for $NH_2 + NO \rightarrow$ products.

Reference	Log(A/cm^3 mol^{-1} s^{-1} K^{-B})	B	θ/K	T range
Lesclaux *et al.* (1975)	16.1	−1.25	0	300–500
Hack *et al.* (1979)	17.43	−1.85	0	209–505
Silver *et al.* (1980, 1981)	20.07	−2.46	938	294–1200
Roose (1981)	13.6	0	8 000	1560–2850
This evaluation (Silver *et al.* expression	20.07	−2.46	938	294–1200

Table 18b. Rate constant data for $NH_2 + NO \rightarrow N_2O + H_2$.

Reference	Log(A/cm^3 mol^{-1} s^{-1})	θ/K	T range
Roose (1981)	13.7	12 400	1560–2850
This evaluation	No recommendation		

a branching ratio α

$$\alpha = k_1/(k_1 + k_2) \simeq k_1/k_{ov}$$

in the range $0.25 < \alpha < 0.5$.

Silver (1981; Silver *et al.*, 1980) carried out measurements of k_{ov} over the temperature range 294–1200 K using a high-temperature flow reactor constructed to allow both spectroscopic and mass spectrometric analysis of species at the reactor exit. The experiments were conducted at low pressure (typically 1 Torr) in NH_2/NO/He mixtures with NO in excess. The NH_2 was produced by reacting F atoms with a slight excess of NH_3. NH_2 was monitored using laser-induced fluorescence, and the decay of NH_2 as a function of distance downstream from the movable injector was converted to a value of k_{ov}. Results for k_{ov} as a function of temperature were best fit by the expression (see Fig. 18a)

$$k_{ov} = 10^{20.07}\, T^{-2.46} \exp(-938/T)\ \text{cm}^3\ \text{mol}^{-1}\ \text{s}^{-1}$$

Although no particular conclusion should be drawn from the specific form of Silver's expression, the agreement in the overall inverse temperature dependence observed in the three-flow reaction studies strongly suggests that the reaction proceeds via a long-lived collision complex. At still higher temperatures, one might expect other reaction channels to become important, leading to a change in the temperature dependence of k_{ov}, and indeed some evidence for such a change has been reported by Roose *et al.* (1981) and Roose (1981).

Silver attempted to determine the product distribution of the reaction using mass spectrometry and fluorescence techniques (for OH and H). Mass spectrometric data at 300 K indicated that 50–70% of the reactions produce $N_2 + H_2O$ directly, while OH data (also at 300 K) suggested that $50 \pm 30\%$ of the reactions produce OH. Taken together, in the context of a model with reaction channels (1) and (2), the data are consistent with α in the range 0.3–0.5. The fact that no H atoms were observed, however, raises prospects for another reaction channel, possibly $N_2H + OH$, as suggested by Miller *et al.* (1981), although no N_2H was observed in the mass spectrometer data and there is reason to believe N_2H should decompose in the exhaust gas stream to $N_2 + H$.

As part of his investigation, Silver (1981) also modeled laboratory data of Muzio *et al.* (1976) on Thermal DeNox using a 54-reaction mechanism, including the two effective channels (1) and (2), and found that $\alpha = 0.4$ provided a best overall fit at $T = 1250$ K. In work of a similar nature, Salimian and Hanson (1980) also modeled results of Muzio *et al.* (1977) and inferred $\alpha = 0.3$ for $T = 1100$–1300 K.

Kinetic data at higher temperatures have been obtained by Roose (Roose *et al.*, 1980; Roose, 1981) using shock tube techniques. NO/NH_3 and $NO/NH_3/H_2$ mixtures dilute in Ar were heated by incident shock waves, and the species NO, NH_3, NH_2, NH, and H_2O were monitored using a variety of spectroscopic techniques. The data were interpreted both by detailed modeling and by a simplified approach, with similar results. The observation of significant N_2O production led to inclusion of a third reaction channel [in addition to reactions (1) and (2)]

$$NH_2 + NO \rightarrow N_2O + H_2 \qquad \Delta H_0^\circ = -198 \text{ kJ} \tag{3}$$

The overall rate constant ($k_{ov} = k_1 + k_2 + k_3$) was determined from the measured NH_2 concentration and the rate of removal of NO, with the result shown in Fig. 18a. Although the magnitude of k_{ov} is in reasonable agreement with Silver's high-temperature data, the observed activation energy is quite different, suggesting a change in the dominant reaction channel.

Roose was able to infer the individual rate constant k_3 from N_2O time histories using different $NH : NH_2$ ratios to distinguish between the $NH_2 + NO$ and $NH + NO$ paths for forming N_2O. The results for k_3 (see Fig. 18b and Table 18b) indicate a reasonable activation energy, about 25 kcal, and show that this channel may generally be neglected at temperatures below about 1600 K. It is worth noting that Roose's data for N_2O do not actually confirm that reaction (3) is an elementary reaction with the products $N_2O + H_2$, and could probably also have been modeled satisfactorily using other products, for example, $HNNO + H$, which could quickly form N_2O.

Roose's (1981) results for α (see Fig. 18c) were obtained by detailed modeling of his data (primarily NO, NH_2, and NH time histories). Also shown in Fig. 18c are the values of α inferred by Silver (1981) and by Salimian and Hanson (1980). These results for α are evidence that the radical-producing path increases in importance with increasing temperature and is dominant above about 1700 K.

Further work on this interesting and important reaction is currently in progress in several laboratories, and hence we anticipate that new results regarding the active reaction channels and the associated rate constants will be reported in the near future. With regard to recommendations for this reaction, we suggest use of Silver's (1981) expression for k_{ov} in the temperature range 298–1200 K, with uncertainty factors of about $f = 0.4$ and $F = 2$, and note that extrapolation of this expression to about 2000 K provides at least crude agreement with the magnitude of Roose's k_{ov}. Further work is needed to document k_{ov} and its temperature dependence above about 1200 K before a recommendation can be made for high temperatures. Since data for k_3 are available from only one source, and the determination was indirect, we prefer to make no recommendation of k_3 at the present time. Finally, with regard to the branching ratio α, we also feel unable to make a specific recommendation with confidence. At temperatures in the approximate range 1100–1800 K, simple interpolation between the limited (and indirect) data for α (see Fig. 18c) may suffice for many purposes, and above about 1800 K it may be reasonable to neglect k_2 relative to k_1. At even higher temperatures, of course, reaction (3) may be important, and the approximation that $k_1 + k_2 \cong k_{ov}$ (used in defining α) no longer applies and should be replaced by $k_1 + k_3 \cong k_{ov}$.

6. N/H/O rate constant compilation

Current efforts in modeling combustion phenomena indicate an increasing use of extensive reaction mechanisms, often up to the limits of computational capacity, with the objective of achieving a fuller and more correct kinetic description of the problem. Although there are hazards associated with this approach, in that larger mechanisms indiscriminately formulated do not necessarily increase the accuracy of a kinetic calculation, it is nevertheless clear that the trend toward the use of large reaction sets will continue. Eventually, one can imagine the formation and use of reliable libraries of reaction mechanisms (or submechanisms) and rate constants. Composites of established mechanisms will be assembled, as needed, much the same as one would assemble and utilize files of thermodynamic data. The establishment of correct mechanisms and accurate rate constants even for the most basic and common reaction mechanisms will occupy kineticists for some time, but the scope of such an effort should not keep us from beginning. Indeed, Westbrook and Dryer (1981) have already advanced this concept and initiated work along these lines.

It is in this spirit that we provide the following compilation of N/H/O reactions and rate constants (Tables 19–21). Although the list of reactions is not complete and the rate constants listed, and indeed even the products of the reactions in some cases are often not well known, we believe that such a compilation can be of value to combustion modelers and kineticists. The rate constants listed are intended for use only at high temperatures, e.g., $T > 1000$ K, and have been drawn from a variety of sources. They therefore should be used with considerable caution. In cases where the rate constant is based on a critical evaluation of experimental data, for example, the evaluations of Baulch *et al.* (1973, 1976) or the 18 reactions evaluated in this study, the confidence level should be fairly high. In other cases, and these are in the majority, there is little or no experimental data and only estimates, often quite crude, are provided. These estimates were made in a variety of ways. Whenever possible, the specified rate constant was based on data, even if it was indirect or obtained at low temperatures. In the absence of such data, the expression was generally selected by comparison with results for one or more similar reactions. Obviously, considerable errors can be made by this

Table 19. Compilation of N/O reaction rate constants.

Reaction	$\log(A/cm^3\ mol^{-1}\ s^{-1}\ K^{-B})$	B	θ/K	References	Comments
$NO + M \rightarrow N + O + M$ $M = Ar, N_2, O_2$	20.6	−1.5	75 490	Wray and Teare, 1962	See discussion in N/O evaluations
$N_2O + M \rightarrow N_2 + O + M$ $M = Ar, Kr$	23.84	−2.5	32 710	This evaluation	$k(M = Ar) = 0.7k(M = N_2)$. See discussion in Endo *et al.* (1979).
$NO_2 + M \rightarrow NO + O + M$ $M = Ar$	16.04	0.0	33 000	Baulch *et al.*, 1973	
$N_2 + M \rightarrow N + N + M$ $M = N_2$	21.57	−1.6	113 200	Baulch *et al.*, 1973	$k(M = Ar) = 0.2k(M = N_2)$
$O_2 + M \rightarrow O + O + M$ $M = Ar$	18.26	−1.0	59 380	Baulch *et al.*, 1976	
$N_2 + O \rightarrow NO + N$	14.26	0.0	38 370	This evaluation	
$NO + O \rightarrow N + O_2$	9.58	1.0	20 820	This evaluation	
$N_2O + O \rightarrow NO + NO$	13.84	0.0	13 400	This evaluation	
$N_2O + O \rightarrow N_2 + O_2$	14.0	0.0	14 100	This evaluation	
$NO_2 + O \rightarrow NO + O_2$	13.0	0.0	300	Baulch *et al.*, 1973	No high-temperature data
$NO_2 + N \rightarrow NO + NO$	12.6	0.0	0	Phillips and Schiff, 1965	Limited data. See discussion Baulch *et al.*, 1973
$NO_2 + N \rightarrow N_2O + O$	12.7	0.0	0	Phillips and Schiff, 1965	Limited data. See discussion in Baulch *et al.*, 1973
$N_2O + N \rightarrow N_2 + NO$	13.0	0.0	10 000	Estimate	No data. Spin hindered
$NO + N_2O \rightarrow N_2 + NO_2$	14.0	0.0	25 000	Estimate	Limited data
$NO_2 + NO_2 \rightarrow NO + NO + O_2$	12.30	0.0	13 500	Baulch *et al.*, 1973	See discussion in Baulch *et al.*, 1973

Table 20. Compilation in N/H reaction rate constants.

Reaction	Log(A/cm^3 mol^{-1} s^{-1} T^{-B})	B	θ/K	References	Comments
$N_2H_4 + M \rightarrow NH_2 + NH_2 + M$	15.60	0	20 600	Baulch *et al.*, 1973	
$M = Ar$					
$N_2H_4 + M + N_2H_3 + H + M$	15.0	0	32 000	Estimate	No data
$N_2H_3 + M \rightarrow N_2H_2 + H + M$	16.0	0	25 000	Estimate	No Data
$N_2H_2 + M \rightarrow NH_2 + NH + M$	16.0	0	21 000	Estimate	No Data
$N_2H_2 + M \rightarrow N_2H + H + M$	16.0	0	25 000	Estimate	No data
$N_2H_2 + M + NH + NH + M$	16.5	0	50 000	Estimate	No data
$N_2H + M \rightarrow N_2 + H + M$	14.3	0	10 000	Miller *et al.*, 1982	Estimate, No data
$N_2H + M \rightarrow NH + N + M$	15.0	0	35 000	Estimate	No data
$NH_3 + M \rightarrow NH_2 + H + M$	16.40	0	47 200	This evaluation	
$M = Ar$					
$NH_3 + M \rightarrow NH + H_2 + M$	14.8	0	47 000	This evaluation	See discussion in N/H evaluations
$M = Ar$					
$NH_2 + M \rightarrow NH + H + M$	23.5	−2	46 000	Estimate	No data
$NH + M \rightarrow N + H + M$	21.5	−2	42 000	Estimate	No data
$H_2 + M \rightarrow H + H + M$	14.34	0	48 300	Baulch *et al.*, 1972	
$M = Ar$					
$N_2H_4 + H \rightarrow N_2H_3 + H_2$	13.11	0	1 260	Baulch *et al.*, 1973	No high-temperature data
$N_2H_4 + H \rightarrow NH_2 + NH_3$	9.65	0	1 560	Gehring *et al.*, 1971	No high-temperature data
$N_2H_3 + H \rightarrow N_2H_2 + H_2$	12.0	0	1 000	Estimate	No data
$N_2H_3 + H \rightarrow NH_2 + NH_2$	12.2	0	0	Gehring *et al.*, 1971	No high-temperature data
$N_2H_3 + H \rightarrow NH + NH_3$	11.0	0	0	Estimate	No data
$N_2H_2 + H \rightarrow N_2H + H_2$	13.0	0	500	Estimate	No data
$N_2H + H \rightarrow N_2 + H_2$	13.6	0	1 500	Estimate	No data
$NH_3 + H \rightarrow NH_2 + H_2$	14.1	0	10 820	This evaluation	

Table 20 (*continued*).

Reaction	Log(A/cm^3 mol^{-1} s^{-1} T^{-B})	B	θ/K	References	Comments
$NH_2 + H \rightarrow NH + H_2$	13.28	0	0	This evaluation	See discussion in N/H evaluations
$NH + H \rightarrow N + H_2$	13.7	0	1 000	Estimate	Limited data. See Morley, 1981
$N_2H_4 + NH \rightarrow NH_2 + N_2H_2$	12.0	0.5	1 000	Estimate	No data
$N_2H_2 + NH \rightarrow N_2H + NH_2$	13.0	0	500	Estimate	No data
$N_2H + NH \rightarrow N_2 + NH_2$	11.3	0.5	1 000	Estimate	No data
$NH + NH \rightarrow NH_2 + N$	11.3	0.5	1 000	Estimate	No data
$NH + NH \rightarrow N_2H + H$	11.9	0.5	500	Estimate	No data
$N_2H_4 + NH_2 \rightarrow N_2H_3 + NH_3$	11.6	0.5	1 000	Estimate	Limited data. See discussion in Baulch *et al.*, 1973
$N_2H_3 + NH_2 \rightarrow N_2H_2 + NH_3$	11.0	0.5	0	Estimate	Limited data
$N_2H_2 + NH_2 \rightarrow N_2H + NH_3$	13.0	0	2 000	Estimate	No data
$N_2H_2 + NH_2 \rightarrow NH + N_2H_3$	11.0	0.5	17 000	Estimate	No data
$N_2H + NH_2 \rightarrow N_2 + NH_3$	13.0	0	0	Miller *et al.*, 1981	Estimate. No data
$NH_3 + NH_2 \rightarrow N_2H_3 + H_2$	11.90	0.5	10 850	Dove and Nip, 1979	Estimate obtained by computer simulation
$NH_2 + NH_2 \rightarrow NH_3 + NH$	12.8	0	5 000	Michel, 1965	Limited data. See discussion in Baulch *et al.*, 1973
$NH_2 + NH_2 \rightarrow N_2H_2 + H_2$	13.6	0	6 000	Michel, 1965	Limited data
$NH + NH_2 \rightarrow N_2H_2 + H$	13.5	0	500	Estimate	No data
$N_2H_4 + N_2H_2 \rightarrow N_2H_3 + N_2H_3$	10.4	0.5	15 000	Estimate	No data
$N_2H_3 + N_2H_2 \rightarrow N_2H_4 + N_2H$	13.0	0	5 000	Estimate	No data
$N_2H_2 + N_2H_2 \rightarrow N_2H + N_2H_3$	13.0	0	5 000	Estimate	No data
$N_2H + N_2H \rightarrow N_2H_2 + N_2$	13.0	0	5 000	Estimate	No data
$NH + N \rightarrow N_2 + H$	11.80	0.5	0	Benson *et al.*, 1975	No data
$N_2H + N \rightarrow NH + N_2$	13.5	0	1 000	Estimate	No data

Table 21. Compilation of N/H/O reaction rate constants.

Reaction	Log(A/cm^3 mol^{-1} s^{-1} K^{-B})	B	θ/K	References	Comments
$N_2H_4 + O \rightarrow N_2H_2 + H_2O$	13.8	0	600	Gehring *et al.*, 1973	No high-temperature data
$N_2H_4 + O \rightarrow N_2H_3 + OH$	12.4	0	600	Estimate	No data. See discussion in Baulch *et al.*, 1973
$N_2H_3 + O \rightarrow N_2H_2 + OH$	11.5	0.5	0	Estimate	No data
$N_2H_3 + O \rightarrow N_2H + H_2O$	11.5	0.5	0	Estimate	No data
$N_2H_2 + O \rightarrow N_2H + OH$	11.0	0.5	0	Estimate	No data
$N_2H + O \rightarrow N_2 + OH$	13.0	0	2 500	Miller *et al.*, 1981	Estimate. No data
$N_2H + O \rightarrow N_2O + H$	13.0	0	1 500	Miller *et al.*, 1981	Estimate. No data
$NH_3 + O \rightarrow NH_2 + OH$	13.3	0	4470	This evaluation	
$NH_2 + O \rightarrow NH + OH$	14.1	−0.5	0	Miller *et al.*, 1982	No high-temperature data
$NH_2 + O \rightarrow HNO + H$	14.8	−0.5	0	Miller *et al.*, 1982	No high-temperature data
$NH + O \rightarrow N + OH$	11.8	0.5	4000	Benson *et al.*, 1975	Estimate. No data
$NH + O \rightarrow NO + H$	11.8	0.5	0	Benson *et al.*, 1975	Estimate. No data
$HNO + O \rightarrow NO + OH$	11.7	0.5	1 000	Estimate	Limited data
$H_2H_4 + OH \rightarrow N_2H_3 + H_2O$	13.6	0	0	Harris *et al.*, 1979	No high-temperature data
$N_2H_3 + OH \rightarrow N_2H_2 + H_2O$	13.0	0	1 000	Estimate	No data
$N_2H_2 + OH \rightarrow N_2H + H_2O$	13.0	0	1000	Estimate	No data
$N_2H + OH \rightarrow N_2 + H_2O$	13.5	0	0	Miller *et al.*, 1981	Estimate. No data
$NH_3 + OH \rightarrow NH_2 + H_2O$	13.76	0	4055	This evaluation	
$NH_2 + OH \rightarrow NH + H_2O$	11.7	0.5	1 000	Estimate	Limited data
$NH + OH \rightarrow N + H_2O$	11.7	0.5	1 000	Estimate	No data
$NH + OH \rightarrow HNO + H$	12.0	0.5	1 000	Estimate	No data
$N_2O + OH \rightarrow N_2 + HO_2$	11.8	0	5 000	Estimate	No high-temperature data

Table 21 (*continued*).

Reaction	log(A/cm^3 mol^{-1} s^{-1} T^{-B})	B	θ/K	References	Comments
$HNO + OH \rightarrow NO + H_2O$	12.1	0.5	1 000	Estimate	Limited data. See discussion in Baulch *et al.*, 1973
$N_2H_4 + HO_2 \rightarrow N_2H_3 + H_2O_2$	13.6	0	1 000	Estimate	No data
$N_2H_3 + HO_2 \rightarrow N_2H_2 + H_2O_2$	13.0	0	1 000	Estimate	No data
$N_2H_2 + HO_2 \rightarrow N_2H + H_2O_2$	13.0	0	1 000	Estimate	No data
$N_2H + HO_2 \rightarrow N_2 + H_2O_2$	13.0	0	1 000	Estimate	No data
$NH_3 + HO_2 \rightarrow NH_2 + H_2O_2$	12.4	0	12 000	Estimate	No data
$NH_2 + HO_2 \rightarrow NH_3 + O_2$	13.0	0	1 000	Estimate	No data
$NH_2 + HO_2 + NH + H_2O_2$	13.0	0	1 000	Estimate	No data.
$NH + HO_2 \rightarrow HNO + OH$	13.0	0	1 000	Estimate	No data
$N + HO_2 \rightarrow NO + OH$	13.0	0	1 000	Estimate	No data
$N + HO_2 \rightarrow NH + O_2$	13.0	0	1 000	Estimate	Limited data
$NO + HO_2 \rightarrow HNO + O_2$	11.3	0	1 000	Estimate	Limited data. See Howard, 1980
$NO + HO_2 \rightarrow NO_2 + OH$	12.32	0	−240	This evaluation	
$HNO + HO_2 \rightarrow NO + H_2O_2$	11.5	0.5	1 000	Estimate	No data
$NO + H \rightarrow N + OH$	14.42	0	25 370	This evaluation	
$NO_2 + H \rightarrow NO + OH$	14.54	0	740	Baulch *et al.*, 1973	Limited high-temperature data
$N_2O + H \rightarrow N_2 + OH$	13.88	0	7 600	This evaluation	
$HNO + H \rightarrow NO + H_2$	13.1	0	2 000	Estimate	Limited data. See discussion in Baulch *et al.*, 1973
$HNO + N \rightarrow NO + NH$	13.0	0	1 000	Estimate	No data
$N0 + NH \rightarrow N_2O + H$	12.03	0	230	Nip (1974)	See discussion in N/H/O evaluation
$NO + NH_2 \rightarrow N_2 + H_2O$	19.8	−2.5	950	Estimate	Based on Silver (1981) expression. See discussion in N/H/O evaluation
$NO + NH_2 \rightarrow N_2 + H + OH$	19.8	−2.5	950	Estimate	Based on Silver (1981) expression. See discussion in N/H/O evaluation

$NO + NH_2 \rightarrow N_2O + H_2$	13.7	0	12 400	Roose 1981	See discussion in N/H/O evaluations
$NO_2 + NH \rightarrow HNO + NO$	11.0	0.5	2 000	Estimate	No data
$NO_2 + NH_2 \rightarrow N_2O + H_2O$	20.3	−3	0	Hack *et al.*, 1979	No high-temperature data. Unlikely as elementary step
$N_2O + NH \rightarrow N_2 + HNO$	12.3	0	3 000	Estimate	Spin hindered, no data
$NH_2 + O_2 \rightarrow HNO + OH$	12.25	0	7 500	Fujii *et al.*, 1981	See discussion in N/H/O evaluations
$NH_2 + O_2 \rightarrow NH + HO_2$	14.0	0	25 000	Fujii *et al.*, 1981	Estimate. No data
$NH_2 + HNO \rightarrow NH_3 + NO$	11.7	0.5	500	Estimate	No data
$NH + O_2 \rightarrow HNO + O$	13.0	0	6 000	Miller *et al.*, 1982	Limited data
$N_2H + O_2 \rightarrow N_2 + HO_2$	12.3	0	4 500	Miller *et al.*, 1981	Estimate. No data
$HNO + M \rightarrow H + NO + M$	16.25	0	24 500	Baulch *et al.*, 1973	Limited high-temperature data

$M = Ar$

procedure, and the user is again cautioned to be wary in using large reaction sets with uncertain rate constants. Once preliminary calculations have been done to establish critical reactions for a given problem, then the modeler would be well advised to try to improve the relevant rate constants through reference to more recent experimental data, if available, or by a more detailed estimate of the rate constant. Modelers should also be aware of other compilations which may provide additional reactions or alternative rate constants. Examples of such current compilations are the works of Engleman (1976) and Westley (1980).

We hope that the value of this compilation, in providing a reasonably complete listing based on our current understanding, outweighs the disadvantages which may accrue from the errors which are present.

7. References

Albers, E. A. *et al.* (1975). 15th Int. Symp. Combust., Combustion Institute, Pittsburgh, pp. 765–773.

Ando, H. & Asaba, T. (1976). Int. J. Chem. Kinet. **8,** 259–275.

Balakhnine, V. P., Vandooren, J., & Van Tiggelen, P. J. (1977). Combust. Flame **28,** 165–173.

Baulch, D. L., Drysdale, D. D., & Horne, D. G. (1969). Report No. 5, Dept. of Physical Chemistry, The University of Leeds.

Baulch, D. L., Drysdale, D. D., & Horne, D. G. (1973). *Evaluated Kinetic Data for High Temperature Reactions*, Vol. 2, Butterworths, London.

Baulch, D. L. *et al.* (1976). *Evaluated Kinetic Data for High Temperature Reactions*, Vol. 3, Butterworths, London.

Benson, S. W. *et al.* (1975). Environmental Protection Agency, Washington, D.C. Report No. EPA-600/2-75-019.

Benson, S. W. (1981). 18th Int. Symp. Combust., Combustion Institute, Pittsburgh, p. 882.

Binkley, J. S. & Melius, C. F. (1982). Paper No. WSS/CI 82-96, Fall Meeting of the Western States Section of the Combustion Institute.

Blauwens, J., Smets, B., & Peeters, J. (1977). 16th Int. Symp. Combust., Combustion Institute, Pittsburgh, pp. 1055–1064.

Bowman, C. T. (1971). Combust. Sci. Technol. **3,** 37–45.

Bradley, J. N. & Craggs, X. (1975). 15th Int. Symp. Combust., Combustion Institute, Pittsburgh, pp. 833–842.

Camac, M. & Feinberg, R. M. (1967). 11th Int. Symp. Combust., Combustion Institute, Pittsburgh, pp. 137–145.

Campbell, I. M. & Thrush, B. A. (1968). Trans. Faraday Soc. **64,** 1265–1274.

Chase, M. W., Jr. *et al.* (1982). J. Phys. Chem. Ref. Data **11,** 695–940.

Clark, T. C., Garnett, S. H., & Kistiakowsky, G. B. (1969). J. Chem. Phys. **51,** 2885–91.

Clyne, M. A. A. & McDermid, J. S. (1975). J. Chem. Soc. Faraday Trans. I **71,** 2189–2202.

Clyne, M. A. A. & Thrush, B. A. (1961a). Proc. Royal Soc. A. **261,** 259–273.

Clyne, M. A. A. & Thrush, B. A. (1961b). Nature **189,** 56–57.

Dean, A. M. (1976). Int. J. Chem. Kinet. **8,** 459–474.

Dean, A. M. & Steiner, D. C. (1977). J. Chem. Phys. **66,** 598–604.

Dean, A. M., Steiner, D. C., & Wang, E. E. (1978). Combust. Flame **32,** 73–83.

Dean, A. M., Hardy, J. E., & Lyon, R. K. (1981). Submitted to 15th Int. Symp. on Free Radicals.

Dove, J. E. & Nip, W. S. (1974). Can. J. Chem. **52,** 1171–1180.

Dove, J. E. & Nip, W. S. (1979). Can. J. Chem. **57,** 689–701.

Dove, J. E., Nip, W. S., & Teitelbaum, H. (1975). 15th Int. Symp. Combust., Combustion Institute, Pittsburgh, pp. 903–916.

Duff, R. E. & Davidson, N. (1959). J. Chem. Phys. **31,** 1018–1027.

Duxbury, J. & Pratt, N. H. (1975). 15th Int. Symp. Combust., Combustion Institute, Pittsburgh, pp. 843–855.

Endo, H., Glänzer, K., & Troe, J. (1979). J. Phys. Chem. **83,** 2083–2090.

Fenimore, C. P. & Jones, G. W. (1962). *8th Int. Symp. Combust.*, Williams and Wilkins, Baltimore. pp. 127–133.

Engleman, V. S. (1976). Environmental Protection Agency, Washington, D. C. Report No. EPA-600/2-76-003.

Flower, W. L., Hanson, R. K., & Kruger, C. H. (1975). 15th Int. Symp. Combust., Combustion Institute, Pittsburgh, pp. 823–832.

Flower, W. L., Hanson, R. K., & Kruger, C. H. (1977). Combust. Sci. Technol. **15,** 115–128.

Freedman, E. & Daiber, J. W. (1961). J. Chem. Phys. **34,** 1271–1278.

Fujii, N. *et al.* (1981). 18th Int. Symp. Combust., Combustion Institute, Pittsburgh, pp. 873–883.

Gehring, V. M. *et al.* (1971). Ber. Bunsenges Phys. Chem. **75,** 1287–1294.

Gehring, V. M. *et al.* (1973). 14th Int. Symp. Combust., Combustion Institute, Pittsburgh, pp. 99–105.

Glänzer, K. & Troe, J. (1975). Ber. Bunsenges Phys. Chem. **79,** 465–469.

Glick, H. S., Klein, J. J., & Squire, W. (1957). J. Chem. Phys. **27,** 850–857.

Gordon, S., Mulac, W., & Nangia, P. (1971). J. Phys. Chem. **75,** 2087–2093.

Hack, V. W., Hoyermann, K., & Wagner, H. Gg. (1974). Ber Bunsenges Phys. Chem. **78,** 386–391.

Hack, V. W. *et al.* (1979). 17th Int. Symp. Combust., Combustion Institute, Pittsburgh, pp. 505–513.

Hancock, G. *et al.* (1975). Chem. Phys. Lett. **33,** 168–172.

Hansen, I. *et al.* (1976). Chem. Phys. Lett. **42,** 370–372.

Hanson, R. K., Flower, W. L., & Kruger, C. H. (1974). Combust. Sci. Technol. **9,** 79–86.

Harris, G. W., Atkinson, R., & Pitts, J. N. (1979). J. Phys. Chem. **83,** 2557–2559.

Harris, R. J., Nasralla, M., & Williams, A. (1976). Combust. Sci. Technol. **14,** 85–94.

Henrici, M. (1966). Ph.D. thesis, University of Gottingen, Gottingen.

Holzrichter, K. (1980). Ph.D. thesis, Göttingen, Germany.

Holzrichter, K. & Wagner, H. Gg. (1981). 18th Int. Symp. Combust., Combustion Institute, Pittsburgh, pp. 769–775.

Howard, C. J. (1980). J. Am. Chem. Soc. **102,** 6937–6941.

JANAF Thermochemical Tables (1971). D. R. Stull & H. Prophet, Project Directors, Nat. Bur. Stand., Washington, D. C.: NBS37, 2nd Ed.

Kaufman, F. & Decker, L. J. (1959). *7th Int. Symp. Combust.*, Butterworths, London, pp. 57–60.

Kaufmann, F. & Kelso, J. (1955). J. Chem. Phys. **23,** 1702–1707.

Koshi, M. *et al.* (1975). 15th Int. Symp. Combust., Combustion Institute, Pittsburgh, pp. 809–822.

Koshi, M. *et al.* (1979). 17th Int. Symp. Combust., Combustion Institute, Pittsburgh, pp. 553–562.

Leonard, P. A., Plee, S. L., & Mellor, A. M. (1976). Combust. Sci. Technol. **14,** 183–193.

Lesclaux, R. *et al.* (1975). Chem. Phys. Lett. **35,** 493–497.

Livesey, J. B., Roberts, A. L., & Williams, A. (1971). Combust. Sci. Technol. **4,** 9–15.

Louge, M. Y. & Hanson, R. K. (1981) (unpublished).

Lyon, R. K. (1975). U. S. Patent No. 3900554.

Lyon, R. K. (1976). Int. J. Chem. Kinet. **8,** 315–318.

McCullough, R. W., Kruger, C. H., & Hanson, R. K. (1977). Combust. Sci. Technol. **15,** 213–223.

Michel, K. W. (1965). 10th Int. Symp. Combust., Combustion Institute, Pittsburgh, p. 351.

Miller, J. A., Branch, M. C., & Kee, R. J. (1981). Combust. Flame **43,** 81–98.

Miller, J. A. (1982). Paper No. WSS/CI 82-93, Fall Meeting of the Western States Section of the Combustion Institute.

Monat, J. P. (1977). Ph.D. thesis. Stanford University, Stanford, California.

Monat, J. P., Hanson, R. K., & Kruger, C. H. (1977). Combust. Sci. Technol. **16,** 21–28.

Monat, J. P., Hanson, R. K., & Kruger, C. H. (1979). 17th Int. Symp. Combust., Combustion Institute, Pittsburgh, pp. 543–552.

Morley, C. (1981). 18th Int. Symp. Combust., Combustion Institute, Pittsburgh, pp. 23–32.

Muzio, L. J. & Arand, J. K. (1976). Electric Power Research Inst., Palo Alto, CA. Report No. FP-253.

Muzio, L. J., Arand, J. K., & Teixeira, D. P. (1977). 16th Int. Symp. Combust., Combustion Institute, Pittsburgh, pp. 199–208.

Myerson, A. L. (1973). 14th Int. Symp. Combust., Combustion Institute, Pittsburgh, pp. 219–228.

Newhall, H. K. & Shahed, S. M. (1970). 13th Int. Symp. Combust., Combustion Institute, Pittsburgh, pp. 381–389.

Niemitz, K. J., Wagner, H. Gg., & Zellner, R. (1981). Z. Phys. Chem. Neue Folge **124,** 155–170.

Nip, W. S. (1974). Ph.D. thesis, University of Toronto, Toronto, Canada.

Olschewski, H. A., Troe, J., & Wagner, H. Gg. (1966). Ber. Bunsenges Phys. Chem. **70,** 450–459.

Perry, R. A., Atkinson, R., & Pitts, J. N. (1976). J. Chem. Phys. **64,** 3237–3239.

Peterson, R. C. (1981). Ph.D. thesis. Purdue University, West Lafayette.

Phillips, L. F. & Schiff, H. I. (1965). J. Chem. Phys. **42,** 3171–3174.

Piper, L. G. (1979). J. Chem. Phys. **70,** 3417–3419.

Pouchan, C. & Chaillet, M. (1982). Chem. Phys. Lett. **90,** 310–316.

Roose, T. R., Hanson, R. K., & Kruger, C. H. (1980). *Proc. 12th Int. Symp. Shock Tubes and Waves*, Magnus, Hebrew Univ. Press, Jerusalem, pp. 476–485.

Roose, T. R. (1981). Ph.D. thesis. Stanford University, Stanford, California.

Roose, T. R., Hanson, R. K., & Kruger, C. H. (1981). 18th Int. Symp. Combust., Combustion Institute, Pittsburgh, pp. 853–862.

Roth, P. & Just, T. (1977). Ber. Bunsenges Phys. Chem. **81,** 572–577.

Salimian, S. & Hanson, R. K. (1980). Combust. Sci. Technol. **23,** 225–230.

Salimian, S., Hanson, R. K., & Kruger, C. H. (1983). Int. J. Chem. Kinet, in press.

Seery, D. J. & Zabielski, M. F. (1980). *Laser Probes for Combustion Chemistry*, American Chem. Society, Washington, D. C., pp. 375–380.

Silver, J. A. & Kolb, C. E. (1980). Chem. Phys. Lett. **75,** 191–195.

Silver, J. A., Gozewski, C. M., & Kolb, C. E. (1980). Western State Section, The Combustion Inst. Paper No. 80-41, Los Angeles, California.

Silver, J. A. (1981). Opt. Eng. **20,** 540–545.

Smith, G. P., Crosley, D. R. (1981). 18th Int. Symposium Combust., Combustion Institute, Pittsburgh, pp. 1511–1520.

Smith, I. W. M. & Zellner, R. (1975). Int. J. Chem. Kinet., Symp. **1,** 341–351.

Sulzmann, K. G. P., Kline, J. M., & Penner, S. S. (1980). *Proc. 12th Int. Symp. Shock Tubes and Waves*, Magnus, Hebrew University Press, Jerusalem, pp. 465–475.

Thompson, D., Brown, T. D., & Beer, J. M. (1972). Combust. Flame **19,** 69–79.

Troe, J. (1975). *Proc. 10th. Int. Symp. Shock Tubes and Waves*, Shock Tube Research Society Japan, Kyoto, pp. 29–51.

Vetter, K. (1949). Z. Electrochem. **53,** 369–376.

Waldman, C. H., Wilson, R. D., & Maloney, K. L. (1974). EPA Report No. EPA-650/2-74-045.

Westbrook, C. K. & Dryer, F. L. (1981). 18th Int. Symposium Combust., Combustion Institute, Pittsburgh, pp. 749–767.

Westley, F. (1980). Nat. Bureau of Standards, Washington D. C. Report No. NSRDS-NBS 67.

Wray, K. L. & Teare, J. D. (1962). J. Chem. Phys. **36,** 2582–2596.

Yumura, M. & Asaba, T. (1981). 18th Int. Symposium Combust., Combustion Institute, Pittsburgh, pp. 863–872.

Zeldovich, Ya. B. (1946). Acta Physiochim. U.R.S.S. **21,** 577–628.

Zellner, R. & Smith, I. W. M. (1974). Chem. Phys. Lett. **26,** 72–74.

Chapter 7

Modeling

Michael Frenklach

> Data themselves cannot produce information; they can only produce information in the light of a particular model (Box and Hunter, 1965)

1. Introduction

Modeling in combustion research may be defined as a *procedure* of deducing the mathematical description of a process, i.e., a model, from experimental observations. This procedure is nontrivial in the sense that there is no "recipe" or algorithm to follow. A strategy for modeling must be developed by the researcher himself based upon various factors involved in a particular study, such as objectives of the investigation, purpose of the modeling, experimental technique employed, quality and quantity of the experimental observations, reference data available, desired degree of reliability of the model, etc. While a mathematical background is an essential requirement, the success of the modeling is mainly determined by having a well-chosen strategy.

The goal of this chapter is to prepare the reader for independent modeling research. The first half (Sections 2 to 6) reviews the subject of modeling in general, while the second half is devoted to detailed kinetic modeling of combustion chemistry. The reader is advised to concentrate more on the qualitative aspects than on mathematical details; thus Subsections 4.2 and 4.3 may be skipped at first reading.

2. Basic concepts and definitions

We define a *model* as a functional relationship or relationships among various quantities involved in a physical process. The relationships may be expressed in the form of algebraic, differential, or integral equations, which need not necessarily possess an explicit analytical solution. In this sense the concept of a model is similar to the concept of a computer subroutine, as shown schematically in the following diagram:

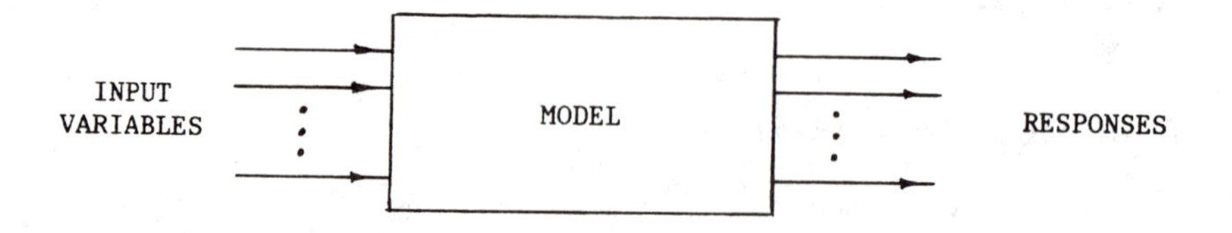

The quantities appearing in the equations are divided into *input* variables and *output* variables or *responses* of the model. The responses would be the predictions of experimentally observed entities or their known functions. *Actual* responses, typical for combustion research, are the intensity of a light beam, the voltage generated by a pressure transducer, etc. The researcher is interested, however, in concentrations of species as well as their logarithms and ratios, pressures, temperatures, ignition delay times, luminosity of flames, amounts of soot formed, etc. They can be taken for responses but only when the *instrumental functions*, i.e., the relationships between the actual responses and the entities considered, are known precisely.

The input variables are subdivided into *controllable* variables and *parameters*. The controllable variables are those quantities whose values can be varied by the experimenter as, for example, initial and boundary conditions of experiments. Empirical and physical "constants" that are assumed to remain unchanged in a given set or a subset of experiments constitute the parameters of the model. To clarify an ambiguity that may arise from the above definitions let us consider a rate constant that is expressed in the Arrhenius form $k = A\exp(-E_a/RT)$. In isothermal modeling the rate constant at the assumed temperature, k, is a parameter. However, if temperature is a controllable variable of the model, then the preexponential factor A and the activation energy E_a are the parameters.

The model concept takes the mathematical form

$$\boldsymbol{\eta} = \mathbf{f}(\boldsymbol{\xi}, \boldsymbol{\theta}) \tag{2.1}$$

where

$\boldsymbol{\xi}$ is a vector of m controllable variables, $\xi_1, \xi_2, \ldots, \xi_m$
$\boldsymbol{\theta}$ is a vector of p parameters, $\theta_1, \theta_2, \ldots, \theta_p$
$\mathbf{y}$ is a vector of l observable quantities, $y_1, y_2, \ldots, y_l$
$\boldsymbol{\eta}$ is a vector of l responses, i.e. expectation values of $\mathbf{y}$, $\eta_1, \eta_2, \ldots, \eta_l$
$\mathbf{f}$ is a vector of l functions, $f_1, f_2, \ldots, f_l$.

In modeling theory the term "vector" is customarily used to denote sets of functions, variables, or parameters (Dorny, 1975). In ordinary mathematical convention, Eq. (2.1) has the form of a vector function with $m + p$ "independent variables" and l "dependent variables".

Of particular interest to combustion modeling are *dynamic* models, which are models formulated in terms of differential equations. For the sake of simplicity in presentation only ordinary differential equations are considered here, that is

$$\frac{d\mathbf{s}}{dt} = \mathbf{h}(t, \mathbf{s}, \boldsymbol{\xi}, \boldsymbol{\theta}) \tag{2.2}$$

with the initial conditions

$$\mathbf{s}(t = 0) = \mathbf{s}(\boldsymbol{\xi})$$

where

t is the elapsed time of reaction–in most combustion modeling t does not appear explicitly on the right-hand side of Eq. (2.2), but in principle it can

$\mathbf{s}$ is a vector of variables (e.g., concentrations, temperature, pressure) describing the state of the system being modeled

$\mathbf{h}$ is a vector of functions.

It is not difficult to extend the principal aspects of the further discussion to partial differential equations.

Equation (2.1) can be thought of as a solution to Eq. (2.2) at a given observation time t_i, hence the additional variables t and $\mathbf{s}$ are actually internal variables of the dynamic model.

The general philosophy of modeling research as an iterative process may be summarized in the following steps:

(a) *Induction of a model*—setting forth a specific but tentative hypothesis as to the explicit form of the model, i.e., writing down the functions $\mathbf{f}(\boldsymbol{\xi}, \boldsymbol{\theta})$ in Eq. (2.1) or the functions $\mathbf{h}(t, \mathbf{s}, \boldsymbol{\xi}, \boldsymbol{\theta})$ in Eq. (2.2);

(b) *Parameter estimation*—fitting the model responses to agree with the experimental observations by adjusting the vector $\boldsymbol{\theta}$;

(c) *Hypothesis testing*—analysis of variances, discrimination between alternative models, or deduction of an improved hypothesis;

(d) *Planning new experiments.*

A general discussion of these four topics is presented in Sections 3–6.

3. Construction of models

3.1. The nature of a model

The inductive part of modeling is based on the researcher's knowledge of the process (here, the process of combustion) and includes elements of his experience, intuition, and guesswork. Perhaps the most important factor at this stage is to conceive clearly the nature of the suggested hypothesis, i.e., whether it is *empirical* or *physical*; this distinction sets the tone of the entire modeling procedure.

3.2. Empirical models

There are situations when the underlying physics and chemistry of a process are completely unknown or purposely assumed so, and one is interested only in the formal relationship between the variables of the process. Models derived under these conditions are called *empirical* models.

The choice of an explicit mathematical form for an empirical model, being arbitrary in principle, is usually made upon consideration of the functional appropriateness and simplicity of the mathematical treatment involved and the minimal experimental efforts required. The simplest possible single-response model would be a linear form

$$\eta = \theta_0 + \theta_1\xi_1 + \theta_2\xi_2 + \cdots + \theta_m\xi_m \tag{3.1}$$

where noting that

$$\frac{\partial\eta}{\partial\xi_i} = \theta_i \qquad \text{for} \quad i = 1, 2, \ldots, m,$$

each parameter θ_i is seen to represent the *effect* of the corresponding controllable variable ξ_i on the model response η. This model has a minimal number of parameters and their values can be determined by standard regression analysis, which will be discussed in Section 4.

If the linear model appears to be inadequate, it can be expanded to the quadratic form that takes interactions between the controllable variables into account

$$\eta = \theta_0 + \sum_{i=1}^{m} \theta_i\xi_i + \sum_{\substack{i,j=1 \\ j\leq i}}^{m} \theta_{ij}\xi_i\xi_j \tag{3.2}$$

which still can be subjected to standard regression analysis since the controllable variables ξ_i and the correlation variables $\xi_{ij} = \xi_i\xi_j$ are linearly independent. It should be noticed, however, that the effects of the controllable variables

$$\frac{\partial\eta}{\partial\xi_i} = \theta_i + 2\theta_{ii} + \sum_{j=1}^{m} \theta_{ij}\xi_j, \tag{3.3}$$

can no longer be associated with a single parameter.

If the quadratic model also appears to be inadequate, it can be expanded further to cubic form and so forth. The expansion procedure must be stopped immediately after adequacy has been achieved, since further increase in the order of the polynomial may lead to serious distortion of the model.

The higher-order polynomial model requires a larger number of parameters to be determined and would therefore necessitate more experiments than if this number is kept small. Thus, if a linear model [Eq. (3.1)] is inadequate, one may prefer to seek a nonpolynomial model. The latter is often chosen upon physical considerations or by following some tradition.

Sometimes it is possible to transform a nonlinear model into a generalized linear form

$$\psi(\eta) = \theta_1 g_1(\xi) + \theta_2 g_2(\xi) + \cdots + \theta_p g_p(\xi) \tag{3.4}$$

where

$\psi(\eta)$ is the expectation value of the transformed response
$g_1(\xi), g_2(\xi), \ldots, g_p(\xi)$ are linearly independent functions of the controllable variables.

For instance, Arrhenius-type expressions (rate constants, induction times, etc.) take the form of model (3.4) after logarithmic transformation.

No matter what specific form empirical models may take, they possess the following limitations:

(a) the model is "reliable" only for the conditions studied experimentally;
(b) empirical models reflect but do not disclose the mechanisms of phenomena. Consequently, the model parameters cannot be interpreted in terms of physical constants.

Within limitation (a), the empirically obtained effects of controllable variables determine the gradient of the response, i.e.,

$$\text{grad}(\eta) = \left(\frac{\partial \eta}{\partial \xi_1}, \frac{\partial \eta}{\partial \xi_2}, \ldots, \frac{\partial \eta}{\partial \xi_m} \right).$$

This result has led to a number of valuable applications of empirical modeling.

3.3. Physical models

Models developed from theoretical considerations, that is, on the basis of physical and chemical laws governing the process of interest, are called *physical* models. When the physical mechanism of the phenomena is felt to be completely understood, this approach leads to *exact physical* models, and the modeling procedure is reduced to the determination of the model's parameters and their statistical uncertainties.

The chemical kinetics of combustion systems is usually not well-established. Nevertheless, it is often possible to construct plausible mechanisms including elementary reactions of various degrees of certainty and even hypothetical ones never identified in direct laboratory experiments. The corresponding model, being physico-empirical in nature, is called a *mechanistic* model. The adequacy of the latter, in a statistical sense, does not necessarily mean that the "true" model is derived, but rather that with a certain probability the model cannot be rejected. Such an interpretation reflects the empiricism of mechanistic modeling and suggests that future experimental discoveries may lead to modification or abandonment of the model.

4. Parameter estimation

4.1. Preliminary remarks

Measurements have errors. When constructing a model, however, one must assume that systematic errors are absent. Whether this assumption is valid or not has to be the subject for later hypothesis testing. However, the presence of only random errors must be postulated *a priori.*

Due to these random errors, the values of the parameters cannot be determined exactly, but rather only estimated within a certain degree of uncertainty. The properly stated result must be that with a given probability, called a *confidence level*, the value of a parameter belongs within a certain *confidence interval.* These intervals are determined by sample statistics, called *estimators*, which will be denoted hereafter by placing ^ on the top of the parameter's symbol, i.e., $\hat{\theta}$ is an estimator of θ. For example, if the weight of an object has to be determined, the arithmetic mean of the results of single weighings is used as an estimator for the "true" weight (Harnett and Murphy, 1975, Chap. 1)

$$\hat{\mu} = \bar{z} = \frac{1}{n} \sum_{i=1}^{n} z_i$$

where

μ is the true weight
$\hat{\mu}$ is the estimator of μ
z_i is the result of the ith weighing
n is the number of weighings and
$\bar{z}$ is the arithmetic mean.

The confidence interval for the true weight is defined by the following statement (Harnett and Murphy, 1975, Chap. 7)

$$\Pr(\hat{\mu} - t_{n-1,\alpha/2}\hat{\sigma}_{\hat{\mu}} < \mu < \hat{\mu} + t_{n-1,\alpha/2}\hat{\sigma}_{\hat{\mu}}) = 1 - \alpha$$

or, in simplified form,

$$\mu = \hat{\mu} \pm t_{n-1,\alpha/2}\hat{\sigma}_{\hat{\mu}}$$

where

$1 - \alpha$ is the confidence level, often assumed to be 0.95 in technical applications
$\hat{\sigma}^2_{\hat{\mu}}$ is the estimate of the variance of $\hat{\mu}$, $\hat{\sigma}^2_{\hat{\mu}} = \hat{\sigma}^2/n$
$\hat{\sigma}^2$ is the estimate of the sample error variance
$t_{n-1,\alpha/2}$ is the upper $\alpha/2$ point of the t distribution with $n - 1$ degrees of freedom
Pr is the "probability of" operator.

Estimators, being functions of random observations, are random variables themselves. Hence, the choice of particular estimators rests upon their statistical properties. Here we will consider two of them, bias and efficiency (Bard, 1974, Chap. 3; Hudson, 1963, Chap. 4; Linnik, 1961, Chap. 3).

(a) Estimator $\hat{\theta}$ is said to be *unbiased* if its expectation is equal to the true value of the parameter, i.e., $E(\hat{\theta}) = \theta$. Thus, the *bias* of an estimator is a measure of the systematic error in an estimator.
(b) An estimator is said to be *efficient* if its variance is the lowest theoretically attainable. Efficiency is a measure of random errors in an estimator.

When the confidence interval is determined by unbiased and efficient estimators, we may say that the parameter is estimated accurately. It is not always possible, however, to satisfy both conditions simultaneously. In the previous example, the arithmetic mean is the unbiased and efficient estimator of the true weight (Hamilton, 1964, p. 39; Johnson and Leone, 1977, p. 214). The estimator of the sample variance, $\hat{\sigma}^2$, can be given by two expressions (Hamilton, 1964, p. 40):

$$\frac{1}{n}\sum_{i=1}^{n}(z_i - \hat{\mu})^2 \quad \text{or} \quad \frac{1}{n-1}\sum_{i=1}^{n}(z_i - \hat{\mu})^2$$

At large n both expressions are practically identical. At small n the former is efficient but biased while the later is unbiased but not efficient. Preference in this case should be given to the unbiased estimator since the efficiency of $\hat{\sigma}^2_{\hat{\mu}}$ is less critical.

The theoretical grounds for parameter estimation are based on the principle of maximum likelihood, which states: if an event has occurred, the maximum probability should have corresponded to its realization. As a consequence, the estimators maximizing the likelihood function possess certain optimal properties (Johnson and Leone, 1977, Chap. 7; Linnik, 1961, Chap. 3). The likelihood function is composed of the distribution functions of the errors. When no information on the type of distribution is available, it is usually assumed to be normal, justified by the results of the central limit theorem (Box *et al.*, 1978, p. 44; Linnik, 1961, pp. 71–74). For normally distributed errors, maximization of the likelihood function is reduced to minimization of the residuals, i.e., to the least-squares methods.

The above material is treated in textbooks on mathematical statistics, and details on these topics can be found in the books and articles cited in the text. The following subsections summarize some of the results relevant to the present discussion.

4.2. Linear models

For a generalized linear model (3.4) the estimates of the parameters and their variances are given by (Draper and Smith, 1981, Chap. 2; Hamilton, 1964, Chap. 4; Linnik, 1961, Chap. 6; Seber, 1977, Chap. 3)

$$\hat{\boldsymbol{\theta}} = (\mathbf{X}'\mathbf{W}\mathbf{X})^{-1}\mathbf{X}'\mathbf{W}\mathbf{Y} \tag{4.1}$$
$$\hat{\mathbf{D}}(\hat{\boldsymbol{\theta}}) = \hat{\sigma}^2(\mathbf{X}'\mathbf{W}\mathbf{X})^{-1}$$

where

$$\mathbf{X} = \begin{pmatrix} g_1(\xi_1) & g_2(\xi_1) & \cdots & g_p(\xi_1) \\ g_1(\xi_2) & g_2(\xi_2) & \cdots & g_p(\xi_2) \\ \vdots & \vdots & & \vdots \\ g_1(\xi_n) & g_2(\xi_n) & \cdots & g_p(\xi_n) \end{pmatrix}; \quad \mathbf{Y} = \begin{pmatrix} \psi(y_1) \\ \psi(y_2) \\ \vdots \\ \psi(y_n) \end{pmatrix}$$

$$\hat{\mathbf{D}}(\hat{\boldsymbol{\theta}}) = \begin{pmatrix} \hat{\sigma}^2_{\hat{\theta}_1} & \mathrm{Cov}(\hat{\theta}_1, \hat{\theta}_2) & \cdots & \mathrm{Cov}(\hat{\theta}_1, \hat{\theta}_p) \\ \mathrm{Cov}(\hat{\theta}_2, \hat{\theta}_1) & \hat{\sigma}^2_{\hat{\theta}_2} & \cdots & \mathrm{Cov}(\hat{\theta}_2, \hat{\theta}_p) \\ \vdots & \vdots & & \vdots \\ \mathrm{Cov}(\hat{\theta}_p, \hat{\theta}_1) & \mathrm{Cov}(\hat{\theta}_p, \hat{\theta}_2) & \cdots & \hat{\sigma}^2_{\hat{\theta}_p} \end{pmatrix}$$

$$\hat{\sigma}^2 = \frac{\mathbf{U}^t\mathbf{W}\mathbf{U}}{n - p} \tag{4.2}$$

$$\mathbf{U} = \mathbf{Y} - \mathbf{X}\hat{\boldsymbol{\theta}}$$

superscript t indicates transposition
$\mathbf{W}$ is the matrix of statistical weights (Hamilton, 1964, pp. 146–149)
$\xi_i, y_i (i = 1, 2, \ldots, n)$ are realizations of the controllable variables and the response, respectively, in the ith experiment
n is the number of experimental "points"
p is the number of parameters estimated
$\mathrm{Cov}(\hat{\theta}_i, \hat{\theta}_j)$ is the covariance of $\hat{\theta}_i$ and $\hat{\theta}_j$.

It should be emphasized that these results do not require an assumption of the type of the error distribution function. Moreover, it can be shown (the Gauss–Markov theorem; Bard, 1974, p. 59; Hamilton, 1964, Chap. 4; Hudson, 1963, Chap. 5; Seber, 1977, Chap. 3) that

(a) $\hat{\boldsymbol{\theta}}$ are unbiased estimates of $\boldsymbol{\theta}$;
(b) among all linear unbiased estimators Eq. (4.1) yields those whose variances are smallest. If the distribution of errors is normal, the estimates $\hat{\theta}$ are efficient.

In contrast to estimators, determination of confidence intervals does require knowledge of the distribution function. In the case of normal distribution, $100(1 - \alpha)\%$ confidence intervals are (Linnik, 1961, Chap. 6)

$$(\hat{\theta}_j - t_{n-p,\alpha/2}\hat{\sigma}_{\hat{\theta}_j}, \hat{\theta}_j + t_{n-p,\alpha/2}\hat{\sigma}_{\hat{\theta}_j}), \; j = 1, 2, \ldots, p \tag{4.3}$$

Generally, estimates $\hat{\boldsymbol{\theta}}$ are correlated and therefore it is more appropriate to consider the $100(1 - \alpha)\%$ *joint confidence region* (Draper and Smith, 1981, Chap. 2; Hamilton, 1964, Chap. 4; Seber, 1977, Chap. 5)

$$(\hat{\boldsymbol{\theta}} - \boldsymbol{\theta})^t\hat{\mathbf{D}}^{-1}(\hat{\boldsymbol{\theta}})(\hat{\boldsymbol{\theta}} - \boldsymbol{\theta}) \le pF_\alpha(p, n - p) \tag{4.4}$$

where $F_\alpha(p, n - p)$ is the upper α point of the F distribution with p and $n - p$ degrees of freedom, that is, with probability $1 - \alpha$ the true values of

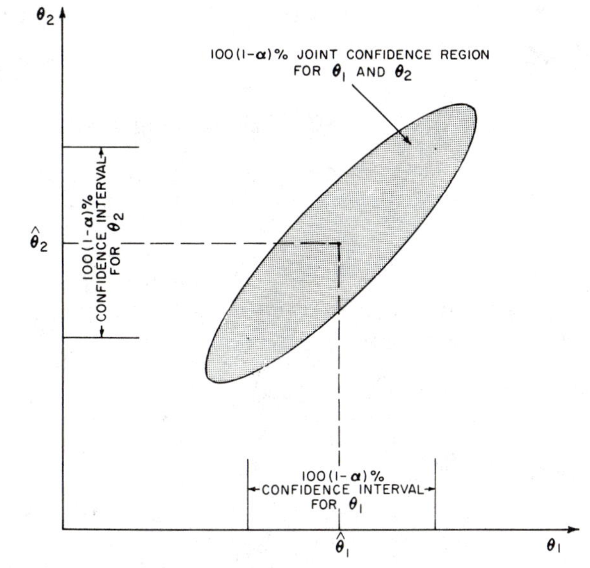

Figure 1. The two-dimensional joint confidence region—an ellipse for a linear model. The projections of this ellipse onto the axes are approximately equal to the individual confidence intervals.

parameters $\boldsymbol{\theta}$ lie inside of p-dimensional ellipsoid [Eq. (4.4)]. The projection of the ellipsoid onto axis θ_j is approximately equal to the individual confidence interval [Eq. (4.3)] (Seber, 1977, Chap. 5), as illustrated for the two-dimensional case in Fig. 1.

4.3. Nonlinear models

4.3.1. Linearizing transformations

The optimal properties of linear regressions encourage transformation, whenever possible, of a nonlinear model into a linear form [Eq. (3.4)]. However, hidden behind the apparent simplicity of such an approach are subtle complications. For example, the two-parameter exponential model

$$\eta = \theta_1 \exp(\theta_2 \xi),$$

after logarithmic transformation takes the linear form

$$\eta' = \theta'_1 + \theta_2 \xi,$$

where $\eta' = \ln \eta$ and $\theta'_1 = \ln \theta_1$. Standard linear regression would result in biased estimates of the original parameters, i.e., $\hat{\theta}'_1 \neq \ln \hat{\theta}_1$. The bias can be removed by adjusting the statistical weights (Cvetanovic and Singleton, 1977). However, the efficiency of the resulting estimators remains unclear.

A detailed analysis of the transformations can be found in Box and Cox (1964) and Draper and Smith (1981), Chap. 5.

4.3.2. Linearizing expansions

Let us consider a single-response nonlinear model

$$\eta = f(\boldsymbol{\xi}, \boldsymbol{\lambda})$$

where $\boldsymbol{\lambda}$ is the vector of parameters and $\boldsymbol{\lambda}^0$ designates the initial guess for $\boldsymbol{\lambda}$. Expanding the model response into Taylor series about $\boldsymbol{\lambda}^0$ and keeping only the first-order terms we obtain

$$\eta - \eta^0 \approx \frac{\partial f}{\partial \lambda_1}(\lambda_1 - \lambda_1^0) + \frac{\partial f}{\partial \lambda_2}(\lambda_2 - \lambda_2^0) + \cdots + \frac{\partial f}{\partial \lambda_p}(\lambda_p - \lambda_p^0)$$

where $\eta^0 = f(\boldsymbol{\xi}, \boldsymbol{\lambda}^0)$.
If the expansion is valid and assuming, in reference to expression (3.4),

$$g_j(\xi) = \frac{\partial f}{\partial \lambda_j}, \quad \theta_j = \Delta\lambda_j^0 \equiv \lambda_j - \lambda_j^0, \quad j = 1, 2, \ldots, p,$$

and

$$\psi(\eta) = \eta - \eta^0,$$

the nonlinear parameter estimation is reduced to an iterative sequence of linear regressions, where

$$\lambda_j^{\kappa+1} = \lambda_j^\kappa + \Delta\lambda_j^\kappa \qquad \text{for } \kappa = 0, 1, 2, \ldots$$

(Bard, 1974, Chap. 5; Chambers, 1977; Draper and Smith, 1981, Chap. 10).

4.3.3. Search methods

Search methods consist of the optimization of an appropriate *objective function*, $\Phi(\boldsymbol{\theta})$, which is usually the likelihood or posterior distribution functions or, for a single-response model with a normal distribution of errors, the sum of squares of the residuals. The latter case is implied in the following discussion. Various gradient and direct search techniques for single-response objective functions are discussed elsewhere (see, e.g., Bard, 1974, Chap. 5; Draper and Smith, 1981, Chap. 10).

The efficiency of these search methods depends on the shape of the objective function. The reader must be aware that the typical shape encountered in chemical kinetics modeling is a so-called *valley*. A particularly troublesome situation arises when the valley is long and very narrow, i.e., when the global minimum lies on a gently sloping locus of "local minima" (Fig. 2). The simple gradient method (e.g., the steepest descent) tends to oscillate around the locus line in a so-called hemstitching pattern (Fig. 3), resulting in

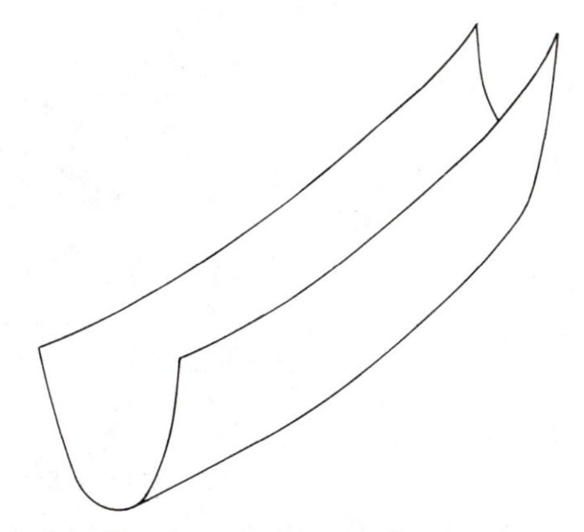

Figure 2. A two-dimensional "valley." Such valley-shaped objective functions are typical for chemical kinetics modeling.

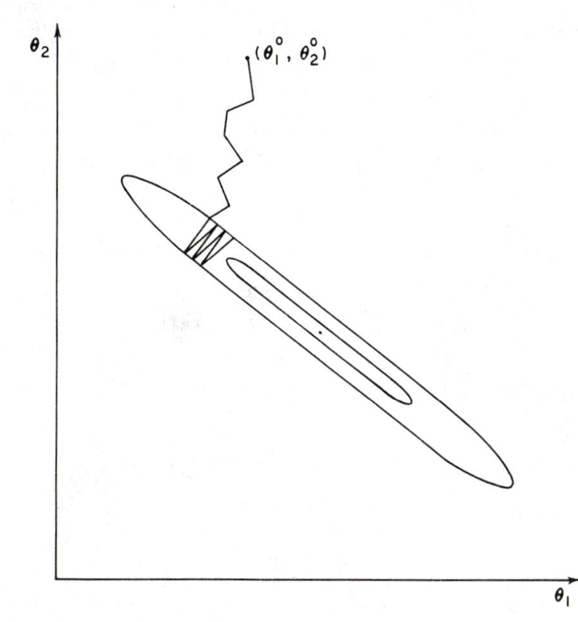

Figure 3. A simple gradient method tends to oscillate around the bottom of the valley in a hemstitching pattern making very slow progress.

very bad convergence. If the search were to be mistakenly halted prior to attainment of the global minimum, one would be trapped in misleading conclusions. While the visual appearance of the "quality" of the fit along the bottom line of the valley can hardly be depicted, the corresponding changes in parameter estimates may be dramatic.

A simple approach to the "valley search" was proposed by Gelfand and Tsetlin (1961). A descent from the initially guessed $\boldsymbol{\theta}^0$ is halted upon reaching

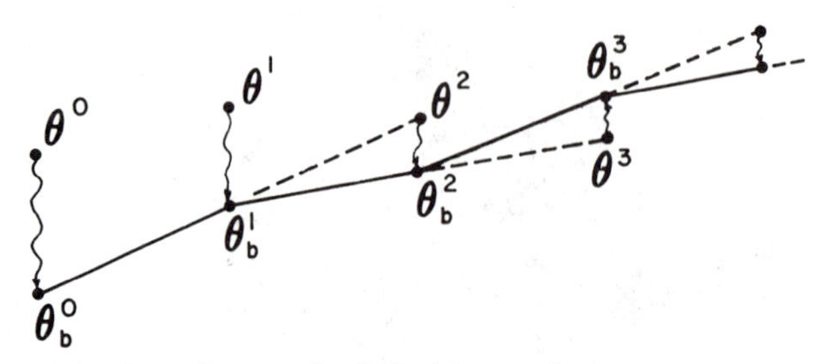

Figure 4. Valley search method (see text).

the bottom of the valley, point $\boldsymbol{\theta}_b^0$ (see Fig. 4). Another "initial guess," $\boldsymbol{\theta}^1$, is generated by a relatively large perturbation in $\boldsymbol{\theta}_b^0$, and the descent from $\boldsymbol{\theta}^1$ is halted again at the bottom of the valley, point $\boldsymbol{\theta}_b^1$. A straight line passing through points $\boldsymbol{\theta}_b^0$ and $\boldsymbol{\theta}_b^1$ provides an approximate local direction of the valley. Therefore, the starting point for the next descent is determined in this direction, towards the decline of the valley, at chosen step size h. Further search is clear from Fig. 4 (Gagarin *et al.*, 1969). The advantage of this method is that any numerical routine can be employed for the descent from $\boldsymbol{\theta}^\kappa$ to $\boldsymbol{\theta}_b^\kappa$.

When the global minimum is found, an approximate $100(1-\alpha)\%$ joint confidence region is given by (Box and Hunter, 1965)

$$\Phi(\boldsymbol{\theta}) - \Phi(\hat{\boldsymbol{\theta}}) \le p\hat{\sigma}^2 F_\alpha(p, n-p)$$

where

$\Phi(\hat{\boldsymbol{\theta}})$ is the minimum value of the objective function
$\hat{\sigma}^2$ is the estimate of the error variance
n is the number of observations
p is the number of independent parameters estimated.

Since

$$\hat{\sigma}^2 = \Phi(\hat{\boldsymbol{\theta}})/(n-p) \tag{4.5}$$

we obtain

$$\Phi(\boldsymbol{\theta}) \le \Phi(\hat{\boldsymbol{\theta}}) \left\{1 + \frac{p}{n-p} F_\alpha(p, n-p)\right\} \tag{4.6}$$

The equality in expression (4.6) specifies a contour of the objective function which is, generally, not ellipsoidal and frequently is crescent-shaped (Draper and Smith, 1981, Chap. 10).

For the multiresponse model, Box and Draper (1965) proposed to optimize the posterior distribution of $\boldsymbol{\theta}$ using uniform or "noninformative" prior distributions of the parameters. They wrote:

> "In considering a parameter like the specific rate [constant] φ which is essentially positive, it is probably most realistic to take $\theta = \ln \varphi$, $-\infty \le \theta \le \infty$, as locally uniform *a priori*. This would mean, for example, that having guessed a value for φ, an experimenter would be about equally prepared to accept a value twice as big as he would to accept a value one-half as big."

When errors are distributed normally and there are no correlations between the responses of model (2.1) (Box *et al.*, 1973), the above criterion leads to minimization of the determinant of matrix $\mathbf{V}$, that is,

$$\Phi(\boldsymbol{\theta}) = |\mathbf{V}|$$

where

$\mathbf{V} = \{v_{ru}\}$, $v_{ru} = \sum_{i=1}^{n} [y_{ri} - f_r(\xi_i, \boldsymbol{\theta})][y_{ui} - f_u(\xi_i, \boldsymbol{\theta})]$,
y_{ri} is realization of the rth response in the ith set of observations, and
n is the number of sets of observations.

The approximate $100(1 - \alpha)\%$ joint confidence region is given by

$$\Phi(\boldsymbol{\theta}) \leq \Phi(\hat{\boldsymbol{\theta}}) \exp(\chi^2_{p,\alpha}/n)$$

where

$\Phi(\hat{\boldsymbol{\theta}}) = |\mathbf{V}|_{\min}$
$\chi^2_{p,\alpha}$ is the upper α point of chi-square distribution with p degrees of freedom
p is the number of parameters estimated.

In this approach the variances and covariances (within the response) are assumed to be unknown, which is a typical situation in combustion research. When the former is known and the latter is zero, the determinant is reduced to the weighted sum of the squares of the responses' residuals (see also Box, 1970; Box and Draper, 1972; Singh and Rao, 1981).

Box and Draper warn, however, that "the investigator should *not* resort immediately to the joint analysis of responses. Rather he should ... consider the consistency" in the minima of individual and combined (the minors of $\mathbf{V}$) responses. Yet a formal lack of fit test has not been proposed.

5. Adequacy of fit

On the *a priori* assumption that only random errors are present, Eq. (4.5), or (4.2) for linear models, provides an unbiased (or asymptotically unbiased) estimator of the error variance σ^2. If the calculated estimate proves to be unbiased, we say that the model adequately fits the experimental data or simply the model is adequate. Otherwise, the bias in the variance estimate indicates the presence of systematic errors. This means that the model is inadequate and neither the parameter estimates nor the confidence region are valid.

In order to test the unbiasedness of the estimate, the exact value of σ^2 must be known. In practice, it is not. The usual approach is to estimate the error variance σ^2 by replicated experiments, designed so that systematic errors of instrumentation (called instrumental errors, such as unbalanced scales, wrong absorption coefficients, etc.) are absent (Box *et al.*, 1978, p. 319). Then, the formalism of the test for adequacy consists of setting forth a so-called null hypothesis stating that the estimators for error variance in the model considered and in the replicated runs both estimate the same variance. Comparing the ratio of these estimators with the corresponding value of the F distribution, we accept or reject the null hypothesis and hence the adequacy of the model. For further discussion on the analysis of variances the reader is referred to Box and Hunter (1965), Box *et al.* (1978), Cvetanovic *et al.* (1979), and Draper and Smith (1981).

There are situations when several rival models are suggested to explain the same experimental facts. Certain discriminatory criteria can be derived for choosing the most probable model. The two most commonly used approaches, the likelihood and the Bayesian, are reviewed by Singh and Rao (1981).

It should be emphasized that the results of the inference tests must be interpreted in their statistical sense. Thus, when a model passes the F test or is chosen by a discriminatory criterion, we should say that with a given probability this model *cannot be rejected* and by no means does it prove the uniqueness of the model. When the results of a test are marginal, the experimenter should abstain from making any conclusions. Further experimental research directed towards an increase in the power of the criterion is required.

6. Design of experiments

The goal of the researcher in conducting experiments is to acquire desired information. It is possible to increase the "amount of information obtained per data point" if the experiments are carried out in accordance with a prearranged plan. Let us consider the following example borrowed from Box *et al.* (1954), pp. 445–446.

Suppose we wish to determine the weights of three objects, A, B and C. The plan of the traditional "one-variable-at-a-time" approach is shown in Table 1,

Table 1. Traditional, "one-variable-at-a-time" plan: three objects A, B, and C are weighed separately.

Object			Observation	
A	B	C		
−	−	−	y_0	Zero error
+	−	−	y_1	A only
−	+	−	y_2	B only
−	−	+	y_3	C only

Table 2. Alternative plan to determine weights of three objects A, B, and C.

Object *A*	*B*	*C*	Observation	
$-$	$-$	$-$	y'_1	Zero error
$+$	$-$	$+$	y'_2	A and C
$-$	$+$	$+$	y'_3	B and C
$+$	$+$	$-$	y'_4	A and B

where $(+)$ and $(-)$ indicate the presence or absence, respectively, of the corresponding object on the scales. The weight of each object is determined then by the difference of the two weighings:

$$\hat{\mu}_A = y_1 - y_0$$
$$\hat{\mu}_B = y_2 - y_0$$
$$\hat{\mu}_C = y_3 - y_0$$

with variances

$$\sigma^2(\hat{\mu}_A) = \sigma^2(\hat{\mu}_B) = \sigma^2(\hat{\mu}_C) = 2\sigma^2$$

where σ^2 is the error variance.

If the same number of weighings is performed according to another plan (Table 2), the weight of each object is determined by the results of all four weighings:

$$\hat{\mu}_A = (-y'_1 + y'_2 - y'_3 + y'_4)/2$$
$$\hat{\mu}_B = (-y'_1 - y'_2 + y'_3 + y'_4)/2$$
$$\hat{\mu}_C = (-y'_1 + y'_2 + y'_3 - y'_4)/2$$

with variances

$$\sigma^2(\hat{\mu}_A) = \sigma^2(\hat{\mu}_B) = \sigma^2(\hat{\mu}_C) = 4\sigma^2/4 = \sigma^2$$

Thus, taking into account that $1/\sigma^2$ provides a measure of information, carrying out the second plan results in twice as much information as carrying out the first one. In other words, the traditional or so-called *passive* experimentation would require twice as many weighings to achieve the same accuracy as that obtained by the alternative or *active* experimentation.

The results of the above example are not happenstance and there exist techniques to design such plans for linear as well as for nonlinear modeling. The subject, however, lies outside the scope of this text; the reader is referred to Box *et al.* (1978) and references therein for further details. The application of active "computer experimentation" to modeling of chemical kinetics is discussed in Section 7.5.

7. Dynamic models in chemical kinetics

7.1. Preliminary remarks

The goal of basic combustion research, as of any basic research, is to elucidate the physical and chemical mechanism of combustion phenomena, i.e., to construct a physical model. Yet the present state of knowledge in combustion chemistry allows one, at best, to deduce only *tentative* mechanisms. We know something of the chemistry, but not all of it. These mechanisms are developed by considering mechanistic models that take the form of differential equations (2.2).

Modeling, rigorously approached, should consist of setting forth a number of alternative kinetic schemes and by proper experimentation choosing the most probable one. However, the present ambiguity of kinetic data would then lead to an enormously large number of close rival models, whereas the design of discriminating experiments is typically constrained into practical impossibility by instrumental limitations. The necessity for numerical integration makes matters even worse. The consequence is that after exhaustive efforts the researcher can find himself with frustratingly inconclusive results.

The complexity of this problem pushes some researchers into another extreme. Combining all conceivable reactions, the results of integration are simply compared with experimental data, upon which the goodness of fit is judged. Although suitable as a computer exercise for poorly established mechanisms, this procedure is insufficient for better known systems. First, for valley-shaped objective functions such comparisons are misleading. Second, such modeling does not disclose the role each reaction plays in the overall mechanism. Questions like what elementary reactions are "responsible" for the observed experimental facts or what elementary reactions can be neglected under what conditions are of importance not only to a kineticist, but also to a practical user of the model.

The experimental design chosen sets a limit on the amount of information that can possibly be extracted from the data obtained. For instance, when measuring the temperature of a flame, fluctuations in the ambient temperature of the room make no difference since, as we say, the temperature fluctuation of the room is below the sensitivity of the instrument. An analogous situation is typical in kinetic modeling where the existence or nonexistence of some reactions cannot be established even in principle within the framework of possible experiments.

Thus, the current objective of chemical kinetics modeling in combustion will be stated as the deduction of a tentative mechanism of the minimal size necessary to explain the observed phenomena. A systematic approach, though informal and still in the stage of development, constitutes the subject of the following discussion.

7.2. Trial models

Initially, a first-trial mechanism is constructed by including all possibly relevant, known reactions as well as some that may be completely hypothetical. Rate constant expressions for these reactions are selected from literature sources, estimated theoretically, or just guessed. At this stage preference is given to experimental rate constants, but when considering them the researcher must scrupulously analyze all the assumptions made and the parameter values used by the authors, for they may be out of date. The researcher must also check the consistency of the authors' results with the reports of others.

As a rule, a rate constant is specified either for the forward or for the reverse direction of the reaction. The other one is computed, in accordance to the principle of detailed balancing, via the equilibrium constant, which is usually much better known than either of the rate constants (Chapter 8).

The mathematical formulation of a kinetic scheme results in a dynamic model, Eqs. (2.2), which in the case of the first mechanism constructed will be called a *trial* model. The numerical solution of the differential equations (2.2) combined with the instrumental functions

$$\boldsymbol{\eta} = \mathbf{I}(\mathbf{s}, \boldsymbol{\nu}) \tag{7.1}$$

constitute model (2.1), where the functions $\mathbf{f}(\boldsymbol{\xi}, \boldsymbol{\theta})$ are defined numerically and the vector $\boldsymbol{\theta}$ includes parameters of the instrumental functions $\mathbf{v}$. Then one could proceed with modeling as outlined in the previous section, parameter estimation followed by statistical analysis of the results, except for the fundamental difficulty that the parameters to be determined in the trial model invariably outnumber the independent experimental observations. That is, the number of degrees of freedom is negative, meaning that the problem may have an infinite number of solutions. Such an answer can hardly satisfy the researcher. Should he attempt to fit the data by random adjustment of the model parameters, he will find himself in the position of a naive burglar guessing the combination of a safe. The sophisticated burglar, however, will try to sense the mechanism of the lock. A similar approach to the trial model is achieved by sensitivity analysis.

7.3. Sensitivity analysis

In a trial kinetic scheme not all reactions "contribute" equally to the chemical history, and hence to the model fit. On the contrary, some of them may contribute significantly, some marginally, and some not at all. Given an overly large first-trial mechanism, one should begin simplifying it immediately by discarding the irrelevant reactions and identifying those which contribute the most. In order to do this systematically we introduce the *fit sensitivity* to parameter θ_j

$$S_j^{\Phi} = \partial\Phi(\boldsymbol{\theta})/\partial\theta_j$$

and the rth *response sensitivity* of parameter θ_j

$$S_j^r = \partial\eta_r/\partial\theta_j$$

Note that the fit and response sensitivities are interrelated, as can be demonstrated in the case of the weighted least squares, when

$$\Phi(\boldsymbol{\theta}) = \sum_{r,i} \omega_{ri}(y_{ri} - \eta_{ri})^2,$$

where ω_{ri} is the statistical weight corresponding to the ith observation of the rth response,

and

$$\partial\Phi(\boldsymbol{\theta})/\partial\theta_j = -2\sum_{r,i} \omega_{ri}(y_{ri} - \eta_{ri})(\partial\eta_r/\partial\theta_j)_i.$$

These sensitivities provide numerical measures for the contribution of the jth reaction when θ_j is associated with the parameter of the corresponding rate expression, namely, with the rate constant k_j for the isothermal case or with the preexponential factor A_j for a nonisothermal model. It is more convenient to deal with nondimensional, or *logarithmic*, measures of sensitivities (Bukhman *et al.*, 1969; Frank, 1978; Gardiner, 1977). Thus, the rth logarithmic response sensitivity of rate constant k_j is given by

$$LS_j^r = \partial \ln \eta_r/\partial \ln k_j = \frac{k_j}{\eta_r}\frac{\partial\eta_r}{\partial k_j}$$

which measures the relative change in response η_r with respect to the relative change in rate constant k_j. Actually, considering $\ln k_j$ as a model parameter, i.e., $\theta_j = \ln k_j$, would be in harmony with the Bayesian approach to parameter estimation discussed earlier (Section 4.3.3).

The evaluation of sensitivities can be approached in a number of ways. The simplest, so-called "brute force" method, consists of the computation of the response at distinct values of the parameter considered while keeping the others constant. Assuming that

$$\partial \ln \eta_r/\partial \ln k_j \approx \Delta \ln \eta_r/\Delta \ln k_j \tag{7.2}$$

we obtain an "interval" or *integral* sensitivity. Another method, called the direct method (Bukhman *et al.*, 1969; Dickinson and Gelinas, 1976), is based on the following considerations. Differentiating instrumental functions [Eq. (7.1)] with respect to k_j yields

$$\frac{\partial\boldsymbol{\eta}}{\partial k_j} = \frac{\partial\mathbf{I}}{\partial\mathbf{s}}\frac{\partial\mathbf{s}}{\partial k_j} \tag{7.3}$$

where $\partial\mathbf{I}/\partial\mathbf{s}$ is the Jacobian matrix of functions $\mathbf{I}$ with respect to $\mathbf{s}$. The elements of the vector $\partial \mathbf{s}/\partial k_j$ are called the *state* sensitivities of rate constant k_j. Differentiating Eq. (2.2) with respect to k_j we obtain

$$\frac{\partial}{\partial k_j}\left(\frac{d\mathbf{s}}{dt}\right) = \frac{\partial \mathbf{h}}{\partial k_j} + \frac{\partial \mathbf{h}}{\partial \mathbf{s}}\frac{\partial \mathbf{s}}{\partial k_j}$$

or

$$\frac{d}{dt}\left(\frac{\partial \mathbf{s}}{\partial k_j}\right) = \frac{\partial \mathbf{h}}{\partial k_j} + \frac{\partial \mathbf{h}}{\partial \mathbf{s}}\frac{\partial \mathbf{s}}{\partial k_j}. \tag{7.4}$$

Simultaneous integration of Eqs. (7.4) and (2.2) yields "point" or *differential* state sensitivities, and if the functions **I** in Eq. (7.1) are known, one obtains differential response and fit sensitivities according to Eq. (7.3). More sophisticated, mathematically involved methods to obtain differential sensitivities are the Green's function method (Demiralp and Rabitz, 1981a,b; Dougherty and Rabitz, 1979; Hwang *et al.*, 1978; Kramer *et al.*, 1981; Rabitz, 1983), the Fourier amplitude sensitivity test (Cukier *et al.*, 1973; Cukier *et al.*, 1978; McRae *et al.*, 1982), the stochastic sensitivity analysis (Constanza and Seinfeld, 1981), and the methods proposed by McKeown (1980), Seigneur *et al.*, (1982) and Hwang (1983). For descriptions of them the reader should consult the literature cited.

A newcomer to kinetic modeling is well advised to begin, and the seasoned modeler to remain, with the "brute force" method for the following reasons. First, it is easy to use since hardly any additional programming is required. Second, the interpretation of sensitivities is direct and any conceivable response can be considered, whereas state sensitivities cannot be easily related, if they can be related at all, to such characteristic combustion measures as induction times. Third, the approximation involved in expression (7.2) is harmless, since for the purpose of sensitivity analysis only relative comparison of sensitivities is of interest. Moreover, this "disadvantage" can easily be removed by empirical computer experimentation (Section 7.5) leading to adequate interval sensitivities far more informative than point measures, and in fact becomes an advantage when, as always happens sooner or later, the computational intervals Δk_j get extended to study wider regions of **k** space. Fourth, the slightly larger computational time requirement, the only objection to the "brute force" method, becomes unessential if efficient numerical codes are employed (Chapter 1), and is probably far less costly than the programming effort required to implement more sophisticated methods of sensitivity analysis and the interpretive effort required to extend the other methods into regions of the **k** parameter space that prove to be interesting.

The set of logarithmic response sensitivities for all rate constants is referred to as a *sensitivity spectrum*. The highest absolute values of the sensitivities in the spectrum pinpoint the most important reactions in the kinetic scheme for determining the chosen set of responses. The lowest absolute values of sensitivities, however, do not necessarily identify unimportant reactions. Let us illustrate this by the following example taken from the modeling of cyanogen oxidation behind reflected shock waves (Lifshitz and Frenklach, 1980). The logarithmic sensitivities for reactions (R3) and (R4)

$$CN + O_2 \xrightarrow{k_3} NCO + O \tag{R3}$$

$$O + C_2N_2 \xrightarrow{k_4} NCO + CN \tag{R4}$$

with regard to ignition delay for a particular set of experimental conditions, which was the only model response considered, were found to be

$$LS_3 \cong 0 \text{ (within truncation error);}$$

$$LS_4 = -0.62.$$

Both reactions proceed with practically identical rates, much higher than the other reactions, and thus determine the instant of ignition. Hence, reaction (R3) can hardly be considered unimportant. Why then is the sensitivity to this reaction so low?

The rate of the chain reaction is (approximately) given by the analytical expression

$$\text{rate(R3 + R4)} \cong \frac{k_3[O_2]k_4[C_2N_2]}{k_3[O_2] + k_4[C_2N_2]}\{[O] + [CN]\}$$

Under the experimental conditions employed

$$\frac{k_3[O_2]}{k_4[C_2N_2]} = \frac{[O]}{[CN]} \cong 25 \gg 1 \tag{7.5}$$

and therefore

$$\text{rate(R3 + R4)} \cong k_4[C_2N_2][O]$$

Thus, the rate of the chain reaction is determined mainly by the rate of reaction (R4), and changing k_3 without upsetting the inequality (7.5) would not much affect the ignition delay. Such behavior manifests the valley character of the objective function as shown in Fig. 5. Indeed, by moving along the bottom line of the valley parallel to axis k_3, *a–b*, without experiencing significant changes in $\Phi(k_3, k_4)$ would mean the fit and the response sensitivities of k_3 are close to zero and those of k_4 are quite large, which explains the results of the sensitivity tests. However, changing k_3 by a large negative amount (say a factor of 100) would completely alter the conclusion. Also, should other responses be considered, the sensitivity with respect to the same rate constant may be different. For instance, the logarithmic state sensitivity of k_3 with respect to the concentration of CN radicals is close to -1.

In order to resolve the ambiguity of "zero" sensitivity one must analyze, in addition to the sensitivities, the reaction rates. A convenient way of doing this is to define (Gardiner, 1977)

$$pR_j = \text{sign} \times \log |(R_j^+ - R_j^-)/\text{molecules cm}^{-3}\,\text{s}^{-1}|$$

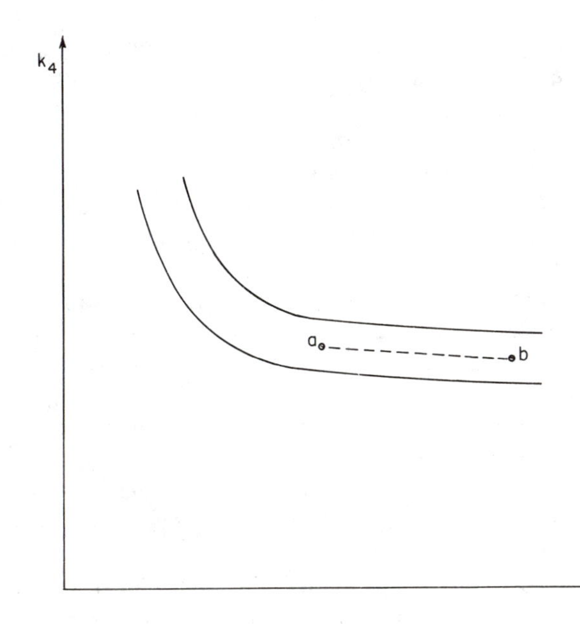

Figure 5. Moving along the valley parallel to axis k_3, e.g., a–b, would not change the value of the objective function significantly and would result in a very low sensitivity with respect to k_3.

where R_j^+ and R_j^- are the forward and reverse rates, respectively, of reaction j; "sign" is the sign of $(R_j^+ - R_j^-)$ and the bars denote absolute value. The values of pR's are easily calculated and conveniently printed at the output points of the integration. Essential reactions which have zero sensitivities due to their fast rates are then immediately apparent.

After a first-trial model has been constructed, modeling may proceed through the following steps.

(a) *First-trial runs.* It may happen that the first-trial model contains "bugs" so large that systematic sensitivity analysis would be premature. For instance, in modeling the ignition delay in COS-O_2-Ar mixtures (Lifshitz *et al.*, 1975), the ignition phenomena could not be simulated at all until the reaction $COS + SO \rightarrow CO_2 + S_2$, whose rate constant was originally overestimated, was removed. When the results of initial computation do not seem chemically realistic, a preliminary analysis of species concentrations, reaction rates, and randomly evaluated sensitivities will probably resolve the problem.

(b) *Discrimination between reactions.* Assuming the initial set of rate constants and varying their values one at a time by a chosen factor, typically 2–10, the sensitivities are determined by expression (7.2). For the slowest reactions, which are the most probable candidates for removal, the variations in rate constants should preferably reach the conceivable upper and lower limit values. Moreover, sensitivity spectra can be remarkably different for various sets of controllable variables and, therefore, they should be determined over the entire range of ξ.

The reactions having the smallest absolute values of both LS_j and pR_j are removed from the scheme and the new responses are computed and compared to the initial ones. If no change is detected, further reactions are removed and the procedure continued until the difference in responses becomes larger than the computational errors. The removed reactions constitute the least important ones, within the range of sensitivity tests. With experience these reactions can be identified by analyzing the pR spectra only and confirmed by a few computer runs.

The sensitivity spectra for the remaining reactions establish the significance of each one of them. Thus, when selecting parameters for optimization, preference should be given to those with the highest absolute values of LS and the largest uncertainty. It should be noted that choosing parameters with low sensitivity would worsen the valley character of the objective function.

The analysis of sensitivities can shed light on the mechanism of the process even when rigorous optimization cannot be justified.

(c) *Optimization.* There are two possible options for optimization of the model: first, to keep all the reactions in the scheme, or second, to remove the insignificant reactions in order to save computational time. In either case, if optimization results in the values of the parameters lying outside the range tested initially, the entire sensitivity analysis should be performed again and if the sensitivity spectra are changed significantly, the model should be reoptimized.

(d) *Interpretation of results*—as required for the problem under study (see the next Section).

The above procedures and their sequence are only recommendations. Common sense, knowledge of chemistry, experience, and the imagination of the researcher are the real guidelines for modeling.

7.4. Optimization and interpretation of results

The classical approach to parameter estimation, optimization of the objective function, encompasses all the experimental observations of the study. The procedures involved have been outlined earlier. Here we are more concerned with the interpretation of results, a subject that causes quite a lot of confusion.

To test the adequacy of a model one needs to determine independently an estimate of the error variance free of systematic errors. The obvious source of systematic errors is the incomplete knowledge of relationships (7.1). Determination of these instrumental functions, assessed by so-called calibrating experiments, constitutes modeling of the instrumental process, but the adequacy of the model [Eq. (7.1)] cannot be easily established, if at all. Let us consider the situation where the state variables (e.g., concentrations) calculated by inadequate instrumental functions are taken as the responses of modeling. Even if the predicted values match these responses perfectly, the kinetic model is inadequate due to the propagating bias of the instrumental functions **I** in Eq. (7.1). This bias will not be detected by any statistical tests as long as the same instrumentation is employed to obtain the data.

For most experimental techniques used in combustion research the problem of adequacy remains unresolved, and it is common practice to compare the results of modeling obtained by various researchers, justifying model adequacy by agreement among their results. The typical approach, however, of comparing values of parameters may be misleading. Suppose that the results of researcher A are compared to those of researchers B and C at point $\boldsymbol{\theta}_A$, $\boldsymbol{\theta}_B$ and $\boldsymbol{\theta}_C$, respectively, in parameter space $\boldsymbol{\theta}$. Assume that points $\boldsymbol{\theta}_A$ and $\boldsymbol{\theta}_B$ are quite removed from one another but still lie in the valley of the "true" objective function or even that both belong to the "true" confidence region I; and point $\boldsymbol{\theta}_C$ is much closer to point $\boldsymbol{\theta}_A$ but lies slightly aside from region I (Fig. 6). Researchers A and B, obtaining almost the same "quality" of fit, would claim their results are contradictory. Researchers A and C may observe a slight difference in the fits but, on the grounds of closeness in parameter values, would report their results are consistent. Nevertheless, the opposite is more probable.

Considering this, it is more appropriate to compare confidence regions, bearing in mind the following possible situations (Fig. 6):

(a) Two regions coincide or cross one another at their centers, as I and II. If the instrumentation used in both studies is identical, this event indicates

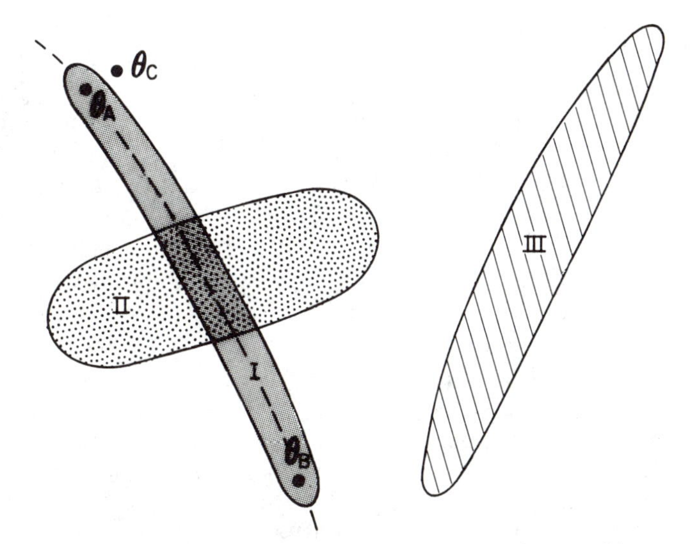

Figure 6. The interpretation of modeling results without considering confidence regions may be misleading (see text).

the consistency of the results but not the adequacy of the model. If very different techniques are employed and there is no reason to suspect bias in either study, the model is assumed to be tentatively adequate and the experimental observations from both studies can be incorporated into multiresponse optimization (Box and Draper, 1965) yielding a tentative joint model.

(b) Two regions are far apart, as I and III. The results are inconsistent and no joint modeling should be considered.

Optimization of the overall objective function places a tremendous demand for computer time. Every iteration requires integration of Eqs. (2.2) for all experimental runs and the number of iterations must be substantial since at least two parameters must be estimated for each reaction considered. Box and Hunter (1962) proposed a simpler method, suitable when experiments are carried out at various exactly known temperatures, each isothermally

$$T_i = \text{const}(i), \qquad i = 1, 2, \ldots, n$$

and the number of degrees of freedom for each experimental run is positive. Under such circumstances, the model is optimized for every run separately, which results in values of rate constants (and confidence regions) at temperatures T_i ($i = 1, 2, \ldots, n$). The null hypothesis consists of assuming the form of the rate constant expression, e.g., Arrhenius. If the rate constants are consistent with this hypothesis, i.e., a straight line in $\log k - 1/T$ coordinates is obeyed, the model is tentatively consistent (with all the reservations discussed above) and the parameters of the rate constant expressions are readily calculated.

The extension of these ideas leads to an even more economical, yet informal, approach (Gardiner, 1979; Gardiner *et al.*, 1981; Lifshitz *et al.*, 1975). An empirical model of the process is derived first. The tentative consistency of the mechanistic model is tested by comparing the parameters of the empirical model with corresponding predictions of the mechanistic one. This method is especially suitable when the trial model is inadequate *a priori* and rigorous modeling is unjustified. Let us consider again the modeling of ignition in COS-O_2-Ar mixtures (Lifshitz *et al.*, 1975). The preliminary analysis of experimental results indicated that COS, the fuel, inhibits the ignition and it was of interest to understand what causes this behavior. Since the gas dynamic aspects of the preignition development are not well understood (Gardiner and Wakefield, 1970; Oppenheim *et al.*, 1975), it was clear that no matter what kinetic scheme were chosen, the trial model would be incomplete and could not, in principle, adequately predict the ignition delay, the response of the model. Under these circumstances the modeling followed the strategy below.

(a) The empirical model was derived as

$$\tau = \theta_1 \exp(\theta_2/T)[\mathrm{COS}]^{\theta_3}[\mathrm{O_2}]^{\theta_4}$$

where τ is the ignition delay (s); T is the initial temperature (K); [COS] and $[O_2]$ are the initial concentrations of COS and O_2, respectively (mol cm^{-3}); $\theta_1 = 2.66 \times 10^{-11}$/s; $\theta_2 = 8505$/K; $\theta_3 = 0.30$; and $\theta_4 = -1.12$.

(b) The first-trial mechanism, with 26 reactions, was reduced to a 14-reaction one by a preliminary sensitivity analysis.

(c) Five computer experiments for the trial model were designed in such a manner that they covered the range of experimental conditions employed. The results of these runs were

$$\theta'_2 = 8706/\mathrm{K}$$
$$\theta'_3 = 0.22$$
$$\theta'_4 = -1.55$$

Even though the computed responses were close to the experimentally observed ignition delays, θ'_1 was not of interest for the reason of unknown gasdynamics.

(d) As can be seen, the computer-simulated parameters values are not far apart from the corresponding empirical ones, in the sense of sign and order of magnitude, and they remained about the same within the uncertainties of the rate constants according to the sensitivity analysis. Further optimization, leading to inconclusive results in any event, would not have justified the computational efforts and, therefore, was not attempted.

(e) Further modeling was directed by the following strategy. The tentatively entertained model is inadequate and cannot be employed to predict the "true" ignition delays. Nevertheless, it is possible to explain the inhibition effect of COS on the *model* response. (Details of the sensitivity tests used can be found in the paper cited above.) Since the simulated inhibition did not disappear over a wide range of parameter values, it is reasonable to speculate a similar explanation for the observed phenomenon. Only further experimental facts can tell us how close we are to reality; for the time being, the tentative results can direct future research.

7.5. Further look at sensitivities

In the "brute force" method the responses are computed at various points of parameter space $\boldsymbol{\theta}$. Following the ideas of Box and Coutie (1956), this procedure can be viewed as computer experimentation for which the integration of Eqs. (2.2) is the "process" and the parameters of functions (2.2) are the controllable variables. The response sensitivities introduced in Section 7.3 become the effects of these variables and the estimation of them constitutes the objective of empirical modeling.

Let us state the problem of deriving an empirical model of responses $\boldsymbol{\eta}$ in variables $\boldsymbol{\theta}$. If the resulting model is adequate in subspace $\boldsymbol{\Omega}$ of space $\boldsymbol{\theta}$, the dynamic model [Eq. (2.2)] can be replaced by the empirical one within $\boldsymbol{\Omega}$.

Then, at each point of $\mathbf{\Omega}$:

(a) the values of $\partial\boldsymbol{\eta}/\partial\boldsymbol{\theta}$, the required sensitivities, can be adequately estimated;
(b) the empirical model itself provides the relationship between various sensitivities;
(c) the evaluation of the objective function $\Phi(\boldsymbol{\theta})$ does not require integration of differential equations.

Let us consider the following example (Miller and Frenklach, 1983). Lifshitz *et al.* (1973) studied decomposition of propane behind reflected shock waves by measuring the concentrations of the products. To interpret the results obtained over the temperature range 1050–1250 K, Lifshitz and Frenklach (1975) proposed a mechanism composed of 11 reactions of 11 species. Under the experimental conditions employed, the pyrolysis can be assumed isothermal; that is, for the actual process, the rate constants are the parameters, the post-shock temperature, pressure and initial concentration of propane are the controllable variables, and the concentrations of products at a given reaction time are the responses.

Let us define a computer experiment as integration of the dynamic model [Eq. (2.2)] corresponding to the 11-reaction mechanism. The outcome of these experiments is the vector of predicted values of the actual responses. Assuming the logarithms of the rate constants as the controllable variables of the computational process, let us begin the empirical modeling of this process (Section 3.2) by considering the linear form

$$\ln C_r = a_{r0} + a_{r1} \ln k_1 + a_{r2} \ln k_2 + \cdots + a_{r11} \ln k_{11} \tag{7.6}$$

where $r = 1, 2, \ldots$; C_r is the concentration of species r at a given reaction time; $k_1, k_2, \ldots, k_{11}$ are the rate constants of the 11-reaction mechanism; and a_{r0}, $a_{r1}, a_{r2}, \ldots, a_{r11}$ are the parameters of the rth model. For the sake of simplicity in presentation the example will be restricted to the following conditions:

(a) 1.6% C_3H_8—argon mixture, $T_5 = 1100$ K, $\rho_5 = 0.06$ kmol/m^3;
(b) two responses: the concentrations of ethane and ethylene at 700 μs (the instrumental functions are assumed to be known exactly);
(c) two adjustable variables, $\ln k_1$ and $\ln k_2$, where k_1 and k_2 are the rate constants of reactions $C_3H_8 \rightarrow CH_3 + C_2H_5$ and $CH_3 + C_3H_8 \rightarrow CH_4 + C_3H_7$, respectively;
(d) the "initial guesses" for k_1 and k_2 at $T_5 = 1100$ K are $k_{1,0} = 0.129$ s^{-1} and $k_{2,0} = 3.225 \times 10^7$ s^{-1} m^3 kmol^{-1}. The rest of the rate constants assumed to be "known" and their values are taken the same as in the work of Lifshitz and Frenklach (1975), except that the rate constant for methyl radical recombination is taken from the works of Skinner and co-workers (Chiang and Skinner, 1981; Rogers and Skinner, 1981).

Thus, Eqs. (7.6) are reduced to

$$\ln C_1 = a_{10} + a_{11} \ln k_1 + a_{12} \ln k_2$$

and

$$\ln C_2 = a_{20} + a_{21} \ln k_1 + a_{22} \ln k_2$$

or, comparing them with Eq. (3.1), to

$$\eta_r = a_{r0} + a_{r1}x_1 + a_{r2}x_2, \qquad r = 1, 2, \tag{7.7}$$

where $x_1 = \ln k_1$ and $x_2 = \ln k_2$; $\eta_1 = \ln C_1$ and $\eta_2 = \ln C_2$; and C_1 and C_2 are the concentrations of ethane and ethylene, respectively, at 700 μs.

As demonstrated in Section 6, it is advantageous to employ active experimentation for model derivation. (Note that the "brute force" method is a "one-variable-at-a-time" design for a linear model.) For this purpose, variables x_1 and x_2 are transformed into new variables

$$X_1 = \frac{x_1 - x_{1,0}}{\Delta x_1} \quad \text{and} \quad X_2 = \frac{x_2 - x_{2,0}}{\Delta x_2},$$

where $(x_{1,0}, x_{2,0})$ is the central point of the experimental design;

$$x_{1,0} = \frac{x_{1\,\max} + x_{1\,\min}}{2} \quad \text{and} \quad x_{2,0} = \frac{x_{2\,\max} + x_{2\,\min}}{2};$$

$$\Delta x_1 = \frac{x_{1\,\max} - x_{1\,\min}}{2} \quad \text{and} \quad \Delta x_2 = \frac{x_{2\,\max} - x_{2\,\min}}{2};$$

Δx_1 and Δx_2 are the half-intervals of variations for variables x_1 and x_2, respectively; $x_{1\,\max}$, $x_{2\,\max}$ and $x_{1\,\min}$, $x_{2\,\min}$ are the upper and lower levels of variables x_1 and x_2, respectively.

In the new variables, the model [Eq. (7.7)] takes from

$$\eta_r = b_{r0} + b_{r1} X_1 + b_{r2} X_2 \tag{7.8}$$

where $b_{r1} = a_{r1}\,\Delta x_1$ and $b_{r2} = a_{r2}\,\Delta x_2$; and $b_{r0} = a_{r0} + a_{r1}\,x_{1,0} + a_{r2}\,x_{2,0}$

A 2^2 factorial design (Box *et al.*, 1978, Chaps. 10, 15) and its realization, i.e., the computed concentrations of ethane and ethylene at 700 μs, are summarized in Table 3. The central point of the design, $(x_{1,0}, x_{2,0})$, is determined by the initial guesses for k_1 and k_2. The half-intervals of variations, Δx_1 and

Table 3. A 2^2 factorial design for computer experimentation with the 11-reaction mechanism for propane decomposition.

Experiment No.	Variables			Observations $y_j = \ln\{C_j\,\mathrm{m}^3/\mathrm{kmol}\}$	
	X_0	X_1	X_2	y_1	y_2
1	+1	+1	+1	−15.15	−10.77
2	+1	+1	−1	−15.03	−13.26
3	+1	−1	+1	−20.27	−13.40
4	+1	−1	−1	−19.98	−16.08

Δx_2, are assumed to be equal to ln5, i.e., the rate constants are varied by a factor of 5. The parameters of model (7.8) are calculated according to (Box *et al.*, 1978, Chap. 15)

$$\hat{b}_{rj} = \frac{\sum_{i=1}^{4} X_{ij} y_{ri}}{4}, \qquad j = 0, 1, 2, \qquad r = 1, 2,$$

where X_{ij} is the value of jth variable in ith experiment and $X_{i0} \equiv 1$. For example,

$$\hat{b}_{12} = \frac{(+1)(-15.15) + (-1)(-15.03) + (+1)(-20.27) + (-1)(-19.98)}{4} = -0.102$$

Thus, we obtain

$$\hat{\eta}_1 = -17.61 + 2.52X_1 - 0.102X_2 \tag{7.9}$$

$$\hat{\eta}_2 = -13.38 + 1.36X_1 + 1.29X_2 \tag{7.10}$$

Due to the absence of "experimental errors" in computer runs, the adequacy of the obtained empirical models (7.9) and (7.10) can be tested by comparing the model predictions with the "experimental" values of the responses at the central point of the factorial design (Box *et al.*, 1978, Chap. 15). The central point is defined by $X_1 = X_2 = 0$ and, therefore, according to Eqs. (7.9) and (7.10), $\hat{\eta}_{1,0} = \hat{b}_{10} = -17.61$ and $\hat{\eta}_{2,0} = \hat{b}_{20} = -13.38$, respectively. The additional computer experiment at these conditions results in $y_{1,0} = -17.25$ and $y_{2,0} = -13.37$. The difference

$$|\hat{\eta}_{2,0} - y_{2,0}| = 0.01$$

means that the corresponding concentrations agree within 1% ($e^{0.01} \cong 1.01$), which is comparable to truncation error of numerical integration. Thus, within

$$\Omega = (k_{1,0}/5 \le k_1 \le 5k_{1,0},\ k_{2,0}/5 \le k_2 \le 5k_{2,0})$$

the linear model (7.10) is adequate. The transformation of the latter to its original form [Eq. (7.7)] results in

$$\ln C_2 = -25.51 + 0.84 \ln k_1 + 0.80 \ln k_2 \tag{7.11}$$

The effect of x_j with respect to η_2 is given by

$$\frac{\partial \eta_2}{\partial x_j} = a_{2j}, \qquad j = 1, 2,$$

and since, by definition,

$$\frac{\partial \eta_r}{\partial x_j} = \frac{\partial \ln C_r}{\partial \ln k_j} \equiv LS_j^r, \qquad j = 1, 2; \qquad r = 1, 2, \tag{7.12}$$

the logarithmic response sensitivities for ethylene within $\mathbf{\Omega}$ can be estimated by the corresponding parameters of empirical model (7.11), i.e.,

$$LS_j^2 = a_{2j}, \qquad j = 1, 2$$

or

$$LS_1^2 = 0.84 \qquad \text{and} \qquad LS_2^2 = 0.80$$

For ethane, however, the difference between $\hat{\eta}_{1,0}$ and $y_{1,0}$ is significant, which indicates inadequacy of the linear model (7.9). The second-order design for the quadratic model (3.2), accomplished by four additional computer experiments (Box *et al.*, 1954, Chap. 11; Box *et al.*, 1978, Chap. 15), leads to

$$\hat{\eta}_1 = -17.25 + 2.51X_1 - 0.102X_2 + 0.043X_1X_2 - 0.288X_1^2 - 0.068X_2^2 \tag{7.13}$$

Based on various criteria (Box *et al.*, 1978, pp. 522–523; Box and Draper, 1982) the quadratic model (7.13) is adequate within Ω. In original variables x_1 and x_2, model (7.13) takes the form

$$\begin{aligned}\ln C_1 = {} & -21.87 + 0.82 \ln k_1 + 0.88 \ln k_2 + 0.0166 \ln k_1 \ln k_2 \\ & - 0.111(\ln k_1)^2 - 0.0262(\ln k_2)^2\end{aligned} \tag{7.14}$$

According to relationships (3.3), (7.12) and (7.14), the logarithmic response sensitivities of ethane within $\mathbf{\Omega}$ are given by

$$LS_1^1 = 0.82 - 0.222 \ln k_1 + 0.0166 \ln k_2$$

and

$$LS_2^1 = 0.88 + 0.0166 \ln k_1 - 0.052 \ln k_2$$

The interrelation of both sensitivities is apparent.

Equations (7.11) and (7.14) predict adequately within $\mathbf{\Omega}$ the results of computer experiments and, therefore, numerical integration of the dynamic model to compute the concentrations of ethane and ethylene can be replaced within $\mathbf{\Omega}$ by the empirical models (7.11) and (7.14). Developing similar equations for all experimental responses, parameter estimation for dynamic models is reduced to a more trivial problem. (Miller and Frenklach, 1983).

8. Closing remark

Most researchers are, justifiably, skeptical of involved computations. Decisive experimentation, disclosing unambiguously the mechanism of phenomcna, is the most desirable research activity and should always have first priority. Even in the most astute experimentation, however, information which could be obtained is often lost by using a merely "good-enough" approach to modeling. Only the right combination of experimentation and modeling will open the "safe" of nature.

9. Acknowledgments

The author wishes to express his thanks to the editor, Professor W. C. Gardiner, Jr., for his valuable aid and encouragement throughout the preparation of this manuscript. Also, appreciation is extended to Mr. D. Miller for preparing the example on propane decomposition, to Ms. N. Duffy for drawing the figures, and to Mrs. J. Easley for typing the manuscript.

10. References

Bard, Y. (1974). *Nonlinear Parameter Estimation*, Academic Press, New York.

Box, G. E. P. & Coutie, G. A. (1956). Proc. I. E. E. **103B** (Suppl. No. 1), 100–107.

Box, G. E. P. & Cox, D. R. (1964). J. Roy. Stat. Soc. **26B,** 211–252.

Box, G. E. P. & Draper, N. R. (1965). Biometrika **52,** 355–365.

Box, G. E. P. & Draper, N. R. (1982). Technometrics **24,** 1–8.

Box, G. E. P. & Hunter, W. G. (1962). Technometrics **4,** 301–318.

Box, G. E. P. & Hunter, W. G. (1965). Technometrics **7,** 23–42.

Box, G. E. P. *et al.* (1954). *The Design and Analysis of Industrial Experiments*, Edited by O. L. Davies, Oliver and Boyd, London.

Box, G. E. P. *et al.* (1973). Technometrics **15,** 33–51.

Box, G. E. P., Hunter, W. G., & Hunter, J. S. (1978). *Statistics for Experimenters. An Introduction to Design, Data Analysis, and Model Building*, Wiley, New York.

Box, M. J. (1970). Technometrics **12,** 219–229.

Box, M. J. & Draper, N. R. (1972). Appl. Statist. **21,** 13–24.

Bukhman, F. A. *et al.* (1969). In *Primenenie Vycheslitel'noi Matematiki v Khimicheskoi i Fizicheskoi Kinetike*, Edited by L. S. Polak, Nauka, Moskow, Chap. 1.

Chambers, J. M. (1977). *Computational Methods for Data Analysis.* Wiley, New York, Chap. 6.

Chiang, C.-C. & Skinner, G. B. (1981). 18th Int. Symp. Combust. The Combustion Institute, Pittsburgh, pp. 915–920.

Constanza, V. & Seinfeld, J. H. (1981). J. Chem. Phys. **74,** 3852–3858.

Cukier, R. I. *et al.* (1973). J. Chem. Phys. **59,** 3873–3878.

Cukier, R. I., Levine, H. B., & Shuler, K. E. (1978). J. Comp. Phys. **26,** 1–42.

Cvetanovic, R. J. & Singleton, D. L. (1977). Int. J. Chem. Kinet. **9,** 481–488.

Cvetanovic, R. J., Singleton, D. L., & Paraskevopoulos, G. (1979). J. Phys. Chem. **83,** 50–60.

Demiralp, M. & Rabitz, H. (1981a). J. Chem. Phys. **74,** 3362–3375.

Demiralp, M. & Rabitz, H. (1981b). J. Chem. Phys. **75,** 1810–1819.

Dickinson, R. P. & Gelinas, R. J. (1976). J. Comp. Phys. **21,** 123–143.

Dorny, C. N. (1975). *A Vector Space Approach to Models and Optimization*, Wiley, New York, Chap. 2.

Dougherty, E. P. & Rabitz, H. (1979). Int. J. Chem. Kinet. **11,** 1237–1248.

Draper, N. R. & Smith, H. (1981). *Applied Regression Analysis*, Wiley, New York.

Frank, P. M. (1978). *Introduction to System Sensitivity Theory*, Academic Press, New York, Chap. 2.

Gagarin, S. G. *et al.* (1969). In *Primenenie Vycheslitel' noi Matematiki v Khimicheskoi i Fizicheskoi Kinetike*, Edited by L. S. Polak, Nauka, Moskow, Chap. 2.

Gardiner, W. C., Jr. (1977). J. Phys. Chem. **81,** 2367–2371.

Gardiner, W. C., Jr. (1979). J. Phys. Chem. **83,** 37–41.

Gardiner, W. C., Jr. & Wakefield, C. B. (1970). Astron. Acta **15,** 399–409.

Gardiner, W. C., Jr., Walker, B. F., & Wakefield, C. B. (1981). In *Shock Waves in Chemistry*, Edited by A. Lifshitz, Marcel Dekker, New York, Chap. 7.

Gelfand, I. M. & Tsetlin, M. L. (1961). Dokl. AN SSSR **137,** 295–298.

Hamilton, W. C. (1964). *Statistics in Physical Science; Estimation, Hypothesis Testing, and Least Squares*, The Ronald Press, New York.

Harnett, D. L. & Murphy, J. L. (1975). *Introductory Statistical Analysis*. Addison-Wesley, Reading.

Hudson, D. J. (1963). *Lectures on Elementary Statistics and Probability*, CERN, Geneva.

Hwang, I.-T. (1983). Int. J. Chem. Kinet. **15,** 959–987.

Hwang, J.-T. *et al.* (1978). J. Chem. Phys. **69,** 5180–5191.

Johnson, N. L. & Leone, F. C. (1977). *Statistics and Experimental Design in Engineering and the Physical Sciences*, Vol. I, Wiley, New York.

Kramer, M. A., Calo, J. M., & Rabitz, H. (1981). Appl. Math. Modelling **5,** 432–441.

Lifshitz, A. & Frenklach, M. (1975). J. Phys. Chem. **79,** 686–692.

Lifshitz, A. & Frenklach, M. (1980). Int. J. Chem. Kinet **12,** 159–168.

Lifshitz, A., Scheller, K., & Burcat, A. (1973). *Recent Developments in Shock Tube Research, Proc. 9th Int. Shock Tube Symp.*, Stanford Univ. Press, Standord, pp. 690–699.

Lifshitz, A. *et al.* (1975). Int. J. Chem. Kinet. **7,** 753–773.

Linnik, Yu. V. (1961). *Method of Least Squares and Principles of the Theory of Observations*, Edited by N. L. Johnson, Pergamon Press, New York.

McKeown, J. J. (1980). In *Numerical Optimization of Dynamic Systems*, Edited by L. C. W. Dixon and G. P. Szegö, North-Holland, Amsterdam, pp. 349–362.

McRae, G. J., Tilden, J. W., Seinfeld, J. H. (1982). Comp. Chem. Eng. **6,** 15–25.

Miller, D. & Frenklach, M. (1983). Int. J. Chem. Kinet. **15,** 677–696.

Oppenheim, A. K. *et al.* (1975). 15th Symposium (International) on Combustion, The Combustion Institute, Pittsburgh, pp. 1503–1513.

Rabitz, H., Kramer, M., & Dacol, D. (1983). Ann. Rev. Phys. Chem. **34,** 419–461.

Rogers, D. & Skinner, G. B. (1981). Int. J. Chem. Kinet. **13,** 741–753.

Seber, G. A. F. (1977). *Linear Regression Analysis*, Wiley, New York.

Scigneur, C., Stephanopoulos, G., & Carr, R. W., Jr. (1982). *Chem. Eng. Sci.* **37,** 845–853.

Singh, S. & Rao, M. S. (1981). Indian Chem. Engr. **23,** 19–26.

Chapter 8

Thermochemical Data for Combustion Calculations

Alexander Burcat

1. Introduction

The thermodynamic and thermochemical properties of molecules, radicals, and atoms are involved one way or another in almost every computational aspect of combustion science. In most chemical kinetics and equilibrium calculations, such as kinetics of reactions behind shock waves or in nozzle flow, adiabatic flame calculations, and detonation processes, to mention a few, these thermodynamic and thermochemical properties must be found at a number of temperatures, usually determined by an automatic iteration process.

There are two ways to handle the task of representing properties of individual species. In the first one the properties of the substances are provided in a tabular form at predetermined temperature intervals and the values needed are calculated by interpolation. This method requires large memory storage, handling thousands of thermodynamic values, and does not permit use beyond the temperature limits of the table. A second and more commonly used technique is representation of the properties of each species by polynomials that allow direct calculation of the thermodynamic properties at any temperature, including limited extrapolation beyond the fitted range of the polynomial.

Polynomials as discussed in this chapter may also include exponential and logarithmic forms.

2. The polynomial representation

We differentiate between two kinds of properties, thermodynamic and thermochemical. The three thermodynamic properties, C_P°—the heat capacity at constant pressure, H°—the enthalpy, and S°—the entropy, can be calculated directly from molecular spectroscopic data using statistical mechanics equations. (The superscript $^\circ$ denotes the ideal gas standard state of 1 atm pressure; the amount of matter is taken as one mole. To conform to the usage in common tables of properties, we use subscript T to denote the value of a property at temperature T.) The three properties are interrelated by

$$H_T^\circ = H_{T\,\mathrm{ref}}^\circ + \int_{T\,\mathrm{ref}}^{T} C_P^\circ \, dT \tag{1}$$

$$S_T^\circ = S_{T\,\mathrm{ref}}^\circ + \int_{T\,\mathrm{ref}}^{T} C_P^\circ \, d\ln T \tag{2}$$

and by definition the Gibbs free energy is derived from them by

$$G_T^\circ = H_T^\circ - TS_T^\circ \tag{3}$$

The other properties are thermochemical ones, those which take cognizance of the chemical reactions undergone by the substance. The basic thermochemical property is the standard heat of formation ΔH_f°, which determines the heat balance when one mole of the substance is formed in its standard state from its constituent elements in their standard states. The heat of formation is used to calculate the standard Gibbs free energy of formation, another thermochemical property

$$\Delta G_{fT}^\circ = \Delta H_{fT}^\circ - T\Delta S_{fT}^\circ \tag{4}$$

The thermochemical standard free energy of formation is calculated practically as a difference of the standard free energy of the compound minus the standard free energy of the constituent elements (McBride and Gordon, 1967)

$$\Delta G_{fT}^\circ = G_T^\circ\,(\text{compound}) - G_T^\circ\,(\text{elements}) \tag{5}$$

From Eq. (3) $\Delta G_{f0}^\circ = \Delta H_{f0}^\circ$. The free energy of formation is also related to the equilibrium constant of formation K_P

$$\Delta G_{fT}^\circ = -RT \ln K_P \tag{6}$$

(But note that

$$G_T^\circ \neq -RT \ln K_P \tag{7}$$

as there is no meaning to K_P except for a chemical reaction.)

Since the value of $H_{T\,\mathrm{ref}}^\circ$ in Eq. (1) is arbitrary, a convention is adopted by which

$$H_{T\,\mathrm{ref}}^\circ = \Delta H_{fT\,\mathrm{ref}}^\circ \tag{8}$$

Thus, engineers refer to a thermochemical rather than a thermodynamic property through the definition

$$H_T^\circ = \Delta H_{f\,T\,\mathrm{ref}}^\circ + \int_{T\,\mathrm{ref}}^{T} C_P^\circ\, dT \tag{9}$$

usually called the "absolute enthalpy" and sometimes called "sensible enthalpy." (Its notation in the Russian literature on thermochemical properties has different symbols. Thus, the thermochemical enthalpy is designated as I_T. But in Eq. (3), the enthalpy that defines G_T° is known only if it is defined through Eq. (9). Therefore, G_T° becomes a thermochemical property also, but still not the ΔG_{fT}° defined by Eq. (4). In the Russian literature it is designated Z_T° to differentiate it from the thermodynamic property.)

Thermodynamic polynomials found their use in engineering practice long ago. Engineers prefer to start with polynomial representations of C_P, essentially because this enables calculation of other properties through simple integration. The use of C_P polynomials is thus found quite commonly in engineering thermodynamics textbooks (e.g., Holman, 1974; Wark, 1977). A number of papers are devoted to different techniques of determining and using C_P polynomials (Huang and Daubert, 1974; Parsut and Danner, 1972; Prothero, 1969; Reid, Prausnitz, and Sherwood, 1977; Tinh *et al.*, 1971; Thompson, 1977; Wilhoit, 1975; Yuan and Mok, 1968; Zeleznik and Gordon, 1960). The main problem with these polynomials is the accuracy with which they reproduce the original values of C_P and other values, particularly of enthalpy and entropy, obtained by integration. While other kinds of polynomials have been suggested and used (Thompson, 1977; Wilhoit, 1975), the most widely used one is the power series $a + bT + cT^2 + dT^3 + \cdots$. It was found long ago that if a large temperature range is covered by a single such polynomial, then even the original data is poorly reproduced. The fundamental reason for this is that the C_P function, and to a lesser extent the other thermodynamic properties, have a characteristic "knee" between 900 and 2000 K. Thus, a single polynomial may fit poorly to the changing slopes.

To overcome this difficulty Duff and Bauer (1962) proposed two different polynomials with overlapping ranges. Later, "pinned polynomials" were suggested by McBride and Gordon (1967) and Zeleznik and Gordon (1960). These polynomials are constrained to fit exactly the C_P value at a temperature that is an endpoint of one polynomial and a starting point to the second, and to give equal values of the other thermodynamic functions at that temperature. McBride and Gordon preferred 1000 K as the common temperature, while Prothero (1969) argues that 2000 K is a better choice.

While most authors (Parsut *et al.*, 1972; Reid *et al.*, 1977; Tinh *et al.*, 1971; Yuan *et al.*, 1968) fit C_P values only and use the C_P polynomial to calculate the rest of the thermodynamic properties, McBride *et al.* (1967) and Zeleznik and Gordon (1960), suggested the simultaneous least squaring of three properties, C_P°, S_T°, and $H_T^\circ - H_{\mathrm{ref}}^\circ$. This procedure generally gives better reproducibility of all the properties and low deviation. For example, the typical fourth-degree least-squares fit to C_P values gives a fit with a maximal error of around 4% in the 1000–5000 K range. In the 300–1000 K range the fitting is easier, normally with lower error percentages; the rest of the thermodynamic properties calculated by integration of the C_P values will have a still lower percent error. Simultaneous least squaring of three properties, however, causes the maximal error of C_P to be the same as before but the other thermodynamic properties to be fit with usually half the error of the C_P-alone method.

The polynomials from simultaneous fitting are known as the NASA thermodynamic polynomials because of their use in a variety of NASA computer programs (Bittker and Scullin, 1972; 1984; Gordon and McBride, 1971; McLain and Rao, 1976; Svehla and McBride, 1973). In Appendix A a program written according to Zeleznik and Gordon (1961) to calculate NASA polynomials by simultaneous least squaring is presented.

The NASA polynomials are usually fitted in the temperature range 300 to 5000 K. The reason for choosing this range is practical. Combustion calculations require thermodynamic and thermochemical properties between room temperature and 3000 or (for special fuels or detonations) 4000 K. In the course of automatic calculations, as well as in some exotic conditions such as spaceship reentry, knowledge of properties to 6000 K is required. Thus, the polynomials discussed here follow the bulk of existing tables (such as JANAF and TSIV as discussed later) by being fit in the range 300–5000 K. Extrapolation to 6000 K is easily done with little error. Extrapolation below 300 K, seldom needed in combustion research, is less accurate. In some cases the polynomials were fit up to 3000 K only.

3. Extrapolation

Extrapolating a polynomial outside the temperature range where it was fitted calls for caution. Not only does the uncertainty increase, but the curve often deviates in a direction opposite to the normal trend. This may be a serious drawback, since most thermodynamic data tabulations, as discussed in the next section, cover temperature ranges too low for combustion calculations. Thus, when these data have to be extrapolated to higher temperatures, the polynomials usually used for this purpose may give improper results (see Fig. 1).

To overcome this inconvenience, different types of polynomials have been developed. The basic concept is to force the C_P° curve to approach asymptotically the correct classical upper value $C_P^\circ(\infty)$. Although this offers problems for molecules with hindered internal rotations and electronically excited species (radicals and ions), it is satisfactory for most molecules, and even for the exceptional species it provides an adequate approximation.

One method, proposed by Wilhoit (1975), uses the following fitting function for the heat capacity

$$C_P^\circ = C_P^\circ(0) + [C_P^\circ(\infty) - C_P^\circ(0)]y^2\left[1 + (y-1)\sum_{i=0}^{n} a_i y^i\right] \qquad (10)$$

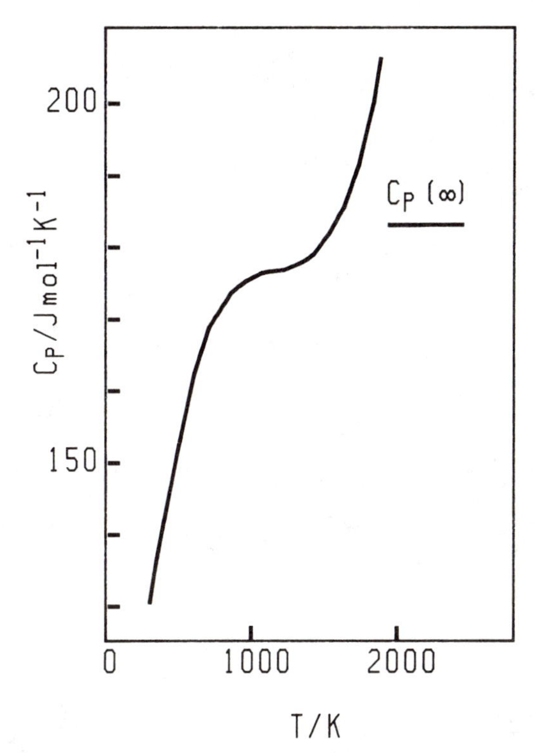

Figure 1. Heat capacity polynomial fit for $CCl_2F\text{-}CF_2Cl$ (Freon 113) from the data compilation of Reid, Prausnitz, and Sherwood (1977) showing the unrealistic upwards curvature above 1300 K. Data points were fitted up to 1500 K.

where $y = T/(T + B)$ varies from 0 to 1. The symbol B represents a scaling factor that determines how rapidly C_P° approaches $C_P^\circ(\infty)$. For most polyatomic molecules it is in the range 300–1000 K, with small molecules near the high end and larger ones in the 300–500 K range. $C_P^\circ(0)$ is the low-temperature limit heat capacity, $(7/2)R$ for linear molecules (except hydrogen) and $4R$ for nonlinear molecules; $C_P^\circ(\infty)$ is the classical high-temperature value, $(3N\text{-}3/2)R$ for linear molecules and $(3N\text{-}2)R$ for nonlinear ones; and N is the number of atoms in the molecule.

The coefficients a_i can be calculated by a least-squares fit of the function

$$\frac{C_P^\circ - C_P^\circ(0) - y^2[C_P^\circ(\infty) - C_P^\circ(0)]}{[C_P^\circ(\infty) - C_P^\circ(0)]y^2(y-1)} = \sum_{i=0}^{n} a_i y^i \tag{11}$$

Heat capacity values and an assumed B are used to compute the left-hand side of Eq. (11) for a range of y values. The results for different values of B are computed and B modified until the best agreement is obtained. The enthalpy

Table 1. Parameters for Wilhoit Polynomials.

Compound	Range (K)	$C_P(\infty)/R$	10^3 RMSD	B(K)	I/R(kK)	J/R	a_0	a_1	a_2	a_3	Notes
O_2	50–3000	4.5	22	1500	−167.575	−2.59	8.594	−84.33	166.26	−107.80	a
H_2	50–3000	4.5	38	500	−86.682	−5.50	4.280	−126.19	326.53	−234.56	a
H_2O	50–3000	7	7	1000	−25.868	−20.84	0.502	6.59	−24.64	14.30	a
F_2	50–3000	4.5	35	700	−16.719	−2.20	6.755	−52.84	79.98	−39.54	a
HF	50–3000	4.5	15	1000	−47.191	−5.98	−1.090	14.57	−19.71	2.60	a
Cl_2	50–3000	4.5	25	700	77.789	0.06	7.966	−111.71	252.84	−168.06	a
HCl	50–3000	4.5	13	1000	−14.11	−4.39	−1.292	23.63	−60.54	34.16	a
N_2	50–3000	4.5	14	1000	37.957	−3.80	−0.745	24.94	−84.07	63.37	a
NO	50–3000	4.5	94	500	25.041	−1.57	−52.396	292.64	−507.88	275.32	a, b
NO_2	50–3000	7	27	1000	−143.24	−14.79	5.708	−61.55	116.27	−69.34	a, b
N_2O	50–3000	7	45	700	−26.207	−42.76	3.684	−39.48	70.03	−38.03	a
N_2O_4	50–3000	16	88	500	58.984	−65.66	−8.828	27.33	−43.36	24.44	a, b
NH_3	50–3000	10	19	1000	−24.11	−41.14	2.447	−12.95	10.35	−2.55	a
CO	50–3000	4.5	20	1500	135.959	−3.48	3.694	−15.32	−22.26	44.28	a
CO_2	50–3000	7.5	28	700	−40.905	20.80	4.520	39.40	66.77	36.12	a
CH_4	60–3000	13	53	1500	291.547	−66.23	5.017	−46.20	76.38	−40.15	c
CH_3F	100–1500	13	31	500	36.772	−52.06	−0.195	14.72	−41.24	26.03	d
CH_3Cl	100–1500	13	37	500	22.202	−50.79	1.444	1.84	−16.82	12.33	e
CH_3NO_2	298–1000	19	3	335	1.23390	−78.377	0.8484	1.291	−6.719	2.673	f
CH_3OH	100–1500	16	45	500	67.414	−69.72	−3.042	23.35	−49.21	28.71	g
C_2H_2	50–2500	9.5	84	700	−149.907	−42.36	4.480	−44.45	88.47	−55.17	h
C_2H_4	50–1500	16	62	500	34.557	−71.31	1.23	3.46	−20.04	14.28	i
C_2H_5F	100–1500	22	43	500	39.983	−104.68	−0.125	2.71	−12.87	9.00	j
C_2H_5Cl	100–1500	22	42	500	27.789	−103.65	−0.023	0.04	−6.79	5.27	j
C_2H_5OH	273–1000	25	15	500	54.945	−121.66	0.021	0.23	−6.34	4.93	a, e
$C_2H_5NO_3$	298–1000	31	5	300	−0.709	−139.90	1.0257	−4.4261	5.702	−4.287	f
CCl_2FCF_2Cl	298–1000	22	25	500	197.499		−16.91	54.94	−83.20	46.70	k
C_2H_6	50–1500	22	35	500	0.111	108.26	0.682	0.94	−8.60	4.98	l
C_3H_6	50–1500	25	42	500	4.959	−123.55	−0.094	1.09	−7.24	4.32	i
C_3H_8	50–1500	31	94	500	−12.832	−161.14	0.077	−0.76	−2.59	1.12	l
C_3H_6O(m)	273–1000	28	15	500	63.804	−140.04	−1.866	8.70	−19.72	11.74	a
C_3H_8O(n)	273–1000	34	23	500	106.470	−175.26	−2.419	10.35	−21.42	12.63	a, e

C_3H_8O(o)	273–1000	34	47	500	50.803	−176.12	0.014	−3.11	1.16	0.48	a, e
1-n-C_4H_8	298–1500	34	15	500	−11.629	−176.31	0.3198	−3.942	3.079	−1.583	a
n-C_4H_{10}	50–1500	40	195	500	52.689	−215.88	−3.029	12.30	−22.62	11.82	e
i-C_4H_{10}	50–1500	40	143	500	−1.783	−216.21	−0.556	−0.10	−2.51	1.15	e
1-n-C_5H_{10}	298–1500	43	10	350	1.374	−216.22	0.7411	−3.058	6.754	−6.368	a
1-n-C_5H_{12}	200–1500	49	100	350	42.984	−255.18	−4.2985	20.66	−32.005	14.108	a
1-n-C_6H_{12}	298–1500	52	15	300	13.846	−259.39	−0.03395	−0.1838	2.63	−4.248	a
n-C_6H_{14}	200–1000	58	60	350	170.074	−306.40	−5.785	28.76	−47.74	24.30	a
1-n-C_7H_{14}	298–1500	61	25	300	21.795	−308.81	−0.5105	1.49876	0.283	−3.053	a
n-C_7H_{16}	200–1000	67		350	227.261	−358.23	−6.212	30.58	−50.97	26.36	a
1-n-C_8H_{16}	298–1500	70	25	300	28.141	−358.06	−0.7283	2.114	−0.481	−2.66	a
n-C_8H_{18}	298–1000	76	25	335	15.981	−400.25	−0.7887	1.95133	−1.1749	−1.8469	f
i-C_8H_{18}	200–1000	76	35	350	−108.576	−407.60	−0.347	−0.29	2.33	−4.43	a
cis-$C_{10}H_{18}$	298–1000	82	25	500	183.861	−470.17	1.193	−5.322	−1.15	3.80	p
trans-$C_{10}H_{18}$	298–1000	82	45	500	127.054	−471.33	0.4636	−1.39	−8.01	7.702	p
c-C_5H_{10}	300–1500	43	20	500	−87.506	−230.86	4.426	−18.96	21.59	−9.78	a
c-C_6H_{12}	298–1500	52	15	400	−14.663	−278.86	0.442	3.96	−15.58	9.05	a
c-C_5H_8(w)	298–1500'	37	20	500	−39.257	−193.87	4.155	−16.95	17.07	−6.66	a
c-C_6H_{10}(x)	298–1500	46	15	500	−105.678	−249.28	4.059	−21.67	28.71	−13.91	a
C_6H_6(q)	273–1500	34	15	500	−113.424	−177.84	5.932	−30.19	41.04	−20.00	a
$C_{10}H_8$	50–1500	52		1000	−2658.86	−320.92	−0.959	−32.67	83.39	−59.78	r
C_7H_8(s)	273–1500	43	20	500	−103.336	−230.22	2.846	−16.34	20.72	−10.08	a
C_6H_5OH(t)	50–1500	37	35	700	367.423	−205.54	1.792	−28.06	53.40	−31.56	p
$(C_6H_5)_2$(u)	298–1000	64		500	−47.371	−356.03	2.491	−17.42	22.81	−10.12	f

[a] API tables.
[b] Chao *et al.* (1974).
[c] McDowell and Kruse (1963).
[d] Rogers *et al.* (1977).
[e] Chen *et al.* (1975).
[f] Stull *et al.* (1969); reconstructed enthalpy is H_T-H_{298}.
[g] Chen *et al.* (1977).
[h] Chao (unpublished).
[i] Chao and Zwolinsky (1975).
[j] Chao *et al.* (1974).
[k] Freon 113; Reid *et al.* (1977); entropy not available.
[l] Chao *et al.* (1973).
[m] Acetone.
[n] 1-Propanol; Wilhoit and Zwolinsky (1973).
[o] 2-Propanol; Wilhoit and Zwolinsky (1973).
[p] Decalin; Miyazawa and Pitzer (1958).
[q] Benzene; API tables.
[r] Napthalene; Chen *et al.* (1979).
[s] Toluene; API tables.
[t] Phenol; Kudchadker *et al.* (1978).
[u] Biphenyl; Stull *et al.* (1969).
[w] Cyclopentene.
[x] Cyclohexene.

and entropy are obtained by integration of the C_P equation

$$\frac{H_T^\circ - H_0^\circ}{T} = \frac{I}{T} + C_P^\circ(0) - [C_P^\circ(\infty) - C_P^\circ(0)]\left(2 + \sum_{i=0}^{n} a_i\right) \times \left[y/2 - 1 + \left(\frac{1}{y} - 1\right)\ln\frac{T}{y}\right] + y^2 \sum_{i=0}^{n} \frac{y^i}{(i+2)(i+3)}\left(\sum_{j=1}^{n} f_{ij} a_j\right) \tag{12}$$

in which $f_{ij} = 3 + j$ for $i = j$ and $f_{ij} = 1$ for $j > i$. I is an integration constant set by comparing an enthalpy value given in the literature at some selected temperature with that obtained by integration of C_P°. The entropy is calculated from

$$S_T^\circ = J + C_P^\circ(\infty)\ln T - [C_P^\circ(\infty) - C_P^\circ(0)] \times \left[\ln y + \left(1 + y\sum_{i=0}^{n} \frac{a_i y^i}{2+i}\right) y\right] \tag{13}$$

J is an integration constant set by comparing an entropy value given in the literature at some selected temperature to that obtained by integration of C_P°.

Table 1 shows some polynomial coefficients obtained by Wilhoit (1975) and the present author. (Additional ones can be obtained from a data bank at the Thermodynamic Research Center (TRC), Texas A & M University, College Station, Texas.) In Appendix B a program is given for evaluating the a_i, B, I, and J.

Thompson's method (1977) provides a different polynomial based on similar principles. It also approaches $C_P(\infty)$ asymptotically. It has fewer parameters and is therefore less accurate, as noted by Thompson himself. Yuan and Mok (1968) propose extrapolation of data using the formula $C_P^\circ = A + B\exp(-C/T^n)$. The four parameters to be fit are A, B, C, and n, where n differs from unity only when data to 6000 K are available.

4. Thermochemical data sources

In order to calculate polynomial coefficients, the thermodynamic and thermochemical property values must be found in tables or calculated from molecular properties by the methods of statistical thermodynamics. The available tables will be described in the following section.

A popular and often quoted source is the "JANAF Thermochemical Tables" (1971), prepared by a team of thermodynamicists at the Dow Chemical Company headed by D. R. Stull and M. W. Chase. This source has thermochemical values for small, mostly inorganic molecules, radicals, and ions up to 6000 K. The calculations are mainly done using the rigid rotator harmonic oscillator (RRHO) approximation. Values for some species are calculated using anharmonic corrections. Four additional groups of tables were issued since the main volume was published in 1971, with both additions and revisions (JANAF, 1974; 1975; 1978; 1982).

Most of the JANAF tables have been fitted to polynomials by the NASA thermodynamics group. Polynomial coefficients for the species included in the main JANAF volume (1971) have been published (Gordon and McBride, 1971) and are supplied with the NASA SP-273 computer program.

A second source comparable to the JANAF tables published in the Soviet Union by Gurvitch *et al.* (1967) is not well known in the Western hemisphere. Gurvitch *et al.* published with their tables a set of nine-term polynomials that cover the temperature range 0–5000 K. This set of polynomials was reprinted in a later publication of Alemasov *et al.* (1974). Unfortunately, the explanation of how to use these polynomials is not clear, making their use guesswork.

The latest edition of the "TSIV" publication of Gurvitch *et al.* (1978) has gained much prestige although only the Russian version is as yet available. The spectroscopic data are well documented and a good discussion of the thermochemical data is provided. The tables list C_P°, $H_T^\circ - H_0^\circ$, $\phi = (G_T^\circ - H_0^\circ)/T$, S_T°, and $\log K_P$. However, K_P is calculated for equilibrium with gaseous atoms rather than reference elements as in most sources (JANAF, 1971; 1974; 1975; 1978; Amer. Petrol Inst.; Stull, Westrum, and Sinke, 1969). ΔH_f° is given only at 0 and 298. 15 K. A seven-coefficient polynomial is fitted for ϕ only in the temperature range 500–6000 K. This polynomial is unsuitable for most combustion research since the room-temperature end is missing and extrapolation into this range introduces a large error.

A publication devoted to stable organic molecules is the API-TRC Project 44. This is a reliable thermodynamic data source calculated either by statistical mechanics formulas or by approximation methods developed by Pitzer, Rossini, and their co-workers over the years. The main problems of this source are the following: First, it covers temperatures up to 1000 K only (for some older calculations up to 1500 K). In most cases this range is too low for combustion calculations; Wilhoit's extrapolation method is recommended if these tables are to be used at higher temperatures. Also, this source, owing to its high cost, is not available in many libraries. Where it is available it is usually not complete, and because of its complex indexing system, updating the looseleaf supplements and finding the desired table are difficult. In addition, it is almost impossible to determine which data sources were used for the calculation of thermochemical values, and the spectroscopic values used are not quoted as they are in the JANAF tables. (In 1984 the first n-alkyl radical tables of thermodynamic properties were published by the API-TRC Project).

Most of these drawbacks were corrected by Stull, Westrum, and Sinke (1969). In a regular-size volume the authors have included the ideal gas thermodynamic functions of the API Project 44 as well as others that were not included in this publication. It contains data for 918 compounds. All values are given, in the JANAF format, for the temperature range 298–1000 K. A complete list of references and comments regarding the calculations and the validity of the spectroscopic and thermal data are appended to each table. It contains an additional data table for organic molecules on which limited thermodynamic data is available.

Reid, Prausnitz, and Sherwood (1977) provide ideal gas C_P data for 468 organic molecules as four-term polynomials. The valid temperature range of these polynomials is not mentioned, nor is the data source, presumably mostly Thinh *et al.* (1971). The absolute enthalpy polynomial (probably in the range 273–1000 or 298–1500 K) can also be found since the value of ΔH°_{f298} is mentioned. The entropy cannot be calculated since the integration constant is not given. However, as mentioned earlier, these polynomials are useless anyway for high-temperature combustion calculation since they cannot be extrapolated far outside the range where they were fitted.

When making combustion calculations one usually needs in addition to the thermochemical properties of the stable organic molecules those of organic radicals or even ions. Unfortunately, in this domain there are very few sources of data.

Duff and Bauer (1961) calculated some thermodynamic properties of complex organic radicals. The values were given only in polynomial form because of the uncertainties in the data obtained. As mentioned earlier, the values were fitted to two polynomials with overlapping temperature ranges: 300–2000 K and 1500–6000 K. (Note: in the J. Chem. Phys. publication, Parts I and II of Table I were interchanged. The original report gives more details and is therefore recommended.) This publication should be considered as a pioneer work, and many of the spectroscopic values recommended in it are still valid today.

The only other publication in which calculations of the thermodynamic properties of a large number of organic radical species are reported is the NASA report of G. S. Bahn (1973), which includes many serious mistakes. These tables were later published as coefficients for NASA-type polynomials by Wakelyn *et al.* (1975). Although the approximation idea used by Bahn is interesting, ΔH°_{f0} values are sometimes used in place of ΔH°_{f298} values as if they were the same, which usually causes the "absolute enthalpy" values to be inconsistent; in most cases it is unclear to which of the possible isomers the calculated species belong or even whether the species mentioned is a chain or a ring compound; some values are simply wrong. It can be shown that these tables or polynomials may lead to serious mistakes if used in chemical kinetics or equilibrium calculations (Van Zeggeren and Storey, 1970).

In Appendix C a table of polynomial coefficients and ancillary information is given. Most of the entries are based on the most accurate data currently available. Some of them, although approximate, were included to provide better representation than the values given by Bahn (1973) or Wakelyn and McLain (1975).

When properties of deuterated species are needed, the monograph of Haar, Friedman, and Beckett (1961) is almost the only existing compilation. The Russian monographs (Gurvitch, 1967; 1978) present data for some deuterated and tritiated species, as do the recent JANAF supplement (1982) and a report of the author (Burcat, 1980). Thermodynamic properties for other isotopic species have to be searched for in the primary literature (Chen, Wilhoit, and Zwolinski, 1977; Chen, Kudchadker, and Wilhoit, 1979).

Finally, a word about statistical thermodynamic calculation methods. Most of the published tables used the rigid rotor harmonic oscillator (RRHO) approximation method. These calculations are accurate for most molecules up to 1500 K. The JANAF (1971) calculations were based mainly on the RRHO approximation. When values at temperatures above 3000 K are desired, however, the RRHO values are too low. Unfortunately, anharmonicity constants are still known only for very few molecules. Some publications do include values obtained using anharmonicity corrections (Burcat, 1980; McBride *et al.*, 1963; McDowell and Kruse, 1963). There are still uncertainties regarding the best way to calculate anharmonic corrections. McBride and Gordon (1967) discuss the alternatives, of which NRRAO2 appears to be the best.

5. Approximation methods

In many cases it may be found that the molecule of interest has not been spectroscopically investigated or that the spectroscopic knowledge is too limited to calculate the thermodynamic properties by ordinary statistical mechanics methods. Approximation methods have to be used.

Benson and Buss (1958) have classified the different approximations possible through additivity of properties. They called the roughest approximation possible, approximation of molecular thermodynamic property through summation of the thermodynamic properties of the individual atoms in the molecule, a "zero additivity rule." The first additivity rule is then summation of the properties of the bonds in the molecule. Graphical extrapolations and interpolations of thermodynamic properties based on the chemical formula, as done by Bahn (1973), fit in between the zero and first additivity rules.

The second additivity rule (also used in Benson, 1976; Benson and O'Neal, 1970; O'Neal and Benson, 1973) utilizes the contribution of groups of atoms. A second-order group is a central atom and its attached ligands, at least one of which must be polyvalent. Benson and co-workers evaluated a large number of group properties, mainly by averaging properties of many molecules or by using kinetic information and least-squares-fitting the values rather than relying on any single input. The properties of the entire molecule or radical are evaluated by summing up all the group contributions. In addition, some third-order corrections have to be added for specific molecules such as ring compounds or those having gauche or cis interactions or other steric factors. The cited publications include tables of thermodynamic and thermochemical properties. Most values are compiled or calculated for ΔH_f°, S°, and C_P° at 298 K. For some species the calculation of C_P was continued for temperatures up to 1500 K.

Benson's method, although supposedly simple and easy to use, turns out in practice to be so complex that few have discovered how to use it. It has been programmed so that properties of unknown molecules can be evaluated automatically for any temperature between 300 and 1500 K in a code named CHETAH, the ASTM Chemical Thermodynamic and Energy Release Evaluation Program (Seaton, Freedman, and Treveek, 1974). This program calculates the thermodynamic properties ΔH_f°, S°, C_P°, and the heat of combustion ΔH_c of the requested molecule as a side procedure. (Its main objective is evaluation of possible energy release of a given compound in order to characterize the relative hazard of the substance in industrial use.) Properties of radicals cannot be evaluated with this program; they have to be done separately (Benson, 1976, and the cited references above).

A higher quality approximation method for radicals was proposed by Forgeteg and Berces (1967). This method uses the spectroscopic assignment of the parent molecule, from which the vibrations relevant to the atom missing in the radical are deleted. Other vibrations may then be adjusted according to known ratios between bond lengths or force constants in the molecule and the radical. Thereafter the ordinary statistical mechanics formulas can be used to calculate the thermodynamic properties. Benson's additivity method or other estimates can be used to obtain the enthalpy of formation.

Reid *et al.* (1977) present an evaluation of most of the approximation methods available. These methods should only be used, however, as a last resort, after conventional methods have failed.

What should be done when needed thermodynamic properties cannot be found in one of the standard thermodynamic compilations?

First, the literature should be searched for spectroscopic data, and if these are sufficient, the thermodynamic properties can be calculated by statistical mechanics formulas. McBride and Gordon's program (1967) is recommended for this purpose. The latest verson, PAC3, includes, among many possible calculation methods, an accurate calculation method for internal rotation contributions, which are important when organic species are involved, and a subroutine which automatically calculates the coefficients of the NASA polynomials. Wilhoit's extrapolation method was recently included in the code.

Among the parameters that have to be known for the calculation of thermodynamic properties are the rotational constants. If these are unknown then the three principal moments of inertia of the molecule have to be calculated from assumed molecular bond lengths and angles. For this purpose there are two documented computer programs (Brinkmann and Burcat, 1979; Ehlers and Cowgill, 1964).

For radicals, chances are that no spectroscopic data can be found for the radical itself, and therefore the parent molecule has to be investigated. Afterwards, the procedure of Forgeteg and Berces (1967) can be applied as described above.

In case no spectroscopic information at all is available about a molecule, then the properties must be estimated by analogies or through Benson's method.

6. Thermochemical polynomials in combustion chemistry

Before dealing with the different uses of thermochemical polynomials in combustion, the relevance of using thermochemical properties for elevated temperatures must be discussed. The relevance of the enthalpy or entropy of an organic molecule at 4000 K may be questioned (Chase, Downey, and Syvernd, 1979) since these molecules can hardly exist even at 1000 K for more than a few milliseconds. Modern technical and experimental devices, however, require engineers and scientists to consider the behavior of fuel molecules that have been brought in microseconds or less from ambient temperature to 2000–3000 K or more. It is clear that at such elevated temperatures large molecules dissociate quickly. Some combustion processes, however, proceed on a microsecond time scale. Thus, although large molecules cannot survive a few seconds at combustion temperatures, they can survive a few microseconds, and for studying their behavior over these limited time scales their thermodynamic properties are important.

One of the simplest uses of thermodynamic polynomials is in the calculation of steady-flow gas properties, such as behind a shock wave, assuming "frozen chemistry." In this case the specific enthalpy of the gaseous mixture must be found for the calculation of the flow properties (pressure, temperature, Mach Number, density, etc.). To find the molar enthalpy of the mixture, the polynomial expressions for H_T° are multiplied by the mole fractions of each component and added. More sophisticated calculations are involved in studying the equilibrium chemistry of a flame or behind a shock wave. In this case the calculation of equilibrium composition is required, using one of the methods that will be discussed in the next section. To find the Chapman-Jouguet detonation velocity, the equilibrium chemistry of the mixture has to be evaluated as well as the specific enthalpy. The equilibrium composition is also needed for calculating adiabatic flame temperatures, exhaust properties of rocket engine nozzles, and stagnation processes in gasdynamic combustion lasers.

In kinetic modeling, one must calculate reverse rate constants of elementary reactions from forward rate constants. This is easily achieved from

$$k_f/k_r = K_c \tag{14}$$

The equilibrium constant calculated from thermochemical polynomials [Eq. (6)] is K_P. Misidentifying K_c with K_P can cause confusion (Golden, 1971) since $K_c = K_P$ only when the number of moles of products is equal to the number of moles of reactants i.e., when $\Delta n = 0$. When this is not the case

$$K_c = K_P(R'T)^{-\Delta n} \tag{15}$$

where $R' = 82.06\ \text{cm}^3\ \text{atm/mol K}$ or $0.08206\ \text{l atm/mol K}$. K_P is calculated from the free energy of reaction ΔG_r°

$$\Delta G_r^\circ = \sum G_T^\circ(\text{products}) - \sum G_T^\circ(\text{reactants}) \tag{16}$$

and

$$K_P = \exp(-\Delta G_r^\circ/RT) \tag{17}$$

or from the equilibrium constants for the formation of each substance participating in the reaction, as defined in Eq. (6), through

$$\ln K_P = \sum \ln K_P(\text{products}) - \sum \ln K_P(\text{reactants}) \tag{18}$$

Reviews of the different methods for calculation of equilibrium composition are given by Zeleznik and Gordon (1968), Van Zeggeren and Storey (1970), and by Alemasov (1974). Basically, there are two methods: those based on the calculation of the equilibrium constant K, and those based on the free energy minimization. Superficially, both would appear to be the same since the two properties are interconnected by Eq. (17). They differ in the method used to solve the system of nonlinear equations. Minimization of free energy is usually preferred for complex systems. The NASA program NASA SP-273 (Gordon *et al.*, 1971) was noted by the reviewers (Alemasov, 1974; Van Zeggeren and Storey, 1970) to be one of the best codes available for calculating equilibrium conditions in a variety of practical situations arising in combustion science. There are others.

7. Required accuracy of thermochemical information

Opinions about the accuracy of thermochemical data needed for combustion research are educated guesses rather than facts. Unfortunately, no systematic investigation has been carried out yet on this question.

The "well-determined species," such as those appearing in JANAF tables or API Project 44 tables or those published in the Journal of Physical and Chemical Reference Data, are calculated to the third or fourth digit after the decimal point. In the majority of cases the third digit after the decimal point can be well reproduced but the fourth is sometimes doubtful. However, for most calculations even the second digit after the decimal point is more than is ever necessary. Thus, C_P and S may be known to two digits after the decimal point while there is seldom any need for this degree of accuracy. The value of H_T°, however, is seldom known to better than 0.01 kJ while it would be preferable to know it to 0.001 kJ. Whether this type of accuracy is really needed in calculations is not clear, although general opinion holds that it is usually not necessary.

One property we would want to know to at least two digits after the decimal point is log K_P. The reason for this is that K_P is the property used to calculate back reaction rate constants when forward reaction rate constants are known or assumed. In chemical kinetics calculations this is of primary importance and a difference of one unit in the second digit of log K_P causes a 2% difference in the value of the calculated reverse reaction rate constant. On the other hand, in the calculation of shock tube experimental properties where reactive species are highly diluted in a noble gas or nitrogen, there is hardly any importance to knowledge of the exact properties of any species other than the diluent itself.

Even the thermochemical values for "well-determined species," such as those appearing in JANAF tables, are changed from time to time as physical constants or atomic weights change or mistakes are found. This should concern combustion science only if the changes affect the second digit after the decimal point. On the other hand, it is important to use reliable sources as much as possible, and if there is a choice not to mix different thermodynamic sources since they tend to differ at temperatures higher than 3000 K, where anharmonicity effects may be important.

One must also realize that except for elements, stable compounds which burn cleanly in oxygen, and small molecules or radicals whose electronic spectra in dissociative regions have been thoroughly analyzed, standard enthalpies of formation must be evaluated through difficult experiments subject to interpretive difficulties. This means that the largest errors—by far—in thermochemical calculations arise from uncertain enthalpies of formation. Fortunately, it is a fairly straightforward matter to modify the polynomial representations to change enthalpies of formation when new experimental results become available or sensitivity checks on ΔH_{f298}° are to be made (cf. comments in Appendix C).

There are also some technical uncertainties about whether values calculated by the RRHO method and values calculated including anharmonicity corrections should be used in the same calculations. These uncertainties are under consideration by the JANAF, NBS, and NASA thermodynamics teams. Special care should also be taken when dealing with species such as CH_3, CH_4, CD_4, CD_3H, CD_2H_2, and CDH_3 at temperatures beyond 3000 K. The mixed isotopic species have been calculated in this compilation using only the RRHO method, while CH_3, CH_2, etc., were calculated using RRHO plus electronic excitation (Appendix C). On the other hand, CH_4 and CD_4 were calculated using both anharmonic and simple RRHO methods. It is the author's recommendation to use together species that were calculated by the same method. CD_3, etc., should be used together with CH_4 and CD_4 of the anharmonic method, while CD_3H, CD_2H_2 and CDH_3 should be used with CH_4 and CD_4 calculated by the simple RRHO method.

For the nonspecialist user of Appendix C who wants to know by which method a species was calculated, the following should be noted:

(1) Anharmonic corrections were used if anharmonicity constants are given together with the other molecular constants;
(2) RRHO formulas plus electronic excitation contributions were used if electronic levels are given;
(3) In all other cases the calculations were done using the RRHO formulas.

As a closing word of caution it should be noted that the values of the polynomial coefficients generally have no meaning by themselves and that virtually the same property values can be generated by different sets of coefficients. Hence, judgment about the actual values of the coefficients is usually not possible, but it is possible to judge how well they collectively represent the thermodynamic functions themselves.

8. Acknowledgment

The author is grateful to Professor W.C. Gardiner for his effort in bringing this chapter to press.

9. References

Alemasov, V.E. *et al.* (1974). *Thermodynamic and Thermophysical Properties of Combustion Products*, Vol. I., V.P. Glushko, Ed., Keter, Jerusalem.

American Petroleum Institute Research Project 44, Selected Values of Physical and Thermodynamic Properties of Hydrocarbons and Related Compounds. Thermodynamic Research Center, Texas A & M University, College Station, TX, (looseleaf sheets published in different years).

Astholz, D.C., Durant, J., & Troe, J. (1981). 18th Combustion Symposium, 855.

Bahn, G.S. (1973). Approximate Thermochemical Tables for Some C-H and C-H-O Species, NASA CR-1278.

Benson, S.W. & Buss, J.H. (1958). J. Chem. Phys. **29,** 546.

Benson, S.W., O'Neal, H.E. (1970). Kinetic Data on Gas Phase Unimolecular Reactions, NSRDS-NBS-21.

Benson, S.W. (1976). *Thermochemical Kinetics*, Wiley, New York.

Bittker, D.A. & Scullin, V.J. (1972). General Chemical Kinetics Computer Program for Static and Flow Reactions with Application to Combustion and Shock Tube Kinetics, NASA TN-D 6586. Revised (1984) NASA TP-2320.

Brinkmann, U. & Burcat, A. (1979). A Program for Calculating the Moments of Inertia of a Molecule, TAE Report 382, Technion, Haifa.

Brown, J. M. & Ramsey D. A., (1975). Canad. J. Phys. **53,** 2232.

Burcat, A. & Kudchadker, S. A. (1979). Acta Chimica Hung. **101,** 249.

Burcat, A., (1980). Ideal Gas Thermodynamic Functions of Hydrides and Deuterides, Part I, TAE Report No. 411, Technion, Haifa.

Burcat, A. (1982). Ideal Gas Thermodynamic Properties of C_3 Cyclic Compounds, TAE Report No. 476, Technion, Haifa.

Burcat, A., Miller, D., & Gardiner, W. C. (1983). Ideal Gas Thermodynamic Properties of C_2H_nO Radicals, TAE Report No. 504, Technion, Haifa.

Burcat A., Zeleznik, F.J. & McBride, B. (1984). Ideal Gas Thermodynamic functions of phenyl, deuterophenyl and biphenyl radicals. NASA Rept to be published.

Carnaham, B., Luther, H. A., & Wilkes, J. O. (1969). *Applied Numerical Methods*, Wiley, New York.

Chao, J., Wilhoit, R. C., & Zwolinksi, B. J. (1973). J. Phys. Chem. Ref. Data **2,** 427.

Chao, J. *et al.*, (1974). J. Phys. Chem. Ref. Data **3,** 141.

Chao, J., Wilhoit, R. C., & Zwolinski, B. J. (1974). Thermochimica Acta **10,** 359.

Chao, J. & Zwolinski, B. J. (1975). J. Phys. Chem. Ref. Data **4,** 251.

Chao, J. & Zwolinski, B. J. (1978). J. Phys. Chem. Ref. Data **7,** 363.

Chao, J., Wilhoit, R. C., & Hall, K. R. (1980). Thermochimica Acta **41,** 41.

Chase, M. W., Downey, J. R., & Syvernd, A. N. (1979). Evaluation and Compilation of the Thermodynamic Properties of High Temperature Species, in 10th Materials Research Symposium on Characterization of High Temperature Vapors and Gases, NBS SP-561, p. 1581, and discussion to this paper, p. 1595.

Chen, S. S., Wilhoit, R. C., & Zwolinski, B. J. (1975). J. Phys. Chem. Ref. Data **4,** 859.

Chen, S. S., Wilhoit, R. C., & Zwolinski, B. J. (1977). J. Phys. Chem. Ref. Data **6,** 105.

Chen, S. S., Kudchadker, S. A., & Wilhoit, R. C. (1979). J. Phys. Chem. Ref. Data **8,** 527.

Czuchajowski, L. & Kucharski, S. A. (1972). Bull. Acad. Pol. Sci., Ser. Sci. Chim. **20,** 789.

Dewar, M. J. S. & Rzepa, H. S., (1977). J. Mol. Struct. **40,** 145.

Draeger, J.A., Harrison, R.H. & Good, W.D. (1983). J. Chem. Thermo. **15,** 367.

Draeger, J.A. & Scott, D.W. (1981). J. Chem. Phys. **74,** 4748.

Duff, R. E. & Bauer, S. H. (1961). The Equilibrium Composition of the C/H System at Elevated Temperatures, Atomic Energy Commission Report, Los Alamos 2556.

Duff, R. E. & Bauer, S. H. (1962). J. Chem. Phys. **36,** 1754.

Duncan, J.L. & Burns, G.R. (1969). J. Molec. Spectr. **30,** 253.

Ehlers, J. G. & Cowgill, G. R. (1964). Fortran Program for Computing the Principal Moments of Inertia of a Rigid Molecule, NASA TN D-2085.

Forgeteg, S. & Berces, T. (1967). Acta Chimica Hung. **51,** 205.

Friedman, A. S. & Haar, L. (1954). J. Chem. Phys. **22,** 2051.

Golden, D. M. (1971). J. Chem. Ed. **48,** 235.

Gordon, S. (1970). Complex Chemical Equilibrium Calculations, in Kinetics and Thermodynamics in High Temperature Gases, NASA SP-239.

Gordon, S. & McBride, B. J. (1971). A Computer Program for Complex Chemical Equilibrium Compositions—Incident and Reflected Shocks and Chapman Jouguet Detonations, NASA SP-273.

Green, J. H. S. (1961). Trans. Faraday Soc. **57,** 2132.

Gurvitch L. V. *et al.* (1967). Thermodynamic Properties of Individual Substances," Handbook in 2 vols., 2nd Ed., Moskva, AN SSSR, 1962. Reports NTIS AD-659660, AD-659659, and AD-659679.

Gurvitch L. V. *et al.* (1978). Thermodynamic Properties of Individual Substances, Vols. 1, 2 and 3, Nauka Moskva (in Russian).

Haar, L., Friedman, A. S., & Beckett, C. W. (1961). Ideal Gas Thermodynamic Functions and Isotope Exchange Functions for the Diatomic Hydrides, Deuteides and Tritides, NBS Monograph 20.

Hitchcock, A. P. & Laposa, J. B. (1975) J. Mol. Spectr. **54,** 223.

Holman, J. P. (1974). *Thermodynamics*, 2nd Edn, McGraw Hill, New York.

Holtzclaw, J. R., Harris, W. C., & Bush, S. F. (1980). J. Raman Spectr. **9,** 257.

Huang, P. K. & Daubert, T. E. (1974). Ind. Eng. Chem. Process, Des. Develop. **13**(2), 193.

Huber, K.P., & Herzberg, G. (1979). *Molecular Spectra and Molecular Structure IV. Constants of Diatomic Molecules*, Van Nostrand, Toronto.

JANAF Thermochemical Tables (1971). Edited by D. R. Stull and H. Prophet, NSRDS NBS-37.

JANAF Thermochemical Tables (1974). Edited by M. W. Chase *et al.*; J. Phys. Chem. Ref. Data. **2** (Supplement 1), 1.

JANAF Thermochemical Tables (1975). Edited by M. W. Chase *et al.*; J. Phys. Chem. Ref. Data. **4,** 1.

JANAF Thermochemical Tables (1978). Edited by M. W. Chase *et al.*; J. Phys. Chem. Ref. Data. **7,** 793.

JANAF Thermochemical Tables (1982). Edited by M. W. Chase *et al.*; J. Phys. Chem. Ref. Data **11,** 695.

Kanazawa, Y. & Nukada, K. (1962). Bull. Chem. Soc. Jpn. **35,** 612.

Katon, J. E. & Lippincott, E. R. (1959). Spectrochim. Acta **11,** 627.

Kovats, E., Günthard, Hs. H., & Plattner, Pl. A. (1955). Helv. Chem. Acta. **27,** 1912.

Kudchadker, S. A. *et al.* (1978). J. Phys. Chem. Ref. Data **7,** 417.

Levin, I. W. & Pearce, R. A. R. (1978). J. Chem. Phys. **69,** 2196.

Lewis Research Center, NASA (unpublished).

McBride, B. J. *et al.* (1963). Thermodynamics Properties to 6000 K for 210 Substances, NASA SP-3001.

McBride, B.J. & Gordon, S. (1967). Fortran IV Program for Calculation of Thermodynamic Data, NASA TN-D, 4097.

McDowell, R. S. & Kruse, F. H. (1963). J. Chem. Eng. Data **8,** 547.

McLain, A. G. & Rao, C. S. R. (1976). A Hybrid Computer Program for Rapidly Solving Flowing or Static Chemical Kinetic Problems Involving Many Chemical Species, NASA TM-X-3403.

Miyazawa, T. & Pitzer, S. (1958). J. Am. Chem. Soc. **80,** 60.

Moore, C.B. & Pimentel, G.C. (1963). J. Chem. Phys. **38,** 2816.

Nielsen, J. R., El-Sabban, M. Z., & Alpert, M. (1955). J. Chem. Phys. **23,** 324.

O'Neal, H. E. & Benson, S. W. (1973). *Free Radicals*, Edited by J. K. Kochi, Chap. 17, Wiley, New York.

Pacansky, J. & Chang, J.S. (1981). J. Chem. Phys. **74,** 5539.

Pamidimukkala, K. M. Rogers, D., & Skinner, G. B. (1982). J. Phys. Chem. Ref. Data **11,** 83.

Parsut, C. A. & Danner, R. P. (1972). Ind. Eng. Chem. Process, Des. Develop. **11,** 543.

Prothero, A. (1969). Comb. Flame **13,** 399.

Reid, R. C., Prausnitz, J. M. & Sherwood, T. K. (1977). *The Properties of Gases and Liquids*, 3rd Ed., *et al.*, McGraw-Hill, New York.

Rodgers, A. S. *et al.* (1974). J. Phys. Chem. Ref. Data **3,** 141.

Seaton, W. H., Freedman, E., & Treveek, B. (1974). CHETAH—The ASTM Chemical Thermodynamic and Energy Release Evaluation Program, ASTM D51, American Society for Testing and Materials, 1916 Race Street, Philadelphia, PA 19103.

Shimanouchi, T. (1972). Tables of Molecular Vibrational Frequencies NSRDS-NBS-39.

Shimanouchi, T. (1974). J. Phys. Chem. Ref. Data **3**, 304.

Spangenberg, H. J., Borger, I., & Schirmer, W. (1974). Z. Phys. Chem. Leipzig **255,** 1.

Stull, D. R., Westrum, E. F., & Sinke, G. C. (1969). *The Chemical Thermodynamics of Organic Compounds*, Wiley, New York.

Svehla, R. A. & McBride, B. J. (1973). Fortran IV Computer Program for Calculation of Thermodynamic and Transport Properties of Complex Chemical Systems, NASA N-D 7056.

Swalen, J. D. & Herschbach, D. R. (1957). J. Chem. Phys. **27,** 100.

Tinh, T. P. *et al.* (1971). Hydrocarbon Processing **50**(1), 98.

Thompson, P. A. (1977). J. Chem. Eng. Data **22,** 431.

TSIV—see Gurvitch, L. V., *et al.* (1978).

Van Zeggeren, F. & Storey, S. H. (1970). *The Computation of Chemical Equilibria*, Cambridge University Press.

Wakelyn, N. T. & McLain, A. G. (1975). Polynomial Coefficients of Thermochemical Data for C-H-O-N Systems, NASA TM X-72657.

Wark, K. (1977). *Thermodynamics*, 3rd Edn., McGraw-Hill, New York.

Wilhoit, R.C. & Zwolinski, B.J. (1973). J. Phys. Chem. Ref. Data **2,** Supplement # 1.

Wilhoit, R. C. (1975). Thermodynamics Research Center Current Data News, Vol. 3, # 2 Texas A & M University.

Wurrey, C. J., Bucy, W. E., & Durig, J. R. (1977). J. Chem. Phys. **67,** 2765.

Yuan, S. C. & Mok, Y. I. (1968). Hydrocarbon Processing **47**(3), 133.

Yost, D. M., Osborne, D. W., & Garner, C. S. (1942). J. Am. Chem. Soc. **63,** 3492.

Yum, T. Y. & Eggers, D. F. (1979). J. Phys. Chem. **83,** 501.

Zeleznik, F. J. & Gordon, S. (1960). An Analytical Investigation of Three General Methods of Calculating Chemical Equilibrium Compositions, NASA TN D-473.

Zeleznik, F. J. & Gordon, S. (1961). Simultaneous Least-Square Approximation of a Function and its First Integrals with Application to Thermodynamic Data, NASA TN D-767.

Zeleznik, F. J. & Gordon, S. (1968). Ind. Eng. Chem. **60**(6), 27.

Appendix A

Program for Finding Coefficients of NASA Polynomials

The coding follows NASA TN D-767. The values of all thermochemical functions are forced to fit the input values at 1000 K. The units of the input data are interpreted to be those of the input value of the gas constant R. A sample input data deck is included at the end of the listing.

```
      PROGRAM NASA
      DIMENSION T(50),DELH(50),STO(50),CPO(50),KP(10),ITITLE(20),ARG(50)
     ..A(11,11),C(10),X(10),BB(50),CC(50),DD(50),ER1(50),ER2(50),ER3(50)
      COMMON X,R
C*****************************************************************
C     SET PINNING TEMPERATURE AT 1000 K AND TSCALE AT 5000 K.   *
C     NOTE THAT THE FOLLOWING LINE- AND MANY OTHERS LATER IN    *
C     THE PROGRAM - HAS TWO FORTRAN STATEMENTS SEPARATED BY A   *
C     DOLLAR SIGN.  IF THE FORTRAN COMPILER YOU ARE USING WILL  *
C     NOT RECOGNIZE THIS SEPARATOR, THEN YOU WILL HAVE TO GO    *
C     THROUGH THE ENTIRE PROGRAM AND PLACE THESE STATEMENTS     *
C     ON SUCCESSIVE LINES OF THE PROGRAM.                       *
C*****************************************************************
      TO=1000.  $ TSCALE=5000.
C***************************************************************************
C     THE NEXT CARDFUL OF 80 CHARACTERS MAY HAVE  ANY INFORMATION AT ALL.  *
C     IT WILL BE PRINTED AS AN IDENTIFIER OF THE RUN.                      *
C***************************************************************************
      READ(5,98) ITITLE
98    FORMAT(20A4)
C***********************************************************************
C     READ IN THE NUMBER OF TEMPERATURES AT WHICH DATA POINTS WILL BE  *
C     READ IN (M),  THE VALUE OF THE GAS CONSTANT TO BE USED (R) AND   *
C     THE ENTHALPY OF FORMATION AT 298 K (HF298) IN UNITS THAT ARE     *
C     1000 TIMES LARGER THAN THE UNITS OF THE GAS CONSTANT.  EXAMPLE - *
C     IF THE GAS CONSTANT IS IN JOULES, THEN THE ENTHALPY OF FORMATION *
C     MUST BE IN KILOJOULES.                                           *
C***********************************************************************
      READ(5,*) M, R, HF298
C***********************************************************************
C     READ IN THE MOLAR HEAT CAPACITY, ENTROPY AND (H - H298) VALUES   *
C     AT THE PINNING TEMPERATURE OF 1000 K.  VALUES OF (H - HO) MAY BE *
C     USED AS WELL - THERE WILL BE VERY SLIGHT DIFFERENCES IN THE      *
C     COMPUTED COEFFICIENTS.  IT IS ASSUMED THAT THE ENTHALPY VALUES   *
C     ARE IN UNITS THAT ARE 1000 TIMES LARGER THAN THE UNITS OF THE    *
C     HEAT CAPACITY AND THE ENTROPY.                                   *
C***********************************************************************
      READ(5,*) CPOATO,STOATO,DELHTO
C*************************************************
C     READ IN THE LIST OF TEMPERATURES, FORMAT-FREE.  *
C*************************************************
      READ(5,*) (T(I),I=1,M)
```

```
C**************************************************
C     READ IN THE LIST OF MOLAR HEAT CAPACITIES.  *
C**************************************************
      READ(5,*) (CPO(I),I=1,M)
C*******************************************
C     READ IN THE LIST OF MOLAR ENTROPIES.  *
C*******************************************
      READ(5,*) (STO(I),I=1,M)
C*************************************************************
C     READ IN THE LIST OF MOLAR (H-H298) OR (H-H0) VALUES.   *
C     IT IS ASSUMED THAT THE UNITS ARE 1000 TIMES LARGER     *
C     THAN THE UNITS USED FOR HEAT CAPACITY AND ENTROPY.     *
C*************************************************************
      READ(5,*) (DELH(I),I=1,M)
C***************************************************************
C     WRITE OUT THE TITLE SO THAT NO MATTER WHAT HAPPENS THERE *
C     WILL BE AN OUTPUT RECORD SHOWING WHAT WAS TRIED.         *
C***************************************************************
      WRITE(6,99) ITITLE
99    FORMAT(1H1,5X,20A4/////)
C****************************************************************
C     SCALING THE INPUTS.  NOTE THAT THE PINNING TEMPERATURE TO  *
C     HAS BEEN ASSUMED TO BE 1000 K.  IF THIS IS NOT THE CASE,  *
C     THEN 1000.  IN THE NEXT STATEMENT HAS TO BE REPLACED BY   *
C     THE VALUE OF THE PINNING TEMPERATURE CHOSEN.  THE ENTHALPY *
C     VALUES ARE ALSO BROUGHT TO A COMMON ENERGY UNIT WITH THE  *
C     OTHER VARIABLES AT THIS POINT.                             *
C****************************************************************
      TO=1000./TSCALE $ DELHTO=DELHTO*1000./TSCALE
      DO 48 I=1,M $ T(I)=T(I)/TSCALE
48    DELH(I)=DELH(I)*1000./TSCALE
C****************************************************************
C     CALCULATING THE COEFFICIENT MATRIX.  SEE NASA TN D-767 FOR *
C     EXPLANATION OF THIS MATRIX.                                *
C****************************************************************
      DO 21 I=1,10 $ C(I)=0.0
      DO 21 J=1,10
21    A(I,J)=0.0
      DO 22 I=1,M $ ARG1=ALOG(T(I)) $ A(1,1)=A(1,1)+(2.+ARG1*ARG1)
      A(1,2)=A(1,2)+(3./2.+ARG1)*T(I) $ T2=T(I)*T(I)
      A(1,3)=A(1,3)+(4./3.+0.5*ARG1)*T2 $ T3=T(I)*T2
      A(1,4)=A(1,4)+(5./4.+(1./3.)*ARG1)*T3 $ T4=T(I)*T3
      A(1,5)=A(1,5)+(6./5.+(1./4.)*ARG1)*T4 $ TINV=1./T(I)
      A(1,6)=A(1,6)+TINV $ A(1,7)=A(1,7)+ARG1 $ A(2,2)=A(2,2)+T2
      A(2,3)=A(2,3)+T3 $ A(2,4)=A(2,4)+T4 $ T5=T(I)*T4
      A(2,5)=A(2,5)+T5 $ A(2,7)=A(2,7)+T(I) $ T6=T(I)*T5
      A(3,5)=A(3,5)+T6 $ T7=T(I)*T6 $ A(4,5)=A(4,5)+T7 $ T8=T(I)*T7
      A(5,5)=A(5,5)+T8 $ A(6,6)=A(6,6)+1./T2 $ SS1=CPO(I)/R
      SS2=(DELH(I))/(R*T(I)) $ SS3=STO(I)/R $ C(1)=C(1)+SS1+SS2+SS3*ARG1
      C(2)=C(2)+(T(I))*(SS1+0.5*SS2+SS3)
      C(3)=C(3)+T2*(SS1+(1./3.)*SS2+0.5*SS3)
      C(4)=C(4)+T3*(SS1+0.25*SS2+(1./3.)*SS3)
      C(5)=C(5)+T4*(SS1+0.2*SS2+0.25*SS3)
      C(6)=C(6)+TINV*SS2 $ C(7)=C(7)+SS3
22    CONTINUE
      A(1,8)=1.0 $ A(1,9)=1.0 $ A(1,10)=ALOG(TO) $ SUM22=A(2,2)
      A(2,2)=(9./4.)*SUM22 $ SUM23=A(2,3) $ A(2,3)=(5./3.)*SUM23
      SUM24=A(2,4) $ A(2,4)=(35./24.)*SUM24 $ SUM25=A(2,5)
      A(2,5)=(27./20.)*SUM25 $ P=FLOAT(M) $ A(2,6)=P/2.
      A(2,8)=TO $ A(2,9)=TO/2.  $ A(2,10)=TO $ A(3,3)=(49./36.)*SUM24
      A(3,4)=(5./4.)*SUM25 $ SUM35=A(3,5) $ A(3,5)=(143./120.)*SUM35
      A(3,6)=(1./3.)*A(2,7) $ A(3,7)=0.5*SUM22 $ TOSQ=TO*TO
      A(3,8)=TOSQ $ A(3,9)=TOSQ/3.  $ A(3,10)=TOSQ/2.
      A(4,4)=(169./144.)*SUM35 $ A(4,5)=(17./15.)*A(4,5)
      A(4,6)=(1./4.)*SUM22 $ A(4,7)=(1./3.)*SUM23 $ TOQUB=TO*TOSQ
      A(4,8)=TOQUB $ A(4,9)=TOQUB/4.  $ A(4,10)=TOQUB/3.
      A(5,5)=(441./400.)*A(5,5) $ A(5,6)=0.2*SUM23
      A(5,7)=0.25*SUM24 $ TOFORT=TOSQ*TOSQ $ A(5,8)=TOFORT
      A(5,9)=TOFORT/5.  $ A(5,10)=TOFORT/4.  $ A(6,9)=1./TO
      A(7,7)=P $ A(7,10)=1.0 $ C(8)=CPOATO/R
      C(9)=(DELHTO)/(R*TO) $ C(10)=STOATO/R
      DO 70 I=1,10
      DO 60 J=1,10
60    A(J,I)=A(I,J)
70    CONTINUE
```

```
C*******************************************************
C       FORMING THE AUGMENTED MATRIX.  SEE NASA TN D-767 *
C       FOR AN EXPLANATION OF THIS MATRIX.               *
C*******************************************************
        DO 91 I=1,10
91      A(I,11)=C(I)
C*********************************************
C       SET UP AND CALL THE LINEAR EQUATIONS SOLVER *
C       SUBROUTINE SIMUL.                           *
C*********************************************
        N=10 $ EPS=1.E-10 $ CALL SIMUL(N,A,X,EPS,NER)
C**********************************************************
C       THE EQUATIONS HAVE BEEN SOLVED.  NOW RECONVERT THE  *
C       VALUES OF EVERYTHING BACK TO THE UNSCALED ONES THAT *
C       WE STARTED WITH, PRINTING OUT AS WE GO - -          *
C**********************************************************
        TO=TSCALE*TO $ DELHTO=TSCALE*DELHTO $ WRITE(6,1010)
1010    FORMAT(5X,'INPUT VALUES',/5X,12('-'))
        WRITE(6,45) M,R,HF298,CPOATO,DELHTO,STOATO
45      FORMAT(//1X,'NO.POINTS =',I5,4X,'GAS CONSTANT=',F12.1,5X,
       .'ENTHALPY OF FORMATION=',F11.3////1X,'CPO AT TO=',F14.3,
       .5X,'H-H298 AT TO=',F14.3,5X,'STO AT TO=',F14.3////)
C************************************************************
C       THERE MAY HAVE BEEN AN ERROR IN THE LINEAR EQUATIONS   *
C       SOLVER SUBROUTINE.  IF SO, JUMP OUT OF THE REST OF THE *
C       PRINTING AT THIS POINT.                                *
C************************************************************
        IF(NER.EQ.0)GO TO 222
C**********************************************************
C       THE LINEAR EQUATIONS SUBROUTINE FOUND SOME ANSWERS, *
C       SO NOW WE CAN GO AHEAD AND PRINT OUT THE RESULTS.   *
C**********************************************************
        WRITE(6,100)
100     FORMAT(//5X,'TEMP.',10X,'H-H298',10X,'PRED/CORRECT',
       .10X,'STO',10X,'PRED/CORRECT'/5X,5('-'),10X,6('-'),10X,
       .12('-'),10X,3('-'),10X,12('-'),10X,3('-'),10X,12('-'))
30      DO 8 I=1,M $ T(I)=TSCALE*T(I) $ DELH(I)=TSCALE*DELH(I)
8       CONTINUE
        X(2)=X(2)/TSCALE $ X(3)=(X(3))/(TSCALE**2)
        X(4)=(X(4))/(TSCALE**3) $ X(5)=(X(5))/(TSCALE**4)
C***************************************************************
C       SO FAR WE HAVE NOT ADJUSTED THE SCALE OF ENTHALPY TO FIX *
C       THE REFERENCE AT THE HEAT OF FORMATION AT 298 K, AND     *
C       THEREBY ADOPT THE SO-CALLED ABSOLUTE ENTHALPY SCALE.     *
C       THE NEXT STATEMENT DOES THIS ADJUSTMENT.                 *
C***************************************************************
        X(6)=TSCALE*X(6)+1000.*HF298/R
        X(7)=X(7)-X(1)*ALOG(TSCALE)
C*****************************************************************
C       CHECK NOW TO SEE HOW WELL THE POLYNOMIAL MATCHES THE DATA.  *
C*****************************************************************
        DO 9 I=1,M $ ER1(I)=CSUPO(T(I))/CPO(I)
        ER2(I)=DEHSO(T(I))/DELH(I) $ ER3(I)=STSUPO(T(I))/STO(I)
9       CONTINUE
C***************************************************************
C       PRINT A COMPARISON OF INPUT DATA TO THE COMPUTED VALUES.  *
C***************************************************************
        WRITE(6,101)(T(I),DELH(I),ER2(I),STO(I),ER3(I),CPO(I),
       .ER1(I),I=1,M)
101     FORMAT(/(1X,F12.1,3X,F13.4,7X,F12.4,5X,F13.4,5X,F12.4,5X,
       .F13.4,5X,F12.4))
        WRITE(6,102)
102     FORMAT(5X,100('-'))
C*******************************************************
C       PRINT OUT THE RESULTS OF THE FITTING PROCEDURE.  *
C*******************************************************
        WRITE(6,110)
110     FORMAT(///5X,'LEAST SQUARE COEFFICIENTS',/5X,25('-'))
        WRITE(6,112)(I,X(I),I=1,5)
112     FORMAT(5X,'A(',I1,')=',E15.8/)
        WRITE(6,113)(X(I),I=6,10)
113     FORMAT(5X,'A6=',E15.8//5X,'A7=',E15.8//5X,
       .'LAMBDAO=',E15.8//5X,'LAMBDA1=',E15.8//5X,'LAMBDA2=',E15.8)
C********************************************************
C       FOR EXPLANATIONS OF LAMBDAO, LAMBDA1, AND LAMBDA2, *
C       SEE NASA TN D-767.                                 *
C*******************************************************************
```

```
C     THE FOLLOWING STATEMENTS DEAL WITH ERROR MESSAGES FROM SIMUL.  *
C*******************************************************************
222   WRITE(6,102) $ WRITE(6,221)NER
221   FORMAT(5X,'NER=',I2)
      WRITE(6,102) $ WRITE(6,223)
223   FORMAT(1H1)
      WRITE (6,224) ITITLE,T(M)
C*****************************************************************
C     PRINT OUT THE NASA POLYNOMIAL COEFFICIENTS AGAIN IN A COUPLE *
C     OF USEFUL FORMATS.  YOU WILL NEED THESE FORMATS TO CONSTRUCT *
C     NASA-TYPE DATA CARDS HOLDING TWO SETS OF COEFFICIENTS, I.E.  *
C     ONE SET FOR 300 TO 1000 K AND ONE SET FOR ABOVE 1000 K.      *
C*****************************************************************
      WRITE (6,226) (X(I),I=1,7) $ WRITE (6,225) (X(I),I=1,3)
      WRITE (6,226) (X(I),I=4,7)
224   FORMAT(20A4,/F10.0)
225   FORMAT(30X,3E15.8)
226   FORMAT(5E15.8)
      STOP
      END



      SUBROUTINE SIMUL(N,A,X,EPS,NER)
C*****************************************************************
C     THIS IS A STANDARD LIBRARY SUBROUTINE FOR SOLVING SYSTEMS OF *
C     LINEAR EQUATIONS.  SEE YOUR COMPUTER CENTER LIBRARY FOR AN   *
C     EXPLANATION OF HOW IT WORKS.                                 *
C*****************************************************************
      DIMENSION IROW(10),JCOL(10),JORD(10),Y(10),A(11,11),X(10)
      MAX=N+1
5     DETER=1.
      DO 18 K=1,N $ KM1=K-1 $ PIVOT=0.
      DO 11 I=1,N
      DO 11 J=1,N $ IF(K.EQ.1)GO TO 9
      DO 8 ISCAN=1,KM1
      DO 8 JSCAN=1,KM1 $ IF(I.EQ.IROW(ISCAN))GO TO 11
      IF(J.EQ.JCOL(JSCAN))GO TO 11
8     CONTINUE
9     IF(ABS(A(I,J)).LE.ABS(PIVOT))GO TO 11
      PIVOT=A(I,J) $ IROW(K)=I $ JCOL(K)=J
11    CONTINUE
      IF(ABS(PIVOT).GT.EPS)GO TO 13 $ NER=0
      RETURN
13    IROWK=IROW(K)
15    NER=1
      JCOLK=JCOL(K)
      DO 14 J=1,MAX
14    A(IROWK,J)=A(IROWK,J)/PIVOT
      A(IROWK,JCOLK)=1./PIVOT
      DO 18 I=1,N
      AIJCK=A(I,JCOLK) $ IF(I.EQ.IROWK)GO TO 18
      A(I,JCOLK)=-AIJCK/PIVOT
      DO 17 J=1,MAX
17    IF(J.NE.JCOLK)A(I,J)=A(I,J)-AIJCK*A(IROWK,J)
18    CONTINUE
      DO 20 I=1,N $ IROWI=IROW(I) $ JCOLI=JCOL(I)
      JORD(IROWI)=JCOLI
20    X(JCOLI)=A(IROWI,MAX)
      RETURN
      END



      FUNCTION CSUPO(T)
C*************************************************************
C     COMPUTE THE MOLAR HEAT CAPACITY FROM THE POLYNOMIAL.   *
C*************************************************************
      COMMON X,R
      DIMENSION X(10)
      CSUPO=R*(X(1)+X(2)*T+X(3)*(T*T)
     .+X(4)*T**3+X(5)*T**4)
      RETURN
      END
```

```
      FUNCTION DEHSO(T)
C**********************************************************
C     COMPUTE THE ENTHALPY FUNCTION FROM THE POLYNOMIAL.   *
C**********************************************************
      COMMON X,R
      DIMENSION X(10)
      DEHSO=R*T*(X(1)+X(2)*T/2.+X(3)*(T*T/3.)
     .+X(4)*(T**3/4.)+X(5)*(T**4/5.)+X(6)/T)
      RETURN
      END

      FUNCTION STSUPO(T)
C*****************************************************
C     COMPUTE THE MOLAR ENTROPY FROM THE POLYNOMIAL.   *
C*****************************************************
      COMMON X,R
      DIMENSION X(10)
      STSUPO=R*(X(1)*ALOG(T)+X(2)*T+X(3)*(T*T/2.)
     .+X(4)*(T**3/3.)+X(5)*(T**4/4.)+X(7))
      RETURN
      END
```

SAMPLE DATA DECK. IT WAS PUT TOGETHER TO FIT THE TSIV THERMOCHEMICAL DATA FOR THE OD RADICAL FROM 1000 TO 2000 K, WITH THE VALUES OF ALL PROPERTIES AT 1000 K FIXED TO THE TSIV DATA.

```
TRIAL OF NASA PROGRAM FOR TSIV DATA ON OD RADICAL.
10, 8.314, 37.214
32.263,226.206,30.420
1100.,1200.,1300.,1400.,1500.,1600.,1700.,1800.,1900.,2000.
32.831,33.357,33.838,34.273,34.666,35.021,35.342,35.633,35.898,36.141
229.308,232.186,234.877,237.401,239.779,242.028,244.161,246.189,
248.123,249.971
33.675,36.985,40.345,43.751,47.198,50.683,54.201,57.750,61.327,64.929
```

NOTES ON THE PROGRAM -

1) IT IS NOT NECESSARY TO USE THIS PARTICULAR LINEAR EQUATIONS SUBROUTINE (SIMUL) TO SOLVE FOR THE COEFFICIENTS. IF THERE IS ALREADY ONE LIKE IT IN YOUR LIBRARY, YOU ONLY HAVE TO CHANGE THE CALL STATEMENT ACCORDINGLY IN ORDER TO USE THE SUBROUTINE YOU ALREADY HAVE.

2) IF THERE IS A CONVERGENCE DIFFICULTY, THE MOST LIKELY EXPLANATION IS THAT YOUR DATA POINTS ARE TOO FEW IN NUMBER OR TOO CLOSE TO A CURVE THAT CAN BE FIT VERY ACCURATELY WITH A POLYNOMIAL OF DEGREE LOWER THAN FOUR. TRY ADDING SOME MORE DATA POINTS, OR CHANGING SOME OF THEM SLIGHTLY.

3) IN ANY EVENT, THE ACCURACY OF THE POLYNOMIAL BETWEEN THE INPUT DATA POINTS IS VERY IMPORTANT IN AN ACTUAL APPLICATION, AND SO IF ACCURACY IS IMPORTANT TO YOU, BE SURE TO HAVE AS MANY POINTS FITTED AS YOU ARE WILLING TO COMPUTE INPUT DATA FOR.

4) AGAIN FOR PURPOSES OF ACCURATE FITTING, THE TEMPERATURE RANGE FITTED SHOULD NOT BE ANY WIDER THAN REQUIRED FOR THE INTENDED APPLICATION.

5) REMEMBER THAT THE COEFFICIENTS THEMSELVES HAVE NO MEANING - DO NOT TRY TO COMPARE COEFFICIENTS BETWEEN RUNS OF THIS PROGRAM ON DIFFERENT COMPUTERS. THE ONLY VALID COMPARISON IS OF THE MATCHES BETWEEN COMPUTED AND INPUT DATA.

Appendix B

Program Written by A. Lifshitz and A. Burcat for Evaluating the Coefficients of the Wilhoit Polynomials

The inputs include the number of atoms in the molecule and values of heat capacity, enthalpy and entropy at given temperatures. The heat capacity limits are computed from the number of atoms assuming a nonlinear molecule unless a linear one is specified in the input namelist. An optimum (to 50 units) value of B is found by minimizing the standard deviation of the match to the input data. The polynomial coefficients are obtained from the heat capacity fit. Using these coefficients the constants of integration I and J are found by comparing computed to input values of enthalpy and entropy.

For FORTRAN compilers that do not recognize the NAMELIST declaration, appropriate formatted input is required. For compilers that do not accept the dollar sign $ separator, the statements have to be moved to successive lines.

```
C   PROGRAM FOR EVALUATING THE COEFFICIENTS OF WILHOIT POLYNOMIALS.
C
C    CPO=7/2*R FOR LINEAR MOLECULES, 4*R FOR NONLINEAR MOLECULES.
C    CPI=(3*N-3/2)*R FOR LINEAR MOLECULES, (3*N-2)*R FOR NONLINEAR MOLECULES.
C
C   NAMELIST INPUT -
C
C   VARIABLE  DIM.     DEFAULT           DEFINITION
C    NAME      4        ' '          1 TO 16 CHARACTERS
C    R         1     1.987192        GAS CONSTANT
C    T        200       0.           TEMPERATURE, K
C    CP       200                    CP OR CP/R
C    H        200                    H-HO, H-H298, H-HO/RT, H-H298/RT, H,
C                                       H-HO/T, OR H-H298/T
C    S        200                    S OR S/R
C    CLABEL    1       'CP'          'CP' OR 'CP/R'
C    HLABEL    2        ' '          PAC LABEL FOR H INPUT
C    SLABEL    1       'S'           'S' OR 'S/R' DEPENDING ON S
C    LINEAR    1      FALSE          TRUE IF MOLECULE IS LINEAR
C    NATOMS    1                     NUMBER OF ATOMS IN MOLECULE
C    ID        2        ' '          1 TO 6 CHARACTERS TO BE PUT IN CC 1-6
C
```

```
      REAL NAME(4),ID(2)
      LOGICAL LINEAR
      DIMENSION M(15),L(15),HLABEL(2)
      DOUBLE PRECISION  T(200),CP(200),CCAL(200),HCAL(200),SCAL(200),
     . C(200),H(200),HDIF(200),S(200),SCSI(200),SHPT(8),SPT(8),A(8),
     . Y(200),TS(8,8),SUM,HCONS,SCONS,SD,R,SDMIN,HERR,SERR,THERR,TSERR,
     . HX,SX,AS(4),SUMS
      DATA HT/2H/T/,RT/'RT'/,BK/' '/,SR/'S/R'/,SLAB/'S'/,CLAB/'CP'/
      DATA CPR/'CP/R'/,PXNL/'NON-'/
      NAMELIST /INPUT/NAME,R,T,CP,H,S,HLABEL,SLABEL,LINEAR,ID,NATOMS
     . R=1.987192 $ NT=0 $ ID(1)=BK $ ID(2)=BK $ CLABEL=CLAB
      SLABEL=SLAB $ HLABEL(2)=BK
      DO 3 I=1,200
    3 T(I)=0.
      DO 4 I=1,4
    4 NAME(I)=BK
      LINEAR=.FALSE.  $ READ (5,INPUT,END=1000)
      DO 7 I=1,200 $ IF (T(I).EQ.0.) GO TO 20 $ NT=I
      IF (CLABEL.EQ.CPR) CP(I)=CP(I)*R
      IF (SLABEL.EQ.SR) S(I)=S(I)*R
      IF (HLABEL(2).EQ.RT) H(I)=H(I)*R*T(I)
      IF (HLABEL(2).EQ.HT) H(I)=H(I)*T(I)
    7 CONTINUE
   20 IF (NT.GT.0 .AND.  NATOMS.GT.0) GO TO 25
      WRITE (2,22) NATOMS,NT $ GO TO 1
   22 FORMAT (' ERROR - NATOMS=',I4,6X,'NT =',I5)
   25 IF (LINEAR) GO TO 30
      CPO=4.0*R $ CPI=(3*NATOMS-2)*R $ PREFIX=PXNL $ GO TO 40
   30 PREFIX=BK
      CPO=3.5*R $ CPI=R*(FLOAT(3*NATOMS)-1.5)
   40 B=0.0 $ SDMIN=1.E+10
      DO 280 IS=1,20
      B=B+50.0 $ SD=0.
      DO 50 K=1,7 $ SPT(K)=0.0
      DO 50 I=1,NT $ Y(I)=(T(I)/(T(I)+B))
   50 SPT(K)=SPT(K)+(Y(I)*10.)**(K-1)
      DO 60 KR=1,4
      DO 60 KC=1,4 $ K=KR+KC-1
   60 TS(KC,KR)=SPT(K)
      DO 70 KR=1,4 $ SHPT(KR)=0.0
      DO 70 I=1,NT $ C(I)=(CP(I)-CPO-((Y(I)**2)*(CPI-CPO)))/
     .((CPI-CPO)*(Y(I)**2)*(Y(I)-1.))
   70 SHPT(KR)=SHPT(KR)+C(I)*(Y(I)*10.)**(KR-1)
      DO 80 J=1,4 $ L(J)=J
   80 M(J)=J
      DO 160 J=1,4 $ BIGEST=0.0
      DO 100 IL=J,4
      DO 90 IM=J,4
      KL=L(IL) $ KM=M(IM)
      IF (BIGEST.GE.DABS(TS(KL,KM))) GO TO 90
      BIGEST=DABS(TS(KL,KM)) $ NL=IL $ NM=IM
   90 CONTINUE
  100 CONTINUE
      IF (BIGEST.NE.0.0) GO TO 120
      WRITE (6,110) IJ $ GO TO 1
  110 FORMAT (1H1,23HTHE MATRIX IS SINGULAR.//46HUNABLE TO COMPUTE THE C
     .OEFFICIENTS FOR SET NO ,I1,1H.)
  120 IT=L(J) $ L(J)=L(NL) $ L(NL)=IT $ IT=M(J) $ M(J)=M(NM)
      M(NM)=IT $ KL=L(J) $ KM=M(J) $ SHPT(KL)=SHPT(KL)/TS(KL,KM)
      DO 130 IA=1,4 $ KLA=L(IA) $ IF (KLA.EQ.KL) GO TO 130
      SHPT(KLA)=SHPT(KLA)-SHPT(KL)*TS(KLA,KM)
  130 CONTINUE $ IF (J.EQ.4) GO TO 170 $ JP=J+1
      DO 140 IB=JP,4 $ KMA=M(IB)
  140 TS(KL,KMA)=TS(KL,KMA)/TS(KL,KM)
      DO 160 IA=1,4 $ KLA=L(IA) $ IF (KLA.EQ.KL) GO TO 160
      DO 150 IB=JP,4 $ KMA=M(IB)
  150 TS(KLA,KMA)=TS(KLA,KMA)-TS(KL,KMA)*TS(KLA,KM)
  160 CONTINUE
  170 DO 180 I=1,4 $ KLA=L(I) $ KMA=M(I)
  180 A(KMA)=SHPT(KLA)
      SUM=0.
      DO 190 J=1,4 $ A(J)=A(J)*10.0**(J-1)
  190 SUM=SUM+A(J)
      HERR=0.  $ SERR=0.
       DO 220 I=1,NT
      CALL WCALC(A,Y(I),T(I),CPO,CPI,SUM,CCAL(I),HCAL(I),SCAL(I))
      HDIF(I)=HCAL(I)-H(I) $ SCSI(I)=SCAL(I)-S(I)
      HX=DABS(HDIF(I)-HDIF(1)) $ IF(HX.LE.HERR)GO TO 210
      HERR=HX $ THERR=T(I)
```

```
210 SX=DABS(SCSI(I)-SCSI(1))
    IF(SX.LE.SERR)GO TO 220
    SERR=SX $ TSERR=T(I)
220 SD=SD+(CP(I)-CCAL(I))**2
    SD=DSQRT(SD/NT) $ IF(SD.GT.SDMIN)GO TO 228
    SDMIN=SD $ BMIN=B $ SCONS=-SCSI(NT) $ HCONS=-HDIF(NT)
    SUMS=SUM
    DO 224 J=1,4
224 AS(J)=A(J)
228 WRITE (6,230) NAME
230 FORMAT (1H1,3X,'WILHOIT COEFFICIENTS FOR ',4A4)
    WRITE (6,250) ((A(J),J),J=1,3),A(4)
250 FORMAT (1H0,3X,'A(0) = ',3(E16.9,5X,'A(',I1,') = '),E16.9)
    WRITE (6,260) B,CPO,CPI,PREFIX,NATOMS
260 FORMAT (1H0,'B =',F7.1,5X,'CPO =',F9.4,5X,'CPI =',F9.4,5X,A4,
   . 'LINEAR',5X, 'NO.  ATOMS =',I4/)
    WRITE(6,262) SD,HERR,THERR,SERR,TSERR
262 FORMAT(1H ,'STD DEV =',F9.4,5X,'MAXERR H =',F11.4,2X,'T =',F9.2,
   . 5X,'MAXERR S =',F6.4,2X,'T =',F9.2)
    WRITE(6,265)
265 FORMAT(//9X,1HT,6X,'CP CALC',4X,'CP INPUT',9X,'H CALC',14X,
   . 'H INPUT',12X,'HCAL-HINP',6X,6HS CALC,5X,'S INPUT',3X,'SCAL-SINP'/)
    WRITE (6,270) (T(I),CCAL(I),CP(I),HCAL(I),H(I),HDIF(I),SCAL(I),
   . S(I),SCSI(I),I=1,NT)
270 FORMAT (1X,3F11.3,3E20.7,3F11.3)
280 CONTINUE
C  EXTRAPOLATION
    NT1=NT+1 $ T(NT1)=T(NT)+100.
290 DO 300 I=NT1,200  $ Y(I)=T(I)/(T(I)+BMIN)
    CALL WCALC(AS,Y(I),T(I),CPO,CPI,SUMS,CP(I),H(I),S(I))
    H(I)=H(I)+HCONS $ S(I)=S(I)+SCONS
    IF(T(I).GE.5000.)GO TO 320
    NTOT=I+1 $ T(NTOT)=T(I)+100.
300 CONTINUE
320 WRITE(6,230)NAME
    WRITE (6,250) ((AS(J),J),J=1,3),AS(4)
    WRITE (6,322) HCONS,SCONS
322 FORMAT('0INTEGRATION CONSTANTS      H V',E16.9,5X,'S =',E16.9)
    WRITE (6,260) BMIN,CPO,CPI,PREFIX,NATOMS
    WRITE (6,325)  NAME
325 FORMAT ('0PAC INPUT FOR ',4A4//)
    DO 330 I=1,NTOT
    IF (CLABEL.EQ.CPR)CP(I)=CP(I)/R
    IF (SLABEL.EQ.SR)S(I)=S(I)/R
    IF (HLABEL(2).EQ.RT)H(I)=H(I)/(R*T(I))
    IF (HLABEL(2).EQ.HT)H(I)=H(I)/T(I)
330 CONTINUE
    WRITE(6,335) (ID,T(I),CLABEL,CP(I),HLABEL,H(I),SLABEL,S(I),I=1,
   . NTOT)
    WRITE(8,340) (ID,T(I),CLABEL,CP(I),HLABEL,H(I),SLABEL,S(I),I=1,
   . NTOT)
335 FORMAT(1X,A4,A2,'T',F15.3,2X,A4,F12.4,2X,A4,A2,G12.8,A4,2X,F12.5)
340 FORMAT(A4,A2,'T',F15.3,2X,A4,F12.4,2X,A4,A2,G12.8,A4,2X,F12.5)
    GO TO 1
1000 STOP
    END
```

```
      SUBROUTINE WCALC(A,Y,T,CPO,CPI,SUM,C,H,S)
      DOUBLE PRECISION A(4),Y,T,CPO,CPI,SUM,C,H,S,SF(4),SC,SY,SS
      DO 200 J=1,4 $ SF(J)=FLOAT(J+1)*A(J)
      DO 200 N=J,4
  200 SF(J)=SF(J)+A(N)
      SC=(((A(4)*Y+A(3))*Y+A(2))*Y+A(1))*Y
      SS=(((A(4)*Y/5.+A(3)/4.)*Y+A(2)/3.)*Y+A(1)/2.)*Y
      SY=(((SF(4)*Y/3.+SF(3)/2.)*Y/5.+SF(2)/6.)*Y+SF(1)/3.)*Y/2.
      C=CPO+(CPI-CPO)*Y*(Y+(Y-1.)*SC)
      H=(CPO-(CPI-CPO)*((2.+SUM)*(Y/2.-1.+(1./Y-1.)*DLOG(T/Y))+Y*SY))
      H=H*T $ S=CPI*DLOG(T)-(CPI-CPO)*(DLOG(Y)+(1.+SS)*Y)
      RETURN
      END
```

```
$INPUT NATOMS=21,NAME='C7H14',
 T=298.15, 300, 400, 500, 600, 700, 800, 900,1000, 1100,
 1200,1300,1400,1500,CP=37.10,37.27,46.97,55.72,63.24,69.62,
 75.08,79.81,83.90,87.40,90.5, 93.2, 95.5, 97.5,
 H=7221.4,7287,11504,16655,22608,29246,36488,44244,52430,61006.,
 69924,79118,88564,98190,
 S=101.24,101.48,113.55,125.01,135.82,146.04,155.71,164.81,173.43,181.60,
 189.36,196.7,203.71,210.39,HLABEL='H-HO',LINEAR=F,ID='C7H14'$END
$INPUT NAME='1-C3H7OH',HLABEL='H-HO/T',T=298.15,300,400,500,600,700,800,
 900,1000,CP=20.82,20.91,25.86,30.51,34.56,38.03,41.04,43.65,45.93,
 H=14.48,14.52,16.74,19.04,21.29,23.43,25.46,27.32,29.08,
 S=77.61,77.71,84.31,90.54,96.44,102.01,107.29,112.26,116.99,NATOMS=12,
 ID='1C3H8O'$END
$INPUT NAME='C2N2',NATOMS=4,LINEAR=T,HLABEL='H-HO',ID='C2N2',T=298.15,
 300,400,500,600,700,800,900,1000,1500,2000,2500,3000,3500,4000,4500,
 CP=13.5815,13.6085,14.7827,15.6233,16.3108,16.9040,17.4190,17.8629,
 18.2425,19.4441,20.0087,20.3058,20.4794,20.5895,20.664,20.7173,
 H=3027.1,3052.3,4475.9,5998.,7595.6,9257,10973.8,12738.5,14544.2,
 24005.5,33884.9,43970.7,54170.5,64439.7,74754.2,85100.2, S=57.7889,
 57.8730,61.9604,65.3533,68.2643,70.8242,73.1158,75.1938,77.096,
 84.752,90.4324,94.9324,98.6512,101.817,104.5715,107.0086,$END
```

Appendix C

Table of Coefficient Sets for NASA Polynomials

The NASA thermochemical polynomials have the form

$$C_P^\circ/R = a_1 + a_2T + a_3T^2 + a_4T^3 + a_5T^4$$

$$H^\circ/RT = a_1 + a_2T/2 + a_3T^2/3 + a_4T^3/4 + a_5T^4/5 + a_6/T$$

$$S^\circ/R = a_1 \ln T + a_2T + a_3T^2/2 + a_4T^3/3 + a_5T^4/4 + a_7$$

One mole of matter is assumed, and unless explicitly noted on the first data card the reference state is the ideal gas at one atmosphere pressure. The enthalpy reference is chosen such that H° includes the enthalpy of formation as in Eq. (9) of the text. This means that values of H°/RT calculated from the polynomials can be used directly to compute enthalpies of reaction. (See text for discussion of absolute enthalpy). The usual thermochemical manipulations can be done on these polynomials to compute other thermochemical quantities such as standard molar free energies.

It should be emphasized that the accuracy of the fit given by the coefficients tabulated here varies considerably from one species to the next. For essentially all purposes in combustion modeling, however, the accuracy of the polynomials as given by these coefficients is much better than the uncertainties of the modeling introduced from other sources.

For species that are not to be found in standard thermochemical tabulations, enough information is given prior to the coefficient cards to indicate the nature of the procedures used to generate the input data for the fitting program. The corresponding information for the species for which data was taken from tabulations can be found in the tabulations themselves.

The arrangement of the coefficients on the card images is this. On the second card are listed a_1 through a_5, on the third card a_6 and a_7 for the temperature range over 1000 K. Continuing on the third card are a_1 through a_3 and then on the fourth card a_4 through a_7 for the temperature range 300 to 1000 K. The first card contains the name of the species, information about the source of the data, (see below) some formula composition information required by certain equilibrium-solving programs, and the temperature range over which the polynomials were fitted. Some data sets contain in addition to the NASA-format information the molecular weight on card 1 and $(H298\text{-}H_0)/R$ on card 4.

The coding of data sources is J = JANAF Tables L = NASA Lewis Tables R = TSIV Tables T = Technion Reports U = LSU Calculations

While the coefficients collected here form a set that is complete enough for modeling a broad variety of combustion processes, including combustion of hydrocarbons up to C_8, individual users must expect to have to generate some of their own thermochemical data and fit it to polynomials by a program such as the one given in Appendix A. We note also that some of the information provided here, especially for deuterated and unusual hydrocarbon species, has not been presented in the literature before and is included for archival purposes more than to suggest special importance in combustion science.

```
AR                L 5/66AR  100             0G   300.      5000.             1
 0.25000000E 01 0.             0.             0.             0.              2
-0.74537502E 03 0.43660006E 01 0.25000000E 01 0.             0.              3
 0.             0.            -0.74537498E 03 0.43660006E 01                 4

BR                J 6/74BR  10    00    00  0G   300.      5000.             1
 0.20843207E+01 0.71949483E-03-0.27419924E-06 0.42422650E-10-0.23791570E-14  2
 0.12858837E+05 0.90838003E+01 0.24611551E+01 0.33319275E-03-0.10080655E-05  3
 0.12262126E-08-0.44283510E-12 0.12711920E+05 0.69494733E+01                 4

BR2               J12/61BR  20    00    00  0G   300.      5000.             1
 0.44479495E 01 0.10051208E-03-0.16393816E-07 0.22685621E-11-0.10236774E-15  2
 0.23659941E 04 0.40888431E 01 0.38469580E 01 0.26111841E-02-0.40034147E-05  3
 0.28120689E-08-0.73256202E-12 0.24846984E 04 0.69696985E 01                 4

C(S)              J 3/78C   1     0     0   0S   300.      5000.             1
 0.16272659E+01 0.13726102E-02-0.47427392E-06 0.78269058E-10-0.46932334E-14  2
-0.75550659E+03-0.94597750E+01-0.39358151E+00 0.50322227E-02 0.21335575E-06  3
-0.44809667E-08 0.22281673E-11-0.10043521E+03 0.14578543E+01                 4

C                 J 3/78C   100             0G   300.      5000.             1
 0.25769424E+01-0.13903944E-03 0.69481807E-07-0.67414021E-11-0.43389004E-16  2
 0.85425220E+05 0.43358122E+01 0.25279476E+01-0.12519400E-03 0.22544496E-06  3
-0.18489024E-09 0.57291741E-13 0.85448374E+05 0.46274790E+01                 4

CD  METHYLIDENE-D RADICAL  SIGMA=1  TE=0(2)  WE=2099.7  WEXE=34.02  BE=7.806
ALPHAE=0.208  TE=23184.   WE=2203.3  WEXE=78.5  BE=8.032  ALPHAE=0.26
TE=26043.(2)  WE=1652.5  WEXE=123.8  BE=7.104  ALPHAE=0.341  TE=31818(2)
WE=2081.3 WEXE=66.79 WEYE=5.364 WEZX=-1.15 BE=7.879 ALPHAE=0.283 TE=59039.(2)
WE=2025.  BE=7.425 REF=HORNE AND HERZBERG MAX ERROR CP AT 1300 K .3%
CD                T 2/80C   1D    1     0   0G   300.      5000.             1
 0.26841459E 01 0.18855776E-02-0.48628311E-06 0.38441708E-10 0.64605384E-15  2
 0.70531750E 05 0.70191174E 01 0.35427971E 01-0.47720969E-03 0.10656331E-05  3
 0.73458772E-09-0.74328873E-12 0.70311938E 05 0.26285248E 01                 4

CDH3  METHANE-D  STATWT=1  SIGMA=3  A0=C0=5.25  B0=3.878  NU=2945,2200,1300,
3017(2),1471(2),1155(2)  REF=BURCAT  MAX ERROR CP AT 1300 K .9%
CDH3 G            T05/79C   1D    1H    3   0G   300.      5000.             1
 0.29389458E 01 0.84684640E-02-0.28219238E-05 0.41319725E-09-0.21738508E-13  2
-0.10964586E 05 0.42649403E 01 0.26380539E 01 0.41823313E-02 0.72133871E-05  3
-0.58564389E-08 0.79961938E-12-0.10462590E 05 0.74964495E 01                 4
```

```
CDO  FORMYL-D RADICAL  STATWT=2  SIGMA=1  A0=14.69  B0=1.281375  C0=1.171465
NU=1937,847,1800  X12=-13.1,X22=-2.44,Y222=.0096  ALFA1=.006  ALFA2=.002
ALFA3=.02  T0=9162  B0=1.1011  REF=BROWN AND RAMSAY  MAX ERROR CP AT 1300 K .6%
CDO               L11/83D   1C   10   1    0G   300.      5000 0  MW=30.02     1
 0.42518167E 01 0.25338491E-02-0.91200246E-06 0.15268785E-09-0.96286403E-14    2
-0.70481602E 04 0.21420174E 01 0.35113592E 01 0.22066066E-02 0.21039923E-05    3
-0.20886284E-08 0.28339429E-12-0.66476914E 04 0.67500134E 01          1213.5   4

CD2  METHYLENE-D2 RADICAL SIGMA=2  T0=0(3)=30300(1) IA=.0744 IB=.60878 IC=.6831
NU=2115,767,2345   T0=2600(1)  IA=.24223  IB=.50049  IC=.74272  NU=2209,926,2273
T0=9700(1)  IA=.06458  IB=.6511  IC=.71567  NU=2093,545,2338  REF=BURCAT
MAX ERROR CP AT 1300 K .7%
CD2               T05/80C   1D   2    0    0G   300.      5000.                1
 0.36602430E 01 0.33572798E-02-0.12381643E-05 0.20197106E-09-0.12083819E-13    2
 0.44684898E 05 0.25554295E 01 0.38409843E 01 0.12651016E-02 0.18910869E-05    3
-0.77415541E-09-0.25377709E-12 0.44799531E 05 0.22202606E 01                   4

CD2H2  METHANE-D2  STATWT=1  SIGMA=2  A0=4.303  B0=3.506  C0=3.05  NU=2974,
2202,1435,1033,1331,3013,1090,2234,1234  REF=BURCAT  MAX ERROR CP AT 1300 K .9%
CD2H2             T05/79C   1H   2D   2    0G   300.      5000.                1
 0.35087013E 01 0.81863180E-02-0.27852266E-05 0.41648370E-09-0.22470558E-13    2
-0.11595125E 05 0.18748598E 01 0.23866291E 01 0.57553649E-02 0.64751221E-05    3
-0.67107635E-08 0.13974620E-11-0.10846535E 05 0.94474039E 01          1220.9   4

CD2O  METHANAL-D2 (FORMALDEHYDE-D2)  STATWT=1  SIGMA=2  IA=.59244  IB=2.5995
IC=3.2048  NU=2056,1700,1106,2160,990,938  REF=CHAO,WILHOIT,HALL MAX ERROR CP AT
1300 K .8%
CD2O              T 8/81C   1D   2O   1    0G   300.      5000.                1
 0.46622076E+01 0.50203055E-02-0.18413848E-05 0.29739944E-09-0.17558905E-13    2
-0.15805738E+05-0.17819729E+01 0.25921259E+01 0.59901401E-02 0.39293818E-05    3
-0.62653172E-08 0.18746567E-11-0.14881922E+05 0.10376950E+02                   4

CD3 METHYL-D3-RADICAL  STATWT=1  SIGMA=6  IA=IB=.596 IC=1.191 T0=0(2),46200(2)
NU=2153,463,2381(2),1026(2)  REF=BURCAT  MAX ERROR CP AT 1300 K .7%
CD3               T11/79C   1D   3    0    0G   300.      5000.                1
 0.44567032E 01 0.49626939E-02-0.17476059E-05 0.27139846E-09-0.15351469E-13    2
 0.14782500E 05-0.23942318E 01 0.34687710E 01 0.49496330E-02 0.19827057E-05    3
-0.36768906E-08 0.12036257E-11 0.15276805E 05 0.35894156E 01                   4

CD3H  METHANE-D3  STATWT=1  SIGMA=3  AE=CE=2.62  BE=3.27  NU=2993,2142,1003,
2263(2),1291(2),1036(2)  REF=BURCAT  MAX ERROR CP AT 1300 K .9%
CD3H              T05/79C   1H   1D   3    0G   300.      5000.                1
 0.40764599E 01 0.79434291E-02-0.27834194E-05 0.42990389E-09-0.24151396E-13    2
-0.12245391E 05-0.13617935E 01 0.21469107E 01 0.74287578E-02 0.56749586E-05    3
-0.77548528E-08 0.21464679E-11-0.11265895E 05 0.10438148E 02                   4

CD4   METHANE-D4  STATWT=1  SIGMA=12  A0=B0=C0=2.634  NU=2109,1092(2),2259(3),
996(3)  REF=BURCAT  MAX ERROR CP AT 1300 K .9%
CD4    RRHO       T05/79C   1D   4    0    0G   300.      5000.                1
 0.47153826E 01 0.75838268E-02-0.2712 208E-05 0.4266 048E-09-0.2442063 E-13    2
-0.12937410E 05-0.62130547E 01 0.19176292E 01 0.91806799E-02 0.47843714E-05    3
-0.88772119E-08 0.29830964E-11-0.11712711E 05 0.10117057E 02                   4

CD4  METHANE-D4 ANHARMONIC.  DATA AS FOR RRHO.   X11=-13.6,X12=-1.54,X13=-40.6,
X14=-2.2,X22=-.2,X23=-4.8,X24=-10.9,X33=-9.6,X34=-12.7,X44=-6.4  ALFA1=.07
ALFA2=-.06  ALFA3=.03  ALFA4=.05  REF=TSIV(CH4)  MAX ERROR CP AT 1300 K .9%
CD4*  ANHARMONIC  T06/81C   1D   4    0    0G   300.      5000.                1
 0.44482183E+01 0.81195608E-02-0.27020378E-05 0.43419712E-09-0.24605867E-13    2
-0.12860102E+05-0.47993603E+01 0.19425707E+01 0.89269280E-02 0.54267666E-05    3
-0.89088488E-08 0.28879408E-11-0.11714484E+05 0.10023487E+02                   4

CH  METHYLIDENE RADICAL  SIGMA=1  T0=0(2)  WE=2861.6  WEXE=64.3  BE=14.455
ALFAE=.5339  RHO=1.9E-5  T0=17.9(2)  T0=4500(4)  T0=23150(4)  T0=25949(2)
WE=2921  WEXE=90.4  BE=14.912  ALFAE=.67  T0=3182(2)  WE=2824.1  WEXE=105.8
BE=14.629  ALFAE=.744  REF=BURCAT  MAX ERROR CP AT 5000 K .6%
CH                T 2/80C   1H   1    0    0G   300.      5000.                1
 0.24161863E 01 0.19084704E-02-0.43044935E-06 0.20121099E-10 0.22894041E-14    2
 0.70784500E 05 0.79757013E 01 0.36162109E 01-0.60296291E-03 0.66297542E-06    3
 0.65568906E-09-0.41529920E-12 0.70400250E 05 0.15434847E 01                   4

CHO               J12/70H   1C   1O   1O   0G   300.      5000.                1
 0.34738348E+01 0.34370227E-02-0.13632664E-05 0.24928645E-09-0.17044331E-13    2
 0.39594005E+04 0.60453340E+01 0.38840192E+01-0.82974448E-03 0.77900809E-05    3
-0.70616962E-08 0.19971730E-11 0.40563860E+04 0.48354133E+01                   4
```

```
CHO+                J12/70H   1C   10   1E  -1G    300.      5000.              1
 0.37411880E+01 0.33441517E-02-0.12397121E-05 0.21189388E-09-0.13704150E-13    2
 0.98884078E+05 0.20654768E+01 0.24739736E+01 0.86715590E-02-0.10031500E-04    3
 0.67170527E-08-0.17872674E-11 0.99146608E+05 0.81625751E+01                   4

CH2                 J12/72C   1H   2    0    0G    300.      5000.              1
 0.30643921E 01 0.33640424E-02-0.10989143E-05 0.15985906E-09-0.84323282E-14    2
 0.45435059E 05 0.49476233E 01 0.36884661E 01 0.14331874E-02 0.57268682E-06    3
-0.99654077E-10-0.11374164E-12 0.45305152E 05 0.18445559E 01                   4

CH2O                J 3/61C   1H   2O   1O   0G    300.      5000.              1
 0.28364249E 01 0.68605298E-02-0.26882647E-05 0.47971258E-09-0.32118406E-13    2
-0.15236031E 05 0.78531169E 01 0.37963783E 01-0.25701785E-02 0.18548815E-04    3
-0.17869177E-07 0.55504451E-11-0.15088947E 05 0.47548163E 01                   4

CH2O2  METHANOIC(FORMIC) ACID  HCOOH MONOMER  STATWT=1  SIGMA=1  IA=1.0953
IB=6.9125  IC=8.0078  IR(OH)==0.1122  POTENTIAL  BARRIER V(3)=12200.
NU=3570,2943,1770,1387,1229,1105,625,1033,(610 TORSION)  REF=CHAO AND ZWOLINSKI
CH2O2               L 5/80C   1H   2O   2    0G    300.      5000.              1
 0.57878771E 01 0.75539909E-02-0.30995161E-05 0.54494809E-09-0.34704210E-13    2
-0.48191230E 05-0.65299015E 01 0.21183796E 01 0.11175469E-01 0.26270773E-05    3
-0.81816403E-08 0.30133404E-11-0.46669293E 05 0.14480175E 02                   4

CH3 METHYL RADICAL  STATWT=1  SIGMA=6 IA=IB=.2923  IC=.5846  T0=0(2),46205(2)
NU=3002,580,3184(2),1383(2)  REF=BURCAT  MAX ERROR CP AT 1400 K .6%
CH3                 T11/79C   1H   3    0    0G    300.      5000.              1
 0.32985334E 01 0.51838532E-02-0.15955029E-05 0.21366862E-09-0.99468265E-14    2
 0.16425031E 05 0.29979439E 01 0.35155468E 01 0.34882184E-02 0.18435312E-05    3
-0.27320166E-08 0.97533353E-12 0.16448859E 05 0.22105637E 01                   4

CH3NO2 NITROMETHANE   EXTRAPOLATED DATA THROUGH WILHOIT POLYNOMIAL FROM STULL,
SINKE,WESTRUM  MAX ERROR CP AT 1300 K .8%
CH3NO2              T 1/81C   1H   3N   1O   2G    300.      5000.              1
 0.72579584E+01 0.97425506E-02-0.32998314E-05 0.49756554E-09-0.27304449E-13    2
-0.12274141E+05-0.12664469E+02 0.16658440E+01 0.18138818E-01 0.12019345E-05    3
-0.12436082E-07 0.56004723E-11-0.10272887E+05 0.18221619E+02                   4

CH3NO3  METHYLNITRATE  STATWT=1  SIGMA=1  IA=7.11599  IB=17.8297  IC=24.4114
IRCH3=.28  IRNO2=1.05  NU=2992,2991,2905,1635,1463,1438,1431,1283,1177,1089,
1008,863,760,630,540,262  POTENTIAL BARRIERS  V(3)=2320.  (CH3)  V(3)=9100.
(NO2)  REF=CZUCHAJOWSKI AND KUCHARSKI  MAX ERROR CP AT 1300 K .8%
CH3NO3              T09/81C   1H   3N   1O   3G    300.      5000.              1
 0.10090198E+02 0.10290518E-01-0.37422951E-05 0.59494920E-09-0.34445171E-13    2
-0.18800801E+05-0.26184555E+02 0.30306816E+01 0.21521591E-01 0.11596558E-05    3
-0.16239241E-07 0.77263126E-11-0.16334395E+05 0.12570032E+02                   4

CH3O METHYLOXIDE  STATWT=2 SIGMA=3 IA=.511 IB=IC=3.830 NU=3000,2844,1477,1455,
1060,1033,2960,1477,1165 REF=BURCAT AND KUDCHADKER MAX ERROR CP AT 400 K .5%
CH3O                U10/77C   1H   3O   1    0G    300.      3000.      MW=31.03  1
 0.37707996E+01 0.78714974E-02-0.26563839E-05 0.39444314E-09-0.21126164E-13    2
 0.12783252E+03 0.29295750E+01 0.21062040E+01 0.72165951E-02 0.53384720E-05    3
-0.73776363E-08 0.20756105E-11 0.97860107E+03 0.13152177E+02          1223.4  4

H3CO  HYDROXYMETHYLENE RADICAL (CH2OH)  STATWT=2  SIGMA=1  IA=.4041  IB=3.0848
IC=3.1518 IR=.79 NU=3681,2844,1455,1345,1060,1119,2960,1165  POTENTIAL BARRIER
B=35.4411  V2=2.     REF=BURCAT AND KUDCHADKER  MAX ERROR CP AT 1300 K .7%
H3CO                U 3/78H   3C   1O   1    0G    300.      3000.              1
 0.47235041E+01 0.61020441E-02-0.19132094E-05 0.27607427E-09-0.14548367E-13    2
-0.39329165E+04-0.85243821E-01 0.33368406E+01 0.65881237E-02 0.29979328E-05    3
-0.58719714E-08 0.21229572E-11-0.33168267E+04 0.80668154E+01                   4

CH3O2  METHYLPEROXIDE RADICAL (CH3OO)  SIGMA=1  STATWT=2  IA*IB*IC=7.203E-116
NU=2930(3),1400(2),1350,1100,960,900(2),450,300 (FREE INTERNAL ROTATION)
REF=W.TSANG, NBS, UNPUBLISHED CALCULATIONS.   MAX ERROR CP AT 1300 K .7%
CH3O2               L 1/84C   1H   3O   2    0G    300.      5000.    MW=47.03    1
 0.66812963E 01 0.80057271E-02-0.27188507E-05 0.40631365E-09-0.21927725E-13    2
 0.52621851E 03-0.99423847E 01 0.20986490E 01 0.15786357E-01 0.75683261E-07    3
-0.11274587E-07 0.56665133E-11 0.20695879E 04 0.15007068E 02          1471.7  4

CH4     RRHO        J 5/61C   1H   4    0    0G    300.      5000.              1
 0.23594046E 01 0.87309405E-02-0.28397053E-05 0.40459835E-09-0.20527095E-13    2
-0.10288820E 05 0.60290012E 01 0.29283962E 01 0.25691092E-02 0.78437060E-05    3
-0.49102979E-08 0.20380030E-12-0.10054172E 05 0.46342220E 01                   4

CH4 METHANE  STATWT=1 SIGMA=12  A0=B0=C0=5.2410356  NU=2916.7,1533.295(2),
3019.491(3),1310.756(3)  X11=-26,X12=-3,X13=-75,X14=-4,X22=-.4,X23=-9,X24=-20,
X33=-17,X34=-17,X44=-11  ALFA1=.01,ALFA2=-.09,ALFA3=.04,ALFA4=.07 D0=1.10864E-4
REF=TSIV  MAX ERROR 1300 K .8%
```

```
CH4*  ANHARMONIC  T05/79C   1H   4    0    0G   300.      5000.                   1
 0.20916119E+01 0.91304034E-02-0.27888073E-05 0.40232462E-09-0.20205537E-13    2
-0.10161687E+05 0.75069170E+01 0.29824648E+01 0.22276966E-02 0.83464556E-05    3
-0.48251323E-08 0.83841408E-13-0.10026887E+05 0.44046726E+01                   4

CH4O  METHANOL (CH3OH)  STATWT=1  SIGMA=3  SIGMA BARRIER=1.  IA=.6578
IB=3.4004  IC=3.5306  IR=.0993  NU=3681,3000,2844,1477,1455,1345,1060,1033,2960,
1477,1165,298  POTENTIAL BARRIER  V3=1.06706  V6=-.00149  REF=CHEN,WILHOIT,
ZWOLINSKI  MAX ERROR CP AT 1300 K  .8%
CH4O                T 4/82C   1H   40   1    0G   300.      5000.      MW=32.04  1
 0.40290613E+01 0.93765929E-02-0.30502542E-05 0.43587933E-09-0.22247232E-13    2
-0.26157910E+05 0.23781958E+01 0.26601152E+01 0.73415078E-02 0.71700506E-05    3
-0.87931937E-08 0.23905704E-11-0.25353484E+05 0.11232631E+02             1375.3  4

CH5N  METHYLAMINE (CH3NH2)  STATWT=1  SIGMA=1  IA=.81375  IB=3.8663  IC=3.7089
IR=0.5288  POTENTIAL BARRIER V(3)=1980.   SIGMAR=6
NU=3361,2961,2820,1623,1473,1430,1130,1044,780,3427,2985,1485,1419,1195
REF=DEWAR AND RZEPA      MAX ERROR CP AT 1300K .8%
CH5N G              T09/81C   1H   5N   1    0G   300.      5000.                1
 0.44235811E+01 0.11449948E-01-0.36999727E-05 0.52389848E-09-0.26375054E-13    2
-0.49847539E+04-0.42785645E+00 0.27267694E+01 0.10014653E-01 0.67409546E-05    3
-0.98750093E-08 0.30637376E-11-0.40688989E+04 0.10201913E+02                   4

CN                  J 6/69C   1N   10   00   0G   300.      5000.                1
 0.36036285E 01 0.33644390E-03 0.10028933E-06-0.16318166E-10-0.36286722E-15    2
 0.51159833E 05 0.35454505E 01 0.37386307E 01-0.19239224E-02 0.47035189E-05    3
-0.31113000E-08 0.61675318E-12 0.51270927E 05 0.34490218E 01                   4

CNN                 J 6/66C   1N   200       0G   300.      5000.                1
 0.48209077E 01 0.24790014E-02-0.94644109E-06 0.16548764E-09-0.10899129E-13    2
 0.68685948E 05-0.48484039E 00 0.35077779E 01 0.72023958E-02-0.75574589E-05    3
 0.42979217E-08-0.94257935E-12 0.68994281E 05 0.60234964E 01                   4

CN2     (NCN)       J12/70C   1N   20   00   0G   300.      5000.                1
 0.55626268E+01 0.20860606E-02-0.88123724E-06 0.16505783E-09-0.11366697E-13    2
 0.54897907E+05-0.55989355E+01 0.32524003E+01 0.70010737E-02-0.22653599E-05    3
-0.28939808E-08 0.18270077E-11 0.55609085E+05 0.66966778E+01                   4

CO                  J 9/65C   10   100       0G   300.      5000.                1
 0.29840696E 01 0.14891390E-02-0.57899684E-06 0.10364577E-09-0.69353550E-14    2
-0.14245228E 05 0.63479156E 01 0.37100928E 01-0.16190964E-02 0.36923594E-05    3
-0.20319674E-08 0.23953344E-12-0.14356310E 05 0.29555351E 01                   4

COS                 J 3/61C   10   1S   100  0G   300.      5000.                1
 0.52392000E 01 0.24100584E-02-0.96064522E-06 0.17778347E-09-0.12235704E-13    2
-0.18480455E 05-0.30910517E 01 0.24625321E 01 0.11947992E-01-0.13794370E-04    3
 0.80707736E-08-0.18327653E-11-0.17803987E 05 0.10792556E 02                   4

CO2                 J 9/65C   10   200       0G   300.      5000.                1
 0.44608041E 01 0.30981719E-02-0.12392571E-05 0.22741325E-09-0.15525954E-13    2
-0.48961442E 05-0.98635982E 00 0.24007797E 01 0.87350957E-02-0.66070878E-05    3
 0.20021861E-08 0.63274039E-15-0.48377527E 05 0.96951457E 01                   4

CS                  J12/76C   1S   100       0G   300.      5000.                1
 0.36826012E+01 0.90473203E-03-0.36436374E-06 0.63854294E-10-0.36933982E-14    2
 0.32497490E+05 0.38850496E+01 0.34039344E+01-0.65773308E-03 0.61712157E-05    3
-0.73689604E-08 0.27346738E-11 0.32689393E+05 0.58977001E+01                   4

CS2                 J12/76C   1S   20   00   0G   300.      5000.                1
 0.59252610E+01 0.18252996E-02-0.75585380E-06 0.14605073E-09-0.10438595E-13    2
 0.12048071E+05-0.60723317E+01 0.28326013E+01 0.13290791E-01-0.18144694E-04    3
 0.12831681E-07-0.36800609E-11 0.12766782E+05 0.92087947E+01                   4

C2                  J12/69C   20   00   00   0G   300.      5000.                1
 0.40435359E 01 0.20573654E-03 0.10907575E-06-0.36427874E-10 0.34127865E-14    2
 0.99709486E 05 0.12775158E 01 0.74518140E 01-0.10144686E-01 0.85879735E-05    3
 0.87321100E-09-0.24429792E-11 0.98911989E 05-0.15846678E 02                   4

C2D2  ACETYLENE-D2  STATWT=1  SIGMA=2  IB=3.2838  NU=2701,1762,2439,505(2),
537(2)  X11=15.43,X12=12.1,X13=58.78,X14=10.87,X15=6.92,X22=6.31,X23=.91,X24=
8.34,X25=.56,X34=5.54,X35=3.13,X44=-3.66,X45=7.7,X55=1.24,X33=14.3,G44=-0.75,
G55=-1.36  REF=SHIMANOUCHI  MAX ERROR CP AT 1300 K .6%
C2D2                T 8/80C   2D   2    0    0G   300.      5000.                1
 0.57631445E+01 0.39823391E-02-0.14399011E-05 0.21952536E-09-0.12146185E-13    2
 0.24641469E+05-0.92923393E+01 0.37629929E+01 0.83192550E-02-0.22101658E-05    3
-0.40820787E-08 0.27229842E-11 0.25258297E+05 0.13225250E+01                   4
```

```
C2D2O  KETENE-D2  SIGMA=2  IA=.5974  IB=9.1958  IC=9.7932  NU=2267,2120,1228,
927,2383,855,371,542,432  REF=MOORE AND PIMENTEL  MAX ERROR CP AT 1300 K .7%
C2D2O              T10/82C   2D   2O   1   0G    300.       5000.      MW=44.05   1
 0.68584700E+01 0.55908523E-02-0.19912059E-05 0.31183456E-09-0.17762101E-13   2
 0.21307729E+04-0.11535155E+02 0.34471798E+01 0.11882458E-01-0.17057137E-05   3
-0.64614767E-08 0.35897769E-11 0.32729224E+04 0.69507427E+01          1489.9   4

C2D4   ETHYLENE-D4  STATWT=1  SIGMA=4  IA=1.1487  IB=3.793  IC=4.942  NU=2247,
1515,981,728,2289,1009,720,780,2345,586,2200,1078  REF=BURCAT  MAX ERROR CP AT
1300 K .9%
C2D4               T12/79C   2D   4    0   0G    300.       5000.                 1
 0.67207203E 01 0.84912479E-02-0.30327419E-05 0.47564219E-09-0.27109157E-13   2
 0.6275380 E 03-0.14438146E 02 0.132 4621E 01 0.17719518E-01-0.13082199E-05   3
-0.10431190E-07 0.53182406E-11 0.24874675E 04 0.15012101E 02                  4

C2D4O  ETHANAL-D4 (ACETALDEHYDE-D4)   STATWT=1  SIGMA=1  IA=2.4015  IB=9.7752
IC=11.109  IR=.64048  POTENTIAL BARRIER V(3)=1161.   NU=2265,2130,2060,1737,
1045,938,1028,1151,747,436,2225,1028,573,670  REF=CHAO,WILHOIT,HALL
MAX ERROR CP AT 1300 K .9%
C2D4O              T 8/81C   2D   4O   1   0G    300.       5000.      MW=48.08   1
 0.85226345E+01 0.92743672E-02-0.33571869E-05 0.53372684E-09-0.30898383E-13   2
-0.25431613E+05-0.19842667E+02 0.24537258E+01 0.18615011E-01 0.81830109E-06   3
-0.12927025E-07 0.59826883E-11-0.23262375E+05 0.13635018E+02          1688.8   4

C2D6   ETHANE-D6  STATWT=1  SIGMA=6  SIGMA BARRIER=3  IA=2.0942  IB=IC=6.0986
NU=2083,1155,843,2087,1077,2226(2),1041(2),970(2),2235(2),1081(2),594(2)
POTENTIAL BARRIER VO=2.87  IR=.5235  REF=BURCAT  MAX  ERROR CP AT 1300 K .9%
C2D6               T 5/80C   2D   6    0   0G    300.       5000.      MW=36.11   1
0.87366476E+01 0.11772312E-01-0.42297552E-05 0.66704353E-09-0.38247847E-13    2
-0.17392641E 05-0.25933151E 02 0.81539208E 00 0.24633620E-01 0.28606987E-07   3
-0.16559884E-07 0.79903445E-11-0.14620465E 05 0.97529633E 02          1591.    4

C2D6O  DIMETHYL-ETHER-D6  SIGMA=2  SIGMA BARRIER=3  IA=3.2656  IB=11.2126
IC=12.3437 IR=9.271 V(3)=2500.  NU=2248(2),2054(2),1059(4),1057(2),1033,827,362,
2202,1162,872,2184,931,950 REF=KANAZAWA AND NUKADA MAX ERROR CP AT 1300 K .9%
C2D6O              T12/82C   2D   6O   1   0G    300.       5000.      MW=52.11   1
 0.10630716E+02 0.12416139E-01-0.44895924E-05 0.71285688E-09-0.41213699E-13   2
-0.29983387E+05-0.32505524E+02 0.16130285E+01 0.27251996E-01 0.22420198E-06   3
-0.19127672E-07 0.92674445E-11-0.26856473E+05 0.16880051E+02          1909.3   4

C2D6N2  AZOMETHANE-D6    STATWT=1  SIGMA=2  IA=3.147 IB=24.215 IC=25.133
IR=0.765  POTENTIAL BARRIER VO=1700.   NU=2234,2127,1569,1122,1044,1034,761,
523,2225,1027,803,2239,1049,896,261,2240,1115,1112,1051,921,900,304,(191,166
TORSIONAL FREQS)  REF=PAMIDIMUKKALA,ROGERS,SKINNER  MAX ERROR CP AT 1300 K .9%
C2D6N2             T 8/81C   2D   6N   2   0G    300.       3000.                 1
 0.14394463E+02 0.70246868E-02 0.21563219E-05-0.22486397E-08 0.39756257E-12   2
-0.37601357E+04-0.50448975E+02 0.28080702E+01 0.28044611E-01 0.24821838E-05   3
-0.21411783E-07 0.98014019E-11 0.11770692E+03 0.12440903E+02                  4

C2H RADICAL   SIGMA=1   STATWT=2  IB=1.9   NU=3612,500(2),1848   TO=4000   STATWT=4
HF298=534.522 KJ   REF=TSIV  MAX ERROR CP AT 5000 K .7%
C2H                R  79 C   2H   1    0   0G    300.       5000.      MW=25.03   1
 0.30690565E 01 0.45962296E-02-0.15024734E-05 0.19119895E-09-0.74086109E-14   2
 0.63322871E 05 0.67282467E 01 0.36717138E 01 0.42759366E-02-0.15715832E-05   3
-0.13222836E-08 0.12928322E-11 0.63021723E 05 0.30995226E 01          1185.4   4

C2HO  KETYL RADICAL  SIGMA=1  STATWT=2  IA=0.1342  IB=7.5956  IC=7.7298
NU=3070.2152,1118,3166,438,525    REF=BURCAT,MILLER,GARDINER    MAX ERROR
CP AT 1400 K .5%
C2HO               T10/82C   2H   1O   1   0G    300.       5000.      MW=41.03   1
 0.45077200E+01 0.43579005E-02-0.13757026E-05 0.18988289E-09-0.92407330E-14   2
 0.92941846E+03 0.26024265E+01 0.39291782E+01 0.56519806E-02-0.14746602E-05   3
-0.16703900E-08 0.12344639E-11 0.11102358E+04 0.56634159E+01          1351.8   4

C2H2               J 3/61C   2H   200      0G    300.       5000.                 1
 0.45751083E 01 0.51238358E-02-0.17452354E-05 0.28673065E-09-0.17951426E-13   2
 0.25607428E 05-0.35737940E 01 0.14102768E 01 0.19057275E-01-0.24501390E-04   3
 0.16390872E-07-0.41345447E-11 0.26188208E 05 0.11393827E 02                  4

C2H2O   KETENE    SIGMA=2  IA=.299  IB=8.1477  IC=8.4466  NU=3070.2152,1388,1118,
3133,977,438,591,525  REF= MOORE AND PIMENTEL  MAX ERROR CP AT 1300 K .6%
C2H2O              T10/82C   2H   2O   1   0G    300.       5000.    MW=42.04     1
 0.60388174E+01 0.58048405E-02-0.19209538E-05 0.27944846E-09-0.14588675E-13   2
-0.85834023E+04-0.76575813E+01 0.29749708E+01 0.12118712E-01-0.23450457E-05   3
-0.64666850E-08 0.39056492E-11-0.76326367E+04 0.86735525E+01          1419.1   4
```

```
C2H3  VINYL-RADICAL  STATWT=2  SIGMA=1  IA=0.2813  IB=2.739  IC=3.0203
NU=3050,3150,3100,1625,1400,1000,1300,950,1000  REF=TSIV  HF298=260 KJ
MAX ERROR CP AT 1400 K .6%
C2H3                T 8/81C   2H   3    0    0G    300.       5000.                1
 0.40913172E+01 0.73747411E-02-0.24263809E-05 0.35108782E-09-0.18219093E-13    2
 0.29399117E+05 0.16772680E+01 0.19295549E+01 0.90625472E-02 0.34292807E-05    3
-0.83374339E-08 0.32886263E-11 0.30275852E+05 0.14064019E+02                   4

C2H3N  METHYLCYANIDE (CH3CN)  STATWT=1  SIGMA=3  A0=5.289  B0=C0=.3068
NU=2954,2268,1389,920,3009(2),1453(2),1041(2),362(2) T0=55     REF=SPANGENBERG
AND BORGER  MAX ERROR CP AT 1300 K .8%
C2H3N               T04/80C   2H   3N   1    0G    300.       5000.                1
 0.54956026E 01 0.89018531E-02-0.29834273E-05 0.43930859E-09-0.23263757E-13    2
 0.81332734E 04-0.55127306E 01 0.28112087E 01 0.11668280E-01 0.18374703E-05    3
-0.75577553E-08 0.30708968E-11 0.92079258E 04 0.97456837E 01                   4

C2H3OA (CH2CHO) RADICAL  SIGMA=1  SIGMA ROT=2  STATWT=2  IA=1.226  IB=7.7552
IC=8.7646  IR=.2902     INT ROT POTENTIAL  V(2)=2000.   NU=3005,2822,1743,1441,
1400,1352,1113,509,2967,867,763  REF= BURCAT,MILLER,GARDINER  MAX ERROR CP AT
1300 K .7%
C2H3OA              T04/830   1H   3C   2    0G    300.       5000.     MW=43.05   1
 0.59756699E+01 0.81305914E-02-0.27436245E-05 0.40703041E-09-0.21760171E-13    2
 0.49032178E+03-0.50452509E+01 0.34090624E+01 0.10738574E-01 0.18914925E-05    3
-0.71585831E-08 0.28673851E-11 0.15214766E+04 0.95582905E+01          1552.7   4

C2H3OB CH3CO RADICAL SIGMA=1  STATWT=2  IA=1.0816 IB=8.2237 IC=8.7646 IR=.3648
INT ROT SYM=3  V(3)=1178  NU=3005,2917,1743,1441,1352,1113,919,509,2967,1420,
867  REF=BURCAT,MILLER,GARDINER  MAX ERROR CP AT 1300 K .8%
C2H3OB              T10/82C   2H   3O   1    0G    300.       5000.     MW=43.05   1
 0.56122789E+01 0.84498860E-02-0.28541472E-05 0.42383763E-09-0.22684037E-13    2
-0.51878633E+04-0.32749491E+01 0.31252785E+01 0.97782202E-02 0.45214483E-05    3
-0.90094616E-08 0.31937179E-11-0.41085078E+04 0.11228854E+02          1520.4   4

C2H4  ETHYLENE  STATWT=1  SIGMA=4.   IA=.5752  IB=2.802  IC=3.372  NU=3021,1623,
1342,1023,3083,1236,949,943,3106,826,2989,1444  REF=CHAO,ZWOLINSKY,BURCAT
MAX ERROR CP AT 1300 K .8%
C2H4                T12/79C   2H   4    0    0G    300.       5000.                1
 0.44007187E 01 0.96285827E-02-0.31700802E-05 0.45826254E-09-0.23716445E-13    2
 0.41129180E 04-0.24797812E 01 0.12187214E 01 0.13023630E-01 0.33733277E-05    3
-0.10929970E-07 0.46081021E-11 0.53369570E 04 0.15472055E 02                   4

C2H4O               J 9/65C   2H   4O   1    0G    300.       5000.                1
 0.59249249E 01 0.11120714E-01-0.37434083E-05 0.55413918E-09-0.29549886E-13    2
-0.93028008E 04-0.93792849E 01-0.24173594E 00 0.20761095E-01 0.21481201E-05    3
-0.16948157E-07 0.81075771E-11-0.71720117E 04 0.24432190E 02                   4

C2H4OA ACETALDEHYDE (CH3CHO)  STATWT=1  SIGMA=1  IA=2.76748  IB=6.9781  IR=.44
IC=9.03498  NU=3005,2917,2822,1743,1441,1400,1352,1113,919,509,2967,1420,867,
1113,919,509,2967,1420,867,763  REF=CHAO,WILHOIT,HALL  MAX ERROR CP AT 1300 K .9%
C2H4OA              T 8/81C   2O   1H   4    0G    300.       5000.     MW=44.05   1
 0.58686504E+01 0.10794241E-01-0.36455303E-05 0.54129123E-09-0.28968442E-13    2
-0.22645687E+05-0.60129461E+01 0.25056953E+01 0.13369907E-01 0.46719533E-05    3
-0.11281401E-07 0.42635661E-11-0.21245887E+05 0.13350887E+02          1551.1   4

C2H4O2  ACETIC ACID   STATWT=1  SIGMA=1  IA=7.40342  IB=8.85376   IC=15.7599
IR(CH3)=.0253924  IR(OH)=.095545  POTENTIAL BARRIERS V(OH)=12200.   V(CH3) =
481   NU=3583,3051,2944,1788,1430,1382,1264,1182,989,847,657,581,2996,1430,1048,
642,(565,75 TORSION)  REF=CHAO AND ZWOLINSKI
C2H4O2              L 5/80C   2H   4O   2    0G    300.       5000.                1
 0.80115623E 01 0.12846351E-01-0.48638476E-05 0.80297502E-09-0.48234623E-13    2
-0.55871125E 05-0.17325912E 02 0.12057524E 01 0.22100847E-01 0.23546336E-05    3
-0.16049182E-07 0.71368233E-11-0.53322684E 05 0.20644348E 02                   4

C2H4O4  METHANOIC(FORMIC) ACID  (HCOOH)2 DIMER  STATWT=1  SIGMA=2  IA=13.615
IB=37.724  IC=51.340  NU=3200,2956,1672,1395,1350,1204,675,232,215,1063,677,519,
1073,917,164,68,3110,2957,1754,1450,1365,1218,697,248  REF=CHAO AND ZWOLINSKI
C2H4O4              L 5/80C   2H   4O   0G    300.       5000.                     1
 0.12233341E 02 0.13638251E-01-0.46521382E-05 0.69714767E-09-0.37684533E-13    2
-0.10397013E 06-0.35852188E 02 0.38265133E 01 0.26852809E-01 0.25594691E-05    3
-0.21561117E-07 0.10201341E-10-0.10105763E 06 0.10259444E 02                   4

C2H5  ETHYL RADICAL  STATWT=2  SIGMA=1  IAIBIC=1.3E117  IR=0.265  SIGMA ROT=3
POTENTIAL BARRIER 658  NU=2925,3000(3),2950,1050(2),1400,1450(3),800,1250,1300
REF=TSIV  MAX ERROR CP AT 1300 K .9%
C2H5                R  79 C   2H   5    0    0G    300.       5000.     MW=29.06   1
 0.41068954E+01 0.12077987E-01-0.40150690E-05 0.58634697E-09-0.30743528E-13    2
 0.10632148E+05 0.23229771E+01 0.19983559E+01 0.10580499E-01 0.79106931E-05    3
-0.10932574E-07 0.31684750E-11 0.11754070E+05 0.15462708E+02          1431.5   4
```

```
C2H5O    (CH2CH2OH) RADICAL   SIGMA=1   STATWT=2   IA=2.359   IB=8.08   IC=9.935
IR(CH2)=0.79   IR(OH)=.1363   INT ROT POTENTIAL   SIGMA ROT(OH)=2   V(2)=201
SIGMA ROT(CH2)=3   V(3)=3000   NU=3689,2989(3),1320,1242,1067,1040,877,427,1456,
1270,1104,801   REF=BURCAT,MILLER,GARDINER   MAX ERROR CP AT 1300 K .7%
C2H5O                 T 4/83H    5C    2O    1    0G    300.       5000.      MW=45.06    1
 0.75944014E+01 0.93229339E-02-0.30303854E-05 0.43216319E-09-0.21970039E-13    2
-0.57727852E+04-0.13968735E+02 0.14019508E+01 0.21543175E-01-0.22326512E-05    3
-0.14464092E-07 0.80488420E-11-0.38464519E+04 0.19135818E+02            1606.9    4

C2H5O   (CH3CHOH)     RADICAL   SIGMA =1   STATWT=2   IA=1941   IB=9.542   IC=10.916
IR(CH3)=.4331   IR(OH)=.1363   INT ROT POTENTIAL   SIGMA ROT(CH3)=3   V(3)=3300.
SIGMA ROT(OH)=2   V(2)=800.    NU=3689,2989(4),1456(3),1391,1242,1067,1040,877,427,
1270 1104   REF=BURCAT,MILLER,GARDINER   MAX ERROR CP AT 1300 K   .8%
C2H5O                 T 4/83C    2O    1H    5    0G    300.       5000.      MW=45.06    1
 0.67665424E+01 0.11634436E-01-0.37790651E-05 0.53828875E-09-0.27315345E-13    2
-0.56092969E+04-0.94112072E+01 0.24813328E+01 0.16790036E-01 0.37755499E-05    3
-0.13923497E-07 0.60095193E-11-0.40120054E+04 0.14568459E+02            1709.7    4

C2H5O   ETHYL-OXIDE RADICAL   (CH3CH2O) SIGMA=1   SIGMA BARRIER=3   STATWT=2
IA=2.327 IB=8.395   IC=9.693   IR=0.4331   V(3)=3300.    NU=2989(3),1456(2),1391,
1320,1126,1040,868,423,2989,1456,1270,1104,801   REF=BURCAT,MILLER,GARDINER
MAX ERROR CP AT 1300 K 0.8%
C2H5O                 T11/82O    1C    2H    5    0G    300.       5000.      MW=45.06    1
 0.60114346E+01 0.12165219E-01-0.40449604E-05 0.59076588E-09-0.30969595E-13    2
-0.49366992E+04-0.68033428E+01 0.17302504E+01 0.16908489E-01 0.39996221E-05    3
-0.13711180E-07 0.57643603E-11-0.32922483E+04 0.17322952E+02            1517.0    4

C2H5O   CH2-O-CH3 RADICAL   SIGMA=1   SIGMA BARRIER=2,3   STATWT=2   IA=2.15 IB=7'.91
IC=9.558   IR(CH2)=3.335   V(2)=2000   IR(CH3)=4.787   V(3)=2720   NU=2996(2),2817,
1464(3),1452,1244,928,418,2952,1150,1227,1102,2925,1179   REF=BURCAT,MILLER,
GARDINER MAX ERROR CP AT 1300 K 0.8%
C2H5O                 T11/82C    2H    5O    1    0G    300.       5000.      MW=45.06    1
 0.65567484E+01 0.12180723E-01-0.40628420E-05 0.59495830E-09-0.31276214E-13    2
-0.40282515E+04-0.81434402E+01 0.35953999E+01 0.13379216E-01 0.53914910E-05    3
-0.10947097E-07 0.38193320E-11-0.27021975E+04 0.92718735E+01            1766.4    4

C2H6   ETHANE   STATWT=1   SIGMA=6   SIGMA BARRIER=3   IA=1.0481   IB=IC=4.22486
IR=.26203   NU=2954,1388,995,2896,1379,2969(2),1468(2),1190(2),2985(2),1469(2),
822(2)   POTENTIAL BARRIER V0=2.96 KCAL   REF=CHAO,WILHOIT,ZWOLINSKI   MAX ERROR
CP AT 1300 K 1.0%
C2H6 G                T12/78C    2H    6    0    0G    300.       4000.                   1
 0.48259382E 01 0.13840429E-01-0.45572588E-05 0.67249672E-09-0.35981614E-13    2
-0.12717793E 05-0.52395067E 01 0.14625387E 01 0.15494667E-01 0.57805073E-05    3
-0.12578319E-07 0.45862671E-11-0.11239176E 05 0.14432295E 02                      4

C2H6N2   AZOMETHANE   STATWT=1   SIGMA=2   IA=2.063   IB=19.082   IC=20.029 IR=0.425
POTENTIAL BARRIER V0=1700.    NU=2989,2926,1583,1437,1381,1179,919,591,2977,
1416,1027,2981,1440,1111,312,2988,2925,1447,1384,1112,1008,353,(2148222 TORSION)
REF=PAMIDIMUKKALA,ROGERS,SKINNER   MAX ERROR CP AT 1300 K .9%
C2H6N2 G              T 8/81C    2H    6N    2    0G    300.       3000.                   1
 0.10862436E+02 0.80828294E-02 0.27653668E-05-0.26001215E-08 0.44496541E-12    2
-0.26410972E+04-0.31821930E+02 0.34451923E+01 0.18929366E-01 0.71458344E-05    3
-0.15136251E-07 0.51713807E-11 0.81479553E+02 0.93749094E+01                      4

C2H6O   ETHANOL (C2H5OH)      STATWT=1   SIGMA=1   SIGMA BARRIER CH3=3   IA=2.3051
IB=9.1718   IC=10.558   IRCH3=.4331   IROH=.1363   NU=3689,2989(3),1456(2),1391,1320
1242,1067,1040,877,2989(2),1456,1270,1104,801   POTENTIAL BARRIERS 3.3(CH3) .8(OH)
REF=GREEN   MAX ERROR CP AT 1300 K 1.0%
C2H6O                 U 5/79C    2H    6O    1    0G    300.       5000.                   1
 0.79087286E+01 0.12227729E-01-0.35144249E-05 0.42572035E-09-0.15468177E-13    2
-0.31943867E+05-0.16426895E+02 0.16487226E+01 0.21139644E-01 0.32672033E-05    3
-0.16375399E-07 0.73521788E-11-0.29673598E+05 0.18271423E+02                      4

C2H6O   DIMETHYL-ETHER SIGMA=2   SIGMA BARRIER=3   IA=2.164   IB=8.347   IC=9.445
IR=4.852   V(3)=2720.    NU=2996(2),2817(2),1464(4),1452(2),1244,928,418,1150,2952,
1227,1102,2925,1179   REF=SHIMANOUCHI   MAX ERROR CP AT 1300 K 0.9
C2H6O                 T11/82C    2H    6O    1    0G    300.       5000.      MW=46.07    1
 0.63939190E+01 0.14928680E-01-0.50054186E-05 0.73737505E-09-0.39071824E-13    2
-0.25358937E+05-0.99885874E+01 0.28876772E+01 0.15701063E-01 0.76460055E-05    3
-0.13923732E-07 0.47045120E-11-0.23739461E+05 0.10834729E+02            1732.4    4

C2N                   J 3/67C    2N  100         0G    300.       5000.                   1
 0.61931308E 01 0.14327539E-02-0.61255161E-06 0.11578707E-09-0.80401339E-14    2
 0.64818372E 05-0.84132298E 01 0.32670394E 01 0.98211307E-02-0.83284733E-05    3
 0.17650559E-08 0.59632768E-12 0.65589057E 05 0.65682304E 01                      4
```

```
C2N2              J 3/61C   2N   200         0G    300.      5000.            1
 0.65968935E 01 0.38694131E-02-0.15516161E-05 0.28141546E-09-0.19069442E-13   2
 0.34883726E 05-0.10001801E 02 0.39141782E 01 0.14011008E-01-0.17404350E-04   3
 0.12012779E-07-0.33565772E-11 0.35514550E 05 0.32384353E 01                  4

C2O               J 9/66C   2O   100         0G    300.      5000.            1
 0.48990313E 01 0.28430384E-02-0.10209669E-05 0.16112165E-09-0.95542914E-14   2
 0.32800545E 05-0.91382280E 00 0.35364815E 01 0.69543872E-02-0.53071374E-05   3
 0.17030470E-08-0.14108072E-13 0.33151572E 05 0.60172370E 01                  4

C3                J12/69C   3O   00    00    0G    300.      5000.            1
 0.36815361E 01 0.24165236E-02-0.84348112E-06 0.14508198E-09-0.95697300E-14   2
 0.97413955E 05 0.68377802E 01 0.57408464E 01-0.84281238E-02 0.18620198E-04   3
-0.14510529E-07 0.39676977E-11 0.97157524E 05-0.23837376E 01                  4

C3D4  CYCLOPROPENE-D4  STATWT=1  SIGMA=2  IA=3.861  IB=4.9423  IC=7.826
NU=2435,2142,1548,1147,1023,639,749,640,2313,885,863(2),637,2262,424
REF=BURCAT(1982)  MAX ERROR CP AT 1300 K .8%
C3D4              T 2/82C   3D   4    0      0G    300.      5000.   MW=44.09 1
 0.89251080E+01 0.92740692E-02-0.33307069E-05 0.52548144E-09-0.30162352E-13   2
 0.27717801E+05-0.24785095E+02 0.87993717E+00 0.25426447E-01-0.47690091E-05   3
-0.14818401E-07 0.86449008E-11 0.30267191E+05 0.18301620E+02           1521.4 4

C3D6  CYCLOPROPANE-D6  STATWT=1  SIGMA=6  IA=IB=6.0672  IC=8.75747  NU=2236,
1274,956,800,870,2336,614,2211(2),1072(2),855(2),717(2),2329(2),940(2),528(2)
REF=DUNCAN AND BURNS  MAX ERROR CP AT 1300 K .9%
C3D6              T12/81C   3D   6    0      0G    300.      5000.            1
 0.10402956E+02 0.12471735E-01-0.44642438E-05 0.70182371E-09-0.40115975E-13   2
-0.77593262E+03-0.35106918E+02-0.79611647E+00 0.35631880E-01-0.75448597E-05   3
-0.20582778E-07 0.12364153E-10 0.27102590E+04 0.24627518E+02        1581.86   4

C3H RADICAL  STATWT=3  SIGMA=1  B0=.4047  NU=3312,2089(2),1230,63(2),2040
REF=SPANGENBERG AND BORGER  MAX ERROR CP AT 1300 K .7%
C3H G             T04/80C   3H   1    0      0G    300.      5000.            1
 0.41682930E 01 0.53355880E-02-0.17796119E-05 0.26075297E-09-0.13723909E-13   2
 0.63058313E 05 0.57827826E 01 0.52461081E 01-0.86180924E-04 0.39655006E-05   3
 0.73884454E-09-0.18929944E-11 0.63032672E 05 0.11971579E 01                  4

C3H2 RADICAL  RECONSTRUCTED FROM DUFF AND BAUER POLYNOMIAL  MAX ERROR CP AT
1700 K .6%
C3H2              U04/78C   3H   2    0      0G    300.      5000.            1
 0.65308533E+01 0.58703162E-02-0.17207767E-05 0.21274979E-09-0.82919105E-14   2
 0.51152137E+05-0.11227278E+02 0.26910772E+01 0.14803664E-01-0.32505513E-05   3
-0.86443634E-08 0.52848776E-11 0.52190719E+05 0.87573910E+01                  4

C3H3 RADICAL  STATWT=2  SIGMA=2  IA=.278  IB=9.2073  IC=9.4853  NU=3429,3061,
2142,615,1410,1070(2),626,945,400,2977,450  REF=DUFF AND BAUER    MAX ERROR
CP AT 3000 K .4%
C3H3              U12/77C   3H   3    0      0G    300.      3000.            1
 0.80916252E+01 0.37372850E-02 0.13886647E-05-0.12298604E-08 0.20681585E-12   2
 0.35437793E+05-0.18204468E+02 0.25097322E+01 0.17103866E-01-0.45710858E-05   3
-0.82841574E-08 0.54362287E-11 0.37040680E+05 0.11011264E+02                  4

C3H4P PROPYNE  STATWT=1  SIGMA=3  IA=.5283  IB=IC=9.8172  NU=3334,2918,2142,
1382,931,3008(2),1452(2),1053(2),633(2),328(2)  REF=SHIMANOUCHI  MAX ERROR
CP AT 1300 K .7%
C3H4P             U05/78H   4C   3    0      0G    300.      5000.            1
 0.65344639E+01 0.10337744E-01-0.34001059E-05 0.49091375E-09-0.25363748E-13   2
 0.20110711E+05-0.12785910E+02 0.16365213E+01 0.19788995E-01-0.20252373E-05   3
-0.11532656E-07 0.56520986E-11 0.20800957E+05 0.98740911E+01                  4

C3H4A ALLENE  STATWT=1  SIGMA=2  IA=.555  IB=IC=9.4389  NU=3015,1443,1073,
865,3007,1957,1398,3086(2),999(2),841(2),355(2)  REF SHIMANOUCHI  MAX ERROR
CP AT 1300 K .3%
C3H4A             U06/78C   3H   4    0      0G    300.      5000.            1
 0.67737303E+01 0.10210991E-01-0.33816414E-05 0.49214188E-09-0.25700138E-13   2
 0.20110711E+05-0.12785910E+02 0.16365213E+01 0.11978895E-01-0.20252373E-05   3
-0.11780120E-07 0.64494235E-11 0.21779836E+05 0.14916451E+02                  4

C3H4C CYCLOPROPENE  STATWT=1  SIGMA=2  IA=2.792  IB=3.846  IC=6.085  NU=3152,
2909,1653,1483,1105,905,996,815,3116,1043,1011,769,2995,1088,569  REF=YUM AND
EGGERS  MAX ERROR CP AT 1300 K .8%
C3H4C             T12/81C   3H   4    0      0G    300.      5000.   MW=40.06 1
 0.66999931E+01 0.10357372E-01-0.34551167E-05 0.50652949E-09-0.26682276E-13   2
 0.30199051E+05-0.13391933E+02-0.24621047E-01 0.23197215E-01-0.18474357E-05   3
-0.15927593E-07 0.86846155E-11 0.32334137E+05 0.22716599E+02           1367.9 4
```

```
C3H5 RADICAL  STATWT=2  SIGMA=2  IA=1.5225 IB=9.8756 IC=11.3981  NU=3019,1342,
950,3075,1288,1500,3105,1236,1400,2989,810,380,950,949,750(2),1443,943  REF=DUFF
AND BAUER  MAX ERROR CP AT 1300 K .8%  HF0=146.1 KJ
C3H5              U12/77C   3H   5    0    0G    300.      5000.              1
 0.79091978E+01 0.12115255E-01-0.41175863E-05 0.61566796E-09-0.33235733E-13    2
 0.12354156E+05-0.19672333E+02-0.54100400E+00 0.27284101E-01-0.96365329E-06    3
-0.19129462E-07 0.98394175E-11 0.15130395E+05 0.26067337E+02          1453.9  4

C3H5NO2  NITROCYCLOPROPANE  STATWT=1  SIGMA=2  IA=10.5515  IB=28.5698  IC=32.482
IR=2.61 POTENTIAL BARRIER  V(2)=3300.  SIGMAR=2 NU=3095,3019,2934,1571,1443,1373,

1325,1202,1110,1075,1042,(880),854,770,483,289,3103,3019,1407,1118,936,921,828,
730,645,309,(70 TORSION)  REF=HOLTZCLAW,HARRIS,BUSH   MAX ERROR CP AT 1300 K .7%
C3H5NO2           T11/81C   3H   5N   10   2G    300.      5000.      MW=87.08 1
 0.13030508E+02 0.15062746E-01-0.51262668E-05 0.76649598E-09-0.41441788E-13    2
-0.25730569E+04-0.44244141E+02-0.53064877E+00 0.42566467E-01-0.55552337E-05    3
-0.29570092E-07 0.16781715E-10 0.15987754E+04 0.28050201E+02          2051.5  4

C3H6  PROPYLENE  STATWT=1  SIGMA=3  IA=1.8133 IB=9.0187 IC=10.317 IR=0.3945
POTENTIAL BARRIER V3=1997.   NU=3091,3022,2991,2973,2932,1653,1459,1414,1378,
1298,1178,935,919,428,2953,1443,1045,990,912,57V,  REF=CHAO AND ZWOLINSKI
MAX ERROR CP AT 1300 K .9%
C3H6              T12/81C   3H   6    0    0G    300.      5000.      MW=42.08  1
 0.67322569E+01 0.14908336E-01-0.49498994E-05 0.72120221E-09-0.37662043E-13    2
-0.92357031E+03-0.13313348E+02 0.14933071E+01 0.20925175E-01 0.44867938E-05    3
-0.16689121E-07 0.71581465E-11 0.10748264E+04 0.16145340E+02          1703.1  4

C3H6  CYCLOPROPANE  STATWT=1  SIGMA=6  IA=IB=4.1766  IC=6.6358  NU=3038,1479,
1188,1126,1070,3102,854,3024(2),1438(2),1029(2),867(2),3082(2),1188(2),739(2)
REF=SHIMANOUCHI  MAX ERROR CP AT 1300 K .9%
C3H6CY            T12/81C   3H   6    0    0G    300.      5000.              1
 0.68271122E+01 0.15246719E-01-0.50710187E-05 0.74134521E-09-0.38929339E-13    2
 0.27764282E+04-0.16869995E+02-0.17865782E+01 0.29799264E-01 0.13068729E-05    3
-0.23190996E-07 0.11576777E-10 0.56478437E+04 0.29963318E+02      1372.6      4

C3H6O  PROPYLENEOXIDE  STATWT=1  SIGMA=1  IA=4.657  IB=12.561  IC=14.103
IR=.53  NU=3065(2),3006,2975,2929,2846,1500,1456(2),1406,1368,1263,1166,1142,
1132,1102,1023,950,896,828,745,416,371 POTENTIAL BARRIER V3=2560 REF=SWALEN AND
HERSCHBACH MAX ERROR CP AT 1300 K .9%
C3H6O             L 4/77C   3H   6O   1    0G    300.      5000.      MW=58.08  1
 0.87072573E+01 0.15987653E-01-0.53762797E-05 0.79422535E-09-0.42212622E-13    2
-0.69370352E+04-0.22579315E+02 0.48378503E+00 0.28574701E-01 0.28022350E-05    3
-0.22371523E-07 0.10581544E-10-0.40665557E+04 0.22616104E+02          1733.7  4

N-C3H7  N-PROPYL RADICAL  STATWT=2 SIGMA=6  IA=9.005953 IB=10.613035 IC=2.67748
IR(CH3)=.26287  V(3)=3400.   IR(CH2)=.4576  V(3)=1980.   NU=3100(2),2960(5),
1440(5),1390,1100,990,980,960,460,380    REF=BURCAT(UNPUBLISHED)
MAX ERROR CP AT 1300 K .8%
N-C3H7            T 8/81C   3H   7    0    0G    300.      5000.              1
 0.79782906E+01 0.15761133E-01-0.51732432E-05 0.74438922E-09-0.38249782E-13    2
 0.75794023E+04-0.19356110E+02 0.19225368E+01 0.24789274E-01 0.18102492E-05    3
-0.17832658E-07 0.85829963E-11 0.97132812E+04 0.13992715E+02                  4

I-C3H7  ISOPROPYL RADICAL STATWT=2  SIGMA=18  IA=10.103 IB=11.3281 IC=2.295433
IR(CH3)=.347966  POTENTIAL BARRIER V(3)=3400  NU=3100,2960(6),1440(6),1300,1200
990(4),950,398,367  REF=BURCAT (UNPUBLISHED)  MAX ERROR CP AT 1300 K .8%
I-C3H7            T 8/81C   3H   7    0    0G    300.      5000.              1
 0.80633688E+01 0.15744876E-01-0.51823918E-05 0.74772455E-09-0.38544221E-13    2
 0.53138711E+04-0.21926468E+02 0.17132998E+01 0.25426164E-01 0.15808082E-05    3
-0.18212862E-07 0.88277103E-11 0.75358086E+04 0.12979008E+02                  4

C3H7N CYCLOPROPYLAMINE (C3H5NH2)  REF=DRAEGER,HARRISON,GOOD   DATA EXTRAPO-
LATED THROUGH WILHOIT POLYNOMIAL   HF298=77.37 KJ  MAX ERROR CP AT 1400 K 1.0%
C3H5NH2           L 2/84C   3H   7N   1    0G    300.      5000.      MW=57.09  1
 0.11077434E 02 0.15626516E-01-0.52517407E-05 0.79408302E-09-0.43887471E-13    2
 0.43691211E 04-0.35471283E 02 0.92693955E 00 0.35704415E-01-0.35520043E-05    3
-0.24779276E-07 0.13902465E-10 0.75181836E 04 0.18755966E 02          2039.3  4

C3H8  PROPANE  STATWT=1  SIGMA=6   SIGMA BARRIER=3  IA=2.8899  IB=IC=10.5472
IR=.44202  NU=2977,2962,2887,1476,1462,1392,1158,869,369,2967,1451,1278,940,
2968,2887,1464,1378,1338,1054,922,2973,2968,1472,1192,748  POTENTIAL BARRIER
V0=3.29  REF=CHAO,WILHOIT,ZWOLINSKI  MAX ERROR CP AT 1300 K .9%
C3H8              L 4/80C   3H   8    0    0G    300.      5000.              1
 0.75252171E 01 0.18890340E-01-0.62839244E-05 0.91793728E-09-0.48124099E-13    2
-0.16464547E 05-0.17843903E 02 0.89692080E 00 0.26689861E-01 0.54314251E-05    3
-0.21260007E-07 0.92433301E-11-0.13954918E 05 0.19355331E 02                  4
```

```
1-C3H8O    1-PROPANOL  C3H7OH  WILHOIT AND ZWOLINSKY DATA EXTRAPOLATED WITH
WILHOIT POLYNOMIAL.   MAX ERROR CP AT 1300 K .8%
1-C3H7OH            L 1/84C   3H   8O   1    0G    300.      5000.   MW=60.10     1
 0.10065836E 02 0.18030774E-01-0.57645793E-05 0.82326701E-09-0.42296153E-13    2
-0.36654254E 05-0.26072647E 02 0.23088903E 01 0.26964691E-01 0.72370322E-05    3
-0.23160375E-07 0.97628529E-11-0.33663242E 05 0.17618973E 02            2172.5  4

C3O2                J 6/68C   3O   2O   0O    0G    300.      5000.                1
 0.81435964E 01 0.54395018E-02-0.22192869E-05 0.40778627E-09-0.27915974E-13    2
-0.14230013E 05-0.15456769E 02 0.37161005E 01 0.19872164E-01-0.20935751E-04    3
 0.11750112E-07-0.26589416E-11-0.13089402E 05 0.69298412E 01                   4

C4                  J12/69C   4O   0O   0O    0G    300.      5000.                1
 0.65602101E 01 0.40985234E-02-0.17000471E-05 0.31615228E-09-0.21842144E-13    2
 0.11430434E 06-0.11820311E 02 0.18432021E 01 0.19343592E-01-0.20627502E-04    3
 0.10822626E-07-0.21289203E-11 0.11550276E 06 0.12006898E 02                   4

C4H  RADICAL  STATWT=2 SIGMA=1  B0=0.1558 NU=3350,2023,700,1700,2089(2),482(2),
220(2)  REF=SPANGENBERG AND BORGER  MAX ERROR CP AT 1300 K .6%
C4H                 T04/80C   4H   1    0     0G    300.      5000.                1
 0.62428818E 01 0.61936826E-02-0.20859316E-05 0.30822034E-09-0.16364826E-13    2
 0.75680188E 05-0.72108059E 01 0.50232468E 01 0.70923753E-02-0.60737619E-08    3
-0.22757523E-08 0.80869941E-12 0.76238125E 05-0.69425941E-01                   4

C4H2  BUTADIYNE  STATWT=1  SIGMA=2  IB=19.054  NU=3329(2),2184,874,2020,627(2)
483(2),630(2),220(2) REF=DUFF AND BAUER MAX ERROR CP AT 400 K .7% HF298=440.5 KJ
C4H2                U06/78C   4H   2    0     0G    300.      5000.                1
 0.90314074E+01 0.60472526E-02-0.19487888E-05 0.27548630E-09-0.13856080E-13    2
 0.52947355E+05-0.23850677E+02 0.40051918E+01 0.19810002E-01-0.98658775E-05    3
-0.66351582E-08 0.60774129E-11 0.54240648E+05 0.18457365E+01            1730.1  4

C4H3 RADICAL  STATWT=2  SIGMA=2  IA=.278  IB=20.2884  IC=20.5664  NU=3429,1450,

220,615,1060,300,629,940,1980,3012,483,2080,3102,510,925   REF=DUFF AND BAUER
MAX ERROR CP AT 1300 K .6%  HF0=429 KJ
C4H3                U06/78C   4H   3    0     0G    300.      5000.                1
 0.84874201E+01 0.86908937E-02-0.28544437E-05 0.41200798E-09-0.21301093E-13    2
 0.47970555E+05-0.19018509E+02 0.35539713E+01 0.19461986E-01-0.48102484E-05    3
-0.97301225E-08 0.62390535E-11 0.49453863E+05 0.70829868E+01                   4

C4H4  BUTHAN-1EN-3YN  STATWT=1  SIGMA=1  IA=1.774  IB=17.301  IC=19.075
NU=3012(2),3102,1410,1288,1090,935,950,678,3305,615,629,875,1600,2099,538,309,
219  REF=DUFF AND BAUER  MAX ERROR CP AT 1300 K .7%  HF0=315 KJ
C4H4                U03/78C   4H   4    0     0G    300.      5000.                1
 0.88921490E+01 0.10908850E-01-0.35949597E-05 0.51934190E-09-0.26808921E-13    2
 0.33284348E+05-0.21726944E+02 0.21403370E+01 0.24605047E-01-0.36391657E-05    3
-0.15304014E-07 0.88964608E-11 0.35374000E+05 0.14282477E+02                   4

C4H6  2-BUTAYN (DIMETHYLACETYLENE) STATWT=1  SIGMA=6  IA=1.0605  IB=IC=25.0029
IR=.530  V(3)=500.  (ONE ROTATION ONLY)  NU=371(2),725,1029(2),1050(2),1380(2),
1126,1448(2),1468(2),2270,2916,2966(2),2976(3),213(2)  REF=YOST,OSBORNE,GARNER
MAX ERROR CP AT 1300 K  .9%
C4H6                T12/82C   4H   6    0     0G    300.      5000.      MW=54.09  1
 0.80465832E+01 0.16485251E-01-0.55222272E-05 0.81235929E-09-0.42950784E-13    2
 0.13701305E+05-0.18004578E+02 0.31971083E+01 0.20255916E-01 0.65101922E-05    3
-0.16584423E-07 0.64002822E-11 0.15715203E+05 0.98956604E+01            2029.7  4

C4H8  1-BUTENE CH2=CH-CH2-CH3        SPECTROSCOPIC DATA NOT AVAILABLE.
REF= CHAO AND HALL   MAX ERROR CP AT 4500 K  1.5%
C4H8                T 6/83C   4H   8    0     0G    300.      5000.   MW=56.10     1
 0.20535841E+01 0.34350507E-01-0.15883197E-04 0.33089662E-08-0.25361045E-12    2
-0.21397231E+04 0.15543201E+02 0.11811380E+01 0.30853380E-01 0.50865247E-05    3
-0.24654888E-07 0.11110193E-10-0.17904004E+04 0.21062469E+02            2036.1  4

C4H8I ISOBUTENE  CH2=C(CH3)2    SPECTROSCOPIC DATA NOT AVAILABLE
REF=CHAO AND HALL   MAX ERROR CP AT 4500 K 1.3 %
C4H8I               T 6/83H   8C   4    0     0G    300.      5000.   MW=56.10     1
 0.44609470E+01 0.29611487E-01-0.13077129E-04 0.26571934E-08-0.20134713E-12    2
-0.50066758E+04 0.10671549E+01 0.26471405E+01 0.25902957E-01 0.81985354E-05    3
-0.22193259E-07 0.88958580E-11-0.40373069E+04 0.12676388E+02            2101.2  4

C4H8T 2-BUTENE-TRANS  CH3-CH=CH-CH3    SPECTROSCOPIC DATA NOT AVAILABLE
REF= CHAO AND HALL   MAX ERROR CP AT 4500 K  1.5%
C4H8T               T 6/83C   4H   8    0     0G    300.      5000.     MW=56.10   1
 0.82797676E+00 0.35864539E-01-0.16634498E-04 0.34732759E-08-0.26657398E-12    2
-0.30521033E+04 0.21342545E+02 0.12594252E+01 0.27808424E-01 0.87013932E-05    3
-0.24402205E-07 0.98977710E-11-0.29647742E+04 0.20501129E+02            2047.4  4
```

```
C4H8C 2-BUTENE-CIS  CH3-CH=CH-CH3   SPECTROSCOPIC DATA NOT AVAILABLE
REF= CHAO AND HALL  MAX ERROR CP AT 4500 K 1.5%
C4H8C              T 6/83C  4H   8    0    0G   300.      5000.   MW=56.10     1
 0.11097383E+01 0.35542578E-01-0.16481703E-04 0.34412202E-08-0.26411468E-12    2
-0.26507607E+04 0.19353516E+02 0.24108791E+01 0.25147773E-01 0.98473047E-05    3
-0.22716758E-07 0.86585895E-11-0.27758694E+04 0.14084535E+02           2071.7  4

C4H8O4  ACETIC ACID DIMER (CH3COOH)2  IA*IB*IC=1.6141 E-112  IR=0.522
POTENTIAL BARRIER V(3)=481  NU=3193,3032,2949,1675,1436,1436,1370,1283,1018,886,
624,448,196,110,3140,3028,2956,1715,1413,1413,1359,1295,1013,886,624,480,188,
2990,1413,1050,934,635,67,47,3000,1436,1112,912,623,115 REF=CHAO AND ZWOLINSKI
C4H8O4             L 5/80C  4H  80    4    0G   300.      5000.                1
 0.16908463E 02 0.24005935E-01-0.80971395E-05 0.11987000E-08-0.63810502E-13    2
-0.11940706E 06-0.56992554E 02 0.38786602E 01 0.43348286E-01 0.58980841E-05    3
-0.35640014E-07 0.16467286E-10-0.11481000E 06 0.14821056E 02                   4

C4H9  T-BUTYL RADICAL  ( C(CH3)3 )  STATWT=2  SIGMA=3  SIGMA BARRIER=3X3
IA=10.8018  IB=10.7942  IC=19.9462  IR(CH3)=0.4896  INTERNAL BARRIER V(3)=500.
NU=2931(6),2825(3),1455(6),1370(3),1279,1252(2),1182(2),1126,992(2),733,541(2),
200  REF= PACANSKY AND CHANG  MAX ERROR CP AT 1300 K 1.0%
C4H9               T 6/83C  4H   9    0    0G   300.      5000.   MW=57.12     1
 0.77541008E+01 0.23770157E-01-0.80376340E-05 0.11938439E-08-0.63866501E-13    2
 0.74535205E+03-0.15286139E+02 0.14004288E+01 0.24845254E-01 0.15668920E-04    3
-0.25635558E-07 0.83376197E-11 0.36863530E+04 0.22517868E+02           2161.6  4

N-C4H10  N-BUTANE   STATWT=1  SIGMA=18   MIXTURE OF ONE TRANS AND TWO IDENTICAL
GAUCHE ISOMERS IN EQUILIBRIUM.     REF=CHEN,WILHOIT,ZWOLINSKI
MAX ERROR CP AT 1300 K .9%
N-C4H10            L 4/80C  4H  10    0    00   300.     5000.                 1
 0.10526502E 02 0.23627248E-01-0.78760077E-05 0.11501951E-08-0.60251446E-13    2
-0.20519313E 05-0.32140045E 02 0.15223389E 01 0.34286838E-01 0.81006638E-05    3
-0.29214874E-07 0.12672835E-10-0.17125832E 05 0.18349014E 02                   4

I-C4H10  I-ISOBUTANE(2-METHYLPROPANE)   STATWT=1  SIGMA=81  IA=18.648  IB=10.777
IC=10.777  IR=3X(0.51364)  POTENTIAL BARRIERS V3=3851.  V6=-150.   NU=2962(2),
2904(2),2880(2),1477(2),1394(2),1177(2),797(2),433(2),2958(2),1450(2),981(2),
2962,2962,2894,1477,1475,1371,1330,1166,966,367,(TORSION 256,220(2))
REF=CHEN,WILHOIT,ZWOLINSKI
I-C4H10            L 5/80C  4H  10    0    0G   300.      5000.                1
 0.10845599E 02 0.23333851E-01-0.77793875E-05 0.11375818E-08-0.59640660E-13    2
-0.21725719E 05-0.35869400E 02 0.50704670E 00 0.38149782E-01 0.46916175E-05    3
-0.29491598E-07 0.13621288E-10-0.18030996E 05 0.21284882E 02                   4

C4F6  PERFLUORO 1-3 BUTADIENE  SIGMA=1  SIGMA BARRIER=2  IA=42.1018 IB=80.1264
IC=116.0475  IR=132.823  V(2)=2850.  NU=1796,1381,1138,933,702,660,529,464,396,
375,329,181,1765,1329,1189,972,633,547,520,422,293,259,204,   REF=WURREY,BUCY,
DURIG  MAX ERROR CP AT 1300 K 0.4%
C4F6               T12/82C  4F   6    0    0G   300.      5000.   MW=162.03    1
 0.20649826E+02 0.63778609E-02-0.24356023E-05 0.40486192E-09-0.24477111E-13    2
-0.12834769E+06-0.75447845E+02 0.61921721E+01 0.40591445E-01-0.14628447E-04    3
-0.22981666E-07 0.15799126E-10-0.12425062E+06 0.14456344E+00           3000.6  4

C4F6  PERFLUOROCYCLOBUTENE  SIGMA=2  IA=53.90  IB=64.95  IC=87.73  NU=1799, 4 8,
1387,1136,966,684,469.2,286,1182,493,337,174,98,1282,638,187,146,1259,1171,983,
579,429,238,217  REF=NIELSEN AND EL-SABEN     MAX ERROR CP AT 1300 K 0.5%
C4F6               T12/82F  6C   4    0    0G   300.      5000.   MW=162.03    1
 0.19723373E+02 0.81368275E-02-0.30685842E-05 0.50541860E-09-0.30311613E-13    2
-0.15313506E+06-0.72036636E+02 0.60944862E+01 0.36527760E-01-0.68788740E-05    3
-0.24970031E-07 0.14493539E-10-0.14896744E+06 0.48356694E+00           3023.0  4

C5                 J12/69C  50  00   00    0G   300.      5000.                1
 0.82067016E 01 0.54889888E-02-0.22694876E-05 0.42073365E-09-0.28981924E-13    2
 0.11463647E 06-0.20246108E 02 0.11012446E 01 0.29513421E-01-0.33754342E-04    3
 0.19056534E-07-0.40989018E-11 0.11637970E 06 0.15360193E 02                   4

C5H  RADICAL   T0=0   STATWT=4   IB=35.5335  NU=712(2),557(2),637(2),843,3329,
2290,586(2),2200,1570   SIGMA=1   T0=4      STATWT=2   REF=DUFF AND BAUER
MAX ERROR CP AT 400 K .9%
C5H                T12/81C  5H   1    0    0G   300.      5000.   MW=61.06     1
 0.86957493E+01 0.60543008E-02-0.20160105E-05 0.28928926E-09-0.14700995E-13    2
 0.90310687E+05-0.21015945E+02 0.16348248E+01 0.25095381E-01-0.12066364E-04    3
-0.10465111E-07 0.88099883E-11 0.92124875E+05 0.15121937E+02           1444.8  4

C5H2 RADICAL   SIGMA=2 T0=0  STATWT=3 IB=37.7286  NU=627(2),350(2),1900,630(2),
3329(2),1800,550(2),1570,450(2),843  T0=1576.4  STATWT=2 T0=2624 STATWT=1
REF=DUFF AND BAUER  MAX ERROR CP AT 400 K .6%
C5H2               T12/81C  5H   2    0    0G   300.      5000.   MW=62.07     1
 0.11329175E+02 0.74240565E-02-0.26281887E 05 0.40825410E-09-0.23013326E-13    2
 0.78787062E+05-0.36171173E+02 0.30623217E+01 0.27099982E-01-0.10091697E-04    3
-0.12727451E-07 0.91672191E-11 0.81149687E+05 0.70710783E+01           1764.9  4
```

```
C5H3 RADICAL  SIGMA=2  STATWT=2  IA=.2813 IB=39.0398 IC=39.3211  NU=3012,3102,
1410,1090,935,3005,615,629,870,1950,2100,1200,480,220,530,350,200,300
REF=DUFF AND BAUER  MAX ERROR CP AT 1300 K .6%
C5H3                  T12/81C   5H   3    0    0G   300.      5000.     MW=63.08   1
 0.10787622E+02 0.95396191E-02-0.32067446E-05 0.47333226E-09-0.25121354E-13    2
 0.63929043E+05-0.30054443E+02 0.43287201E+01 0.23524802E-01-0.58567230E-05    3
-0.12154494E-07 0.77264783E-11 0.65885312E+05 0.41732588E+01            2067.2  4

C5H6  CYCLOPENTADIENE  STATWT=1  SIGMA=2  IA=9.9561  IB=10.198  IC=19.639
NU=709(2),3040,465,3079,1150,1115,3100,1388,1350,3150,730,990,2853,806,960,2959,
1460,911,1085,1316,1496,1248,1207,1531,761,964  REF=DUFF AND BAUER MAX ERROR CP
AT 1300 K .9%
C5H6                  T12/81C   5H   6    0    0G   300.      5000.     MW=66.10   1
 0.96898146E+01 0.18382620E-01-0.62648842E-05 0.93933772E-09-0.50877081E-13    2
 0.11021242E+05-0.31229080E+02-0.31967392E+01 0.40813610E-01 0.68165053E-06    3
-0.31374590E-07 0.15772231E-10 0.15290676E+05 0.38699387E+02            1526.1  4

C5H8  CYCLOPENTENE  REF=DRAEGER,HARRISON,GOOD  DATA EXTRAPOLATED THROUGH
WILHOIT POLYNOMIAL.  HF298=32.34 KJ  MAX ERROR CP AT 1300 K 1.0%
C5H8                  L 2/84C   5H   8    0    0G   300.      5000.     MW=68.12   1
 0.10138640E 02 0.22714138E-01-0.77910463E-05 0.11876522E-08-0.65932448E-13    2
-0.17218359E 04-0.33125885E 02-0.24190111E 01 0.40430389E-01 0.67802339E-05    3
-0.33724742E-07 0.15116713E-10 0.28121887E 04 0.36459244E 02            1812.   4

C5H12  N-PENTANE    API DATA EXTRAPOLATED THROUGH WILHOIT POLYNOMIAL.
MAX ERROR CP AT 1300 K 1.1%
C5H12                 T 5/83C   5H  12    0    0G   300.      4000.   MW=72.15     1
 0.16677979E+02 0.21144830E-01-0.35333214E-05-0.57422023E-09 0.15159483E-12    2
-0.25536699E+05-0.63729401E+02 0.18779078E+01 0.41216455E-01 0.12532337E-04    3
-0.37015369E-07 0.15255686E-10-0.20038156E+05 0.18772568E+02            2908.7  4

C6D5  PHENYL-D5 RADICAL  IA=33.7696  IB=17.9469  IC=15.8227  SIGMA=2  STATWT=2
NU=2293,943,1037,497,2292,969,827,601,1286,824,662(2),2287,1335(2),814,2265(2),
1552(2),867(2),577(2),795,352(2)  REF=BURCAT,ZELEZNIK,MCBRIDE  MAX ERROR CP
AT 1300 K .8%
C6D5                  L 1/84C   6D   5    0    0G   300.      5000.   MW=82.14     1
 0.14720439E 02 0.15228342E-01-0.55355194E-05 0.88271146E-09-0.51227533E-13    2
 0.32547816E 05-0.55710754E 02-0.11891832E 01 0.46892993E-01-0.70647329E-05    3
-0.30889694E-07 0.17495533E-10 0.37568633E 05 0.29493134E 02            1914.6  4

C6D6  BENZENE-D6  IA=35.8938  IB=IC=17.9469  SIGMA=12  NU=2293,943,1037,497,
2292,969,827,601,1286,824,662(2),2287(2),1335(2),814(2),2265(2),1552(2),867(2),
577(2),795(2),352(2)  REF=SHIMANOUCHI  MAX ERROR CP AT 1300 K 0.9%
C6D6                  L 1/84C   6D   6    0    0G   300.      5000.   MW=84.15     1
 0.15617328E 02 0.17128926E-01-0.62044683E-05 0.98573927E-09-0.56961922E-13    2
 0.25859829E 04-0.63879486E 02-0.20221338E 01 0.52606363E-01-0.88031138E-05    3
-0.33621070E-07 0.19310706E-10 0.81311406E 04 0.30485214E 02            1963.4  4

C6H RADICAL RECONSTRUCTED FROM DUFF AND BAUER POLYNOMIAL
MAX ERROR CP AT 1600 K 1.2%
C6H                   U06/78C   6H   1    0    0G   300.      5000.                1
 0.11587352E+02 0.72953627E-02-0.24660085E-05 0.34070458E-09-0.14981855E-13    2
 0.10314481E+06-0.31725784E+02 0.47698479E+01 0.24572790E-01-0.75612525E-05    3
-0.14806908E-07 0.97680536E-11 0.10485231E+06 0.32415304E+01                    4

C6H2  HEXATRIYNE  STATWT=1  SIGMA=2  B0=63.5805  NU=263(2),630(2),2241,504(2),
3329(2),692,472(2),627(2),2119,107(2),1154,2290  REF=DUFF AND BAUER
MAX ERROR CP AT 400 K .7%
C6H2                  U06/78C   6H   2    0    0G   300.      5000.                1
 0.12756519E+02 0.80343808E-02-0.26182151E-05 0.37250603E-09-0.18788509E-13    2
 0.80754687E+05-0.40412628E+02 0.57510853E+01 0.26367199E-01-0.11667596E-04    3
-0.10714498E-07 0.87902975E-11 0.82620125E+05-0.43355322E+01                    4

C6H3 RADICAL  STATWT=2  SIGMA=2  IA=.278 IB=66.1121 IC=66.3901  NU=3012,629,
450,3102,870,230,1410,1580,530,1090,1100,147,935,1950,490,3305,2100,290,615,
107,480 REF=DUFF AND BAUER  MAX ERROR CP AT 1300 K .6%
C6H3                  U05/78C   6H   3    0    0G   300.      5000.                1
 0.12761181E+02 0.10385573E-01-0.34791929E-05 0.51097326E-09-0.26909651E-13    2
 0.74777062E+05-0.38917450E+02 0.50070896E+01 0.26928518E-01-0.59198655E-05    3
-0.15272335E-07 0.94083101E-11 0.77132000E+05 0.22256212E+01                    4

C6H4  BENZYNE  RECONSTRUCTED FROM DUFF AND BAUER POLYNOMIAL
MAX ERROR CP AT 1300 K .6%
C6H4                  U 5/78C   6H   4    0    0G   300.      5000.                1
 0.10062741E+02 0.16903043E-01-0.64730457E-05 0.11240806E-08-0.73075660E-13    2
 0.56453730E+05-0.29693100E+02-0.13004847E+01 0.38664766E-01-0.36439442E-05    3
-0.26685807E-07 0.14509357E-10 0.60029074E+05 0.31249390E+02                    4
```

```
C6H5  PHENYL RADICAL  SIGMA=2  STATWT=2  IA=13.7884 IB=14.8396  IC=28.628
NU=3062,992,1326,673,3068,1010,995,703,1310,1150,849(2),3063,1486(2),1038,
3047(2),1596(2),1178(2),606(2),975,410(2)  REF=BURCAT,ZELEZNIK,MCBRIDE
HF298=325 KJ   MAX ERROR CP 1300 K .9%
C6H5                  L 1/84C   6H   5    0    0G   300.      5000.              1
 0.11433509E 02 0.17014805E-01-0.58358992E-05 0.88020347E-09-0.47983644E-13    2
 0.33941625E 05-0.38578339E 02-0.22878075E 01 0.42416330E-01-0.17666744E-05    3
-0.31422520E-07 0.16505464E-10 0.38370789E 05 0.35398270E 02            1684.4  4

C6H6  BENZENE  IA=IB=14.8396  IC=29.6792  SIGMA=12  NU=3062,992,1326,673,3068,
1010,995,703,1310,1150,849(2),3063(2),1486(2),1038(2),3047(2),1596(2),1178(2),
606(2),975(2),410(2)  REF=SHIMANOUCHI  MAX ERROR CP AT 1300 K 0.9%
C6H6                  L 1/84C   6H   6    0    0G   300.      5000.   MW=78.11   1
 0.11809893E 02 0.19180011E-01-0.65489467E-05 0.98388275E-09-0.53417303E-13    2
 0.40879304E 04-0.43937561E 02-0.32246819E 01 0.47203060E-01-0.21823034E-05    3
-0.34845410E-07 0.18420931E-10 0.89179141E 04 0.37035767E 02            1707.2  4

C6H6O  PHENOL  STATWT=1  SIGMA=1  IA=14.854  IB=32.0450  IC=46.8942  IR=.12236
POTENTIAL BARRIER V(2)=3468.   NU=3656,3087,3063,3027,1603,1501,1261,1176,1168,
1025,999,823,526,958,817,409,973,881,751,686,503,225,3070,3049,1610,1472,1342,
1277,1150,1070,619,403,(309 TORSION) REF=KUDCHADKER,KUDCHADKER,WILHOIT,ZWOLINSKI

MAX ERROR CP AT 1300 K .8%
C6H6O                 T10/81C   6H   6O   1    0G   300.      5000.              1
 0.14943568E+02 0.18317845E-01-0.61606479E-05 0.91048524E-09-0.48410806E-13    2
-0.18386828E+05-0.56096741E+02-0.16677656E+01 0.52091170E-01-0.67895007E-05    3
-0.36257116E-07 0.20586241E-10-0.13287539E+05 0.32421951E+02                    4

C6H10  CYCLOHEXENE  API PROJ 44 VALUES EXTRAPOLATED THROUGH WILHOIT POLYNOMIAL
MAX ERROR CP AT 1300 K 1.0%
C6H10                 U 5/78C   6H  10    0    0G   300.      5000.              1
 0.15927771E+02 0.23744129E-01-0.69086718E-05 0.81097773E-09-0.26831226E-13    2
-0.86426562E+04-0.65251862E+02-0.13942280E+01 0.47206931E-01 0.11960419E-04    3
-0.41628958E-07 0.17403357E-10-0.22177900E+04 0.31296036E+02                    4

C6H12  CYCLOHEXANE  API PROJ 44 VALUES EXTRAPOLATED THROUGH WILHOIT POLYNOMIAL
MAX ERROR CP AT 1300 K 1.3%
C6H12                 U 5/78C   6H  12    0    0G   300.      5000.              1
 0.15942625E+02 0.30796450E-01-0.98351547E-05 0.12947003E-08-0.54355744E-13    2
-0.23763000E+05-0.70234879E+02-0.41429682E+01 0.55591956E-01 0.16841441E-04    3
-0.50317500E-07 0.20171503E-10-0.16109449E+05 0.42525314E+02                    4

C6H12   1-HEXENE   API DATA EXTRAPOLATED THROUGH WILHOIT POLYNOMIAL
MAX ERROR CP AT 1300 K 1.0%
C6H12                 T 5/83C   6H  12    0    0G   300.      4000.   MW=84.16   1
 0.18663635E+02 0.20971451E-01-0.31082809E-05-0.68651618E-09 0.16023608E-12    2
-0.13590895E+05-0.70902695E+02 0.19686203E+01 0.47656231E-01 0.66015373E-05    3
-0.37148173E-07 0.16922463E-10-0.77118789E+04 0.20846069E+02            3134.4  4

C6H14  N-HEXANE   API DATA EXTRAPOLATED THROUGH WILHOIT POLYNOMIAL
MAX ERROR CP AT 1100 K 1.0%
C6H14                 T 5/83C   6H  14    0    0G   300.      4000.   MW=86.17   1
 0.22804718E+02 0.20979892E-01-0.35306739E-05-0.54662452E-09 0.14789499E-12    2
-0.30737566E+05-0.95831619E+02 0.18361740E+01 0.50984614E-01 0.12595857E-04    3
-0.44283624E-07 0.18722371E-10-0.22927496E+05 0.20881454E+02            3452.2  4

C7H7  BENZYL RADICAL  STATWT=2  SIGMA=2  A0=.187 B0=.0906 C0=.0613  NU=3060(2),
3050,2860,1600,1480,1330,1270,1160(2),1010,980(2),810,520,950,820,360(2),890,
750,640,420,170,3040,3030,2930,1550,1440,1350,1300,1070,600,610,(700 TORSION)
REF=ASTHOLZ,DURANT,TROE  MAX ERROR  CP AT 1300 K .8%
C7H7                  T 9/81C   7H   7    0    0G   300.      5000.              1
 0.14765557E+02 0.21834318E-01-0.74497539E-05 0.11167121E-08-0.60388706E-13    2
 0.16991187E+05-0.56287125E+02-0.20902777E+01 0.53650826E-01-0.37096879E-05    3
-0.38106585E-07 0.20462368E-10 0.22393387E+05 0.34406097E+02                    4

C7H8  TOLUENE  STATWT=1  SIGMA=1  IA=14.65  IB=47.993  IC=33.341  IR=.5008
POTENTIAL BARRIER V6=50.     NU=3085,3070,3058,2920,1604,1493,1378,1208,1176,
1028,1002,784,524,973,841,406,2979,1455(3),1040,983,893,734,690,467,217,3037,
3028,2950,1540,1331,1313,1153,1080,1040,6S0,347  REF=HITCHCOCK AND LAPOSA
MAX ERROR CP AT 1300 K .9%
C7H8                  T 9/81C   7H   8    0    0G   300.      5000.              1
 0.13957725E+02 0.24616607E-01-0.83795358E-05 0.12537165E-08-0.67675520E-13    2
-0.10295066E+04-0.52245728E+02-0.25368824E+01 0.52898869E-01 0.14038515E-05    3
-0.40762323E-07 0.20377519E-10 0.44778477E+04 0.37415115E+02                    4

C7H14   1-HEPTENE   API DATA EXTRAPOLATED THROUGH WILHOIT POLYNOMIAL
MAX ERROR CP AT 4000 K .9%
C7H14                 T 5/83C   7H  14    0    0G   300.      4000.   MW=98.18   1
 0.22051590E+02 0.24358388E-01-0.35566181E-05-0.82239082E-09 0.18938377E-12    2
-0.17617004E+05-0.87406464E+02 0.22972660E+01 0.55908546E-01 0.81451562E-05    3
-0.44481435E-07 0.20351013E-10-0.10655879E+05 0.21163208E+02            3634.   4
```

```
C7H16  N-HEPTANE  API PROJ 44 VALUES EXTRAPOLATED THROUGH WILHOIT POLYNOMIAL
MAX ERROR CP AT 1300 K .9%
C7H16                  U05/78C    7H  16     0    0G    300.       5000.             1
 0.22818893E+02 0.32543454E-01-0.11120041E-04 0.17131743E-08-0.96212101E-13     2
-0.33678738E+05-0.94335007E+02 0.30149546E+01 0.54457203E-01 0.21812681E-04     3
-0.54234111E-07 0.20808730E-10-0.26003379E+05 0.17508575E+02                    4

C8H RADICAL  TABLES RECONSTRUCTED FROM DUFF AND BAUER POLYNOMIAL
MAX ERROR CP AT 1600 K 1.2%
C8H                    U06/78C    8H   1     0    0G    300.       5000.             1
 0.14749907E+02 0.99315010E-02-0.33748411E-05 0.46875925E-09-0.20735360E-13     2
 0.13994481E+06-0.48926895E+02 0.44895077E+01 0.35215210E-01-0.10193898E-04     3
-0.21970248E-07 0.14214165E-10 0.14259919E+06 0.39962254E+01                    4

C8H2  OCTATETRAYNE  TABLES RECONSTRUCTED FROM DUFF AND BAUER POLYNOMIAL
MAX ERROR CP AT 1600 K 1.4%
C8H2                   U 6/78C    8H   2     0    0G    300.       5000.             1
 0.15680213E+02 0.11154614E-01-0.37243726E-05 0.51978910E-09-0.23755503E-13     2
 0.10811225E+06-0.55714371E+02 0.46304274E+01 0.39370801E-01-0.11480348E-04     3
-0.25622136E-07 0.16707913E-10 0.11082850E+06 0.80774248E+00                    4

C8H10  1,4-DIMETHYLBENZENE  REF=DRAEGER AND SCOTT  DATA EXTRAPOLATED THROUGH
WILHOIT POLYNOMIAL  HF298=18.03 KJ  MAX ERROR CP AT 1300 K 1.1%
C8H10                  L 2/84C    8H  10     0    0G    300.       5000.   MW=106.17 1
 0.15268401E 02 0.34433573E-01-0.13685810E-04 0.21177802E-08-0.11564062E-12     2
-0.61602461E 04-0.59529587E 02-0.16422014E 01 0.58058664E-01 0.55675910E-05     3
-0.45693085E-07 0.21727536E-10 0.10412372E 03 0.34509140E 02             2642.8 4

C8H16  1-OCTENE  API DATA EXTRAPOLATED THROUGH WILHOIT POLYNOMIAL
MAX ERROR CP AT 1300 K .9%
C8H16                  T 5/83C    8H  16     0    0G    300.       4000.   MW=112.21 1
 0.25359207E+02 0.27841240E-01-0.40735995E-05-0.93202601E-09 0.21516030E-12     2
-0.21615809E+05-0.10340465E+03 0.26774750E+01 0.63882887E-01 0.99165463E-05     3
-0.51358793E-07 0.23292035E-10-0.13626941E+05 0.21278259E+02             4133.5 4

C8H18  NORMAL OCTANE  DATA EXTRAPOLATED THROUGH WILHOIT POLYNOMIAL FROM STULL
MAX ERROR CP AT 1300 K 0.9%
C8H18 G                T 7/80C    8H  18     0    0G    300.       5000.             1
 0.22451614E 02 0.40858209E-01-0.12878297E-04 0.17911266E-08-0.88108611E-13     2
-0.36112590E 05-0.89322800E 02 0.25595255E 01 0.68079829E-01 0.13468589E-04     3
-0.58320591E-07 0.26347424E-10-0.28872742E 05 0.21120300E 02                    4

C8H18 2,2,4 TRIMETHYLPENTANE  DATA EXTRAPOLATED THROUGH WILHOIT POLYNOMIAL
FROM API 44  MAX ERROR CP AT 1300 K .9%
C8H18ISO                 T 7/80C    8H  18     0    0G    300.       5000.           1
 0.21708740E 02 0.45720968E-01-0.14820756E-04 0.21025155E-08-0.10605221E-12     2
-0.36258039E 05-0.92011612E 02 0.11759005E 01 0.73226452E-01 0.11216891E-04     3
-0.55649959E-07 0.24636362E-10-0.28667469E 05 0.22365265E 02                    4

C10D8  NAPHTHALENE-D8  SIGMA=4  IA=32.0228 IB=76.0297 IC=108.0525  NU=2272,2257,
1553,1386,1298,863,835,692,495,785,545,346,875,760,663,410,2302,2275,1604,1330,
1030,929,830,490,800,653,507,177,2286,2258,1545,1258,1050,889,715,328,2289,2258,
1439,1316,1082,880,825,593,791,628,404,163  REF=CHEN,KUDCHADKER,WILHOIT
MAX ERROR CP AT 1300 K .8%
C10D8                  T 9/82C   10D   8     0    0G    300.       5000.   MW=136.23 1
 0.24693802E+02 0.25579888E-01-0.93010221E-05 0.14824513E-08-0.85934623E-13     2
 0.29915154E+04-0.11212885E+03-0.29223614E+01 0.80820084E-01-0.12762395E-04     3
-0.52788202E-07 0.30022318E-10 0.11687422E+05 0.35690598E+02             2844.0 4

C10H8  NAPHTHALENE    SIGMA=4  IA=26.8532 IB=67.4189 IC=94.2721  NU=3060,
3030,1577,1463,1380,1145,1025,761,512,950,725,386,980,876,778,466,3092,3060,
1628,1443,1242,1168,936,506,970,841,581,191,3065,3058,1595,1389,1265,1125,877,
359,3090,3027,1509,1361,1209,1144,1008,617,958,782,472,176  REF=CHEN,KUDCHAKER
AND WILHOIT  MAX ERROR CP AT 1300 K .9%
C10H8                    9/82 C   10H   8     0    0G    300.       5000.   MW=128.17 1
 0.19726501E+02 0.28167639E-01-0.96896429E-05 0.14654855E-08-0.80134955E-13     2
 0.85581797E+04-0.86023239E+02-0.46547480E+01 0.74611127E-01-0.53069607E-05     3
-0.54125859E-07 0.29066527E-10 0.16325324E+05 0.45005463E+02            2491.13 4

C10H8  AZULENE  SIGMA=2  IAIBIC=1.88E- 112 NU=3070,1690,1634,1621,1577,1535,
1482,1442,1389,1367,1295,1290,1265,1201,1150,1114,1055,1045,1036,1007,978,963,
945,899,855,820,787,766,940,724,708,694,671,653,610,510,475,350,280,200,3070(7),
175  REF=KOVATS,GUNTHARD,PLATTNER  MAX ERROR CP AT 1300 K .9%
C10H8A                 T 9/82H    8C  10     0    0G    300.       5000.   MW=128.17 1
 0.19087189E+02 0.28716661E-01-0.98752744E-05 0.14930039E-08-0.81601501E-13     2
 0.24276551E+05-0.81962631E+02-0.48537226E+01 0.73454738E-01-0.38748985E-05     3
-0.53900077E-07 0.28514219E-10 0.31977461E+05 0.46992599E+02             2449.7 4
```

```
C12H9   O-BIPHENYL RADICAL   SIGMA=1  STATWT=2  IA=28.895  IB=148.3008  IC=178.1957
IR=7.22   SIGMAR=2  V(2)=1000.    NU=3083(2),3031,1583,1497,1275,1157,1025,1019,
996,733,841,399,3052,1603,1448,1357,1185,1145,1032,606,302,904,778,695,543,487,
120,955,775,696,531,470,246,970,834,397,3086(2),3067(2),1608,1440,1397,1182,
1162(2),1077,606,140,3038,1583,1497,1012,993,738    REF=BURCAT,ZELEZNIK,MCBRIDE
MAX ERROR CP AT 1300 K .8%
C12H9                   L 1/84C  12H    9    0    0G    300.       5000.      MW=153.20  1
 0.24023865E 02 0.31809665E-01-0.11027002E-04 0.16776169E-08-0.92268857E-13    2
 0.40086453E 05-0.10351401E 03-0.39727516E 01 0.85021615E-01-0.57617590E-05    3
-0.62286745E-07 0.33391831E-10 0.49016336E 05 0.46985733E 02             3175.7  4

C24H18  TRIPHENYLBENZENE   INTERIM TABLE CONSTRUCTED BY GRAPHICAL INTERPOLATION
MAX ERROR CP AT 500 K 3.6%
C24H18                  U10/78C  24H   18    0    0G    300.       5000.                1
 0.49104904E+02 0.64315677E-01-0.23532120E-04 0.39017642E-08-0.23995102E-12    2
 0.20562879E+05-0.23493062E+03-0.12553463E+01 0.17089677E+00-0.51354727E-04    3
-0.11426977E-06 0.89534297E-10 0.38495016E+05 0.37223114E+02                   4

D                       J 3/77D   1    0    0    00G    300.       5000.                1
 0.25        E+01   0.              0.              0.              0.             2
 0.25921761E+05 0.57845408E+00 0.25        E+01   0.              0.             3
 0.              0.              0.25921761E+05 0.57845408E+00                   4

DCL                     J 6/77D   1CL   1    0    0G    300.       5000.                1
 0.29572034E+01 0.15918160E-02-0.63320272E-06 0.11755658E-09-0.81599911E-14    2
-0.12173515E+05 0.58856456E+01 0.38269213E+01-0.25013326E-02 0.60466124E-05    3
-0.44837519E-08 0.11367641E-11-0.12301921E+05 0.18786267E+01                   4

D2                      J 3/77D   2    0    0    0G    300.       5000.                1
 0.28345604E+01 0.12769562E-02-0.35157262E-06 0.47151782E-10-0.21806160E-14    2
-0.83293848E+03 0.10613022E+01 0.36185608E+01-0.62479009E-03 0.90200223E-06    3
 0.91954319E-10-0.18281335E-12-0.10589973E+04-0.30491915E+01                   4

D2O                     J 6/77D   2O   1    0    0G    300.       5000.                1
 0.27264595E+01 0.39845173E-02-0.14932626E-05 0.26349772E-09-0.17649557E-13    2
-0.30902638E+05'0.73050385E+01 0.38541431E+01 0.14712288E-03 0.30069006E-05    3
-0.17747628E-08 0.23018862E-12-0.31151651E+05 0.17202570E+01                   4

D2O2  DEUTERIUM PEROXIDE  A0=4.92  B0=0.788  C0=0.733  NU=2667,1028,869,2661,
947  POTENTIAL BARRIERS V(1)=1029 V(2)=651 V(3)=51 1/CM  X11=-48 X12=-6 X13=-8
X15=-88 X16=-2 X22=-5 X23=-5 X25=-6 X26=-2 X33=-10 X35=-8 X36=-1 X55=-47
X56=-2 X66=-2  ALFAA1=-.09  ALFAA2=.04  ALFAA3=.03  ALFAA5=-.09  ALFAA6=
0.07  ALFAB1=-.001  ALFAB2=-.002  ALFAB3=-.006  ALFAB5=-.001  ALFAB6=-.001
ALFAC1=.0005  ALFAC2=-.002  ALFAC3=-.01  ALFAC5=0.0005  ALFAC6=-0.003
REF=TSIV 1978
D2O2                    R  78 D   2O   2    0    0G    300.       5000.                1
 0.53462811E+01 0.38918597E-02-0.13136314E-05 0.20201987E-09-0.11233771E-13    2
-0.19302543E+05-0.29733829E+01 0.34752569E+01 0.72003081E-02-0.12431274E-05    3
-0.23989009E-08 0.10817658E-11-0.18677613E+05 0.72012196E+01                   4

D2S                     J 6/77D   2S   1    0    0G    300.       5000.                1
 0.36662901E+01 0.34992264E-02-0.14207284E-05 0.26685639E-09-0.18684739E-13    2
-0.42147308E+04 0.37862870E+01 0.38070824E+01 0.37596311E-03 0.57530799E-05    3
-0.53485740E-08 0.14054083E-11-0.40661219E+04 0.38658748E+01                   4

E                       L02/67E   10   00   00    0G    300.       5000.                1
 0.25000000E 01 0.              0.              0.              0.             2
-0.74537496E 03-0.11734026E 02 0.25000000E 01 0.              0.             3
 0.              0.             -0.74537500E 03-0.11734026E 02                   4

H                       J 9/65H   100              0G    300.       5000.                1
 0.25000000E 01 0.              0.              0.              0.             2
 0.25471627E 05-0.46011763E 00 0.25000000E 01 0.              0.             3
 0.              0.              0.25471627E 05-0.46011762E 00                   4

HBR                     J 9/65H   1BR   1    0    0G    300.       5000.                1
 0.27935804E+01 0.15655925E-02-0.56171064E-06 0.95783142E-10-0.61813990E-14    2
-0.52338384E+04 0.76423703E+01 0.36056690E+01-0.59529431E-03 0.65029568E-06    3
 0.93781219E-09-0.71141852E-12-0.54389455E+04 0.34831774E+01                   4

HCN                     L12/69H   1C   1N   10    0G    300.       5000.                1
 0.37068121E 01 0.33382803E-02-0.11913320E-05 0.19992917E-09-0.12826452E-13    2
 0.14962636E 05 0.20794904E 01 0.24513556E 01 0.87208371E-02-0.10094203E-04    3
 0.67255698E-08-0.17626959E-11 0.15213002E 05 0.80830085E 01                   4
```

```
HD                J 6/77H   1D   1    0    0G   300.      5000.              1
 0.28464544E+01 0.10631961E-02-0.24433805E-06 0.29050834E-10-0.11621531E-14    2
-0.76182465E+03 0.96695917E+00 0.34325477E+01 0.65107028E-03-0.19332666E-05    3
 0.24101736E-08-0.86732397E-12-0.10009272E+04-0.24022073E+01                   4

HDO               J 6/77H   1D   10   1    0G   300.      5000.              1
 0.26672688E+01 0.35575209E-02-0.12026003E-05 0.19607209E-09-0.12352620E-13    2
-0.30372869E+05 0.79704328E+01 0.40754422E+01-0.13820285E-02 0.57025534E-05    3
-0.44163646E-08 0.12263062E-11-0.30707608E+05 0.95790167E+00           1193.7  4

HNCO              J12/70H   1N   1C   10   1G   300.      5000.              1
 0.51300390E+01 0.43551371E-02-0.16269022E-05 0.28035605E-09-0.18276037E-13    2
-0.14101787E+05-0.22010995E+01 0.23722164E+01 0.13664040E-01-0.13323158E-04    3
 0.64475457E-08-0.10402894E-11-0.13437059E+05 0.11588263E+02                   4

HNO               J 3/63H   1N   10   10   0G   300.      5000.              1
 0.35548619E 01 0.32713182E-02-0.12734071E-05 0.22602046E-09-0.15064827E-13    2
 0.10693734E 05 0.51684901E 01 0.37412008E 01-0.20067061E-03 0.75409300E-05    3
-0.79105713E-08 0.25928389E-11 0.10817845E 05 0.50063473E 01                   4

HNO2              J 6/63H   1N   10   2    0G   300.      5000.              1
 0.55144941E+01 0.41394403E-02-0.15878702E-05 0.27977639E-09-0.18584209E-13    2
-0.11276885E+05-0.31425253E+01 0.25098874E+01 0.12171605E-01-0.78618375E-05    3
 0.35351571E-09 0.11540858E-11-0.10450008E+05 0.12399634E+02                   4

HNO3              J 6/63H   1N   10   3    0G   300.      5000.              1
 0.70591100E+01 0.56769446E-02-0.22348863E-05 0.40155529E-09-0.27080510E-13    2
-0.18920009E+05-0.10778285E+02 0.14377135E+01 0.20903552E-01-0.14574553E-04    3
 0.11972023E-08 0.19117285E-11-0.17385368E+05 0.18246253E+02                   4

HO2               J 9/78H   10   200       0G   300.      5000.              1
 0.40173060E+01 0.22175883E-02-0.57710171E-06 0.71372882E-10-0.36458591E-14    2
-0.11412445E+04 0.37846051E01 0.35964102E+01 0.52500748E-03 0.75118344E-05     3
-0.95674952E-08 0.36597628E-11-0.89333502E+03 0.66372671E+01                   4

H2                J 3/77H   2    0    0    0G   300.      5000.              1
 0.30667095E+01 0.57473755E-03 0.13938319E-07-0.25483518E-10 0.29098574E-14    2
-0.86547412E+03-0.17798424E+01 0.33553514E+01 0.50136144E-03-0.23006908E-06    3
-0.47905324E-09 0.48522585E-12-0.10191626E+04-0.35477228E+01                   4

H2O(L)            J 3/79H   20   100       0L   273.150    500.0             1
 0.             0.             0.             0.             0.                2
 0.             0.             0.28630800E+02-0.20260986E+00 0.78529479E-03    3
-0.13653020E-05 0.91326966E-09-0.38579539E+05-0.11895046E+03                   4

H2O               J 3/79H   20   1    0    0G   300.      5000.              1
 0.26110472E+01 0.31563130E-02-0.92985438E-06 0.13331538E-09-0.74689351E-14    2
-0.29868167E+05 0.72091268E+01 0.41677234E+01-0.18114970E-02 0.59471288E-05    3
-0.48692021E-08 0.15291991E-11-0.30289969E+05-0.73135474E+00                   4

H2O2              L 2/69H   20   20   00   0G   300.      5000.              1
 0.45731667E 01 0.43361363E-02-0.14746888E-05 0.23489037E-09-0.14316536E-13    2
-0.18006961E 05 0.50113696E 00 0.33887536E 01 0.65692260E-02-0.14850126E-06    3
-0.46258055E-08 0.24715147E-11-0.17663147E 05 0.67853631E 01                   4

H2S               J 6/77H   2S   100       0G   300.      5000.              1
 0.27452199E+01 0.40434607E-02-0.15384510E-05 0.27520249E-09-0.18592095E-13    2
-0.34199444E+04 0.80412439E+01 0.39323476E+01-0.50260905E-03 0.45928473E-05    3
-0.31807214E-08 0.66497561E-12-0.36505359E+04 0.23023599E+01                   4

H2SO4(L)          J 9/77H   2S   10   4    0L   300.      1000.              1
 0.             0.             0.             0.             0.                2
 0.             0.             0.99421525E+01 0.21786369E-01 0.34974458E-05    3
-0.33548857E-08 0.11699586E-11-0.10185979E+06-0.44398695E+02                   4

H2SO4             J 9/77H   2S   10   4    0G   300.      5000.              1
 0.10889532E+02 0.75004178E-02-0.29210478E-05 0.52595513E-09-0.35789415E-13    2
-0.92471364E+05-0.29418038E+02 0.10725680E+01 0.43769226E-01-0.55333243E-04    3
 0.35518253E-07-0.90677358E-11-0.90259758E+05 0.18926326E+02                   4

H3O+              R  79 H   30   1E  -1    0G   300.      5000.     MW=19.02 1
 0.28246822E+01 0.50812885E-02-0.14264715E-05 0.16926313E-09-0.63539296E-14    2
 0.70856812E+05 0.56510124E+01 0.33014021E+01 0.26223860E-02 0.27255019E-05    3
-0.27153022E-08 0.70842800E-12 0.70803750E+05 0.35236931E+01           1208.3  4

HE                L 5/66HE  100            0G   300.      5000.              1
 0.25000000E 01 0.             0.             0.             0.                2
-0.74537498E 03 0.91534888E 00 0.25000000E 01 0.             0.                3
 0.             0.            -0.74537498E 03 0.91534884E 00                   4
```

```
KR                 L 1/84KR  1    0    0    0G    300.     5000.                    1
 0.24999990E 01     0.00000000      0.00000000      0.00000000      0.00000000    2
-0.74537476E 03 0.54777565E 01 0.25000000E 01     0.00000000      0.00000000    3
     0.00000000      0.00000000-0.74537476E 03 0.54777565E 01                     4

N                  J 3/77N   100            0G    300.     5000.                    1
 0.24370811E+01 0.13233886E-03-0.90907754E-07 0.22864054E-10-0.13762291E-14    2
 0.56128585E+05 0.45211111E+01 0.25000004E+01-0.31078154E-08 0.83216097E-11    3
 0.89478278E-14 0.38108039E-17 0.56106975E+05 0.41806431E+01                   4

NCO                J12/70N   1C   10   10    0G    300.     5000.                    1
 0.49964357E+01 0.26250880E-02-0.10928387E-05 0.20309111E-09-0.13915195E-13    2
 0.17379356E+05-0.17325320E+01 0.31092021E+01 0.66201022E-02-0.26070086E-05    3
-0.14966380E-08 0.10922032E-11 0.17977514E+05 0.83561334E+01                   4

ND                 J 6/77N   1D   1    0    0G    300.     5000.                    1
 0.29787664E+01 0.13644909E-02-0.44426554E-06 0.66850067E-10-0.36520334E-14    2
 0.44202715E+05 0.51782103E+01 0.35822582E+01-0.54360228E-03 0.70695836E-06    3
 0.10904153E-08-0.87384879E-12 0.44087676E+05 0.22182770E+01                   4

ND2                J 6/77N   1D   2    0    0G    300.     5000.                    1
 0.13124552E+01 0.73207803E-02-0.37609880E-05 0.84924379E-09-0.67714551E-13    2
 0.21822223E+05 0.15539929E+02 0.37544718E+01 0.55300584E-03 0.28321620E-05    3
-0.12810171E-08-0.20484656E-12 0.21126371E+05 0.28866510E+01                   4

ND3                J 6/77N   1D   3    0    0G    300.     5000.                    1
 0.38974762E+01 0.53573176E-02-0.17675020E-05 0.25481817E-09-0.13516735E-13    2
-0.86511758E+04 0.31102449E+00 0.29169312E+01 0.52944236E-02 0.22344675E-05    3
-0.41223203E-08 0.14051026E-11-0.81626094E+04 0.62506895E+01                   4

NH                 J 6/77N   1H   1    0    0G    300.     5000.                    1
 0.27414945E+01 0.14032028E-02-0.46001046E-06 0.80217694E-10-0.52770870E-14    2
 0.44499551E+05 0.59638059E+01 0.34520634E+01 0.54983583E-03-0.20358506E-05    3
 0.30336916E-08-0.12401129E-11 0.44249545E+05 0.20208741E+01          1036.7   4

NH2                J 6/77N   1H   200           0G    300.     500Y.                    1
 0.27554321E+01 0.32905847E-02-0.11160410E-05 0.20903758E-09-0.15676329E-13    2
 0.22001133E+05 0.69276497E+01 0.41262894E+01-0.16485841E-02 0.56461668E-05    3
-0.40505222E-08 0.10499871E-11 0.21697548E+05 0.16960494E+00                   4

NH3                J 6/77N   1H   300           0G    300.     5000.                    1
 0.23168577E+01 0.62841460E-02-0.21251163E-05 0.34018690E-09-0.21470026E-13    2
-0.64265487E+04 0.82987657E+01 0.37729747E+01-0.82975716E-03 0.11801882E-04    3
-0.12126874E-07 0.41763790E-11-0.66908514E+04 0.14968947E+01                   4

NO                 J 6/63N   10   100           0G    300.     5000.                    1
 0.31890000E 01 0.13382281E-02-0.52899318E-06 0.95919332E-10-0.64847932E-14    2
 0.98283290E 04 0.67458126E 01 0.40459521E 01-0.34181783E-02 0.79819190E-05    3
-0.61139316E-08 0.15919076E-11 0.97453934E 04 0.29974988E 01                   4

NO+                J 6/66N   10   1E  -100    0G    300.     5000.                    1
 0.28885488E 01 0.15217119E-02-0.57531241E-06 0.10051081E-09-0.66044294E-14    2
 0.11819245E 06 0.70027197E 01 0.36685056E 01-0.11544580E-02 0.21755608E-05    3
-0.48227472E-09-0.27847906E-12 0.11803369E 06 0.31779324E 01                   4

NO2                J 9/64N   10   200           0G    300.     5000.                    1
 0.46240771E 01 0.25260332E-02-0.10609498E-05 0.19879239E-09-0.13799384E-13    2
 0.22899900E 04 0.13324138E 01 0.34589236E 01 0.20647064E-02 0.66866067E-05    3
-0.95556725E-08 0.36195881E-11 0.28152265E 04 0.83116983E 01                   4

NO2-               J 6/72N   10   2E   1    0G    300.     5000.                    1
 0.50160903E+01 0.21884463E-02-0.94586144E-06 0.17939789E-09-0.12052428E-13    2
-0.26200160E+05-0.12861447E+01 0.29818036E+01 0.49398681E-02 0.28557293E-05    3
-0.78905297E-08 0.35391483E-11-0.25501540E+05 0.99161680E+01                   4

NO3                J12/64N   10   3    0    0G    300.     5000.                    1
 0.72033289E+01 0.30908791E-02-0.13329045E-05 0.25461601E-09-0.17939047E-13    2
 0.58244016E+04-0.12608119E+02 0.76867377E+00 0.21181075E-01-0.16980256E-04    3
 0.22963836E-08 0.19321041E-11 0.75292921E+04 0.20406284E+02                   4

N2                 J 3/77N   20   00   00   0G    300.     5000.                    1
 0.28532899E+01 0.16022128E-02-0.62936893E-06 0.11441022E-09-0.78057465E-14    2
-0.89008093E+03 0.63964897E+01 0.37044177E+01-0.14218753E-02 0.28670392E-05    3
-0.12028885E-08-0.13954677E-13-0.10640795E+04 0.22336285E+01                   4

N2D2               J 6/77N   2D   2    0    0G    300.     5000.                    1
 0.40114886E+01 0.61707515E-02-0.25716035E-05 0.48376813E-09-0.33904283E-13    2
 0.23210917E+05 0.18402704E+01 0.29546664E+01 0.42616224E-02 0.84357864E-05    3
-0.12016733E-07 0.44251578E-11 0.23786486E+05 0.85980418E+01                   4
```

```
N2H2              J12/65H   2N   2    0    0G   300.      5000.           1
 0.31327846E+01 0.64136459E-02-0.25033625E-05 0.45031298E-09-0.30577306E-13 2
 0.24283790E+05 0.63329308E+01 0.35531690E+01-0.73824010E-03 0.17081641E-04 3
-0.19348477E-07 0.69147109E-11 0.24471653E+05 0.56516775E+01                4

N2H4              J12/65N   2H   400       0G   300.      5000.           1
 0.50947770E 01 0.93296138E-02-0.33626986E-05 0.56308304E-09-0.35859661E-13 2
 0.92996644E 04-0.35950952E 01 0.79803836E 00 0.21788097E-01-0.13456754E-04 3
-0.12698753E-09 0.25865213E-11 0.10379887E 05 0.18248696E 02                4

N2O               J12/64N   2O   100       0G   300.      5000.           1
 0.47306679E 01 0.28258267E-02-0.11558115E-05 0.21263683E-09-0.14564087E-13 2
 0.81617682E 04-0.17151073E 01 0.26189196E 01 0.86439616E-02-0.68110624E-05 3
 0.22275877E-08-0.80650330E-13 0.87590123E 04 0.92266952E 01                4

N2O+              J12/70N   2O   1E  -10   0G   300.      5000.           1
 0.53926946E+01 0.22337196E-02-0.93548832E-06 0.17466166E-09-0.12059043E-13 2
 0.15847633E+06-0.36920186E+01 0.34273064E+01 0.63787690E-02-0.22585149E-05 3
-0.20421800E-08 0.13481477E-11 0.15909237E+06 0.67997616E+01                4

N2O4              J 9/64N   2O   400       0G   300.      5000.           1
 0.10506637E 02 0.58723267E-02-0.24766296E-05 0.46556024E-09-0.32402082E-13 2
-0.28609096E 04-0.26252230E 02 0.36662865E 01 0.23491748E-01-0.16007297E-04 3
 0.11845939E-08 0.20001618E-11-0.90631797E 03 0.93973337E 01                4

N2O5              J12/64N   2O   5    0    0G   300.      5000.           1
 0.14413736E+02 0.40494080E-02-0.17661640E-05 0.33912224E-09-0.23926356E-13 2
-0.38366062E+04-0.43313433E+02 0.32144535E+01 0.37992511E-01-0.36847600E-04 3
 0.12409293E-07 0.24351911E-12-0.98609506E+03 0.13555831E+02                4

N3                J12/70N   3O   00   00   0G   300.      5000.           1
 0.51996828E+01 0.24335678E-02-0.10192340E-05 0.19062350E-09-0.13212412E-13 2
 0.47963131E+05-0.35547759E+01 0.30624389E+01 0.73590658E-02-0.38229374E-05 3
-0.71824202E-09 0.91110236E-12 0.48614547E+05 0.77570129E+01                4

NE                L 5/66NE  100            0G   300.      5000.           1
 0.25000000E 01 0.             0.             0.             0.             2
-0.74537500E 03 0.33420438E 01 0.25000000E 01 0.             0.             3
 0.             0.            -0.74537498E 03 0.33420438E 01                4

NE+               L12/66NE  1E  -100       0G   300.      5000.           1
 0.29285147E 01-0.41229320E-03 0.16341709E-06-0.29554891E-10 0.20056917E-14 2
 0.25015218E 06 0.24159397E 01 0.21006406E 01 0.32416425E-02-0.56265881E-05 3
 0.38693079E-08-0.93291304E-12 0.25029535E 06 0.63098678E 01                4

O                 J 3/77O   100            0G   300.      5000.           1
 0.25342961E+01-0.12478170E-04-0.12562724E-07 0.69029862E-11-0.63797095E-15 2
 0.29231108E+05 0.49628591E+01 0.30309401E+01-0.22525853E-02 0.39824540E-05 3
-0.32604921E-08 0.10152035E-11 0.29136526E+05 0.26099342E+01                4

OD                J 6/77O   1D   1    0    0G   300.      5000.           1
 0.29295740E+01 0.12846475E-02-0.38446586E-06 0.51644730E-10-0.23868544E-14 2
 0.35230369E+04 0.58604975E+01 0.38530350E+01-0.11608871E-02 0.72619503E-06 3
 0.15297850E-08-0.10691248E-11 0.32959333E+04 0.11456270E+01                4

OH                J 6/77O   1H   10   00   0G   300.      5000.           1
 0.28897814E+01 0.10005879E-02-0.22048807E-06 0.20191288E-10-0.39409831E-15 2
 0.38857042E+04 0.55566427E+01 0.38737300E+01-0.13393772E-02 0.16348351E-05 3
-0.52133639E-09 0.41826974E-13 0.35802348E+04 0.34202406E+00                4

O2                J 3/77O   2O   00   00   0G   300.      5000.           1
 0.36122139E+01 0.74853166E-03 0.19820647E-06 0.33749008E-10-0.23907374E-14 2
-0.11978151E+04 0.36703307E+01 0.37837135E+01-0.30233634E-02 0.99492751E-05 3
-0.98189101E-08 0.33031825E-11-0.10638107E+04 0.36416345E+01                4

O3                J 6/61O   3O   00   00   0G   300.      5000.           1
 0.54665239E+01 0.17326031E-02-0.72204889E-06 0.13721660E-09-0.96233828E-14 2
 0.15214096E+05-0.34712616E+01 0.24660617E+01 0.91703209E-02-0.49698480E-05 3
-0.20634230E-08 0.20015595E-11 0.16059556E+05 0.12172130E+02                4

S(S)              J 9/77S   10   00   00   0S   300.       388.360         1
 0.             0.             0.             0.             0.             2
 0.             0.             0.18018871E+01 0.31543881E-02 0.             3
 0.             0.            -0.68614468E+03-0.72259874E+01                4

S(L)              J 9/77S   10   00   00   0L   388.360   5000.           1
 0.33906200E+01 0.71182514E-03-0.39087832E-06 0.87327456E-10-0.68755181E-14 2
-0.63358440E+03-0.14788307E+02-0.38449885E+02 0.25707392E+00-0.55555365E-03 3
-0.51325813E-06-0.17253650E-09 0.42933552E+04 0.16753043E+03                4
```

```
S                         J 9/77S   10   00   00   0G   300.      5000.          1
 0.29171783E+01-0.57194760E-03 0.28890940E-06-0.52911339E-10 0.33647474E-14    2
 0.32484363E+05 0.37503810E+01 0.28695579E+01 0.63609306E-03-0.34420074E-05    3
-0.40332507E-08-0.15123007E-11 0.32453195E+05 0.37536114E+01                   4

SD                        J 6/77S   1D   1    0    0G   300.      5000.          1
 0.33471988E+01 0.12129646E-02-0.47730138E-06 0.88323669E-10-0.60740591E-14    2
 0.15627147E+05 0.48642147E+01 0.47285597E+01-0.50939881E-02 0.99134605E-05    3
-0.73290813E-08 0.19461608E-11 0.15399579E+05-0.15819068E+01                   4

SF4                       J 6/76S   1F   40   00   0G   300.      5000.          1
 0.11124383E+02 0.21457994E-02-0.95452444E-06 0.18746111E-09-0.13535953E-13    2
-0.95581669E+05-0.28888894E+02 0.12819645E+01 0.43569899E-01-0.70125168E-04    3
 0.53677244E-07-0.15914356E-10-0.93586701E+05 0.18406624E+02                   4

SF6                       J 6/76S   1F   600       0G   300.      5000.          1
 0.15162950E+02 0.43842318E-02-0.19486337E-05 0.38247196E-09-0.27605050E-13    2
-0.15226801E+06-0.54428944E+02-0.38388088E+01 0.83221721E-01-0.13181689E-03    3
 0.99636154E-07-0.29248767E-10-0.14836477E+06 0.37147918E+02                   4

SH                        J 6/77S   1H   100       0G   300.      5000.          1
 0.30014537E+01 0.13394957E-02-0.46789663E-06 0.78804015E-10-0.50280453E-14    2
 0.15905320E+05 0.62711902E+01 0.44420322E+01-0.24359197E-02 0.19064576E-05    3
 0.99166630E-09-0.95740762E-12 0.15523258E+05-0.11579273E+01                   4

SN                        J 6/61S   1N   100       0G   300.      5000.          1
 0.38493976E 01 0.72756788E-03-0.29370203E-06 0.55013628E-10-0.38123551E-14    2
 0.30459962E 05 0.44179139E 01 0.39422971E 01-0.20035515E-02 0.73534644E-05    3
-0.75168560E-08 0.25591098E-11 0.30563949E 05 0.45669484E 01                   4

SO                        J 6/77S   10   10   00   0G   300.      5000.          1
 0.40142873E+01 0.27022817E-03 0.82896667E-07-0.34323741E-10 0.31121444E-14    2
-0.71051956E+03 0.34863848E+01 0.31490233E+01 0.11839347E-02 0.25740686E-05    3
-0.44443419E-08 0.18735159E-11-0.40407571E+03 0.83065289E+01                   4

SO2                       J 6/61S   10   200       0G   300.      5000.          1
 0.52451364E 01 0.19704204E-02-0.80375769E-06 0.15149969E-09-0.10558004E-13    2
-0.37558227E 05-0.10873524E 01 0.32665338E 01 0.53237902E-02 0.68437552E-06    3
-0.52810047E-08 0.25590454E-11-0.36908148E 05 0.96513476E 01                   4

SO3                       J 9/65S   10   300       0G   300.      5000.          1
 0.70757376E 01 0.31763387E-02-0.13535760E-05 0.25630912E-09-0.17936044E-13    2
-0.50211376E 05-0.11200793E 02 0.25780385E 01 0.14556335E-01-0.91764173E-05    3
-0.79203022E-09 0.19709473E-11-0.48931753E 05 0.12251863E 02                   4

S2                        J 9/77S   200            0G   300.      5000.          1
 0.39886069E+01 0.55775051E-03-0.50189278E-07-0.15470319E-10 0.26661771E-14    2
 0.14198015E+05 0.44777479E+01 0.28585754E+01 0.51758355E-02-0.65493434E-05    3
 0.33998643E-08-0.40156766E-12 0.14412402E+05 0.98778348E+01                   4

S8                        J 6/64S   8    0    0    0G   300.      5000.          1
 0.20751765E+02 0.14126296E-02-0.61834767E-06 0.11933438E-09-0.84707650E-14    2
 0.55882550E+04-0.67731586E+02 0.10602257E+02 0.44008819E-01-0.68651729E-04    3
 0.48505851E-07-0.12808281E-10 0.75807966E+04-0.19129652E+02                   4

XE                        L 4/70XE  10   00   00   0G   300.      5000.          1
 0.25000000E 01 0.              0.              0.              0.             2
-0.74537501E 03 0.61512740E 01 0.25000000E 01 0.              0.             3
 0.             0.             -0.74537499E 03 0.61512742E 01                  4
```

Index